Microbial and Nanotechnological Approaches for Environmental Remediation

NIPA® GENX ELECTRONIC RESOURCES & SOLUTIONS P. LTD.
New Delhi-110 034

About the Editors

Dr. Namita Ashish Singh is currently working as Assistant Professor at Department of Microbiology, Mohanlal Sukhadia University, Udaipur, Rajasthan. She completed her Masters in Microbiology from Gurukula Kangri Vishwavidyalaya, Haridwar. She earned her doctoral study in Biotechnology from Punjabi University Patiala, Punjab. She has more than fourteen years of research and teaching experience. Dr. Singh's current research focuses on Microbial diversity, Mycotoxin contamination, Antimicrobial Resistance, Food quality & safety. Dr. Singh has supervised 1 Ph.D. and 16 post-graduation students while 4 Ph.D. are ongoing. She is having two Indian patents on Aflatoxin M1 and *Enteroccoci* in milk to her credit. She has published more than 26 research/ review articles in International and National reputed journals along with 1 book and 19 book chapters. She has handled two research projects sanctioned by UGC and RUSA, MHRD, Delhi. She has many prestigious awards to her credit like Best Paper Award by Indian Dairy Association (IDA) along with 20000 Cash Prize, Dr. C.M. Singh Memorial NABS-Best Research Paper Award by NABS, Chennai, India, Certificate of appreciation by DST Rajasthan, Best Inventor award, by Microbiologists Society India etc.

Dr. Girima Nagda is currently working as Assistant Professor at the Department of Zoology, University College of Science, Mohanlal Sukhadia University (MLSU), Udaipur Rajasthan. She completed her M.Sc. in Zoology with specialization in Cell and Molecular Biology in the year 2004 and received Gold Medal for standing first in Mohanlal Sukhadia University. She cleared her CSIR-NET exam in the year 2004. She has been awarded Ph.D. degree by Mohanlal Sukhadia University in the year 2012. She has 16 years of UG and PG teaching experience. Her research areas include Nanotechnology, Toxicology, Biochemistry, Developmental Biology etc. She is handling one RUSA MHRD project on "Pesticide Residue Analysis in Fruits and Vegetables" as a Co-PI. She has published about 15 research/review articles in reputed international and national journals, 14 chapters in edited books and 4 books. She has to her credit awards like Women in Science Award 2023, Prof. R S Paroda Best Presentation Award 2019 by Rajasthan Science Congress, Appreciation Award on Women's Day 2021, ONGC Young Scientist Award 2017 by ONGC at international conference for overall academic excellence.

Dr. Tripta Jain is currently working as Assistant Professor at Department of Botany, Mohanlal Sukhadia University, Udaipur. She got her Ph.D. degree from Mohanlal Sukhadia University and her field of specialization is Plant Pathology and Microbiology. She also holds a Diploma Certificate in Pharmacy. She is actively involved in conducting research since last 15 years. Her research is focused on plant disease management, plant microbe interaction, identification and characterization of plant secondary metabolites and microbial metabolites and isolation of novel PGPR strains especially for their biocontrol potential. She has been associated with the teaching profession for the past 18 years. She has successfully completed one research project funded by DST, New Delhi and currently running two research projects in her credit funded by RUSA. She published more than 30 research/review papers and 15 book chapters in her areas of proficiency and interest. Dr. Jain Co-supervised 4 Ph.D. and more than 45 short term research projects/minor research project in the field of botany and microbiology. She is recipient of Women scientist (WOS-B) fellowship from DST, New Delhi in the year 2009. She also receives Smt. Guman Verma Memorial Best Women Scientist Award best research paper presentation and Women's Day Appreciation Award in 2021.

Microbial and Nanotechnological Approaches for Environmental Remediation

Namita Ashish Singh
Assistant Professor
Department of Microbiology
Mohanlal Sukhadia University
Udaipur - 313001, Rajasthan

Girima Nagda
Assistant Professor
Department of Zoology
Mohanlal Sukhadia University
Udaipur - 313001, Rajasthan

Tripta Jain
Assistant Professor
Department of Botany
Mohanlal Sukhadia University
Udaipur -313001, Rajasthan

NIPA® GENX ELECTRONIC RESOURCES & SOLUTIONS P. LTD.
New Delhi-110 034

NIPA® GENX ELECTRONIC
RESOURCES & SOLUTIONS P. LTD.
101,103, Vikas Surya Plaza, CU Block
L.S.C. Market, Pitam Pura, New Delhi-110 034
Ph : +91-11-43860225, Mob.: +91 9717133558, 9540816132
E-mail: newindiapublishingagency@gmail.com
Website: www.nipaersources.com

Print ISBN: 978-93-58873-28-3

ebook ISBN: 978-93-5887-339-9

Composed and Designed by NIPA®.

Dedicated to Our Parents

&

Family

Who've been there always,
thank you for your unwavering
love and support

Preface

This book, ***"Microbial and Nanotechnological Approaches for Environmental Remediation"*** presents a promising avenue for the degradation and detoxification of environmental contaminants by harnessing the capabilities of microbes. The book highlights the role of microorganisms including bacteria, fungi, and microalgae as well as nanotechnology in bioremediation of different environmental contaminants i.e. heavy metals, pesticides, nanomaterials, pharmaceutical products, plastics, microplastic, radioactive material etc. In recent decades, environmental contamination has emerged as a serious threat to our modern society. Microorganisms can break down, detoxify, and mineralize various synthetic dyes under specific environmental conditions.

The book includes nineteen chapters highlighting various aspects of bioremediation. The chapters within this book reflects the diverse and recent advances in the field of bioremediation of environmental pollutants i.e. heavy metals, pesticides, plastics, textile dyes, oil, microplastic, hydrocarbons etc. in waster water, soil or agro-industrial waste by means of microbes and nanotechnology. The book mainly covers biological solutions to various environmental pollutants from soil, water, air, industrial waste across diverse research fields, fostering innovation and addressing complex challenges with sustainable solutions. Further, the routes and pathways by which human are exposed to various contaminants are discussed. It also highlights the toxicity of these contaminants on human, animals, plants and ecosystem. In this book, we embark on a comprehensive exploration of the dynamic realm of bioremediation, where innovation meets societal impact.

We extend our heartfelt thanks to all the contributing authors who have shared their expertise and insights. The contributors to this book, experts in their respective fields, offer cutting-edge research and real-world applications. Whether you are an experienced researcher, a student eager to delve into the exciting world of microbial bioremediation, or a curious reader interested in understanding its societal implications, this book provides a comprehensive and accessible resource. We are confident that this compilation of knowledge will be an important addition to the scientific literature, inspiring further research & innovation in this exciting field.

Editors

Contents

1

Toxic Footprints: Understanding Environmental Contaminants and Their Toxicity on Living Beings and Ecosystem

Girima Nagda[1#], Tripta Jain[2#], Radhika Soni[1], Pushpa Gehlot[2] Shakshi[3] and Namita Ashish Singh[3*]

[1]*Dept. of Zoology, Mohanlal Sukhadia University, Udaipur-313001, Rajasthan, India*

[2]*Dept. of Botany, Mohanlal Sukhadia University, Udaipur-313001, Rajasthan, India*

[3]*Dept. of Microbiology, Mohanlal Sukhadia University, Udaipur-313001 Rajasthan, India*

[#] *Equally contributed author*

[*]*Corresponding Author Email: namita.singh@mlsu.ac.in*

Abstract

Environmental contamination has become one of the most serious global issues, endangering ecosystem integrity, biodiversity, and human health. This chapter covers various environmental pollutants viz. heavy metals, pesticides, plastics, pharmaceuticals, nanomaterials, and endocrine-disrupting substances, their sources, exposure paths, and toxicity mechanisms. The chapter explores the effects of pollutants on cells and systems, emphasizing mechanisms such as mitochondrial failure, oxidative stress, endocrine disruption, and DNA damage. It also highlights the negative impact on the human respiratory, neurological, immunological, and reproductive systems, as well as the dangers of exposure in the workplace and community. Animals, plants, and microorganisms also suffer from the pollutants that affect ecological functions. This chapter emphasizes how urgently thorough research, strict regulations, sustainable practices, and public awareness are needed to lower the hazards related to environmental toxins and advocates a worldwide coordinated interdisciplinary strategy to preserve environmental balance and guarantee public health resilience.

Keywords: *Environmental contaminants, Food Chain, Nanomaterials, Cellular toxicity, Bioremediation, Phytoremediation*

1. Introduction

Environmental contaminants are defined as substances introduced into the environment, either naturally or anthropogenically, that pose potential harm to living organisms and ecosystems (Masindi and Muedi, 2018). These contaminants include a diverse range of pollutants, such as chemical agents, biological contaminants and physical pollutants (Ahmed *et al.*, 2021). Understanding these contaminants and their toxicity is of paramount importance, given their profound implications on human health, biodiversity and the overall functioning of ecosystems (Samuel *et al.*, 2023).

Environmental contaminations are pervasive in nature and originated from diverse sources. Industrial activities are major contributors, releasing heavy metals, organic solvents, and particulate matter into air, water and soil (Williams, 2019). Similarly, agricultural practices contribute to contamination through excessive use of fertilizers and pesticides, which often runoff into aquatic systems, causing eutrophication and harming aquatic life (Zahoor and Mushtaq, 2023). Urbanization and improper waste management have also intensified the problem with plastics and emerging contaminants finding their way into ecosystems (Singh *et al.*, 2022). Even natural events like volcanic eruptions and forest fires contribute to environmental contamination, though their impacts are often localized and episodic compared to the persistent and widespread pollution caused by human activities (Ernst, 2012).

The study of environmental contaminants and their toxic effects is of vital importance for several reasons in today's context of escalating global pollution (Hussain and Reza, 2023). First, it provides insights into the health risks posed to humans and wildlife (Pereira *et al.*, 2015). Many contaminants, such as lead, mercury and polychlorinated biphenyls (PCBs), are known to have chronic and acute toxic effects on biological systems, affecting neurological, reproductive and immune functions (Priyadarshanee *et al.*, 2022). These substances interact with biological systems in complex ways, often leading to adverse outcomes such as bioaccumulation, oxidative stress, endocrine disruption and ecological imbalances (Satkar *et al.*, 2024). Heavy metals like lead and mercury are known to accumulate in organisms and disrupt critical biological functions, while synthetic chemicals like bisphenol A (BPA) can interfere with hormonal systems, resulting in reproductive and developmental issues (Pan *et al.*, 2024). Such impacts underscore the need to understand how these substances interact with living organisms, identify their exposure thresholds and mechanisms of toxicity to formulate strategies for risk assessment and mitigation (Kumar and Singh, 2023).

Studying contaminant toxicity helps to understand their cumulative and long-term ecological impacts. Many pollutants exhibit persistence in the environment and tend to bioaccumulate in organisms, magnifying their effects across food chains (Edo *et al.*, 2024). For instance, pesticides like DDT and heavy metals like cadmium and arsenic have been linked to population declines in fish, birds and mammals. Such effects threaten biodiversity, disrupt ecosystem balance and reduce the resilience of natural systems, undermining their ability to deliver crucial ecosystem services like pollination, water purification, and carbon sequestration (Chakraborty, 2023).

Emerging contaminants, such as pharmaceuticals and nanomaterials, propose unique challenges due to their novel chemical properties and limited regulatory oversight (Khan *et al.*, 2023). These substances, often found in trace amounts, may have subtle yet significant impacts on living organisms and ecosystems, necessitating advanced analytical techniques and interdisciplinary research to understand their behaviour and effects fully (Samuel *et al.*, 2023).

Contaminants can be categorized into several types, each posing distinct challenges. Chemical contaminants include substances like heavy metals (e.g., lead, mercury, arsenic), organic pollutants (e.g., pesticides, dioxins, and hydrocarbons), and industrial chemicals such as polychlorinated biphenyls (PCBs) (Onyeaka *et al.*, 2024). These chemicals often enter the environment through industrial emissions, agricultural runoff and improper disposal of waste. Biological contaminants such as viruses, bacteria and toxins from algae, primarily affect water and food systems, posing risks to human health and aquatic organisms (Naik *et al.*, 2019). Physical contaminants include microplastics, particulate matter and radioactive particles, which can cause physical and chemical damage to organisms upon ingestion or exposure (Ribeiro *et al.*, 2019). Emerging contaminants, such as pharmaceuticals and endocrine-disrupting compounds, often originate from domestic sewage and industrial effluents and their long-term ecological and health impacts are still under study (Yadav *et al.*, 2021).

According to recent estimates, the society currently uses about 100,000 compounds. Over 40,000 of these compounds have been identified as persistent organic pollutants. This number becomes even more worrying when one considers technological advances and the rampant discharge of these pollutants into the environment (Rodil *et al.*, 2012). The US Environmental Protection Agency defines "emerging contaminant" as a chemical or material with perceived, potential, or real threat to the human health or the environment or for which published health standards are lacking (USEPA, 2008). Emerging contaminants include a long list of routinely employed products such as pharmaceuticals, cosmetics, plasticizers and flame retardants. These contaminants has grown rapidly in recent years, to include an array of compounds with reported presence in the environment. Immune toxicity, neurotoxicity, endocrine disruption and carcinogenicity are among the consequences of exposure to these contaminants (Balducci *et al.* 2012).

The study of environmental contaminants and their toxicity is essential for identifying and mitigating risks to living beings and ecosystems. By understanding the pathways, interactions and effects of these contaminants, researchers and policy makers can devise effective solutions to minimize their impact (Gavrilescu *et al.*, 2015). This research not only safeguards human health and biodiversity but also promotes sustainable development and ecological resilience in the face of rising environmental challenges.

2. Types of Environmental Contaminants

Any physical, chemical, biological, or radioactive material that negatively affects soil, water, air, or living things is considered an environmental contamination. Any material that is present in higher than usual concentrations due to human activities and negatively impacts the environment is considered a pollutant. As diffuse nonpoint sources, pollutants can reach rivers, lakes, streams, and groundwater by runoff and soil percolation, or they can enter directly through dumping, piped outflow, and channeled waste streams. Nonpoint sources, such as runoff from paved roadways and parking lots, agricultural lots, soil erosion from logging, and atmospheric deposition of acidic or toxic air pollutants, are thought to be significant contributors to pollution of the air, water, and soil. Long-lived pollutants may eventually end up in long-term sinks such soils and aquatic sediments (Anderson *et al.*, 2005).

The hydrosphere, lithosphere, and atmosphere are all potential entry points for pollutants into the ecosystem. Anthropogenic activity is a major source of pollutants entering the ecosystem, in addition to the natural pathways mentioned before, such as volcanic activity and rock weathering. They can occur in the intended application of biocides, like vector controls; in waste disposal, like industrial effluents and sewage disposal; or in an unintentional release, such in shipwrecks, oil spills, mining, and fires. Temperature, surface water movement and direction, air mass circulation, wind speed all these affect the migration of heavy metals and other contaminants (Walker *et al.*, 2012).

2.1. Chemical Contaminants

2.1.1. Heavy Metals

Due to their toxicity, propensity to persist in the atmosphere, and capacity for bioaccumulation within the human body, heavy metals are well-known environmental contaminants. Toxic heavy metal poisoning of terrestrial and aquatic ecosystems is a serious environmental issue with health implications. While the majority of heavy metals are found in nature, some are produced by human activity. High atomic mass and toxicity to living things are characteristics of heavy metals. The majority of heavy metals can be fatal to humans and pollute the ecosystem and atmosphere. Through the food chain, people and other living things can be exposed to heavy metals, which can become extremely dangerous when they combine with other environmental components like soil, water, and air. Mercury, cadmium, arsenic, chromium, nickel, copper, and lead are among the most common heavy metals that pollute the environment (Mitra *et al.*, 2022). Animals and humans can be exposed to cadmium in many ways, and it can be released into the atmosphere as a result of natural or man-made activities. Surface runoff into sediments, industrial waste, & absorption all contribute to cadmium poisoning of the aquatic environment. Cadmium poisoning can occur when a person consumes food, breathes in air, or drinks water contaminated with the metal (Hayat *et al.*, 2018).

One of the most dangerous heavy metals that can be present in the biosphere is mercury. When mercury comes into touch with aquatic sediments, it transforms into the extremely hazardous methyl mercury (Gworek *et al.*, 2020). Through the food chain, methylmercury enters the human body through fish, seafood, and wildlife that have been contaminated by consuming harmful microbes. After being absorbed into the human body, it enters the bloodstream and results in a number of neurological issues (Rice *et al.*, 2014). Lead is a metal that is present in nature in comparatively small levels and is not biodegradable. Because of human activities like mining, manufacturing, and burning fossil fuels, atmospheric lead levels are steadily rising. When exposed to levels beyond the recommended threshold, lead is harmful to the human body (Loh *et al.*, 2016).

The most common of the harmful heavy metals, manganese, can be found in nature in a variety of oxidation states. Manganese oxides are released into the atmosphere when methylcyclopentadienyl manganese tricarbonyl (MMT), a gasoline additive, burns. Manganese is necessary for many physiological processes, but too much of it can be extremely harmful (O'Neal and Zheng, 2015). The metal chromium is poisonous and carcinogenic. It can be found in two stable oxidation states in the environment: chromium (III) and chromium

(IV) (VI). A less dangerous version of chromium (VI) is chromium (III). During industrial activities, they have the ability to interconvert to one another. However, because chromium (III) is less poisonous than chromium (VI), its conversion is less detrimental to the environment. Numerous enterprises that employ chromium endanger local climates. Emissions from the ferrochrome business are the highest when compared to environmental natural chromium emissions (Coetzee *et al.*, 2020).

Cobalt is used to create alloys and is abundant in the environment, including plants, soils, rocks, and water. Despite having a modest discharge rate, it poses a serious risk to people. Cobalt has both positive and negative effects on the human body. Large-scale releases of cobalt into the environment can be lethal, while small levels often have no detrimental effects. Nickel has several industrial applications and is a naturally occurring element. It is released into the atmosphere by both man-made and natural sources (Li *et al.*, 2016a). Due to the intake of contaminated air, it has numerous negative consequences on humans, including allergies, lung and nasal cancer, and kidney and cardiovascular disorders (Genchi *et al.*, 2020; Lu *et al.*, 2005).

Copper is an essential micronutrient for living things. It plays a part in the regular physiological processes of plants, including photosynthesis, the synthesis of chlorophyll, and the metabolism of proteins and carbohydrates. Important metabolic pathways are altered by copper shortage, and toxicity results from increased exposure (Schwartz *et al.*, 2003). One essential and ubiquitous metal is zinc. It functions as a cofactor and is linked to numerous enzyme reactions. The type and extent of exposure determine zinc toxicity. The two main sources of zinc are mining and smelting. Mineral processing processes release a significant amount of zinc into the environment, which impacts both ecosystems and living things (Zhang *et al.*, 2012).

2.1.2. Pesticides

According to Aktar *et al.* (2009), a pesticide is a chemical that either kills pests or prevents them from causing harm to the environment. Pesticides are helpful in preventing diseases and pests, but they can also be hazardous to people and other living things. More than 95% of pesticides wind up in areas other than the air, water, and soils. Eight pesticides (CIAT, chlorpyrifos, AMPA, metolachlor, dieldrin, atrazine, de-sulfinyl fipronil, and glyphosate) and two pharmaceutical medications (caffeine, metformin) had detection rates ranging from 66 to 84%, according to a US Geological Survey study on the surface water of 38 streams in the US (Bradley *et al.*, 2017). Among the substances that are quite persistent in soil are pesticides. When these compounds are used excessively, pollutants may accumulate in the soil (Pullagurala *et al.*, 2018).

Despite being prohibited, it has been discovered that the usage of pesticides is rising in some places (Pan *et al.,* 2020).

Among the persistent organic contaminants, the main are organochlorine insecticides (OCPs). High toxicity, bioaccumulation, and biomagnification of OCPs in the environment are well-established (Sparling, 2016). There was a substantial decrease in wildlife populations almost everywhere in the world as a result of the extensive and absurd usage of OCPs throughout the 1960s and 1970s (Kohler *et al.,* 2013). During that time, birds and other wild species were impacted by DDT, dieldrin, and other hazardous OCPs, which have now been banned from use in agriculture. However, since then, populations of aquatic organisms, birds, and wild bees have continued to drop (Kennedy *et al.,* 2013) due to the use of recently developed pesticides, which are found in all types of habitats worldwide. In freshwater, marine, and terrestrial habitats, pesticides continue to be found (Law, 2014). Additionally, residue issues are a consequence of pesticide use. According to the World Health Organization (WHO), pesticide residue is any material or combination of materials found in human or animal food that results from the use of pesticides and any of their designated byproducts, including metabolites, degradation and conversion products, reaction products, and toxic impurities (Mani *et al.,* 2014).

2.1.3. Industrial Chemicals

In a variety of industries, including textiles, insecticides, and refrigeration, PFCs (perfluoroalkyl carboxylic acid) are frequently utilized as industrial chemicals (Prevedouros *et al.,* 2006; Liu *et al.,* 2020). One kind of perfluoroalkyl carboxylic acid is perfluoroalkyl acids (PFAAs). It is well known that PFAAs are highly abundant and widely distributed (Sha *et al.,* 2022). PFAAs have been regarded as a growing pollutant of concern and a threat to both human health and the environment because of their persistence in the environment (Kurtz *et al.,* 2019). PFAAs pollute the soil in a variety of ways, including when biosolids are applied as fertilizer for crop production (Blaine *et al.,* 2013) or reclaimed wastewater is utilized for irrigation (Jurling, 2021). The organic compounds created when sewage sludge is treated are known as biosolids. Usually, they are handled in accordance with the laws of their individual governments (Lu *et al.,* 2012). According to Blaine *et al.* (2013), the application of these materials occupies around 60% of the area utilized for agricultural in the United States. The US Environmental Protection Agency currently enforces a number of laws pertaining to biosolids land use (USEPA, 2013b). On the other hand, PFAAs in biosolids are unregulated. Therefore, frequent usage of biosolids may result in environmental contamination of soil, surface water, and groundwater (Muller *et al.*, 2011).

2.1.4. Pharmaceuticals and personal care products

Products like toothpaste, skincare products, fragrances, antibiotics, pharmaceutical medications used by consumers for their health and beauty, or veterinary medications used by the agroindustry to improve the growth or health of livestock are all considered pharmaceutical and personal care products (PPCPs) (Wang and Huang, 2019; Bishnoi *et al.,* 2022). In settings where conventional wastewater treatment facilities are unable to adequately remove them, PPCPs are becoming increasingly prevalent (Wang and Wang, 2016). PCPs may also be released into soil and aquatic habitats as a result of waste from sewage treatment facilities and animal farms (Ebele *et al.,* 2017). Nevertheless, a number of newly discovered pollutants, including chemicals, engineering nanomaterials, and antibiotics, may also be present in treated sewage sludge (Koumaki *et al.,* 2021). Among the newly discovered pollutants in sewage sludge, PPCPs pose a concern to human health or the environment if they are discharged into ecosystems (Van *et al.*, 2021). The other source of PPCPs in soil is contaminated river water utilized for irrigation or treated wastewater (Gallego *et al.*, 2021). It is common for low-hydrophobic CECs to contaminate soils. Through interactions with organic materials, they can subsequently accumulate in the soil (Beltran *et al.*, 2020). According to Bai *et al.* (2018), water samples contained 76% (109 of 144) of the pharmacological chemicals that were examined.

2.2. Physical Contaminants

2.2.1. Plastics

In nature and chemistry, plastic additives are abundant and have a variety of uses. They are typically embedded in the polymer structure and are present as traces (50% w/w) of these components. Plastics may contain a variety of chemicals, any randomly selected plastic product has roughly 20 additives (Oers *et al.*, 2012). These additives may leak during use or disposal since they are only loosely bonded to the polymers. Furthermore, the additives themselves may break down and produce other harmful substances while still connected to the plastic products or after they leach, which could cause them to remain in the environment and (bio)accumulate in biota.

The possible risks connected to additives persist across the course of plastics' life cycle. Furthermore, a comprehensive perspective of the chemical profile of the finished goods across the value chain is not available due to the lack of transparency by many industrial firms on the chemicals employed and their amounts, as well as the associated health risks (Wagner and Schlummer, 2020). For instance, recycled products such as toys (Miller *et al.*, 2016) and home items (Pivnenko *et al.,* 2017) contain brominated flame retardants, known endocrine disruptors, which pose a serious health risk. Burning plastic garbage does not reduce these health concerns, which is especially troublesome when low-tech incinerators are used or when uncontrolled conditions are present. Hazardous substances, such as dioxins and other persistent organic pollutants (POPs), as well as acid gasses and contaminated ashes, are released during incomplete combustion (Costa *et al.,* 2020).

2.2.2. Microplastics

Synthetic polymers having a particle size of less than 5 mm are known as microplastics. Microplastics are one of the most alarming environmental and worldwide challenges of the past few decades, given the ever-increasing production of plastic and microplastics particles, their inherent toxicity, and their potential for transportation by natural forces (such as wind, water streams, etc.). First, because microplastics can be easily carried by winds and river currents and because their polymeric structures are long-lasting, they can absorb various contaminants on their way to the atmosphere, land, and sea, which could pose health and environmental risks. Second, the long-term negative effects of microplastics on the environment, such as in aquatic systems, endanger the health of people and other marine life. Among all the suggested strategies for getting rid of microplastics in recent years, chemical technologies, photocatalytic degradation, biodegradation, and the usage of hybrid systems have all been well studied (Golmohammadi *et al.,* 2023). Numerous organisms suffer when they consume microplastics and related chemicals since they infiltrate and move up the trophic chain (Enfrin *et al.,* 2019; Talvitie *et al.,* 2017). Even tiny microbes, such as Antarctic krill, can break down polyethylene (PE) microbeads into nanoplastics by ingestion (Shen *et al.,* 2021). The average person consumes 5 g of microplastics each week, according to reports (Rajala *et al.,* 2020).

One thing is certain despite the fact that the effects of microplastics on the human body are not fully understood: the chemical compounds found in microplastics, such as dyes, fillers, additives, surfactants, and polymers, have an undeniable negative influence on human health, especially on the respiratory system (Enfrin *et al.*, 2019; Shen *et al.,* 2022).

2.2.3. Radioactive material

Earlier in World War II, an experiment on mine wastes revealed the existence of radioactive pollutants. The investigation found that the mineral water has a remarkable amount of naturally occurring radio components. Numerous sources, including fertilizers, nuclear businesses, mining effluents, etc., release radioactive and toxic nuclides into the environment (Wang *et al.,* 2017). Numerous radioactive elements have been found in drinking and municipal water, according to the World Health Organization. Radioactive nuclides tend to accumulate in bone tissues when they enter the circulatory system, even at low concentrations (Awual *et al.*, 2014).

These radioactive nuclides have a detrimental effect on both human life and the natural ecology because they are highly soluble in water and readily move into surface and groundwater. Numerous radionuclides are employed extensively in therapeutic and medical settings for sterilizing and cancer treatment. However, there are a number of adverse effects associated with large dosages, including bone abnormalities, low blood pressure, diarrhea, lung irritations, and paralysis. Various mining activities, particularly the extraction of plutonium from radioactive uranium, may be the cause of radioactive contamination in groundwater, particularly by uranium. According to one study, the growing need for uranium in the production of nuclear energy has led to an increase in uranium mining in Africa (Datta *et al.,* 2022).

Weakly radioactive and unstable, radioactive elements like uranium and thorium continuously decay to produce more stable radioisotopes. Heavy rainfall has been shown to promote the increase in radioactive concentrations in ground water (Datta *et al.*, 2022). The permissible limit of uranium in drinking water 30 μg/L (WHO, 2017).

2.3. Emerging Contaminants

2.3.1. Nanomaterials (NMs)

Particle size, surface chemistry, abiotic and biological processes in the environmental matrix all affect the fate and transit of NMs, including nanoparticles (NPs), in the environment. As a result, NPs can aggregate into bigger NMs or exist in suspension as isolated particles. It might also react with natural elements or exist in a dissolving condition (Saleh *et al.*, 2020; Luoma *et al.,* 2008). Because of their small size, NPs settle more slowly, which can lead to prolonged suspension in both water and air. As a result, NPs of the same material can transfer more easily and widely than larger ones (USEPA, 2010). Rainwater runoff or wind can release NMs found in solid wastes and wastewater effluents directly or unintentionally into the aquatic environment (Klaine *et al.,* 2008; Lowry *et al.,* 2010). Additionally, NMs' mobility across porous media is impacted by their small size, which also allows them to firmly adhere to and aggregate with mineral surfaces (Tourinho *et al.,* 2012). For example, NMs' mobility in groundwater aquifers is slowed down when they adhere to mineral surfaces (Wiesner *et al.,* 2012).

More than 36% of NPs in urban atmospheres come from stationary combustion sources and cars powered by gasoline and diesel. Manufactured NPs release more NPs into the atmosphere than do natural NPs. Since the greater toxicity of the finer particles has been revealed, regulatory concerns have shifted from traditional particulate matter (PM10) (< 10 μm) to PM5, PM2.5, and below, and the health effects of such NPs are still being studied (Klaine *et al.*, 2008; Maher *et al.,* 2020). Car exhaust is the primary generator of NPs in the atmosphere. While gasoline engines emit NPs of 20-60 nm, diesel engines emit NPs of 20-130 nm. It has been discovered that carbon nanotubes and fibers are released as byproducts during the burning of gas and diesel. Vehicle pollution is a primary source of nanoparticulate contamination in the urban atmosphere, as diesel-generated particles account for almost 90% of the carbon nanoparticles in the atmosphere (Jeevanandam *et al.,* 2010). Anthropogenic NPs are produced by building destruction and cigarette smoking in addition to exhausts. When a major building is demolished, NPs and microparticulates (~10 μm) are discharged. Lead, glass, respirable asbestos fibers, and other hazardous particles from household goods may be present in the released NPs near the building demolition site (Stefani *et al.,* 2005).

Natural NPs found in soils, including clays, organic matter, iron oxides, and other minerals, are crucial to biogeochemical processes. Contaminants may travel more easily through soils and other porous surfaces when colloids are present. When conditions for colloidal transport are favorable, the pollutants that have been sorbed into colloids can be transferred (Zhai *et al.,* 2020; Lin *et al.,* 2022). The benthic species suffered as a result of the NPs' gradual accumulation in the sediment and bottom-deposition aggregates in freshwater. In contrast, NPs may build up in the marine environment between warm and cold currents (Hyseni *et al.,* 2016). Biota may potentially recycle NPs in marine environments (Klaine *et al.,* 2008). As a result, species like tuna that live at the interface of cold and warm currents may be more at danger (Hyseni *et al.,* 2016). Additionally, Nam *et al.* (2014) demonstrate that manufactured NPs, including Ag and TiO_2 NPs, can pass through aquatic creatures' feeding habits. This study also demonstrated that because NPs adhere to the cell wall, algal cells concentrate NPs.

There are various ways that NMs can enter the air, water, and soil media, and exposure to these substances is now being investigated. Therefore, the next part will outline the ways in which humans are exposed to these materials. NPs can harm DNA or result in cell death since they can be found inside the cell in a variety of places, including the cytoplasm, lipid vesicles, cell membrane, and nucleus (Singh *et al.,* 2020, Cao *et al.*, 2021).

2.3.2. Endocrine-disrupting Chemicals

The ecosystem and human health are seriously threatened by endocrine disrupting chemicals (EDCs), which have been identified as emerging pollutants. These contaminants are directly tied to everyday life and include pesticides, polycyclic aromatic hydrocarbons, dioxins, brominated flame retardants, steroid hormones, and alkylphenols. According to WHO's (2002), EDCs are exogenous molecules that can disrupt the endocrine systems of humans and animals, negatively affecting organisms and even their progeny (Combarnous, 2017; Sabir *et al.,* 2019). According to Alvarez-Munoz *et al.* (2018) these substances are typically divided into three categories: androgenic compounds (EDsA) that mimic or block natural testosterone; thyroid compounds (EDsT) that affect the thyroid gland directly or indirectly; and estrogen compounds (EDsE) that mimic or block natural estrogen. They contribute to the deterioration of aquatic ecosystems and wildlife in addition to having an impact on human health (Aris *et al.,* 2020; Bilal *et al.,* 2022). Malignant tumors, abnormalities in reproduction and development are among the potential outcomes of exposure to EDCs (Wu *et al.,* 2022).

However, recent research also indicates that certain EDCs may have a negative impact on insulin, raising the incidence of type 2 diabetes (Yilmaz *et al.,* 2020). Drugs, detergent surfactants, pesticides, synthetic chemicals used as industrial solvents or lubricants and their byproducts, plastics, plasticizers, and fungicides are the most common sources of EDCs (Kabir *et al.,* 2015; Brander *et al.*, 2016; Vieira *et al.*, 2021). Due to their widespread exposure to humans through diet and personal hygiene products, especially during pregnancy, infancy, and senility, these substances are currently of concern due to their relationship to human metabolic, immune, nervous, and behavioral disorders (Esteban *et al.,* 2016). Despite the fact that concentrations are often found in the environment in the range of a few ng/L to μg/L, they are persistent and bioaccumulative, meaning that prolonged exposure might have side effects (Gonzalez *et al.,* 2022).

3. Routes and Pathways of Exposure

Environmental contaminants are substances that have the potential to negatively impact the ecosystem and human health. These contaminants can enter in the human body through a variety of exposure routes, including air, water, soil, sediments, and biota. Direct exposure with these contaminants occurs via ingestion, inhalation, injection, and dermal contact or exposure through the food chain. It is essential to comprehend the exposure routes and pathways in order to reduce risks and preventing harm. The main pathways of exposure to environmental contaminants include:

3.1. Inhalation or Exposure through Air

Inhalation is the most common and significant route of exposure to a variety of natural and man-made hazardous compounds that pollute the air. Airborne contaminants such as sulfur dioxide, carbon monoxide, nitrogen oxides, heavy metals, biological contaminants, ozone, tobacco smoke, etc. are released into the atmosphere by power plants, burning garbage, automobile emissions, emissions from chemical industries and factories, as well as volcanic eruptions. Chemical components including heavy metals (such as lead, mercury, cadmium, etc.) are released into atmosphere by mining, industrial processes, and automobile emissions; pesticides, especially from agricultural springs; and pharmaceuticals products can enter the atmosphere through evaporation and waste disposal (Manisalidis *et al.*, 2020). Biotoxins from molds and toxic algal blooms, as well as bacteria, viruses, fungus, and other pathogens, are examples of biological contaminants in the air. These biological agents can spread through the air naturally as spores or as a result of human activities like agricultural practices or waste disposal (Hughes *et al.*, 2022; P. Kumar *et al.*, 2021).

There are several health and environmental hazards associated with physical pollutants in the air, such as radioactive elements, plastics, and microplastics. Microscopic plastic particles can be breathed and cause respiratory and inflammatory problems. They can also be consumed through tainted food and water, which can cause microplastics to build up in human tissues and eventually cause health problems (Rochman *et al.*, 2013). Additionally, indoor air quality can be at hazards by radioactive contaminants like radon, a naturally occurring radioactive gas, which can enter the air through soil or building foundations. Inadequate disposal of radioactive waste or accidents at nuclear power plants are additional sources of airborne radioactive elements, which can have detrimental effects on health. In recent times, nanomaterials and endocrine disrupting chemicals have gained attention due to their persistent nature, negative environmental effects and possible health hazards. Despite being employed in a variety of consumer goods and medications, nanoparticles can enter the lungs and may result in oxidative stress, inflammation, and genetic damage because of their small size and large surface area (Oberdörster *et al.*, 2005). Additionally, consumer goods like plastics, solvents, and personal care items discharge a variety of endocrine disruptive chemicals into the atmosphere, which can interfere with the endocrine system and interfere with the control of hormones *(Colborn et al.,* 1993).

Through the respiratory system, all of these pollutants enter the body and interfere with biological functions, leading to diseases such as cancer, heart disease, prenatal central nervous system, respiratory and reproductive health issues (Manisalidis *et al.*, 2020). Health is influenced by the kind of contaminants, the level of exposure, and the individual's sensitivity. Those who are especially vulnerable to these air pollutants include pregnant women, elderly people, and youngsters. In children, developing lungs inhale more air relative to their body size, and thus contaminants can occur in higher-level resulting in developmental and chronic respiratory condition. Specifically in winter, children may suffer from respiratory problem as a result of exposure to oxides of nitrogen. The short-term consequences include eye, skin, nose, throat, coughing, headaches, and serious conditions like bronchitis, asthma, and heart and lung issues. However, prolonged exposure results in neurological, pulmonary, reproductive, and cancerous conditions. When air pollutants combine with pre-existing conditions such as cardiovascular disease, the possibility of developing a more serious unfavourable condition increases (Valavanidis *et al.*, 2013).

3.2. Water

Exposure to polluted water can have serious negative health effects. Pollutants from industrial activities, agriculture activity, floods, and unproper sewage treatment plants are sources of water pollution. Approximately 80% of urban and industrial wastewater worldwide is released into the environment without any prior treatment, endangering ecosystems and human health. As urbanization has accelerated, industrial production of wastewater has steadily increased and the least developed nations, which have inadequate wastewater treatment and sanitation systems, have a larger problem (Wear *et al.*, 2021). Exposure to contaminated water can occur through drinking, recreational activities, or occupational contact with contaminated water. According to the WHO, almost 1.4 million people die each year as a result of poor sanitation and hygiene, or contaminated drinking water, and over 1.7 billion people worldwide drink water containing fecal contaminants (WHO, 2023). Physical, chemical, biological, and emergent contaminants all significantly contribute in the degradation of water quality, which can lead to a variety of health problems, including cancer, brain damage, and gastrointestinal disorders. Many heavy metals like chromium, arsenic, and cadmium are among potentially hazardous substances that are discharged into wastewater from the industrial sector (Jan *et al.*, 2015). Furthermore, water pollution has an immediate connection to agriculture & has severely damaged all freshwater systems in their natural state.

Water pollution from rainwater is mainly caused by organic farm waste, nitrogen fertilizers, and pesticides (Gilden *et al.*, 2010). Waterborne pathogens, such as *Vibrio cholerae* (cholera), *Norovirus*, and *Escherichia coli*, as well as

increases as a result of their consumption of these species (Córdoba-Tovar *et al.*, 2022). The persistent substance may eventually affect the organism's ability to survival as well as reproduction. Additionally, this process can be carried out by trophic amplification, in which chemicals are transported from the fatty tissues of prey to the predator levels. When humans or other top predators consume contaminated meat, the process is more concerning (Ali and Khan, 2019). Prolonged exposure to these bioaccumulated toxins can cause major health problems, including developmental and neurological damage, cancer, and immune system suppression (Mondal *et al.*, 2018).

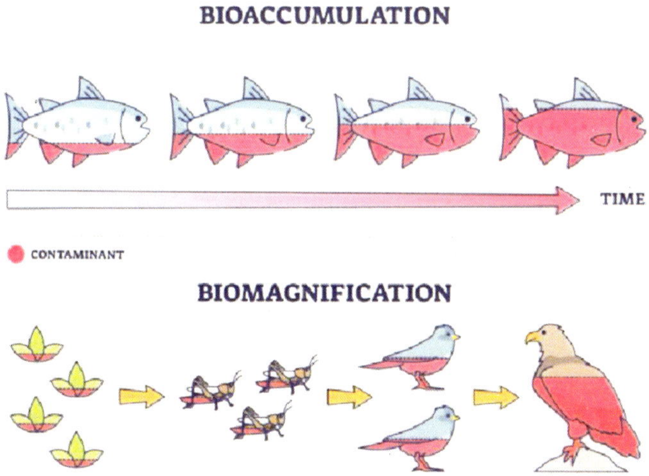

Fig. 1: Difference between bioaccumulation and biomagnification

3.5. Occupational Exposure in Industrial Settings

Industrial work exposure involves the possibility of development of threats to health in connection with potential direct contact with contaminants or risky conditions at the workplace. These could be chemical, physical, biological, or even ergonomic in nature, areas of operation susceptible to these risks include special sectors of manufacturing, construction, mining, and chemical processing (Berglund *et al.*, 2020). Chemical risks are the most widespread risks that are present in working environments of many industries. Employees can be at risk from poisonous substances, mists, fogs, dusts or gases by inhalation, absorption or ingestion. Chemical risk might include: solvents used in, painting and cleaning asbestos in construction and insulation work, pesticides and herbicides used in agriculture and pest control, heavy metals such as lead, mercury, and cadmium employed in manufacturing. The aggregation of these chemicals can cause respiratory diseases, skin ailments, neurological complications, and even cancers, among other diseases, after excessive exploitation (Zunarelli *et al.*, 2021). Physical hazards are dangers associated with aspects of the working environment, such as vibrations, heat, radiation, and noise. Employees who are exposed to high amounts of these elements run the risk of developing various health issues. For example, industrial noise leading to hearing loss, stress, and problems with sleep, which is well observed in industries like construction, mining for instance and the manufacturing industries. Additionally, steel production or construction industry employees may also get problems like heat stroke or hypothermia (Venugopal *et al.*, 2020). Cancer and genetic mutations are common in healthcare fields where workers are exposed to ionizing radiation, such as radiography and nuclear energy. Employees in particular industries like medical keeping, using farming chemicals or handling wastes are likely to experience biological hazards (Rahman *et al.*, 2023) as they come into contact with other people, animals, or environment containing pathogenic microbes or other disease-causing agents. For instance, the caregivers are likely to come across the HIV and hepatitis B virus or the agricultural works being in contact with the years, rodents, or other insects.

Occupational exposure to toxins is a major public health concern, affecting workers across various industries such as construction, mining, manufacturing, and healthcare. Exposure to hazardous substances like asbestos, benzene, lead, pesticides, and silica dust has been linked to severe health effects, including respiratory diseases, neurological impairments, reproductive disorders, and cancers (Smith *et al.*, 2020). Long-term exposure to heavy metals like lead and mercury can cause cognitive decline and kidney damage, while inhalation of silica dust increases the risk of silicosis and lung cancer (Jones & Brown, 2021). Workers in the healthcare sector are at risk of exposure to hazardous drugs, anesthetic gases, and bloodborne pathogens, which can lead to acute and chronic health complications (Doe *et al.*, 2019).

4. Mechanisms of Toxicity

Toxicity arises when a chemical substance interacts with biological systems, disrupting normal physiological functions and potentially leading to cellular or systemic damage. The mechanisms of toxicity depend on factors such as the chemical nature of the toxin, its route and duration of exposure, dose, the affected organ system and individual susceptibility. The general mechanisms by which toxins exert their harmful effects include absorption, distribution, metabolism, and excretion (ADME). For a toxin to exert its effects, it must first enter the body, distribute to target organs, undergo metabolism, and finally be eliminated. Toxins enter the body through inhalation, ingestion, dermal absorption, or injection. Once inside, they are distributed via the bloodstream to target organs, with lipophilic substances accumulating in fatty tissues and hydrophilic substances remaining in circulation (Gallo & Doull, 2019). Distribution depends on lipid solubility, plasma protein binding and blood-brain barrier permeability. After distribution, some toxins undergo biotransformation in the liver via Phase I and Phase II metabolic reactions. While detoxification can occur, some compounds become more toxic through metabolic activation, producing reactive intermediates such as free radicals (Hodgson, 2021). Finally, the toxins are eliminated either through renal excretion via urine (water-soluble toxins e.g., heavy metals); hepatic excretion via bile or feces (lipophilic toxins); pulmonary excretion via exhalation (volatile toxins like alcohol and anesthetic gases are exhaled) or perspiration & breast milk–certain toxins (e.g., DDT, lead) are excreted through sweat and lactation.

4.1. Mechanisms of Cellular Toxicity

Once inside the body, toxins disrupt cellular functions through various mechanisms including generating free radicals and causing oxidative stress, mitochondrial dysfunction and ER stress, DNA damage and mutagenecity, enzyme inhibition and endocrine dysfunction etc.

4.1.1. Oxidative Stress and Cellular Damage

One of the broadly recognized toxicity mechanisms of pollutants is oxidative stress, which is brought on by an imbalance between the generation of reactive oxygen species and antioxidant defense (Samet and Wages, 2018). This imbalance can result in a number of abnormalities, including diabetes, atherosclerosis, cancer, neurological, cardiovascular, respiratory, kidney, and reproductive diseases. Environmental contaminants can induce oxidative stress on cells through mechanisms that are direct, indirect or involve the disruption of metabolic or bioenergetic processes that are regulated by thiol redox switches (Anetor *et al.*, 2022). Numerous pollutants can produce reactive oxygen species (ROS) and reactive nitrogen species (RNS), such as particulate matter (PM), heavy metals (such as lead and mercury), insecticides, and industrial chemicals (Leni *et al.*, 2020). In normal conditions, ROS are essential signalling messengers, maintaining cellular homeostasis between ROS production and antioxidant detoxification. However, an overproduction of ROS or compromised antioxidant defense results in oxidative stress, causing damage to the cells (Niki and Noguchi, 2021). These free radicals are highly unstable and reactive substances that are continuously created as metabolic byproducts or as a result of redox chain reactions and have the ability to damage DNA, proteins, and lipids within cells, which can result in inflammation, cell death, and malfunction (Anetor *et al.*, 2022). It is well established that exposure to fine particulate matter (PM2.5) has increased oxidative stress, but it also activates various inflammatory pathways that further result in chronic respiratory diseases like asthma and COPD (Li *et al.*, 2018). Heavy metals such as cadmium and arsenic cause an imbalance in cellular redox balance, which results in oxidative damage to different cellular components, thereby contributing to carcinogenesis (Lee *et al.*, 2012). Figure 2 shows reactive oxygen species (ROS) or free radical mediated cellular injury which is often initiated by superoxide (O2"-") or hydroxyl radicals (-OH). This ROS initiates lipid peroxidation of cellular, mitochondrial, nuclear and endoplasmic reticulum membranes. This increases cellular permeability to calcium and results in further mitochondrial swelling and damage. Amino acid residues on proteins can be oxidized and degraded by ROS. Nuclear and mitochondrial DNA can also be oxidized, leading to strand breaks and other damage. All of these events ultimately lead to cellular dysfunction and injury.

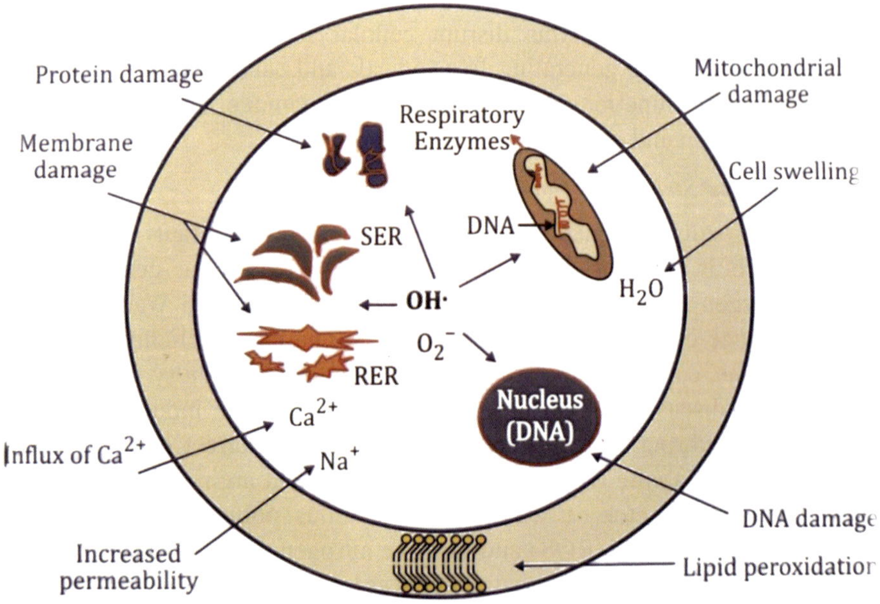

Fig. 2: Cellular injury by ROS or free radical-induced oxidative stress. (adapted by Marks *et al.*, 1996).

4.1.2. Mitochondrial Dysfunction and ER Stress

The mitochondria and endoplasmic reticulum are essential organelles in controlling protein synthesis, energy production, and cellular metabolism. Environmental pollutants and other external stimuli can influence their function and may lead to oxidative stress, ER stress, and mitochondrial damage, in turn causing cellular malfunction and even death (Li *et al.*, 2023). Environmental contaminants such as PM2.5, MPs, and PAHs can increase intracellular Ca^{2+} levels, which may lead to Ca^{2+} overloading (Prokic *et al.*, 2019; Wei *et al.*, 2021). In addition, the main causes of the mPTP opening pathway comprise high Ca^{2+} levels and oxidative stress in the mitochondria (Ichas and Mazat, 1998). Prolonged exposure to these pollutants can inhibit the activities of the enzymes Ca^{2+}-ATPase and Na^{+}-K^{+}-ATPase. By increasing the membrane's permeability to Na^{+}, K^{+}, and $Ca^{2+,}$ the concentration of Ca^{2+} in the mitochondria increases, and the concentration of K^{+} declines. This results in the opening of mPTPs, which causes structural damage, malfunction, and enlargement of the mitochondria (Riva *et al.,* 2011). The mitochondria can be also targeted by other pollutants, such as pesticides (like organophosphates) and heavy metals (including cadmium, lead, and mercury), which can cause oxidative damage and disrupt cellular respiration. For instance, it has been demonstrated that exposure to lead reduces mitochondrial function by causing apoptosis, altering the potential of the mitochondrial membrane, and increasing the production of ROS (Jan *et al.*, 2015)which have widespread environmental distribution and originate from natural and anthropogenic sources, are common environmental pollutants. In recent decades, their contamination has increased dramatically because of continuous discharge in sewage and untreated industrial effluents. Because they are non-degradable, they persist in the environment; accordingly, they have received a great deal of attention owing to their potential health and environmental risks. Although the toxic effects of metals depend on the forms and routes of exposure, interruptions of intracellular homeostasis include damage to lipids, proteins, enzymes and DNA via the production of free radicals. Following exposure to heavy metals, their metabolism and subsequent excretion from the body depends on the presence of antioxidants (glutathione, _-tocopherol, ascorbate, etc.. Pesticides induced mitochondrial dysfunction can lead to the death of neuronal cells as well as the progression of diseases in the case of Parkinson's and Alzheimer's neurodegenerative disease (Mostafalou and Abdollahi, 2013).

4.1.3. DNA Damage and Mutagenesis

Environmental contaminants are also known to cause DNA damage through several mechanisms. The ability of these contaminants to induce mutagenesis is particularly concerning because it increases the risk of genetic alterations that can be passed on to progeny, thereby contributing to the development of various cancers (NCI, 2022). Some are direct intercalators, including benzene and polycyclic aromatic hydrocarbons, which intercalate and produce adducts, disrupting the integrity of DNA and causing mutations, chromosomal abnormalities, and resulting in defective cell division. It results in DNA mutation, which elevates the risk of cancer by causing an accumulation of mutations (Oliveira *et al.,* 2007). Additionally, double-strand breaks in DNA are caused by carcinogens like tobacco smoke and asbestos, which can cause cancer and genomic instability (Msiska *et al.,* 2010).

4.1.4. Endocrine Disruption

Endocrine disruption is another important way by which environmental contaminants impact human health. Endocrine-disrupting chemicals (EDCs) include a variety of exogenous or synthetic substances that interfere with any part of hormone action. Through a variety of processes, including hormone mimicking, disruption of hormone synthesis or breakdown, modification of hormone receptor development, hormone antagonistic action, or altered hormone binding, these compounds can change the hormonal balance in the body *(Kumar et al.,* 2020). Environmental EDCs are mostly released during manufacture and use of human-made materials such as pesticides (methoxychlor, chlorpyrifos and dichlorodiphenyltrichloroethane), plastic/plasticisers (bisphenol A, phthalates), electronic wastes, flame-retardants, metals, food additives, and personal care products. These EDCs can disturb hormonal balance and consequently result in several health disorders, including immune and endocrine system disruption, respiratory problems, metabolic issues, diabetes, obesity, cardiovascular problems, growth impairment, neurological and learning disabilities, and cancer (Li *et al.*, 2023). For instance, BPA, a prevalent EDC in plastics, can alter gene expression and bind to estrogen receptors, resulting in developmental defects, reproductive issues, and even a higher risk of cancer (Cimmino *et al.*, 2020). Heavy metals such as arsenic, cadmium, nickel, copper, and lead, can also act as endocrine disruptors. More specifically, lead affect male fertility and ovarian development, cadmium can affect gonadal steroidogenesis and chromium can affect gonadal structure and function, fertilization rate (Liu *et al.,* 2023)the endocrine-disrupting mechanism of heavy metals is unclear. There are long-term and low-level metal/element exposure scenes for the human body in real life. Therefore, animal models exposed to high doses of heavy metals may not provide key information to elucidate the underlying pathogeny of human diseases. This review collects current knowledge regarding the endocrine-disrupting roles of heavy metals such as lead (Pb. Likewise, organochlorine pesticides like as DDT can disrupt thyroid hormone signaling, which can change development and metabolism (Leemans *et al.,* 2019).

4.2. Impact on Organ Systems

4.2.1. Respiratory System

Air contaminants from the environment constitute a serious risk to human health, particularly to the respiratory system. The negative impacts of air pollutants on the respiratory system might range from temporary irritation to long-term chronic diseases. Among these contaminants include metals, particulate matter (PM), sulfur dioxide (SO_2), nitrogen oxides (NOx), carbon monoxide (CO), ozone (O_3), and volatile organic compounds (VOCs) (Manisalidis *et al.,* 2020). According to recent research, inhaled environmental persistent free radicals (EPFR), which might remain in the air for up to 21 days and are produced when catechols, phenols, and hydroquinones burn, may have a significant effect (Guo *et al.*, 2023). These pollutants can harm the respiratory system in a number of ways, such as oxidative stress, inflammation, and activation of the mitogen-activated protein kinase (MAPK) complex, which promotes the synthesis of RNA and the generation of pro-inflammatory cytokines like TNF-α and IL-8. It has been associated with elevated IgE levels, decreased regulatory T lymphocyte activity, and increased CD4+ and CD8+ T cell production. This leads to a rapidly decline in lung function and chronic respiratory diseases such lung cancer, asthma, and chronic obstructive pulmonary disease (Valavanidis *et al.,* 2013).

4.2.2. Nervous System

Environmental pollutants affect the nervous system through genetic, molecular, cellular, and other pathways, which may damage the brain structure, or directly/indirectly lead to several nervous system diseases. In vulnerable populations, neurological problems are caused by environmental contaminants such as pesticides, smoking, mycotoxins, persistent organic pollutants (POPs), polychlorinated biphenyls, phthalates, and nanomaterials. However heavy metals, especially lead, mercury, and arsenic, are neurotoxic and can contribute to synaptic dysfunction, oxidative stress, and neuroinflammation (Rehman *et al.*, 2021). Through processes like oxidative stress, inflammation, and neurotransmitter disruption, these contaminants can affect brain function and ultimately lead to neurodegenerative diseases like attention-deficit hyperactivity disorder, Parkinson's disease, amyotrophic lateral sclerosis, multiple sclerosis, and Alzheimer's disease (Althobaiti, 2025). Heavy metals have significant effects because they can promote reactive oxygen species, mitochondrial dysfunction, calcium ion efflux, immunogenic response activation, and the inhibition of antioxidants such as glutathione, catalase, and superoxide dismutase (SOD) (Agnihotri and Kesari, 2019). Furthermore, a number of other neurological conditions, including Parkinson's and Alzheimer's diseases, affect memory and create cognitive dysfunctions by depleting brain-derived neurotrophic factor (Gao *et al.*, 2022).

4.2.3. Reproductive System

The reproductive health of both men and women can be adversely affected by environmental pollutants. These pollutants have the potential to cause infertility by interfering with the production of hormones, sperm, and eggs (Pizzorno, 2018). Chemicals that have been linked to interference with hormonal signalling and reproductive development include phthalates, bisphenol A (BPA), heavy metals, pesticides, and EDCs (Cimmino *et al.*, 2020; Cooper *et al.*, 2007). Environmental pollutants, particularly EDCs, can cause reproductive toxicity by disrupting hormonal balance, leading to infertility, developmental abnormalities, and birth defects. These substances have the ability to mimic or inhibit hormones such as testosterone and estrogen, which can result in aberrant reproductive function (Kumar *et al.*, 2020). While exposure to EDCs may induce irregular menstrual periods, polycystic ovarian syndrome (PCOS), and decreased fertility in women, it can also result in decreased sperm quality and motility in males. Additionally, pollutants have the ability to disrupt foetal development, which may result in developmental delays or birth defects (Marlatt *et al.*, 2022). For instance, phthalate exposure has been connected to aberrant fetal development and decreased sperm quality (Almeida-Toledano *et al.*, 2024).

4.2.4. Immune System

Many environmental contaminants also alter immune responses and can provoke immunotoxicity. Chronic exposure to certain contaminants, such as heavy metals, pesticides, and air pollutants, can suppress immune function, making individuals more susceptible to infections. Additionally, environmental contaminants can worsen allergy reactions by making immune cells more sensitive to allergens, which can lead to diseases like asthma and allergic rhinitis (Mitra *et al.*, 2022; Winans *et al.*, 2011). A subset of environmental toxins is sensed and responded to by xenobiotic receptors, such as the aryl hydrocarbon receptor (AHR), which trigger the production of detoxifying enzymes to defend the organism and their chronic activation leads to immunotoxicity. The KEAP1–NRF2 system is another crucial defense against environmental contaminants (Sondermann *et al.*, 2023; Suzuki *et al.*, 2020). In contrast, some contaminants might change immune responses, which can lead to autoimmune diseases. For instance, it has been demonstrated that particulate matter stimulates immunological cells, resulting in systemic inflammation and a higher risk of autoimmune diseases including lupus and rheumatoid arthritis (Adami *et al.*, 2022).

5. Effects on Environmental Contaminants on Different Organisms

Environmental toxicants, including heavy metals, pesticides, industrial chemicals, and air pollutants, pose significant risks to human, animal, plants and microbes. These toxicants enter ecosystems through industrial discharges, agricultural runoff, waste disposal, and air pollution, leading to both acute and chronic effects on living organisms.

5.1. Effects on Humans

Environmental pollutants have a major impact on human health, regardless of whether they come from natural, industrial, or agricultural sources. Depending on the kind, level, and length of exposure, environmental pollutants can have either acute or chronic impacts.

5.1.1. Acute and chronic health effects

Exposure to environmental toxicants can result in both immediate (acute) and long-term (chronic) health effects. Acute exposure often occurs due to accidental spills, inhalation of toxic gases, or ingestion of contaminated food or water. These effects can manifest as nausea, respiratory distress, neurological impairment, and skin irritation (Gallo *et al.,* 2021). Chronic exposure, on the other hand, occurs over an extended period, leading to severe conditions such as cancer, neurodegenerative diseases, and reproductive disorders (Grandjean & Landrigan, 2019).

For example, chronic exposure to heavy metals such as lead and mercury has been linked to developmental delays in children and neurotoxicity in adults (Tchounwou *et al.*, 2012). Similarly, persistent organic pollutants (POPs) like dioxins and polychlorinated biphenyls (PCBs) have been associated with endocrine disruption and increased cancer risks (Bergman *et al.*, 2013). Acute exposure to these environmental pollutants causes food and waterborne diseases, skin irritation, and respiratory issues immediately (Shetty *et al.*, 2023). For example, exposure to chemicals like benzene through the skin can result in rashes or dizziness, while airborne pollutants can induce asthma and chronic obstructive pulmonary disease (COPD) (Kim *et al.*, 2013). Chronic exposure to pollutants can result in more serious long-term health consequences, such as hormone disruption, cancer, respiratory conditions, and neurological diseases. Certain substances, such as heavy metals, asbestos, and benzene, can harm the brain and cause cancer, primarily in children (Mostafalou and Abdollahi, 2013; Rehman *et al.*, 2021). Similar to endocrine disruptors, these chemicals also cause a number of issues with reproductive and developmental health (Marlatt *et al.*, 2022).

5.1.2. Occupational and Community Exposure Risks

Occupational exposure occurs in industries involving chemicals, mining, agriculture, and construction. Workers in these sectors are at higher risk of inhaling or coming into direct contact with toxic substances. For instance, asbestos exposure in construction workers has been linked to mesothelioma, a severe form of lung cancer (Stayner *et al.,* 2013).

Community exposure, on the other hand, arises from environmental contamination, including air and water pollution. Populations living near industrial zones or landfills are more vulnerable to diseases caused by long-term exposure to pollutants. For example, groundwater contamination with arsenic has been a persistent issue in Bangladesh, leading to widespread cases of arsenicosis, a condition characterized by skin lesions and internal organ damage (Smith *et al.,* 2000). Communities surrounding these industrial or agricultural areas are also at risk for respiratory disorders, birth deformities, and malignancies *(*Shankar *et al., 2019;* Zunarelli *et al.*, 2021).

5.2. Effects on Animals

Animal populations are greatly impacted by environmental pollutants, which also have an effect on terrestrial, marine, and wildlife habitats. These pollutants, which include plastics, pesticides, heavy metals, and chemical pollutants, can have both short-term and long-term negative health impacts on animals, frequently interfering with their ability to reproduce, survive, and natural behavior. Exposure to these toxins affects not just the health of individual animals but also the health of entire ecosystems (Reckendorf *et al.*, 2023)pesticides and trace metals. Pesticides like atrazine and heavy metals like mercury can harm aquatic creatures including fish, amphibians, and invertebrates, leading to cognitive impairment, infertility, and even death. Along with air pollutants these contaminants affect terrestrial animal health, behavior, and reproductive success also (Madesh *et al.,* 2024). For example, insecticides could harm bees, while heavy metals like lead may endanger animals and birds. Chemicals such as PCBs, endocrine disruptors, and oil spills pose comparable hazards to wildlife, including birds and marine animals, resulting in health problems, infertility, and even population decreases (Pathak *et al.,* 2022). The disruption of ecosystems and food chains caused by environmental pollution highlights the necessity of conservation and pollution reduction measures to save biodiversity and animal populations.

Aquatic ecosystems are particularly vulnerable to pollution from industrial effluents, oil spills, and agricultural runoff. Heavy metals such as mercury bioaccumulate in fish, leading to toxic effects that impact their growth, reproduction, and neurological functions (Driscoll *et al.,* 2013). Similarly, endocrine-disrupting chemicals (EDCs) found in wastewater disrupt hormone functions in fish, resulting in reproductive abnormalities such as feminization in male fish (Kidd *et al.,* 2007). Terrestrial wildlife is exposed to toxicants through soil contamination, food chain bioaccumulation, and atmospheric pollutants. Pesticides, such as organophosphates and carbamates, have been reported to cause neurological damage and mortality in birds and mammals (Mineau & Palmer, 2013). Additionally, air pollutants such as sulfur dioxide (SO_2) and nitrogen oxides (NOx) contribute to acid rain, which depletes soil nutrients and adversely affects herbivorous species (Driscoll *et al.,* 2001).

Wildlife species often serve as bioindicators of environmental pollution. Exposure to heavy metals, pesticides, and industrial chemicals has been linked to population declines in species such as amphibians and birds of prey. For instance, the widespread use of DDT led to eggshell thinning in birds like the bald eagle, significantly reducing their reproductive success (Carson, 1962). Similarly, polar bears have been found to accumulate high levels of PCBs, which affect their immune and reproductive systems (Sonne, 2010).

5.3. Effects on Plants

Plants are integral to ecosystems, acting as primary producers and serving as a critical link in energy and nutrient cycles. However, they are highly susceptible to environmental contaminants which can impair their growth, reproduction and physiological functions (Khan *et al.*, 2018). The adverse effects of contaminants on growth and development of plants commonly referred as phytotoxicity, highlights the mechanisms by which pollutants adversely impact plants and the broader implications for soil health, ecosystems and agriculture (Anjum *et al.*, 2015).

Phytotoxicity occurs through various mechanisms like heavy metals such as lead, cadmium and arsenic disrupt essential cellular functions by replacing nutrients, impairing enzyme activity and inducing oxidative stress through the generation of reactive oxygen species (ROS) (Dutta *et al.*, 2018). This often leads to visible damage like chlorosis (leaf yellowing), necrosis (tissue death), stunted growth and reduced photosynthesis. Similarly, persistent organic pollutants (POPs), including pesticides and industrial chemicals can interfere with hormonal signalling, thereby affecting germination, flowering and stress responses. These toxic effects not only compromise plant health but also cascade into broader ecological impacts (Gautam *et al.*, 2023). Soil contamination caused by pollutants can lead to reduced crop yields, threatening food security and the economic stability of farming communities (Hou *et al.*, 2020). In natural ecosystems, the loss of sensitive plant species due to contamination can reduce biodiversity and disrupt ecological balances, ultimately affecting ecosystem services like carbon sequestration, nutrient cycling and habitat provision (Power, 2010).

Soil contamination is one of the most direct pathways through which pollutants affect plants. Contaminants can accumulate in the soil, damaging root systems and impeding water and nutrient uptake (Rashid *et al*., 2023). For instance, cadmium is known to inhibit root elongation and reduce photosynthetic efficiency, resulting in diminished plant productivity. POPs and hydrocarbons, commonly found in industrial and urban runoff, persist in soils for long periods, altering its fertility and structure (Aftab and Hakeem, 2023). Furthermore, contaminants disrupt the microbial communities within the soil, which play a vital role in nutrient cycling and organic matter decomposition. These disruptions reduce soil health and the ability of plants to grow sustainably in contaminated areas, posing challenges to agriculture and natural ecosystems (Lipczynska-Kochany, 2018). The broader implications of phytotoxicity extend beyond plant health to entire ecosystems and agriculture. Contaminated plants can serve as conduits for pollutants, transferring toxins up the food chain and posing risks to herbivores, humans and other trophic levels (Rose *et al*., 2023).

Plant species like lichens and mosses are particularly sensitive to airborne pollutants, including sulphur dioxide and heavy metals, making them effective indicators of air quality (Çobanoglu ozyigitoglu, 2020). Observable symptoms like leaf chlorosis, wilting, and necrosis also act as direct indicators of plant stress due to exposure to harmful pollutants, especially in agricultural fields affected by pesticide overuse or industrial effluents (Shafiei-Masouleh, 2021).

5.4. Microorganisms

Among the several ways that environmental pollutants (ECs) can enter the environment, wastewater treatment facilities (WWTPs) are the primary entry site for ECs, mostly in surface waters (Pal *et al.,* 2014). ECs are ineffectively removed during the water treatment process in drinking water treatment facilities (DWTPs) and WWTPs. Therefore, trace amounts of ECs in various ecosystems, including surface water, groundwater, DW, swimming pools, soils, sediments, and irrigation water, have already been detected thanks to advancements in analytical chemistry (Peng *et al.,* 2019). As a result, ECs are constantly present in resident organisms (such as algae, bacteria, aquatic animals, plants, and humans), which may have direct or indirect effects on each class of organism. There is still much to learn about the impacts of such exposure. For example, the primary issue with medications, and especially antibiotics, is the prevalence of antibiotic resistance in bacteria, which has indirect effects on other living things (Lempart *et al.,* 2018). Even at minimal amounts, exposure to ECs may change the microbial community's variety and role in the environment. These changes could have an impact on biodiversity and ecosystem function (Subirats *et al.,* 2019). Numerous studies detail how the behavior of natural microbiomes in various water sources, including river biofilms, marine sediment biofilms, and other aquatic ecosystems, is impacted by newly discovered contamination. Significant changes in the composition and function of the microbial community, including changes in respiration rate, extracellular polymeric substance (EPS) synthesis, and enzymatic activity, were brought about by exposure to ambient relevant concentrations of ECs (Wang *et al.,* 2019).

6. Environmental Contaminants and Ecosystem Health

6.1. Disruption of Ecological Balance; Impact on Food Chains and Ecosystems

Across the globe, anthropogenic activities increase the input of contaminants in ecosystems, disturbing their regime and potentially threaten their health. The delicate balance of nature is intricately intertwined with the diverse array of organisms inhabiting our planet, introduction of contaminants into the environment has posed a significant threat to this delicate balance between ecosystems and organisms. One of the most widely observable impact is the loss of apex predators (Estes *et al*., 2016, Heath *et al*., 2014). As a consequence, there is a surge in attention concerning the effect of contaminants at the food web scale. The response of a food-chain to external disturbances may be gradual, or remains inert until a sudden regime shift, a critical transition, where the food-chain structure and composition are drastically altered (Baudrot *et al.,* 2018). These sudden changes often correspond to abrupt degradation of ecosystem services which are difficult, sometimes impossible, to recover (Rockstrom *et al.,* 2009).

One of the most damaging effects of contaminants is food web which is the "trophic cascade" where the food chain is disturbed by either a change in the abundance of the highest trophic level (i.e., top-down cascade) or by a change in the resources (*i.e.,* bottom-up cascade) (Carpenter *et al*., 1985, Heath *et al*., 2014). For instance, a top-down cascade initiated by a decrease in predator abundance releases the pressure on its prey which consequently increase in density and subsequently induce more pressure on prey resources. The reverse process of a bottom-up cascade in a three-level food chain, may start by an increase/decrease of prey resources favouring/constraining in first prey density and then its predator (Heath *et al*., 2014). Another well-known bottom-up effect is the "paradox of enrichment" where the increase of prey resources (*i.e.,* the carrying capacity) has a destabilizing effect where population dynamics of prey and predators shift from a fixed stable state to an oscillating behaviour

(Rosenzweig *et al.*, 1971). However, recent studies in aquatic systems showed that exposure to contaminants may inhibit the paradox of enrichment by reducing oscillations and driving the system back to a fixed equilibrium (Prosnier*et al.*, 2015, Huang *et al.*, 2015).

6.2. Effects on Biodiversity Like Species Extinction, Habitat Degradation

Environmental contaminants disrupt ecosystem functions, degrade habitats and exacerbate species extinction rates, often through synergistic and cascading effects on ecological networks (Samuel *et al.*, 2023). One of the most critical impacts of environmental contaminants is their role in species extinction. Toxicants exert sub-lethal and lethal effects on organisms across trophic levels. For instance, POPs (Persistent Organic Pollutants) bioaccumulate and biomagnify within food webs, disproportionately affecting apex predators like raptors, cetaceans and large carnivorous fish (Vagi *et al.*, 2021). These chemicals impair physiological functions, including endocrine and immune responses, resulting in reduced reproductive success and increased mortality rates. Amphibians, with their permeable skin and reliance on aquatic habitats, are particularly vulnerable, with contamination contributing to their alarming global population declines (Vos *et al.*, 2000).

Habitat degradation is another profound consequence. Contaminants alter the structural and functional integrity of ecosystems by disrupting biogeochemical cycles and depleting essential resources (Samuel *et al.*, 2023). Eutrophication, driven by agricultural runoff containing nitrogen and phosphorus, creates hypoxic zones in aquatic systems, which are inhospitable to most aerobic organisms (Purohit, 2018). Similarly, soil contamination by heavy metals compromises microbial communities, nutrient availability and the viability of native flora, thereby cascading into declines in dependent faunal populations. Microplastics and other synthetic pollutants further exacerbate habitat degradation by obstructing natural processes and serving as vectors for additional toxicants (Innocent *et al.*, 2024).

The ramifications of biodiversity loss extend beyond individual species, affecting ecosystem services crucial for maintaining planetary health (Sharma *et al.*, 2023). The depletion of keystone species due to contaminants disrupts these services, leading to ecosystem instability. Additionally, ecological homogenization caused by the decline of sensitive species and the proliferation of more tolerant invasive species further diminishes ecosystem resilience (Cinar *et al.*, 2014).

6.3. Alteration of Ecosystem Services

Ecosystem services such as pollination, nutrient cycling, water purification and carbon sequestration are essential for maintaining ecological balance and supporting human well-being (Power, 2010). However, environmental contaminants, including heavy metals, pesticides, plastics and industrial waste, significantly alter these services, compromising ecosystem functionality and resilience (Adla *et al.*, 2022). Pollination is a critical ecosystem service that underpins both natural ecosystems and agricultural productivity. Approximately 75% of flowering plants rely on animal pollinators including bees, butterflies, moths, birds, bats and beetles, to carry out this essential function (Kaur and Kaleka, 2022). Beyond ecological significance, pollination is crucial for the production of many of the crops that form the basis of human diets, such as fruits, vegetables, nuts and oilseeds (Abrol and Shankar, 2012).

Environmental pollutants, however, are severely undermining the process of pollination. Pesticides, particularly neonicotinoids and organophosphates, are among the most significant threats (Abrol and Shankar, 2012). They directly harm pollinators causing neurotoxic effects, disorientation, reduced reproductive success and increased susceptibility to disease. Sub-lethal doses of these chemicals impair memory, navigation and foraging efficiency in pollinators, leading to population declines (Manzoor and Pervez, 2021). Chronic exposure leads to colony collapse disorder (CCD), characterized by the mass disappearance of worker bees, leaving hives unsustainable (Gifford, 2014).

Studies show that ozone and other pollutants chemically alter volatile organic compounds (VOCs) emitted by plants, effectively masking their attractiveness to pollinators (Pinto*et al.*, 2010). Additionally, habitat destruction and fragmentation caused by industrial development further exacerbate the decline of pollinator populations by reducing the availability of flowering plants and nesting sites (Imran *et al.*, 2023). The decline of pollinators has direct implications for biodiversity and food availability. A reduction in pollination services limits the reproductive success of both wild and cultivated plants. This not only diminishes crop yields but also disrupts ecosystems that rely on plant diversity for stability (Allen-Wardell *et al.*, 1998).

Nutrient cycling is particularly vulnerable to contamination. Pollutants can alter soil microbial communities, which play a crucial role in nutrient decomposition and transformation (Morgado *et al.*, 2018). Heavy metals inhibit microbial activity, reducing the availability of nutrients for plants. Similarly, excessive nitrogen from agricultural runoff can lead to eutrophication in aquatic ecosystems, causing algal blooms, oxygen depletion, and loss of aquatic life

(Pérez-de-Mora *et al.*, 2006). These changes not only disrupt local ecosystems but also impair ecosystem services that support agriculture, fisheries and water purification. Persistent organic pollutants (POPs) accumulate in soils and water bodies, altering the balance of nutrient cycles over time (Akhtar *et al.*, 2021). Pharmaceuticals and personal care products often found in wastewater interfere with the nutrient uptake of aquatic plants and animals, further disrupting trophic nutrient dynamics (Yang *et al.*, 2020).

6.4. Contaminants in Sensitive Ecosystems

Sensitive ecosystems such as wetlands and coral reefs are critical to global biodiversity and human well-being. However, these ecosystems are increasingly threatened by a range of environmental contaminants (Chivian, 2002). Wetlands such as marshes, swamps, and bogs, act as natural filters, trapping pollutants and improving water quality (Ramachandra and Solanki, 2007). However, their high nutrient retention capacity makes them particularly vulnerable to nutrient overload, primarily from agricultural runoff rich in nitrogen and phosphorus. This phenomenon, called eutrophication, triggers algal blooms that deplete oxygen levels in water, leading to "dead zones" unsuitable for most aquatic life. For instance, the Gulf of Mexico experiences an extensive dead zone each year due to nutrient runoff from the Mississippi River Basin (Rabalais *et al.*, 2002).

Pesticides and herbicides used in agriculture also enter wetland systems through surface runoff and leaching. These chemicals can be toxic to aquatic organisms, causing developmental, behavioural, and reproductive anomalies in amphibians and fish (Blann *et al.*, 2009). For example, atrazine, a widely used herbicide has been shown to feminize male frogs, disrupting their reproduction and population dynamics (Hayes *et al.*, 2011). Heavy metals, including mercury, lead, and cadmium, pose another significant threat. These metals often enter wetlands through industrial discharge, mining activities and atmospheric deposition (Rai, 2008). Once in the ecosystem, heavy metals bioaccumulate in the food chain, leading to toxic effects on predators and human populations relying on wetlands for subsistence. Mercury, in particular, is converted to its toxic form, methyl mercury, in wetland sediments, posing a direct threat to fish and the larger food web (Ullrich *et al.*, 2001).

Coral reefs, often referred to as the "rainforests of the sea," harbour unparalleled biodiversity and provide ecosystem services such as coastal protection, fisheries and tourism (Sobha *et al.*, 2023). However, they are acutely sensitive to environmental contaminants due to their dependence on clear, nutrient-poor waters. Sedimentation, driven by deforestation, urbanization and agriculture, is a primary concern for coral reefs. Sediments smother corals, block sunlight required for photosynthesis by their symbiotic algae (zooxanthellae), and reduce their growth and reproduction (Selim, 2007). Additionally, sediment can transport pollutants such as pesticides and heavy metals to reef environments. Increased nitrogen and phosphorus levels promote the growth of macroalgae, which compete with corals for space and resources (Zhao *et al.*, 2021). Over time, this shift can lead to algal-dominated systems, fundamentally altering reef ecosystems. The Great Barrier Reef, for instance, has experienced episodes of macroalgal blooms due to nutrient runoff from agricultural lands (DeGeorges, 2017). Chemical pollutants, such as oxybenzone and octinoxate, found in sunscreens, directly harm coral reefs. These chemicals cause coral bleaching, DNA damage and deformities in juvenile corals (Florent *et al.*, 2024). Similarly, hydrocarbons from oil spills coat coral surfaces, impairing gas exchange and leading to widespread coral mortality. Microplastics are an emerging threat to coral reefs (Elumalai *et al.*, 2024). These tiny plastic particles are ingested by reef organisms, leading to physical damage, starvation and exposure to toxic additives. Additionally, microplastics can serve as vectors for pathogens, further stressing coral communities already weakened by climate change (Egbeocha *et al.*, 2018).

The degradation of these ecosystems has profound ecological and socio-economic implications. Loss of biodiversity disrupts food webs, reduces the availability of ecosystem services and increases vulnerability to natural disasters. For human populations, this translates to reduced livelihoods, food insecurity and diminished cultural values associated with these environments (Hernández-Delgado, 2015).

7. Regulation and Management of Environmental Contaminants

Environmental regulations are crucial, and their very existence and implementation act as a potent disincentive, since they have the potential to prevent the most extreme excesses. Even though environmental regulations have made the environment better, they can result in businesses incurring what could be considered excessive costs. One should carefully consider the benefits and drawbacks of regulation in relation to other approaches when determining the best course of action for a given environmental issue. Regulation is one way to improve quality of the environment, but it should be combined with other tactics, such as incentives.

Even though environmental regulation has historically contributed to better environmental outcomes, scientific evidence indicates that many environmental problems have gotten worse. These new issues are being caused by technological advancements in areas such as biotechnology, nuclear energy, and wind power, all of which require some form of regulation. If not controlled, each of these industries might gain more profits, but the environment and society as a whole would suffer significantly. Effective environmental management has far lower economic costs than irreversible environmental devastation to the planet's life systems (Kobayashi, 2019).

Planning controls and soil contaminant estimates are standardized nationally under the National Environmental Standard for Assessing and Managing Contaminants in Soil to Protect Human Health (NESCS) Regulations 2011. It guarantees that land impacted by soil contamination is properly identified and evaluated prior to development; if required, the property is remediated or the toxins are controlled to make the site suitable for human habitation. The Environmental Protection Agency must establish a source reduction program which collects and disseminates information, provides financial assistance to States, and implements the other activities, according to the 1990 Pollution Prevention Act passed by Congress (USEPA, 2010). Pollution prevention is reducing or eliminating waste at the source by modifying production processes, promoting the use of nontoxic or less toxic substances.

7.1. International Treaties and Agreements

7.1.1. Stockholm Convention

This historic agreement was accepted by all nations, this agreement shield the environment and people from the damaging effects of POPs. POPs are semi-volatile, low-water, high fat-soluble substances. The environment, animals, and human health are all being negatively impacted by their toxic and dangerous regional and worldwide dispersion. Climate change is caused by global warming, which is also a result of these harmful chemical substances. Although the Stockholm Convention was ratified in 2001, but it wasn't operative until 2004 with 110 parties worldwide. The main goals of this convention are to control the intentional and unintentional production of POPs, limit storage, stockpiling, and waste generated with POPs, ensure a safe process for adding new POPs during production, enforce general responsibility and use restrictions, and provide financial and technical assistance to the process (Jha *et al.,* 2022).

7.1.2. Basel Convention

Basel Convention is on the control of transboundary movements of hazardous wastes and their disposal, which became operative in 1992. In terms of hazardous wastes and other wastes, it is the most extensive international environmental pact. As of March 31, 2011, it has 175 parties, making its membership almost universal. The Convention seeks to safeguard the environment and human health against the negative consequences of the production, transboundary transportation, and handling of hazardous and other wastes. The Basel Convention requires its Parties to make sure that hazardous wastes and other wastes are managed and disposed of in an environmentally responsible way. It also governs the transboundary flows of these wastes. Toxic, poisonous, explosive, flammable, corrosive, ecotoxic, and infectious wastes are also covered under the Convention. Parties are also required to reduce the amount of trash that is transported, treat and dispose of waste as close to its source as feasible, and avoid or reduce waste creation at the source (UNEP, 2011).

7.1.3. Rotterdam Convention

An international agreement known as the Rotterdam Convention was created to help nations make well-informed decisions about the trafficking of dangerous chemicals. February 24, 2004 marked the Convention's entry into force. To prevent the export of banned hazardous chemicals to nations that do not want them, the Rotterdam Convention institutes a prior informed consent ("PIC") process. Industrial chemicals (such as actinolite asbestos, anthophyllite asbestos, decabromodiphenyl ether, perfluorooctane sulfonic acid, perfluorooctane sulfonates, perfluorooctane sulfonamides, and perfluorooctane sulfonyls, tributyltin compounds, etc.) and pesticides (such as alachlor, 2,4,5-T and its salts and esters, Aldrin, chlorobenzilate, DDT, ethylene oxide, etc.) that meet the requirements for listing under the Rotterdam Convention are covered by the Convention, usually because they are prohibited or severely restricted in party countries, or because they are extremely dangerous pesticide formulations (McDorman, 2004).

7.2. National and Local Regulations

7.2.1. Clean Air Act

The comprehensive federal regulation that governs air pollutants from both fixed and mobile sources is known as the Clean Air Act (CAA). This statute, among other things, gives EPA the authority to set National Ambient Air Quality Standards (NAAQS) in order to control emissions of dangerous air pollutants and safeguard the public's health and welfare. In order to address the hazards to public health and welfare presented by several common air pollutants, one of the Act's objectives was to establish and attain NAAQS in each state by 1975. In order for the states to meet these standards, they were instructed to create state implementation plans (SIPs) that applied to the relevant industrial sources in their states.

The Clean Air Act's Section 112 deals with dangerous air pollution emissions. Only a few standards were created under the risk-based program that CAA had in place prior to 1990. Section 112 was amended by the CAA Amendments of 1990 to first mandate the release of technology-based standards for significant sources and specific area sources. The term "major sources" refers to a stationary source or collection of stationary sources that emit or have the capacity to emit 10 tons or more of a hazardous air pollutant annually, or 25 tons or more of a combination of hazardous air pollutants annually. Any stationary source that is not a large source is referred to as an "area source".

Section 112 mandates the EPA set emission regulations for significant sources that call for the greatest possible reduction in emissions of dangerous air pollutants. The term maximum achievable control technology (MACT) standards are frequently used to describe these emission requirements. For each source category, EPA must assess the technology-based MACT standards eight years after they are released to see if there is any residual risk. If so, the standards must be revised to mitigate the risk (USEPA, 2024).

7.2.2. Water Quality Standards

WHO creates international standards for human health and water quality in the form of guidelines that serve as the foundation for standards and regulations in both developed and developing nations across the globe. One significant environmental predictor of health is the quality of drinking water. Waterborne disease prevention and control are predicated on the assurance of drinking water safety. The WHO Expert Advisory Panel (CPCB, 1974) is consulted along with national authorities, international agencies, and WHO member states (India is one of the member states) in order to draft the guidelines.

The factors that define the quality of water serve as the foundation for water quality standards. Numerous water quality standards establish a maximum concentration of a material in a given medium (such as water, sediment, or biota) that won't be hazardous when the medium is utilized consistently for a single, specific purpose. In order to maintain biological functioning, water quality limits for certain other variables, including dissolved oxygen (DO), are set at the lowest permissible concentration. In order to establish water quality standards for various water bodies, the CPCB has categorized the nation's water resources based on their intended uses. Based on the water quality standards for a certain usage, five classes (A–E) have been established. Surveys have been conducted to compare the actual river quality classification with that needed to maintain the specified best use, and the five classes have been used to define quality objectives for sections of the Yamuna and Ganga rivers. (Sharma and Gulati, 2014).

Table 1. Water quality criteria is mentioned below in the table (CPCB, 1974)

Designated-Best-Use	Class of Water	Criteria
Drinking Water Source without conventional treatment but after disinfection	A	Total Coliforms Organism MPN/100ml shall be 50 or less pH between 6.5 and 8.5 Dissolved Oxygen 6mg/l or more Biochemical Oxygen Demand 5 days 20C 2mg/l or less
Outdoor bathing (Organised)	B	Total Coliforms Organism MPN/100ml shall be 500 or less, pH between 6.5 and 8.5, Dissolved Oxygen 5mg/l or more Biochemical Oxygen Demand 5 days 20C 3mg/l or less
Drinking water source after conventional treatment and disinfection	C	Total Coliforms Organism MPN/100ml shall be 5000 or less, pH between 6 to 9 Dissolved Oxygen 4mg/l or more Biochemical Oxygen Demand 5 days 20C 3mg/l or less
Propagation of Wild life and Fisheries	D	pH between 6.5 to 8.5, Dissolved Oxygen 4mg/l or more Free Ammonia (as N) 1.2 mg/l or less
Irrigation, Industrial Cooling, Controlled Waste disposal	E	pH between 6.0 to 8.5 Electrical Conductivity at 25C micro mhos/cm Max.2250 Sodium absorption Ratio Max. 26 Boron Max. 2mg/l
	Below-E	Not Meeting A, B, C, D & E Criteria

7.3. Remediation and Cleanup Technologies

7.3.1. Bioremediation

The process of removing contaminated materials from the environment by means of bacteria, algae, fungi, and yeast is known as bioremediation (Enerijiofi *et al.,* 2021). Bioremediation was carried out using microbes from diverse settings. Microbes include, among others, *Achromobacter, Alcaligenes, Xanthobacter, Arthrobacter, Pseudomonas, Bacillus, Mycobacterium, Corynebacterium, Flavobacterium,* and *Nitrosomonas* (Sharma, *et al.,* 2021). By giving the organisms, the nutrition and other chemicals they require to function at their best, bioremediation can break down or detoxify toxic pollutants. Every step of the metabolic process depends on enzymes (Malik *et al.,* 2022; Chen *et al.,* 2016). Because of their extensive and pervasive use in the environment, organic compounds (OCs), such as flame retardants and biocides, are today seen as a hazard to almost all living forms on earth. Microbes can break down the majority of OCs in the environment, including polychlorinated biphenyls (PCBs), polybrominated biphenyl ethers (PBEs), and polycyclic aromatic hydrocarbons (PAHs) (Yaashikaa *et al.,* 2022).

The release of hazardous metals into soils and rivers above or below the surface poses intolerable health risks, environmentalists are particularly concerned about toxic elemental pollution (Joutey *et al.,* 2013). It's important to understand that microbes can only alter the oxidation states of metals in order to stabilize them; they cannot break down metal ions (Thakare *et al.,* 2021). Like any other nutrient in the cells, they are able to detoxify and digest metals. It has been observed that a variety of microorganisms can bioremediate both organic and inorganic contaminants which are given in Table 2.

7.3.2. Phytoremediation

Utilizing hyperaccumulator plants and their rhizosphere microbes, phytoremediation is a green technique that stabilizes, transfers, or breaks down contaminants in soil, water, and the environment (Liu *et al.,* 2020). This approach is regarded as being very efficient, economical, and ecologically beneficial (Ashraf *et al.,* 2010; Nedjimi, 2020). There are several phytoremediation techniques that can be applied to the remediation of heavy metal-contaminated soils, depending on the soil conditions, pollutant, and plant species. These include: phytostabilization, which uses plants to lower the bioavailability of heavy metals in soil, Using hydroponically grown plants to absorb or adsorb heavy metal ions from groundwater and aqueous waste is known as phytoextraction. Using plants to extract and remove heavy metals from soil is known as phytovolatilization. Using plants to absorb heavy metals from soil and release them into the atmosphere as volatile compounds is known as phytofiltration (Ernst, 2005; Marques *et al.,* 2009). Phytodegradation and rhizodegradation are two more phytoremediation techniques that break down organic contaminants.

Table 2. Potentially hazardous organic and inorganic pollutants and their degrading microbes

Substrate	Compound	Microorganisms	References
Organic substrate	Chlorobenzenes	*P. putida* (GJ31)	Kunze *et al.*, 2006
	Azo dyes effluents	*Exiguobacterium indicum, B. cereus, E. aurantiacums* and *A. baumanii*	Raquel *et al.*, 2013
	Sulfonate benzene	*A. radiobacter* (S2)	Tripathi *et al.*, 2014
	Aromatic hydrocarbon	*Acinetobacter* sp., *Microbacterium* sp., *Pseudomonas* sp. and *Ralstonia* sp.	Basu *et al.*, 2018
	Phenol	*Alcaligenes odorans, Corynebacterium propinquum, B. subtilis*, and *P. aeruginosa*	Gaur *et al.*, 2018
	Oils	*Alcaligenes odorans, Corynebacterium propinquum, P. aeruginosa* and *Fusarium* sp.	Pande *et al.*, 2020
	Polycyclic aromatic hydrocarbon	*Burkholderia* sp., *Myceliophthorathermophila*	Mohapatra *et al.*, 2021
Inorganic substrate	Hg^{2+}	*Cyclotella cryptica, Chlamydomonas reinhardtii, Pseudochlorococcum typicum, Spirogyra hyaline*	Shah and Jain, 2020
	Heavy metals, mercury, nickel and lead	*Saccharomyces cerevisiae* and, *Cunninghamella elegans*	Duc *et al.*, 2021
	Cobalt, chromium, copper, and lead	*Lysinibacillus sphaericus* CBAM5	Kharangate-lad *et al.*, 2021
	Uranium, copper, nickel, chromium	*P. aeruginosa, Aeromonas* sp.	Gaur *et al.*, 2022

7.4. Role of Environmental Monitoring and Risk Assessment

Monitoring the environment is essential to safeguarding both the environment and human health. Pollution will always be generated because of the growing human population, the expansion of industry and energy use, and the advancements made in pollution control. Therefore, there is still a huge demand for environmental monitoring. To improve the precision and economic viability of monitoring programs, more developments in the creation, use, and automation of monitoring devices are required (Artiola and Brusseau, 2019).

Average concentrations of contaminants in the environmental media i.e. (soil, water, air, and sediment) are typically revealed by environmental monitoring. When the measurement is done at the discharge point (air or water), peak concentrations are obtained. Because the chemical can go through a number of changes and transfer between media before reaching the point of measurement, these data are crucial for estimating the order of magnitude of emissions, but they typically do not provide a comprehensive picture of the concentrations in the environment.

In order to record concentrations in the environmental compartments, monitoring entails a variety of tasks, including creating a statistically sound sampling plan, choosing a sampling technique, transporting and storing the sample, meeting analytical and laboratory requirements for the analysis, and handling data quality and reporting concerns. Lastly, monitoring data can be used to "calibrate" or "validate" Environmental Functional Materials (EFMs) by providing a picture of the pollution of a particular ecosystem or compartment. Environmental risk assessment (ERA) is the term used to describe the detrimental health impacts that arise from ecological and human exposure to environmental hazards (Abbassi *et al.*, 2012; Dadashzadeh *et al.*, 2014). Over the past three decades, ERA has received a lot of attention due to growing public awareness about the effects of chemicals and the risks they pose to wildlife and the ecosystem.

Perrodin *et al.* (2011) state that the majority of risk assessment methods developed at the international level are divided into four main phases: (i) hazard identification (i.e., substances that can have negative effects based on exposure levels), (ii) dose-response relationship evaluation (i.e., the relationship between exposure to a hazardous substance and the incidence on health and the environment), (iii) exposure estimation (i.e., the environmental fate of substances and exposure pathways for populations or ecosystems), and (iv) risk characterization (the likelihood of hazard occurrence for an individual in a given population or an organism in a given ecosystem).

There are two primary benefits to risk assessment. First of all, it is customarily focused on the local level and considers territorial features when determining the vulnerability of particular targets. Although it can be expanded to the scale of a region or even a nation, its scope is typically restricted to the study of a single site. However, because they seek to offer a thorough understanding of numerous stressors, exposures, and receptors, regional environmental risk assessments are more intricate (Sleeswijk, 2010). A risk assessment study's choice of contaminants (such as radionuclides and other inorganic and organic compounds) to examine is a crucial component. The processes that are thought to be significant and pertinent in describing the behaviour of the environment are known as other features. Last but not least, models that can replicate the pertinent processes across time must be found. Simulating long-term behaviour over thousands or tens of thousands of years is undoubtedly a unique problem for risk assessment studies involving contaminated sites (Ugbede *et al.*, 2020).

8. Conclusion

Environmental pollutants are growing more and more common in our ecosystems. These contaminants range from heavy metals, pesticides, plastics, and medications to new dangers like nanomaterials and compounds that disrupt hormones. These contaminants have harmful effects at the molecular, cellular, organismal, and ecological levels and reach the environment through a variety of routes. They damage DNA, cause oxidative stress, interfere with hormones, upset physiological systems, and are linked to both acute and long-term health concerns, such as cancer, neurological conditions, respiratory diseases, and reproductive troubles. Additionally, bioaccumulation, decreased fertility, growth inhibition, and changed ecological functions affect microbes, plants, and animals.

These disturbances have an impact on biodiversity and ecosystem functions including pollination, nutrient cycling, water purification, and climate regulation as they ripple through food webs. Particularly at risk are sensitive ecosystems such as agricultural soils, wetlands, and coral reefs; pollution can result in habitat loss and the extinction of species. Their long-term, sub-lethal impacts on the environment and human health are still being discovered, underscoring the necessity of multidisciplinary study and sophisticated monitoring.

A globally coordinated, policy-supported, and science-driven action is necessary to reduce these hazards. This includes more stringent laws, better waste disposal, eco-friendly technology, and sustainable methods. In the end, maintaining ecological balance and safeguarding human health are closely related objectives. In addition to being a scientific need, preventing environmental contamination is also a moral obligation that calls for swift, coordinated international action to create a healthier, more resilient planet.

References

Abbassi, R., Dadashzadeh, M., Khan, F., & Hawboldt, K. (2012). Risk-based prioritisation of indoor air pollution monitoring using computational fluid dynamics. Indoor and Built Environment, 21(5), 663-673.

Abrol, D. P., & Shankar, U. (2012). Pollination in oil crops: recent advances and future strategies. Technological Innovations in Major World Oil Crops, 2, 221-267.

Adami, G., Pontalti, M., Cattani, G., Rossini, M., Viapiana, O., Orsolini, G., & Fassio, A. (2022). Association between long-term exposure to air pollution and immune-mediated diseases: a population-based cohort study. RMD open, 8(1), e002055.

Adla, K., Dejan, K., Neira, D., & Dragana, S. (2022). Degradation of ecosystems and loss of ecosystem services. In One health. Academic Press, 281-327.

Aftab, T., & Hakeem, K. R. (2023). Environmental Pollution Impact on Plants: Survival Strategies Under Challenging Conditions. CRC Press.

Agnihotri, S. K., & Kesari, K. K. (2019). Mechanistic effect of heavy metals in neurological disorder and brain cancer. Networking of Mutagens in Environmental Toxicology, 25-47.

Ahmed, S. F., Mofijur, M., Nuzhat, S., Chowdhury, A. T., Rafa, N., Uddin, M. A. & Show, P. L. (2021). Recent developments in physical, biological, chemical, and hybrid treatment techniques for removing emerging contaminants from wastewater. Journal of hazardous materials, 416, 125912.

Akhtar, A. B. T., Naseem, S., Yasar, A., & Naseem, Z. (2021). Persistent organic pollutants (POPs): sources, types, impacts, and their remediation. Environmental Pollution and Remediation, 213-246.

Aktar, M. W., Sengupta, D., & Chowdhury, A. (2009). Impact of pesticides use in agriculture: their benefits and hazards. Interdisciplinary toxicology, 2(1), 1.

Alegbeleye, O. O., Singleton, I., & Sant'Ana, A. S. (2018). Sources and contamination routes of microbial pathogens to fresh produce during field cultivation: A review. Food microbiology, 73, 177-208.

Ali, H., & Khan, E. (2019). Trophic transfer, bioaccumulation, and biomagnification of non-essential hazardous heavy metals and metalloids in food chains/webs-Concepts and implications for wildlife and human health. Human and Ecological Risk Assessment: An International Journal, 25(6), 1353-1376.

Allen-Wardell, G., Bernhardt, P., Bitner, R., Burquez, A., Buchmann, S., Cane, J. & Walker, S. (1998). The potential consequences of pollinator decline on the conservation of biodiversity and stability of food crop yields. Conservation biology, 8-17.

Alloway, B. J. (Ed.). (2012). Heavy metals in soils: trace metals and metalloids in soils and their bioavailability, Springer Science & Business Media, 22.

Almeida-Toledano, L., Navarro-Tapia, E., Sebastiani, G., Ferrero-Martinez, S., Ferrer-Aguilar, P., García-Algar, OO, & Gomez-Roig, M. D. (2024). Effect of prenatal phthalate exposure on fetal development and maternal/neonatal health consequences: A systematic review. Science of The Total Environment, 175080.

Althobaiti, N. A. (2024). Heavy metals exposure and Alzheimer's disease: Underlying mechanisms and advancing therapeutic approaches. Behavioural Brain Research, 115212.

American Bird Conservancy, Mineau, P., & Palmer, C. (2013). The impact of the nation's most widely used insecticides on birds. American Bird Conservancy.

Anderson, B., Peyster, A. D., Gad, S. C., Hakkinen, P. J., & Kamrin, M. (2005). Encyclopedia of toxicology. Academic Press.

Anetor, G. O., Nwobi, N. L., Igharo, G. O., Sonuga, O. O., & Anetor, J. I. (2022). Environmental pollutants and oxidative stress in terrestrial and aquatic organisms: examination of the total picture and implications for human health. Frontiers in Physiology, 13, 931386.

Anjum, N. A., Singh, H. P., Khan, M. I. R., Masood, A., Per, T. S., Negi, A. & Ahmad, I. (2015). Too much is bad-an appraisal of phytotoxicity of elevated plant-beneficial heavy metal ions. Environmental Science and Pollution Research, 22, 3361-3382.

Artiola, J. F., & Brusseau, M. L. (2019). The role of environmental monitoring in pollution science. In Environmental and pollution science. Academic Press, 149-162.

Ashraf, M., Ozturk, M., & Ahmad, M. S. A. (2010). Toxins and their phytoremediation. Plant adaptation and phytoremediation, 1-32.

Awual, M. R., & Hasan, M. M. (2014). Novel conjugate adsorbent for visual detection and removal of toxic lead (II) ions from water. Microporous and mesoporous materials, 196, 261-269.

Bai, X., Lutz, A., Carroll, R., Keteles, K., Dahlin, K., Murphy, M., *et al.* (2018). Occurrence, Distribution, and Seasonality of Emerging Contaminants in Urban Watersheds. Chemosphere 200, 133-142.

Balducci, C., Perilli, M., Romagnoli, P., & Cecinato, A. (2012). New developments on emerging organic pollutants in the atmosphere. Environmental Science and Pollution Research, 19, 1875-1884.

Basu, S., Rabara, R. C., Negi, S., & Shukla, P. (2018). Engineering PGPMOs through gene editing and systems biology: a solution for phytoremediation?. Trends in Biotechnology, 36(5), 499-510.

Baudrot, V., Fritsch, C., Perasso, A., Banerjee, M., & Raoul, F. (2018). Effects of contaminants and trophic cascade regulation on food chain stability: Application to cadmium soil pollution on small mammals–raptor systems. Ecological Modelling, 382, 33-42.

Beltran, E. M., Pablos, M. V., Fernandez Torija, C., Porcel, M. A., and González Doncel, M. (2020). Uptake of Atenolol, Carbamazepine and Triclosan by Crops Irrigated with Reclaimed Water in a Mediterranean Scenario. Ecotoxicology and Environment Safety, 191, 110171.

Berglund, L. G., Rissanen, S., Jussila, K., Kaufman, J. W., Piirila, P., Savolainen, K. M., & Juvonen, R. (2020). Physiological and toxicological considerations. In Industrial ventilation design guidebook (pp. 111-226). Academic Press.

Bergman, Å., Heindel, J. J., Jobling, S., Kidd, K. A., & Zoeller, R. T. (2013). Endocrine disrupting chemicals-2012. Geneva: United Nations Environment Programme and the World Health Organization.

Bishnoi, M. M., Verma, A., Kushwaha, A., and Goswami, S. (2022). "Social Factors Influencing Household Waste Management," in Emerging Trends to Approaching Zero Waste, Elsevier, 197-213.

Blaine, A. C., Rich, C. D., Hundal, L. S., Lau, C., Mills, M. A., Harris, K. M., *et al.* (2013). Uptake of Perfluoroalkyl Acids into Edible Crops via Land Applied Biosolids: Field and Greenhouse Studies. Environmental Science and Technology, 47, 14062-14069.

Blann, K. L., Anderson, J. L., Sands, G. R., & Vondracek, B. (2009). Effects of agricultural drainage on aquatic ecosystems: a review. Critical reviews in environmental science and technology, 39(11), 909-1001.

Bradley, P. M., Journey, C. A., Romanok, K. M., Barber, L. B., Buxton, H. T., Foreman, W. T., & Villeneuve, D. L. (2017). Expanded target-chemical analysis reveals extensive mixed-organic-contaminant exposure in US streams. Environmental science & technology, 51(9), 4792-4802.

Briffa, J., Sinagra, E., & Blundell, R. (2020). Heavy metal pollution in the environment and their toxicological effects on humans. Heliyon, 6(9).

Carpenter, S. R., Kitchell, J. F., & Hodgson, J. R. (1985). Cascading trophic interactions and lake productivity. BioScience, 35(10), 634-639.

Chakraborty, B. K. (2023). Effect of pesticide and heavy metal toxicants on fish and human health. Journal of Crop and Weed, 19(1), 110-118.

Chen, B. Y., Ma, C. M., Han, K., Yueh, P. L., Qin, L. J., & Hsueh, C. C. (2016). Influence of textile dye and decolorized metabolites on microbial fuel cell-assisted bioremediation. Bioresource technology, 200, 1033-1038.

Chen, Y., Yang, J., Yao, B., Zhi, D., Luo, L., & Zhou, Y. (2022). Endocrine disrupting chemicals in the environment: Environmental sources, biological effects, remediation techniques, and perspective. Environmental Pollution, 310, 119918.

Chivian, E. (2002). Biodiversity: its importance to human health. Center for Health and the Global Environment, Harvard Medical School, Cambridge, MA, 23.

Cimmino, I., Fiory, F., Perruolo, G., Miele, C., Beguinot, F., Formisano, P., & Oriente, F. (2020). Potential mechanisms of bisphenol A (BPA) contributing to human disease. International journal of molecular sciences, 21(16), 5761.

Çinar, M. E., Arianoutsou, M., Zenetos, A., & Golani, D. (2014). Impacts of invasive alien marine species on ecosystem services and biodiversity. a pan-European review. Aquatic Invasions, 9(4), 391-423.

Cobanoglu Ozyigitoglu, G. (2020). Use of lichens in biological monitoring of air quality. Environmental Concerns and Sustainable Development. Air, Water and Energy Resources, 1, 61-95.

Coetzee, J. J., Bansal, N., & Chirwa, E. M. (2020). Chromium in environment, its toxic effect from chromite-mining and ferrochrome industries, and its possible bioremediation. Exposure and health, 12, 51-62.

Colborn, T., Vom Saal, F. S., & Soto, A. M. (1993). Developmental effects of endocrine-disrupting chemicals in wildlife and humans. Environmental health perspectives, 101(5), 378-384.

Cooper, R. L., Laws, S. C., Das, P. C., Narotsky, M. G., Goldman, J. M., Lee Tyrey, E., & Stoker, T. E. (2007). Atrazine and reproductive function: mode and mechanism of action studies. Birth Defects Research Part B: Developmental and Reproductive Toxicology, 80(2), 98-112.

Cordoba-Tovar, L., Marrugo-Negrete, J., Baron, P. R., & Díez, S. (2022). Drivers of biomagnification of Hg, As and Se in aquatic food webs: a review. Environmental Research, 204, 112226.

D'Surney, S. J., & Smith, M. D. (2005). Chemicals of environmental concern,100.

Da Costa, J. P., Rocha-Santos, T., & Duarte, A. C. (2020). The environmental impacts of plastics and micro-plastics use, waste and pollution: EU and national measures.

Dadashzadeh, M., Khan, F., Abbassi, R., & Hawboldt, K. (2014). Combustion products toxicity risk assessment in an offshore installation. Process Safety and Environmental Protection, 92(6), 616-624.

Datta, S., Radhapyari, K., Saha, N., & Samanta, S. K. (2022). Recent trends in the application of biowaste for hazardous radioactive waste treatment. In Environmental Sustainability and Industries. Elsevier, 159-192.

DeGeorges, P. A. (2017). An historical overview of impacts from land-based pollution on community based natural resource management (CBNRM) as it applies to marine fisheries & coral reefs in the tropics. J Fish Res., 1 (1), 15-27.

Doe, J., Smith, R., & Lee, P. (2019). Occupational health risks and preventive strategies in hazardous workplaces. Occupational Safety Journal, 45(3), 120-135.

Driscoll, C. T., Han, Y. J., Chen, C. Y., Evers, D. C., Lambert, K. F., Holsen, T. M., & Munson, R. K. (2007). Mercury contamination in forest and freshwater ecosystems in the northeastern United States. BioScience, 57(1), 17-28.

Duc, H. D., Hung, N. V., & Oanh, N. T. (2021). Anaerobic Degradation of Endosulfans by a Mixed Culture of Pseudomonas sp. and Staphylococcus sp., Applied Biochemistry and Microbiology, 57(3), 327-334.

Dutta, S., Mitra, M., Agarwal, P., Mahapatra, K., De, S., Sett, U., & Roy, S. (2018). Oxidative and genotoxic damages in plants in response to heavy metal stress and maintenance of genome stability. Plant signaling and behavior, 13(8), 1460048.

Ebele, A. J., Abou-Elwafa Abdallah, M., and Harrad, S. (2017). Pharmaceuticals and Personal Care Products (PPCPs) in the Freshwater Aquatic Environment. Emerging Contaminants,

Edo, G. I., Samuel, P. O., Oloni, G. O., Ezekiel, G. O., Ikpekoro, V. O., Obasohan, P. & Agbo, J. J. (2024). Environmental persistence, bioaccumulation and ecotoxicology of heavy metals. Chemistry and Ecology, 40(3), 322-349.

Egbeocha, C. O., Malek, S., Emenike, C. U., & Milow, P. (2018). Feasting on microplastics: ingestion by and effects on marine organisms. Aquatic Biology, 27, 93-106.

El-Kalliny, A. S., Abdel-Wahed, M. S., El-Zahhar, A. A., Hamza, I. A., & Gad-Allah, T. A. (2023). Nanomaterials: a review of emerging contaminants with potential health or environmental impact. Discover Nano, 18(1), 68.

Elumalai, P., Parthipan, P., Gao, X., Cui, J., Kumar, A. S., Dhandapani, P. & Choi, M. Y. (2024). Impact of petroleum hydrocarbon and heavy metal pollution on coral reefs and mangroves: a review. Environmental Chemistry Letters, 22(3), 1413-1435.

Enerijiofi K.E. Bioremediation for Environmental Sustainability. Elsevier; Amsterdam, The Netherlands: 2021. Bioremediation of environmental contaminants: A sustainable alternative to environmental management, 461-480.

Enerijiofi, K. E., Musa, S. I., Okolafor, F. I., Igiebor, F. A., Odozi, E. B., & Ikhajiagbe, B. (2023). Sustainable Approaches for the Remediation of Agrochemicals in the Environment. In One Health Implications of Agrochemicals and their Sustainable Alternatives, Singapore: Springer Nature Singapore, 511-543.

Enfrin, M., Dumee, L. F., & Lee, J. (2019). Nano/microplastics in water and wastewater treatment processes–origin, impact and potential solutions. Water research, 161, 621-638.

Ernst, W. G. (2012). Overview of naturally occurring Earth materials and human health concerns. Journal of Asian Earth Sciences, 59, 108-126.

Ernst, W.H. (2005). Phytoextraction of mine wastes–options and impossibilities. Geochemistry, 65, 29-42.

Estes, J. A., Heithaus, M., McCauley, D. J., Rasher, D. B., & Worm, B. (2016). Megafaunal impacts on structure and function of ocean ecosystems. Annual Review of Environment and Resources, 41(1), 83-116.

Florent, Y., Nemoto, M., Rowley, A., Villegas, M. I., & Pratt, V. (2024). Dangers of Oxybenzone in Sunscreens on Coral Reefs: Proposed Policy Approaches. Journal of Science Policy and Governance, 24(01).

Gallego, S., Montemurro, N., Béguet, J., Rouard, N., Philippot, L., Pérez, S., *et al.* (2021). Ecotoxicological Risk Assessment of Wastewater Irrigation on Soil Microorganisms: Fate and Impact of Wastewater-Borne Micropollutants in Lettuce-Soil System. Ecotoxicology and Environmental Safety, 223, 112595.

Gallo, J. M., Lee, R., Bhatia, S., & Brown, T. (2021). Acute toxic effects of environmental pollutants: A systematic review. Toxicology Reports, 8, 1034-1045.

Gallo, M. A., & Doull, J. (2019). Toxicology: Mechanisms and risk assessment. Journal of Environmental Science, 45(2), 120-134.

Gao, L., Zhang, Y., Sterling, K., & Song, W. (2022). Brain-derived neurotrophic factor in Alzheimer's disease and its pharmaceutical potential. Translational neurodegeneration, 11(1), 4.

Gaur V.K., Tripathi V., Manickam N (2022). Development in Wastewater Treatment Research and Processes. Elsevier; Amsterdam, The Netherlands. Bacterial-and fungal-mediated biodegradation of petroleum hydrocarbons in soil, 407-427.

Gaur, N., Narasimhulu, K., & PydiSetty, Y. (2018). Recent advances in the bio-remediation of persistent organic pollutants and its effect on environment. Journal of cleaner production, 198, 1602-1631.

Gautam, H., Singh, S., Prashad, H., Kumar, A., Choudhary, A., Kaur, H., & Mehta, S. (2023). Pesticide toxicity and their impact on plant growth, active constituents and productivity. In Plants and Their Interaction to Environmental Pollution, 231-252.

Gavrilescu, M., Demnerova, K., Aamand, J., Agathos, S., & Fava, F. (2015). Emerging pollutants in the environment: present and future challenges in biomonitoring, ecological risks and bioremediation. New biotechnology, 32(1), 147-156.

Gifford, C. (2014). Colony Collapse Disorder: The Vanishing Honeybee (Apis mellifera).

Gilden, R. C., Huffling, K., & Sattler, B. (2010). Pesticides and health risks. Journal of Obstetric, Gynecologic & Neonatal Nursing, 39(1), 103-110.

Giulivo, M., de Alda, M. L., Capri, E., & Barceló, D. (2016). Human exposure to endocrine disrupting compounds: Their role in reproductive systems, metabolic syndrome and breast cancer. A review. Environmental research, 151, 251-264.

Golmohammadi, M., Musavi, S. F., Habibi, M., Maleki, R., Golgoli, M., Zargar, M., & Razmjou, A. (2023). Molecular mechanisms of microplastics degradation: A review. Separation and Purification Technology, 309, 122906.

Gonzalez-Gonzalez, R. B., Parra-Arroyo, L., Parra-Saldívar, R., Ramirez-Mendoza, R. A., & Iqbal, H. M. (2022). Nanomaterial-based catalysts for the degradation of endocrine-disrupting chemicals are a way forward to environmental remediation. Materials Letters, 308, 131217.

Grandjean, P., & Landrigan, P. J. (2019). Neurobehavioral effects of developmental toxicity. The Lancet Neurology, 18(8), 721-731.

Guo, H., Wang, Y., Yao, K., Zheng, H., Zhang, X., Li, R., Wang, N., & Fu, H. (2023). The overlooked formation of environmentally persistent free radicals on particulate matter collected from biomass burning under light irradiation. Environment international, 171, 107668.

Gworek, B., Dmuchowski, W., & Baczewska-Dąbrowska, A. H. (2020). Mercury in the terrestrial environment: a review. Environmental Sciences Europe, 32(1), 128.

Hayat, M. T., Nauman, M., Nazir, N., Ali, S., & Bangash, N. (2019). Environmental hazards of cadmium: past, present, and future. In Cadmium toxicity and tolerance in plants. Academic Press, 163-183.

Hayes, T. B., Anderson, L. L., Beasley, V. R., De Solla, S. R., Iguchi, T., Ingraham, H., & Willingham, E. (2011). Demasculinization and feminization of male gonads by atrazine: consistent effects across vertebrate classes. The Journal of steroid biochemistry and molecular biology, 127, 64-73.

Heath, M. R., Speirs, D. C., & Steele, J. H. (2014). Understanding patterns and processes in models of trophic cascades. Ecology letters, 17(1), 101-114.

Hernandez-Delgado, E. A. (2015). The emerging threats of climate change on tropical coastal ecosystem services, public health, local economies and livelihood sustainability of small islands: Cumulative impacts and synergies. Marine Pollution Bulletin, 101(1), 5-28.

Hodgson, E. (Ed.). (2011). A textbook of modern toxicology. John Wiley & Sons.

Hou, D., O'Connor, D., Igalavithana, A. D., Alessi, D. S., Luo, J., Tsang, D. C., & Ok, Y. S. (2020). Metal contamination and bioremediation of agricultural soils for food safety and sustainability. Nature Reviews Earth and Environment, 1(7), 366-381.

Huang, Q., Wang, H., & Lewis, M. A. (2015). The impact of environmental toxins on predator–prey dynamics. Journal of theoretical biology, 378, 12-30.

Hughes, K. M., Price, D., Torriero, A. A., Symonds, M. R., & Suphioglu, C. (2022). Impact of fungal spores on asthma prevalence and hospitalization. International Journal of Molecular Sciences, 23(8), 4313.

Hussain, S., & Reza, M. (2023). Environmental damage and global health: understanding the impacts and proposing mitigation strategies. Journal of Big-Data Analytics and Cloud Computing, 8(2), 1-21.

Hyseni, S. (2016). Toxicological effects of nanomaterials on aqueous and terrestrial ecosystems. Center for Development and Strategy, 2016(1).

Ichas, F., & Mazat, J. P. (1998). From calcium signaling to cell death: two conformations for the mitochondrial permeability transition pore. Switching from low-to high-conductance state. Biochimica et Biophysica Acta (BBA)-Bioenergetics, 1366(1-2), 33-50.

Imran, M., Sheikh, U. A. A., Rahim, J., & Nasir, M. (2023). A Threat to Pollinators: Understanding the Impacts and Solutions: A Review. Planta Animalia, 2(2), 79-87.

Innocent, M. O., Mustapha, A., Abdulsalam, M., Livinus, M. U., Samuel, J. O., Elelu, S. A., & Muhammad, A. S. (2024). Soil Microbes and Soil Contamination. In Soil Microbiome in Green Technology Sustainability. Cham: Springer Nature Switzerland, 3-35.

Jan, A. T., Azam, M., Siddiqui, K., Ali, A., Choi, I., & Haq, Q. M. R. (2015). Heavy metals and human health: mechanistic insight into toxicity and counter defense system of antioxidants. International journal of molecular sciences, 16(12), 29592-29630.

Jeevanandam, J., Barhoum, A., Chan, Y. S., Dufresne, A., & Danquah, M. K. (2018). Review on nanoparticles and nanostructured materials: history, sources, toxicity and regulations. Beilstein journal of nanotechnology, 9(1), 1050-1074.

Jha, R., Dwivedi, S., & Modhera, B. (2022). Measurement and practices for hazardous waste management. In Hazardous waste management, Elsevier, 89-115.

Jones, M., & Brown, K. (2021). Exposure to toxic substances in industrial settings: Long-term health effects. Journal of Workplace Safety, 58(4), 200-215.

Joutey, N. T., Bahafid, W., Sayel, H., & El Ghachtouli, N. (2013). Biodegradation: involved microorganisms and genetically engineered microorganisms. Biodegradation-life of science, 1, 289-320.

Jurling, H. (2021). Uptake of Perfluorinated Alkyl Acids by Crops: Results from a Field Study. Environmental Science: Processes & Impacts.

Kalia, A., & Gosal, S. K. (2011). Effect of pesticide application on soil microorganisms. Archives of Agronomy and Soil Science, 57(6), 569-596.

Kaur, N., & Kaleka, A. S. (2022). Diversity, importance and decline of pollinating insects in present era.100.

Kennedy, C. M., Lonsdorf, E., Neel, M. C., Williams, N. M., Ricketts, T. H., Winfree, R., & Kremen, C. (2013). A global quantitative synthesis of local and landscape effects on wild bee pollinators in agroecosystems. Ecology letters, 16(5), 584-599.

Khan, M. N., Mobin, M., Abbas, Z. K., & Alamri, S. A. (2018). Fertilizers and their contaminants in soils, surface and groundwater. Encyclopedia of the Anthropocene, 5, 225-240.

Khan, N. A., Lopez-Maldonado, E. A., Majumder, A., Singh, S., Varshney, R., Lopez, J. R., & Aljundi, I. H. (2023). A state-of-art-review on emerging contaminants: Environmental chemistry, health effect and modern treatment methods. Chemosphere, 344, 140264.

Kharangate-Lad, A., & D'Souza, N. C. (2021). Current approaches in bioremediation of toxic contaminants by application of microbial cells; biosurfactants and bioemulsifiers of microbial origin. Rhizobiont in Bioremediation of Hazardous Waste, 217-263.

Kidd, K. A., Blanchfield, P. J., Mills, K. H., Palace, V. P., Evans, R. E., Lazorchak, J. M., & Flick, R. W. (2007). Collapse of a fish population after exposure to a synthetic estrogen. Proceedings of the National Academy of Sciences, 104(21), 8897-8901.

Kim, K. H., Jahan, S. A., & Kabir, E. (2013). A review on human health perspective of air pollution with respect to allergies and asthma. Environment international, 59, 41-52.

Klaine S.J., et al. (2008). Nanomaterials in the environment: behavior, fate, bioavailability, and efects. Environ Toxicol Chem., 27(9),1825-51.

Kobayashi, A. (2019). International encyclopedia of human geography. Elsevier.

Kohler, H. R., & Triebskorn, R. (2013). Wildlife ecotoxicology of pesticides: can we track effects to the population level and beyond?. science, 341(6147), 759-765.

Koumaki, E., Noutsopoulos, C., Mamais, D., Fragkiskatos, G., and Andreadakis, A. (2021). Fate of Emerging Contaminants in High-Rate Activated Sludge Systems. International Journal of Environmental Research and Public Health, 18, 400.

Kumar, K., & Singh, D. D. P. (2023). Assessing the Dynamics of Urbanization: A Comprehensive Review of Associated Risks and Mitigation Strategies. Journal of Research in Infrastructure Designing, 6(3).

Kumar, M., Sarma, D. K., Shubham, S., Kumawat, M., Verma, V., Prakash, A., & Tiwari, R. (2020). Environmental endocrine-disrupting chemical exposure: role in non-communicable diseases. Frontiers in public health, 8, 553850.

Kumar, P., Kausar, M. A., Singh, A. B., & Singh, R. (2021). Biological contaminants in the indoor air environment and their impacts on human health. Air Quality, Atmosphere & Health, 14(11), 1723-1736.

Kunze, M., Zerlin, K. F., Retzlaff, A., Pohl, J. O., Schmidt, E., Janssen, D. B., & Reineke, W. (2009). Degradation of chloroaromatics by Pseudomonas putida GJ31: assembled route for chlorobenzene degradation encoded by clusters on plasmid pKW1 and the chromosome. Microbiology, 155(12), 4069-4083.

Kurtz, A. E., Reiner, J. L., West, K. L., and Jensen, B. A. (2019). Perfluorinated Alkyl Acids in Hawaiian Cetaceans and Potential Biomarkers of Effect: Peroxisome Proliferator-Activated Receptor Alpha and Cytochrome P450 4A. Environmental Science and Technology, 53, 2830-2839.

Law, R. J. (2014). An overview of time trends in organic contaminant concentrations in marine mammals: going up or down? Marine pollution bulletin, 82(1-2), 7-10.

Lee, H., Jun, Z., & Zahra, Z. (2021). Phytoremediation: the sustainable strategy for improving indoor and outdoor air quality. Environments, 8(11), 118.

Lee, J. C., Son, Y. O., Pratheeshkumar, P., & Shi, X. (2012). Oxidative stress and metal carcinogenesis. Free radical biology and medicine, 53(4), 742-757.

Leemans, M., Couderq, S., Demeneix, B., & Fini, J. B. (2019). Pesticides with potential thyroid hormone-disrupting effects: a review of recent data. Frontiers in endocrinology, 10, 743.

Lempart, A., Kudlek, E., & Dudziak, M. (2018). Determination of micropollutants in water samples from swimming pool systems. Water, 10(8), 1083.

Leni, Z., Künzi, L., & Geiser, M. (2020). Air pollution causing oxidative stress. Current opinion in toxicology, 20, 1-8.

Li, K., Geng, Y., Lin, B., & Xi, Z. (2023). Molecular mechanisms underlying mitochondrial damage, endoplasmic reticulum stress, and oxidative stress induced by environmental pollutants. Toxicology Research, 12(6), 1014-1023.

Li, M., Pi, H., Yang, Z., Reiter, R. J., Xu, S., Chen, X., & Zhou, Z. (2016). Melatonin antagonizes cadmium-induced neurotoxicity by activating the transcription factor EB-dependent autophagy-lysosome machinery in mouse neuroblastoma cells. Journal of Pineal Research, 61(3), 353-369.

Li, T., Hu, R., Chen, Z., Li, Q., Huang, S., Zhu, Z., & Zhou, L. F. (2018). Fine particulate matter (PM2. 5): The culprit for chronic lung diseases in China. Chronic diseases and translational medicine, 4(03), 176-186.

Li, Y., Ortiz, G. G. R., Uyen, P. T. M., Cong, P. T., Othman, S. I., Allam, A. A., & Afridi, H. I. (2023). Environmental impact of endocrine-disrupting chemicals and heavy metals in biological samples of petrochemical industry workers with perspective management. Environmental Research, 231, 115913.

Lim, C. C., Yoon, J., Reynolds, K., Gerald, L. B., Ault, A. P., Heo, S., & Bell, M. L. (2023). Harmful algal bloom aerosols and human health. EBioMedicine, 93.

Lin, J., Ma, K., Chen, H., Chen, Z., & Xing, B. (2022). Influence of different types of nanomaterials on soil enzyme activity: A global meta-analysis. Nano Today, 42, 101345.

Lipczynska-Kochany, E. (2018). Humic substances, their microbial interactions and effects on biological transformations of organic pollutants in water and soil: A review. Chemosphere, 202, 420-437.

Liu, B., Zhang, H., Yu, Y., Xie, L., Li, J., Wang, X., *et al.* (2020). Perfluorinated Compounds (PFCs) in Soil of the Pearl River Delta, China: Spatial Distribution, Sources, and Ecological Risk Assessment. Archives of Environmental Contamination and Toxicology, 78, 182-189.

Liu, D., Shi, Q., Liu, C., Sun, Q., & Zeng, X. (2023). Effects of endocrine-disrupting heavy metals on human health. Toxics, 11(4), 322.

Liu, S., Yang, B., Liang, Y., Xiao, Y., & Fang, J. (2020). Prospect of phytoremediation combined with other approaches for remediation of heavy metal-polluted soils. Environmental Science and Pollution Research, 27, 16069-16085.

Loh, N., Loh, H. P., Wang, L. K., & Wang, M. H. S. (2016). Health effects and control of toxic lead in the environment. Natural Resources and Control Processes, 233-284.

Loranger, S., & Zayed, J. (1995). Environmental and occupational exposure to manganese: a multimedia assessment. International archives of occupational and environmental health, 67, 101-110.

Lowry, G. V., Hotze, E. M., Bernhardt, E. S., Dionysiou, D. D., Pedersen, J. A., Wiesner, M. R., & Xing, B. (2010). Environmental occurrences, behavior, fate, and ecological effects of nanomaterials: an introduction to the special series. Journal of environmental quality, 39(6), 1867-1874.

Lu, H., Shi, X., Costa, M., & Huang, C. (2005). Carcinogenic effect of nickel compounds. Molecular and cellular biochemistry, 279, 45-67.

Lu, Q., He, Z. L., and Stoffella, P. J. (2012). Land Application of Biosolids in the USA: A Review. Applied and Environmental Soil Science, 1-11.

Luoma, S. N. (2008). Silver nanotechnologies and the environment. The project on emerging nanotechnologies report, 15, 12-13.

Madesh, S., Gopi, S., Sau, A., Rajagopal, R., Namasivayam, S. K. R., & Arockiaraj, J. (2024). Chemical Contaminants and environmental stressors induced teratogenic effect in aquatic ecosystem-A Comprehensive Review. Toxicology Reports, 101819.

Maher, B. A., Gonzalez-Maciel, A., Reynoso-Robles, R., Torres-Jardon, R., & Calderon-Garciduenas, L. J. E. R. (2020). Iron-rich air pollution nanoparticles: An unrecognised environmental risk factor for myocardial mitochondrial dysfunction and cardiac oxidative stress. Environmental research, 188, 109816.

Malik S., Dhasmana A., Kishore S., Kumari M. Bioremediation and Phytoremediation Technologies in Sustainable Soil Management. Apple Academic Press; New York, NY, USA: 2022. Microbes and Microbial Enzymes for Degradation of Pesticides, 95-127.

Mani, M., Shivaraju, C., Kulkarni, N. S., Mani, M., Shivaraju, C., & Kulkarni, N. S. (2014). Pesticide residue management in grapes. The grape entomology, 181-186.

Manisalidis, I., Stavropoulou, E., Stavropoulos, A., & Bezirtzoglou, E. (2020). Environmental and health impacts of air pollution: a review. Frontiers in public health, 8, 14.

Manzoor, F., & Pervez, M. (2021). Pesticide impact on honeybees declines and emerging food security crisis. In Global Decline of Insects. IntechOpen.100.

Marlatt, V. L., Bayen, S., Castaneda-Cortès, D., Delbès, G., Grigorova, P., Langlois, V. S., & Van Der Kraak, G. (2022). Impacts of endocrine disrupting chemicals on reproduction in wildlife and humans. Environmental Research, 208, 112584.

Marques, A. P., Rangel, A. O., & Castro, P. M. (2009). Remediation of heavy metal contaminated soils: phytoremediation as a potentially promising clean-up technology. Critical Reviews in Environmental Science and Technology, 39(8), 622-654.

Marrone, P. G. (2019). Pesticidal natural products–status and future potential. Pest Management Science, 75(9), 2325-2340.

Masindi, V., & Muedi, K. L. (2018). Environmental contamination by heavy metals. Heavy metals, 10(4), 115-133.

McDorman, T. L. (2004). The Rotterdam Convention on the prior informed consent procedure for certain hazardous chemicals and pesticides in international trade: some legal notes. Review of European Community & International Environmental Law, 13(2), 187-200.

Miller, G. Z., Tighe, M. E., Peaslee, G. F., Peña, K., & Gearhart, J. (2016). Toys, décor, and more: evidence of hazardous electronic waste recycled into new consumer products. Journal of Environmental Protection, 7(03), 341.

Miller, M. E., Hamann, M., & Kroon, F. J. (2020). Bioaccumulation and biomagnification of microplastics in marine organisms: A review and meta-analysis of current data. PloS one, 15(10), e0240792.

Miranda, R. A., Silva, B. S., de Moura, E. G., & Lisboa, P. C. (2023). Pesticides as endocrine disruptors: programming for obesity and diabetes. Endocrine, 79(3), 437-447.

Mitra, S., Chakraborty, A. J., Tareq, A. M., Emran, T. B., Nainu, F., Khusro, A., & Simal-Gandara, J. (2022). Impact of heavy metals on the environment and human health: Novel therapeutic insights to counter the toxicity. Journal of King Saud University-Science, 34(3), 101865.

Mohapatra, B., & Phale, P. S. (2021). Microbial degradation of naphthalene and substituted naphthalenes: metabolic diversity and genomic insight for bioremediation. Frontiers in Bioengineering and Biotechnology, 9, 602445.

Mondal, K., Ghosh, S., & Haque, S. (2018). A review on contamination, bioaccumulation and toxic effect of cadmium, mercury and lead on freshwater fishes. Int. J. Zool. Stud, 3(2), 153-159.

Morgado, R. G., Loureiro, S., & González-Alcaraz, M. N. (2018). Changes in soil ecosystem structure and functions due to soil contamination. In Soil pollution, Academic Press, 59-87.

Mostafalou, S., & Abdollahi, M. (2013). Pesticides and human chronic diseases: evidences, mechanisms, and perspectives. Toxicology and applied pharmacology, 268(2), 157-177.

Msiska, Z., Pacurari, M., Mishra, A., Leonard, S. S., Castranova, V., & Vallyathan, V. (2010). DNA double-strand breaks by asbestos, silica, and titanium dioxide: possible biomarker of carcinogenic potential?. American journal of respiratory cell and molecular biology, 43(2), 210-219.

Muller, C. E., De Silva, A. O., Small, J., Williamson, M., Wang, X., Morris, A., *et al.* (2011). Biomagnification of Perfluorinated Compounds in a Remote Terrestrial Food Chain: Lichen-Caribou-Wolf. Environmental Science & Technology, 45, 8665-8673.

Munzel, T., Hahad, O., Daiber, A., & Landrigan, P. J. (2023). Soil and water pollution and human health: what should cardiologists worry about?. Cardiovascular research, 119(2), 440-449.

Naik, R. K., Naik, M. M., D'Costa, P. M., & Shaikh, F. (2019). Microplastics in ballast water as an emerging source and vector for harmful chemicals, antibiotics, metals, bacterial pathogens and HAB species: A potential risk to the marine environment and human health. Marine Pollution Bulletin, 149, 110525.

Nam, D. H., Lee, B. C., Eom, I. C., Kim, P., & Yeo, M. K. (2014). Uptake and bioaccumulation of titanium and silver-nanoparticles in aquatic ecosystems. Molecular & cellular toxicology, 10, 9-17.

Nedjimi, B. (2020). Germination characteristics of Peganum harmala L. (Nitrariaceae) subjected to heavy metals: Implications for the use in polluted dryland restoration. International Journal of Environmental Science and Technology, 17(4), 2113-2122.

Niki, E., & Noguchi, N. (2021). Antioxidant action of vitamin E in vivo as assessed from its reaction products with multiple biological oxidants. Free radical research, 55(4), 352-363.

O'Neal, S. L., & Zheng, W. (2015). Manganese toxicity upon overexposure: a decade in review. Current environmental health reports, 2, 315-328.

Oberdorster, G., Oberdorster, E., & Oberdörster, J. (2005). Nanotoxicology: an emerging discipline evolving from studies of ultrafine particles. Environmental health perspectives, 113(7), 823-839.

Oliveira, P. A., Colaço, A., Chaves, R., Guedes-Pinto, H., De-La-Cruz, L. F. P., & Lopes, C. (2007). Chemical carcinogenesis. Anais da academia brasileira de ciencias. 79(4).

Onyeaka, H., Ghosh, S., Obileke, K., Miri, T., Odeyemi, O. A., Nwaiwu, O., & Tamasiga, P. (2024). Preventing chemical contaminants in food: Challenges and prospects for safe and sustainable food production. Food Control, 155, 110040.

Osuoha, J. O., Anyanwu, B. O., & Ejileugha, C. (2023). Pharmaceuticals and personal care products as emerging contaminants: Need for combined treatment strategy. Journal of hazardous materials advances, 9, 100206.

Pal, A., He, Y., Jekel, M., Reinhard, M., & Gin, K. Y. H. (2014). Emerging contaminants of public health significance as water quality indicator compounds in the urban water cycle. Environment international, 71, 46-62.

Pan, D., He, M., & Kong, F. (2020). Risk attitude, risk perception, and farmers' pesticide application behavior in China: A moderation and mediation model. Journal of Cleaner Production, 276, 124241.

Pan, J., Liu, P., Yu, X., Zhang, Z., & Liu, J. (2024). The adverse role of endocrine disrupting chemicals in the reproductive system. Frontiers in Endocrinology, 14, 1324993.

Pande, V., Pandey, S. C., Sati, D., Pande, V., & Samant, M. (2020). Bioremediation: an emerging effective approach towards environment restoration. Environmental Sustainability, 3, 91-103.

Pathak, V. M., Verma, V. K., Rawat, B. S., Kaur, B., Babu, N., Sharma, A., ... & Cunill, J. M. (2022). Current status of pesticide effects on environment, human health and it's eco-friendly management as bioremediation: A comprehensive review. Frontiers in microbiology, 13, 962619.

Peng, Y., Gautam, L., & Hall, S. W. (2019). The detection of drugs of abuse and pharmaceuticals in drinking water using solid-phase extraction and liquid chromatography-mass spectrometry. Chemosphere, 223, 438-447.

Pereira, L. C., de Souza, A. O., Bernardes, M. F. F., Pazin, M., Tasso, M. J., Pereira, P. H., & Dorta, D. J. (2015). A perspective on the potential risks of emerging contaminants to human and environmental health. Environmental Science and Pollution Research, 22, 13800-13823.

Perez-de-Mora, A., Burgos, P., Madejón, E., Cabrera, F., Jaeckel, P., & Schloter, M. (2006). Microbial community structure and function in a soil contaminated by heavy metals: effects of plant growth and different amendments. Soil biology and biochemistry, 38(2), 327-341.

Perrodin, Y., Boillot, C., Angerville, R., Donguy, G., & Emmanuel, E. (2011). Ecological risk assessment of urban and industrial systems: A review. Science of the Total Environment, 409(24), 5162-5176.

Pinto, D. M., Blande, J. D., Souza, S. R., Nerg, A. M., & Holopainen, J. K. (2010). Plant volatile organic compounds (VOCs) in ozone (O_3) polluted atmospheres: the ecological effects. Journal of chemical ecology, 36, 22-34.

Pivnenko, K., Granby, K., Eriksson, E., & Astrup, T. F. (2017). Recycling of plastic waste: Screening for brominated flame retardants (BFRs). Waste Management, 69, 101-109.

Pizzorno, J. (2018). Environmental toxins and infertility. Integrative Medicine: A Clinician's Journal, 17(2), 8.

Power, A. G. (2010). Ecosystem services and agriculture: tradeoffs and synergies. Philosophical transactions of the royal society B: biological sciences, 365(1554), 2959-2971.

Prevedouros, K., Cousins, I. T., Buck, R. C., and Korzeniowski, S. H. (2006). Sources, Fate and Transport of Perfluorocarboxylates. Chem Inform, 37.

Priyadarshanee, M., Mahto, U., & Das, S. (2022). Mechanism of toxicity and adverse health effects of environmental pollutants. In Microbial biodegradation and bioremediation, Elsevier, 33-53.

Prokic, M. D., Radovanović, T. B., Gavric, J. P., & Faggio, C. (2019). Ecotoxicological effects of microplastics: Examination of biomarkers, current state and future perspectives. TrAC Trends in analytical chemistry, 111, 37-46.

Prosnier, L., Loreau, M., & Hulot, F. D. (2015). Modeling the direct and indirect effects of copper on phytoplankton–zooplankton interactions. Aquatic toxicology, 162, 73-81.

Pullagurala, V. L. R., Adisa, I. O., Rawat, S., Kalagara, S., Hernandez-Viezcas, J. A., Peralta-Videa, J. R., & Gardea-Torresdey, J. L. (2018). ZnO nanoparticles increase photosynthetic pigments and decrease lipid peroxidation in soil grown cilantro (Coriandrum sativum). Plant physiology and biochemistry, 132, 120-127.

Purohit, R. (2018). Eutrophication: harmful algal blooms. Int. J. Multidiscip. Res. Sci., 1, 267-278.

Rabalais, N. N., Turner, R. E., & Wiseman Jr, W. J. (2002). Gulf of Mexico hypoxia, aka "The dead zone". Annual Review of ecology and Systematics, 33(1), 235-263.

Rahman, M. M., Rahman, M. S., Khan, M. H. R., & Yeasmin, S. (2023). Assessment of radiation level and potential risk to public living around major hospitals in central and western Bangladesh. Heliyon, 9(9).

Rai, P. K. (2008). Heavy metal pollution in aquatic ecosystems and its phytoremediation using wetland plants: an ecosustainable approach. International journal of phytoremediation, 10(2), 133-160.

Rajala, K., Gronfors, O., Hesampour, M., & Mikola, A. (2020). Removal of microplastics from secondary wastewater treatment plant effluent by coagulation/flocculation with iron, aluminum and polyamine-based chemicals. Water Research, 183, 116045.

Ramachandra, T. V., & Solanki, M. (2007). Ecological assessment of lentic water bodies of Bangalore. The Ministry of Science and Technology, 25(96), 2.

Raquel, S., Natalia, G., Luis Fernando, B., & Maria Carmen, M. (2013). Biodegradation of high-molecular-weight polycyclic aromatic hydrocarbons by a wood-degrading consortium at low temperatures. FEMS microbiology ecology, 83(2), 438-449.

Rashid, A., Schutte, B. J., Ulery, A., Deyholos, M. K., Sanogo, S., Lehnhoff, E. A., & Beck, L. (2023). Heavy metal contamination in agricultural soil: environmental pollutants affecting crop health. Agronomy, 13(6), 1521.

Ray, P. C., Yu, H., & Fu, P. P. (2009). Toxicity and environmental risks of nanomaterials: challenges and future needs. Journal of Environmental Science and Health Part C, 27(1), 1-35.

Reckendorf, A., Siebert, U., Parmentier, E., & Das, K. (2023). Chemical pollution and diseases of marine mammals. In Marine mammals: a deep dive into the world of science (pp. 63-78). Cham: Springer International Publishing.

Rehman, Q., Rehman, K., & Akash, M. S. H. (2021). Heavy metals and neurological disorders: from exposure to preventive interventions. In Environmental contaminants and neurological disorders (pp. 69-87). Cham: Springer International Publishing.

Ribeiro, F., O'Brien, J. W., Galloway, T., & Thomas, K. V. (2019). Accumulation and fate of nano-and micro-plastics and associated contaminants in organisms. Trends in analytical chemistry, 111, 139-147.

Rice, K. M., Walker Jr, E. M., Wu, M., Gillette, C., & Blough, E. R. (2014). Environmental mercury and its toxic effects. Journal of preventive medicine and public health, 47(2), 74.

Riva, D., Magalhaes, C. B., Lopes, A., Lanças, T., Mauad, T., Malm, O., ... & Zin, W. A. (2011). Low dose of fine particulate matter (PM2. 5) can induce acute oxidative stress, inflammation and pulmonary impairment in healthy mice. Inhalation toxicology, 23(5), 257-267.

Rochman, C. M., Browne, M. A., Halpern, B. S., Hentschel, B. T., Hoh, E., Karapanagioti, H. K., ... & Thompson, R. C. (2013). Classify plastic waste as hazardous. Nature, 494(7436), 169-171.

Rockstrom, J., Steffen, W., Noone, K., Persson, A., Chapin III, F. S., Lambin, E., & Foley, J. (2009). Planetary boundaries: exploring the safe operating space for humanity. Ecology and society, 14(2).

Rodil, R., Quintana, J. B., Concha-Grana, E., Lopez-Mahía, P., Muniategui-Lorenzo, S., & Prada-Rodríguez, D. (2012). Emerging pollutants in sewage, surface and drinking water in Galicia (NW Spain). Chemosphere, 86(10), 1040-1049.

Rose, P. K., Yadav, S., Kataria, N., & Khoo, K. S. (2023). Microplastics and nanoplastics in the terrestrial food chain: Uptake, translocation, trophic transfer, ecotoxicology, and human health risk. Trends in Analytical Chemistry, 117249.

Rosenzweig, M.L. (1971) Paradox of enrichment: destabilization of exploitation ecosystems in ecological time. Science, 171, 385-387.

S.G. Poulopoulos (2016). Atmospheric Environment, Environment and Development, Elsevier, 45-136.

Saleh, T. A. (2020). Nanomaterials: Classification, properties, and environmental toxicities. Environmental Technology & Innovation, 20, 101067.

Samet, J. M., & Wages, P. A. (2018). Oxidative stress from environmental exposures. Current opinion in toxicology, 7, 60-66.

Samuel, P. O., Edo, G. I., Oloni, G. O., Ugbune, U., Ezekiel, G. O., Essaghah, A. E. A., & Agbo, J. J. (2023). Effects of chemical contaminants on the ecology and evolution of organisms a review. Chemistry and Ecology, 39(10), 1071-1107.

Satkar, S. G., Kumar, A., Bhusare, S., Sahu, A., & Gautam, R. K. (2024). Impact of Pollution and Toxic Stress on Fish Health: Mechanisms, Consequences, and Mitigation Strategies. Uttar Pradesh Journal of Zoology, 45(6), 29-45.

Schneider, T., & Jensen, K. A. (2009). Relevance of aerosol dynamics and dustiness for personal exposure to manufactured nanoparticles. Journal of nanoparticle research, 11, 1637-1650.

Schwartz, G. G., Il'yasova, D., & Ivanova, A. (2003). Urinary cadmium, impaired fasting glucose, and diabetes in the NHANES III. Diabetes care, 26(2), 468-470.

Selim, A. E. M. (2007). Sedimentation threats to Red Sea corals: an ecological study of reefs in the Hurghada region, Egypt (Doctoral dissertation, University of Hull).

Sha, B., Johansson, J. H., Tunved, P., Bohlin-Nizzetto, P., Cousins, I. T., and Salter, M. E. (2022). Sea Spray Aerosol (SSA) as a Source of Perfluoroalkyl Acids (PFAAs) to the Atmosphere: Field Evidence from Long-Term Air Monitoring. Environmental Science & Technology, 56, 228-238.

Shafiei-Masouleh, S. S. (2021). Con tents of Organic Solutes in Horticultural Plants and Crops under Environmental Pollutants. In Organic Solutes, Oxidative Stress and Antioxidant Enzymes Under Abiotic Stressors, CRC Press, 105-122.

Shah, H., & Jain, S. (2020). Bioremediation: An approach for environmental pollutants detoxification. Waste to energy: Prospects and applications, 121-142.

Shankar, A., Dubey, A., Saini, D., Singh, M., Prasad, C. P., Roy, S., & Rath, G. K. (2019). Environmental and occupational determinants of lung cancer. Translational lung cancer research, 8(Suppl 1), S31.

Sharma, A. K., Sharma, M., Sharma, A. K., & Sharma, M. (2023). Mapping the impact of environmental pollutants on human health and environment: A systematic review and meta-analysis. Journal of Geochemical Exploration, 107325.

Sharma, P., Pandey, A. K., Kim, S. H., Singh, S. P., Chaturvedi, P., & Varjani, S. (2021). Critical review on microbial community during in-situ bioremediation of heavy metals from industrial wastewater. Environmental Technology & Innovation, 24, 101826.

Sharma, R. K., Gulati, S., & Puri, A. (2014). Green chemistry solutions to water pollution. In Water Reclamation and Sustainability, Elsevier, 57-75.

Shen, M., Hu, T., Huang, W., Song, B., Zeng, G., & Zhang, Y. (2021). Removal of microplastics from wastewater with aluminosilicate filter media and their surfactant-modified products: Performance, mechanism and utilization. Chemical Engineering Journal, 421, 129918.

Shen, M., Zhang, Y., Almatrafi, E., Hu, T., Zhou, C., Song, B., & Zeng, G. (2022). Efficient removal of microplastics from wastewater by an electrocoagulation process. Chemical Engineering Journal, 428, 131161.

Shetty, S. S., Deepthi, D., Harshitha, S., Sonkusare, S., Naik, P. B., & Madhyastha, H. (2023). Environmental pollutants and their effects on human health. Heliyon, 9(9).

Singh, A., Gautam, P. K., Verma, A., Singh, V., Shivapriya, P. M., Shivalkar, S., & Samanta, S. K. (2020). Green synthesis of metallic nanoparticles as effective alternatives to treat antibiotics resistant bacterial infections: A review. Biotechnology Reports, 25, e00427.

Singh, N., Poonia, T., Siwal, S. S., Srivastav, A. L., Sharma, H. K., & Mittal, S. K. (2022). Challenges of water contamination in urban areas. In Current directions in water scarcity research, Elsevier, 6, 173-202.

Smith, A. H., Lingas, E. O., & Rahman, M. (2000). Contamination of drinking-water by arsenic in Bangladesh: A public health emergency. Bulletin of the World Health Organization, 78(9), 1093-1103.

Smith, A., Johnson, L., & Williams, T. (2020). Chemical hazards in occupational environments: A comprehensive review. Environmental Health Perspectives, 128(7), 1-12.

Sobha, T. R., Vibija, C. P., & Fahima, P. (2023). Coral reef: a hot spot of marine biodiversity. In Conservation and sustainable utilization of bioresources Singapore: Springer Nature Singapore.171-194.

Sondermann, N. C., Faßbender, S., Hartung, F., Hatala, A. M., Rolfes, K. M., Vogel, C. F., & Haarmann-Stemmann, T. (2023). Functions of the aryl hydrocarbon receptor (AHR) beyond the canonical AHR/ARNT signaling pathway. Biochemical pharmacology, 208, 115371.

Sonne, C. (2010). Health effects from long-range transported contaminants in Arctic top predators: an integrated review based on studies of polar bears and relevant model species. Environment international, 36(5), 461-491.

Sparling, D. W. (2016). "Organochlorine Pesticides," in Chapter 4 - Organochlorine Pesticides Ecotoxicology Essentials. Academic Press, 69-107.

Sponsler, D. B., Grozinger, C. M., Hitaj, C., Rundlöf, M., Botías, C., Code, A., & Douglas, M. R. (2019). Pesticides and pollinators: A socioecological synthesis. Science of the Total Environment, 662, 1012-1027.

Stayner, L. T., Dankovic, D. A., & Lemen, R. A. (2013). Occupational exposure to asbestos and risk of lung cancer: A review of the epidemiological evidence. Occupational and Environmental Medicine, 70(5), 343-349.

Stefani, D., Wardman, D., & Lambert, T. (2005). The implosion of the Calgary General Hospital: ambient air quality issues. Journal of the Air & Waste Management Association, 55(1), 52-59.

Subirats, J., Timoner, X., Sànchez-Melsió, A., Balcázar, J. L., Acuña, V., Sabater, S., & Borrego, C. M. (2018). Emerging contaminants and nutrients synergistically affect the spread of class 1 integron-integrase (intI1) and sul1 genes within stable streambed bacterial communities. Water research, 138, 77-85.

Suzuki, T., Hidaka, T., Kumagai, Y., & Yamamoto, M. (2020). Environmental pollutants and the immune response. Nature Immunology, 21(12), 1486-1495.

Talvitie, J., Mikola, A., Setälä, O., Heinonen, M., & Koistinen, A. (2017). How well is microlitter purified from wastewater?- A detailed study on the stepwise removal of microlitter in a tertiary level wastewater treatment plant. Water research, 109, 164-172.

Tchounwou, P. B., Yedjou, C. G., Patlolla, A. K., & Sutton, D. J. (2012). Heavy metal toxicity and the environment. Molecular, Clinical and Environmental Toxicology, 101, 133-164.

Thakare, M., Sarma, H., Datar, S., Roy, A., Pawar, P., Gupta, K.,& Prasad, R. (2021). Understanding the holistic approach to plant-microbe remediation technologies for removing heavy metals and radionuclides from soil. Current research in biotechnology, 3, 84-98.

The Water (Prevention and Control of Pollution) Act, 1974.

Tourinho, P. S., Van Gestel, C. A., Lofts, S., Svendsen, C., Soares, A. M., & Loureiro, S. (2012). Metal-based nanoparticles in soil: Fate, behavior, and effects on soil invertebrates. Environmental Toxicology and Chemistry, 31(8), 1679-1692.

Tripathi, M., & Kumar Garg, S. (2014). Response surface modeling for co-remediation of Cr6+ and pentachlorophenol by Bacillus cereus RMLAU1: bioreactor trial and structural and functional characterization by SEM-EDS and FT-IR analyses. Bioremediation journal, 18(4), 328-344.

U.S.EPA (2010) U.S. Environmental Protection Agency (EPA) Decontamination Research and Development Conference. U.S. Environmental Protection Agency, Washington, DC, EPA/600/R-11/052, 2011.

U.S.EPA (2010). Emerging Contaminants-Nanomaterials.

U.S.EPA. (2008). Nanotechnology for site remediation fact sheet. EPA, Zeist.

Ugbede, F. O., Aduo, B. C., Ogbonna, O. N., & Ekoh, O. C. (2020). Natural radionuclides, heavy metals and health risk assessment in surface water of Nkalagu river dam with statistical analysis. Scientific African, 8, e00439.

Ullrich, S. M., Tanton, T. W., & Abdrashitova, S. A. (2001). Mercury in the aquatic environment: a review of factors affecting methylation. Critical reviews in environmental science and technology, 31(3), 241-293.

United Nations Environment Programme (UNEP), (2011) Towards a Green Economy: Pathways to Sustainable Development and Poverty Eradication. Report, United Nations Environment Programme, Nairobi.

United States Environmental Protection Agency (USEPA), (2024). Environmental Justice.

USEPA (2013b). Waste, Chemical, and Cleanup Enforcement. Available at: https://www.epa.gov/enforcement/waste-chemical-and-cleanup-enforcement.

USEPA-United States Environment Protection Agency (2008) Emerging contaminants—polybrominated diphenyl ethers (PBDE) and polybrominated biphenyls (PBB).

USGS (2017). Pesticides | US Geological Survey. Available at: https://www.usgs. gov/centers/ohio-kentucky-indiana-water-science-center/science/pesticides.

Vagi, M. C., Petsas, A. S., & Kostopoulou, M. N. (2021). Potential effects of persistent organic contaminants on marine biota: a review on recent research. Water, 13(18), 2488.

Valavanidis, A., Vlachogianni, T., Fiotakis, K., & Loridas, S. (2013). Pulmonary oxidative stress, inflammation and cancer: respirable particulate matter, fibrous dusts and ozone as major causes of lung carcinogenesis through reactive oxygen species mechanisms. International journal of environmental research and public health, 10(9), 3886-3907.

Van Oers, L., van der Voet, E., & Grundmann, V. (2012). Additives in the plastics industry. Global Risk-Based Management of Chemical Additives I: Production, Usage and Environmental Occurrence, 133-149.

Van, D.-A., Ngo, T. H., Huynh, T. H., Nakada, N., Ballesteros, F., and Tanaka, H. (2021). Distribution of Pharmaceutical and Personal Care Products (PPCPs) in Aquatic Environment in Hanoi and Metro Manila. Environmental Monitoring and Assessment, 193, 847.

Venugopal, V., Latha, P. K., Shanmugam, R., Krishnamoorthy, M., & Johnson, P. (2020). Occupational heat stress induced health impacts: A cross-sectional study from South Indian working population. Advances in climate change research. 11(1), 31-39.

Vos, J. G., Dybing, E., Greim, H. A., Ladefoged, O., Lambré, C., Tarazona, J. V., & Vethaak, A. D. (2000). Health effects of endocrine-disrupting chemicals on wildlife, with special reference to the European situation. Critical reviews in toxicology, 30(1), 71-133.

Wagner, S., & Schlummer, M. (2020). Legacy additives in a circular economy of plastics: Current dilemma, policy analysis, and emerging countermeasures. Resources, Conservation and Recycling, 158, 104800.

Walker, C.H., Sibly, R.M., Sibly, R.M., & Peakall, D.B. (2012). Principles of Ecotoxicology (4th ed.). CRC Press.

Wang, H., Hu, C., Shen, Y., Shi, B., Zhao, D., & Xing, X. (2019). Response of microorganisms in biofilm to sulfadiazine and ciprofloxacin in drinking water distribution systems. Chemosphere, 218, 197-204.

Wang, J., and Wang, S. (2016). Removal of Pharmaceuticals and Personal Care Products (PPCPs) from Wastewater: A Review. Journal of Environmental Management, 182, 620-640.

Wang, L. K., Eroglu, V., & Erturk, F. (2009). Management and Treatment of Acid Pickling Wastes Containing Heavy Metals. Heavy Metals in the Environment, 293.

Wang, Y., and Huang, H. (2019). "Carbon Nanotube Composite Membranes for Microfiltration of Pharmaceuticals and Personal Care Products," in Advanced Nanomaterials For Membrane Synthesis And its Applications Micro and Nano Technologies. Elsevier, 183-202.

Wear, S. L., Acuña, V., McDonald, R., & Font, C. (2021). Sewage pollution, declining ecosystem health, and cross-sector collaboration. Biological Conservation, 255, 109010.

Wegener Sleeswijk, A. (2010). Regional LCA in a global perspective. Leiden University.

Wei, H., Yuan, W., Yu, H., & Geng, H. (2021). Cytotoxicity induced by fine particulate matter (PM 2.5) via mitochondria-mediated apoptosis pathway in rat alveolar macrophages. Environmental Science and Pollution Research, 28, 25819-25829.

Wiesner, M. R., Lowry, G. V., Alvarez, P., Dionysiou, D., & Biswas, P. (2006). Assessing the risks of manufactured nanomaterials.

Williams, E. T. (2019). Environmental pollution by heavy metal: an overview. International Journal of Environmental Chemistry, 12(2), 72-82.

Winans, B., Humble, M. C., & Lawrence, B. P. (2011). Environmental toxicants and the developing immune system: a missing link in the global battle against infectious disease?. Reproductive toxicology, 31(3), 327-336.

World Health Organization (WHO) (2017). World Health Statistics 2017: Monitoring Health for the SDGs, Sustainable Development Goals. WHO, Geneva.

World Health Organization. (2023). WHO Global water, sanitation and hygiene: annual report 2022. World Health Organization.

Yaashikaa, P. R., Kumar, P. S., Jeevanantham, S., & Saravanan, R. (2022). A review on bioremediation approach for heavy metal detoxification and accumulation in plants. Environmental Pollution, 301, 119035.

Yadav, D., Rangabhashiyam, S., Verma, P., Singh, P., Devi, P., Kumar, P., & Kumar, K. S. (2021). Environmental and health impacts of contaminants of emerging concerns: Recent treatment challenges and approaches. Chemosphere, 272, 129492.

Yang, H., Chen, G., & Wang, J. (2021). Microplastics in the marine environment: sources, fates, impacts and microbial degradation. Toxics, 9(2), 41.

Yang, H., Lu, G., Yan, Z., Liu, J., Dong, H., Bao, X., & Sun, Y. (2020). Residues, bioaccumulation, and trophic transfer of pharmaceuticals and personal care products in highly urbanized rivers affected by water diversion. Journal of hazardous materials, 391, 122245.

Yasunari, T. J., Stohl, A., Hayano, R. S., Burkhart, J. F., Eckhardt, S., & Yasunari, T. (2011). Cesium-137 deposition and contamination of Japanese soils due to the Fukushima nuclear accident. Proceedings of the National Academy of Sciences, 108(49), 19530-19534.

Zahoor, I., & Mushtaq, A. (2023). Water pollution from agricultural activities: A critical global review. Int. J. Chem. Biochem. Sci, 23(1), 164-176.

Zhai, Y., Wang, Z., Wang, G., Peijnenburg, W. J., & Vijver, M. G. (2020). The fate and toxicity of Pb-based perovskite nanoparticles on soil bacterial community: Impacts of pH, humic acid, and divalent cations. Chemosphere, 249, 126564.

Zhang, X., Yang, L., Li, Y., Li, H., Wang, W., & Ye, B. (2012). Impacts of lead/zinc mining and smelting on the environment and human health in China. Environmental monitoring and assessment, 184, 2261-2273.

Zhao, H., Yuan, M., Strokal, M., Wu, H. C., Liu, X., Murk, A., & Osinga, R. (2021). Impacts of nitrogen pollution on corals in the context of global climate change and potential strategies to conserve coral reefs. Science of the Total Environment, 774, 145017.

Zunarelli, C., Godono, A., Visci, G., Violante, F. S., & Boffetta, P. (2021). Occupational exposure to asbestos and risk of kidney cancer: an updated meta-analysis. European Journal of Epidemiology, 36(9), 927-936.

2

Bioremediation: A Pathway to Sustainable Development

Ghanshyam Kachhawa**, Priyanka Verma and Ritul Kachhawa

Department of Zoology, University of Rajasthan, Jaipur-302004, Rajasthan, India

**Corresponding Author Email: garistotlekachhawa@gmail.com*

Abstract

Bioremediation refers to the process of addressing pollutants using biological agents. This approach is cost-efficient, employing microorganisms to break down or neutralize pollutants, transforming them into less harmful substances. In the current context, where developing nations are increasingly focused on industrial growth, urban expansion, and the use of agrochemicals in agriculture, the importance of bioremediation has grown significantly, garnering widespread attention. It offers a means to achieve a cleaner and healthier environment while providing a sustainable solution for eliminating contaminants from ecosystems. The use of microorganisms for bioremediation can be traced back to ancient times when natural processes were harnessed to purify water bodies and wetlands. However, the scientific exploration and systematic application of these processes developed much later. Bioremediation primarily involves three key processes: bioaccumulation, biodegradation, and biotransformation. In bioaccumulation, microorganisms absorb heavy metals from polluted environments, while biodegradation refers to the breakdown of contaminants into simpler molecules with the help of microorganisms and biotransformation refers to the conversion of contaminants into non-toxic forms by microbial activity. Bioremediation has a wide range of applications, including the treatment of both domestic and industrial waste. It is particularly effective in removing toxic metals from soil and degrading agrochemicals like pesticides. This technique has been successfully employed in various settings, such as groundwater, surface water, soils, and sediments. Despite its potential, bioremediation faces certain challenges. For instance, incomplete biodegradation can sometimes result in the production of harmful by-products. Additionally, the efficiency of bioremediation may be reduced under extreme environmental conditions, such as high contaminant concentrations or low temperatures. Enhancements in bioremediation efficacy can be achieved through techniques like biostimulation and bioaugmentation. Bioaugmentation involves introducing specific microbial strains into contaminated sites, whereas biostimulation focuses on optimizing environmental conditions to support the growth and activity of indigenous microorganisms.

Keywords: *Biodegradation, Bioremediation, Detoxification, Genetic engineering, Sustainable development*

1. Introduction

Bioremediation is a process that utilizes biological agents, such as microorganisms, to eliminate pollutants, making it a natural and efficient method for environmental cleanup. This approach is considered cost-effective as it transforms harmful pollutants into less toxic substances through microbial degradation or detoxification. Currently, many developing nations are prioritizing industrial growth, urbanization, and the use of agrochemicals in agriculture, which has heightened the importance of bioremediation and drawn significant attention to this practice. As a sustainable solution, bioremediation offers a way to restore healthier ecosystems by removing pollutants from the environment (Fig. 1).

Fig. 1: Bioremediation (Adopted from Matilda *et al.*, 2024)

Historically, the use of microorganisms for environmental clean-up has existed for centuries. In ancient times, biological agents were employed to purify water bodies and wetlands. However, the systematic study and application of these natural processes did not begin until much later. During the 19th century, researchers began to acknowledge the crucial role of microorganisms in decomposing organic matter. The groundbreaking contributions of scientists such as Louis Pasteur and Robert Koch established the foundations of microbiology, showcasing microorganisms' metabolic capabilities and their potential for waste treatment applications (Pasteur, 1879; Koch, 1881).

The early 20th century marked a pivotal moment in wastewater management with the development of microbial-based treatment systems. In 1914, the activated sludge process was introduced in England, revolutionizing wastewater treatment. This method aerates wastewater to encourage the growth of microorganisms that degrade organic pollutants, effectively reducing the harmful effects of sewage disposal (Ardern and Lockett, 1914).

The term "bioremediation" gained prominence in the 1980s, alongside growing environmental awareness and the need for sustainable pollution control. During this time, environmental biotechnology gained traction, and researchers explored the potential of microorganisms to clean up environmental contaminants such as oil spills, industrial waste, and polluted soils.

A turning point in the development of bioremediation occurred following the 1989 Exxon Valdez oil spill, which underscored the need for effective clean-up methods. The event prompted significant research into bioremediation, leading to the deployment of hydrocarbon-degrading bacteria to mitigate the environmental damage caused by the spill (Atlas and Hazen, 2011).

The 1990s and early 2000s saw further advancements in bioremediation through genetic engineering and molecular biology. These developments allowed scientists to enhance the capabilities of microorganisms, enabling them to degrade pollutants that were previously resistant to natural biodegradation. Notably, the creation of genetically engineered bacteria capable of breaking down polychlorinated biphenyls (PCBs) and other persistent organic pollutants represented a major leap forward in the field (Wackett, 2004).

2. Bioremediation Mechanisms

Bioremediation operates through several mechanisms, such as biodegradation, bioaccumulation, and biotransformation (Fig. 2). These processes can take place in both aerobic and anaerobic environments, depending on the microorganisms present and the characteristics of the pollutants.

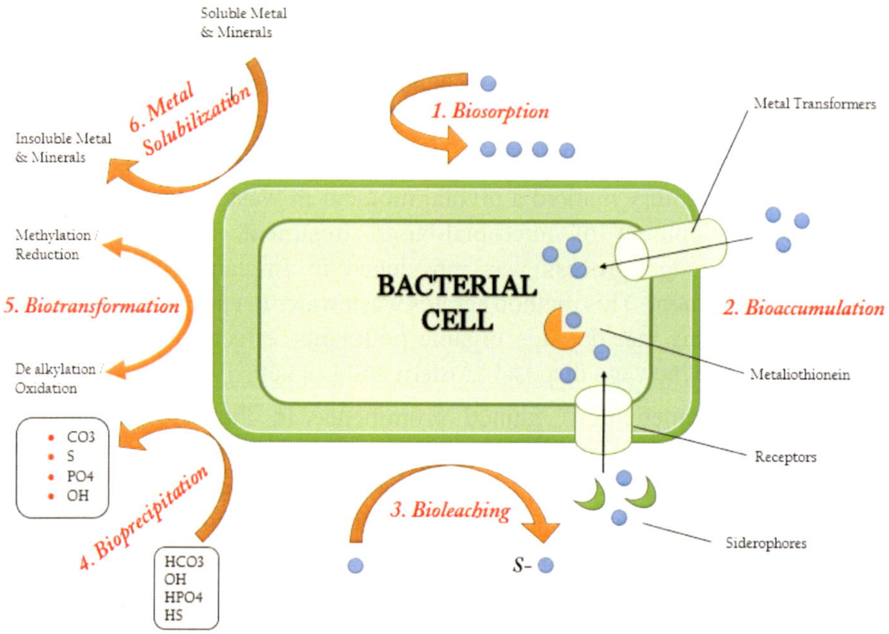

Fig. 2: Bioremediation mechanisms (Adopted from Ma *et al.*, 2016)

2.1. Biodegradation

Biodegradation is the process by which microorganisms break down organic pollutants into simpler, non-toxic substances. For example, bacteria such as *Pseudomonas* and *Alcanivorax* species can degrade hydrocarbons present in oil spills (Head *et al.*, 2006). Biodegradation involves the enzymatic breakdown of organic compounds by microorganisms, including bacteria, fungi, and algae, into less harmful substances. This process plays a vital role in recycling organic matter within ecosystems and reducing pollution (Alexander, 1999).

Several factors influence the efficiency of biodegradation, such as the chemical structure of the pollutant, environmental factors like temperature, pH, and oxygen availability, as well as the presence of specific microorganisms that can break down the target compound (Leahy and Colwell, 1990). Pollutants with simpler, non-branched molecular structures tend to degrade more easily, while more complex compounds like halogenated or aromatic substances are more resistant to microbial degradation (Pignatello, 2013).

Microorganisms use various metabolic pathways to degrade contaminants. Aerobic biodegradation, which takes place in the presence of oxygen, is typically faster and more efficient because oxygen serves as the terminal electron acceptor. In contrast, anaerobic biodegradation, occurring in environments with little to no oxygen, relies on alternative electron acceptors such as nitrate, sulphate, or carbon dioxide, often resulting in slower degradation (Wackett, 2004).

Recent advances in molecular biology have deepened our understanding of the genetic and enzymatic mechanisms involved in biodegradation. For instance, the discovery of genes responsible for producing enzymes like monooxygenases and dioxygenases has underscored their role in initiating the breakdown of complex hydrocarbons (Van Beilen and Funhoff, 2007). Additionally, biotechnology innovations, such as genetically engineered microorganisms (GEMs), have shown promise in addressing pollutants that are difficult to degrade (Cases and de Lorenzo, 2005).

Despite these advances, challenges remain in optimizing biodegradation for practical use. Issues such as microbial competition, the availability of contaminants, and the potential formation of toxic intermediate products need careful management to ensure effective remediation (Amran *et al.*, 2022). Biodegradation continues to be a critical approach for sustainable environmental management, offering environmentally friendly solutions for waste treatment and pollution control. Ongoing research is essential to improve its application, particularly for emerging pollutants like microplastics and pharmaceutical residues.

2.2. Bioaccumulation

Bioaccumulation refers to the accumulation of substances, such as heavy metals or persistent organic pollutants (POPs), within the tissues of organisms over time. Unlike biodegradation, where substances are broken down, bioaccumulation results in the concentration of chemicals, often to levels higher than those found in the surrounding environment (Connolly and Pedersen, 1988). Microorganisms such as algae, fungi, and bacteria have been found to accumulate metals like cadmium, lead, and mercury from contaminated sites, reducing their toxicity and bioavailability (Gadd, 2004).

The main factors driving bioaccumulation include the chemical properties of the pollutants and the biological characteristics of the organisms involved. Lipophilic compounds like polychlorinated biphenyls (PCBs) and DDT tend to accumulate in organisms due to their low water solubility and high affinity for lipid-rich tissues (Mackay and Fraser, 2000). Substances that resist degradation, such as Methylmercury, are particularly prone to bioaccumulation (Boening, 2000).

Bioaccumulation is linked to biomagnification, a process in which the concentration of pollutants increases as they move up the food chain. In aquatic systems, for example, phytoplankton may absorb small amounts of contaminants, which then magnify as they are passed on to zooplankton, small fish, larger fish, and eventually to the top predators, including humans (Lavoie *et al.*, 2013).

The effects of bioaccumulation can be significant at both the organismal and ecosystem levels. At the organism level, the accumulation of toxic substances can interfere with vital physiological functions like reproduction, growth, and immunity. At the ecosystem level, bioaccumulation poses a risk to biodiversity and food web stability, as seen with elevated mercury levels in fish that impact their predators, including birds and mammals (Eisler, 2000). Bioaccumulation through dietary exposure, especially from seafood, raises serious public health concerns, as highlighted by studies on mercury and other POPs (UNEP, 2013).

Advancements in analytical chemistry have made it easier to detect bioaccumulative substances at trace levels, improving monitoring and risk assessment. Regulatory measures, such as the Stockholm Convention on Persistent Organic Pollutants, aim to control and reduce the release of bioaccumulative chemicals. However, new challenges have emerged, including the rise of bioaccumulative substances like certain per- and polyfluoroalkyl substances (PFAS) (Ahrens and Bundschuh, 2014).

Understanding bioaccumulation is crucial for assessing the long-term environmental and health risks of contamination. Ongoing research is essential for identifying and managing the potential dangers posed by both legacy and newly emerging bioaccumulative pollutants.

2.3. Biotransformation

Biotransformation refers to the modification of toxic compounds into less harmful substances by microorganisms. For example, the bacterium *Dehalococcoides* can transform toxic chlorinated solvents into non-toxic Ethene (Löffler *et al.*, 1999). This process involves the chemical alteration of compounds by biological systems, primarily through enzymatic actions in microorganisms, plants, and animals. It plays an essential role in detoxifying xenobiotics, metabolizing natural compounds, and breaking down pollutants in the environment.

Biotransformation occurs in two stages: Phase I and Phase II. In Phase I, reactions such as oxidation, reduction, and hydrolysis introduce or expose functional groups, increasing the substrate's reactivity. The cytochrome P450 enzymes are central to this phase, catalyzing the oxidation of a wide variety of substrates, including drugs and environmental pollutants (Guengerich, 2008). Phase II involves conjugation reactions, where functional groups from Phase I products are linked with endogenous molecules like glucuronic acid, sulphate, or glutathione, making them more water-soluble for easier excretion (Testa and Kramer, 2008).

The effectiveness of biotransformation depends on the physicochemical properties of the substrate, the type of enzyme involved, and environmental factors such as temperature, pH, and the presence of cofactors. Although biotransformation typically results in detoxification, it can also lead to bioactivation, where non-toxic compounds are converted into harmful metabolites. For example, polycyclic aromatic hydrocarbons (PAHs) are metabolized into reactive epoxide intermediates, which can contribute to carcinogenesis (Shimada and Fujii-Kuriyama, 2004).

In addition to its detoxifying function, biotransformation has significant industrial and biotechnological uses. It is employed in the pharmaceutical industry to produce chiral intermediates and active pharmaceutical ingredients. In environmental science, biotransformation underpins many bioremediation strategies for removing pollutants such as hydrocarbons and pesticides from contaminated environments (Van der Meer, 2006). Furthermore, advances in metabolic engineering and synthetic biology have led to the creation of specialized microbial strains tailored for specific biotransformation processes (Keasling, 2010).

Despite its utility, scaling biotransformation for industrial and environmental applications presents challenges. Issues such as enzyme stability, substrate specificity, and the production of undesirable by-products must be addressed. Innovations in enzyme engineering and computational modeling offer potential solutions to these challenges (Bornscheuer *et al.*, 2012).

Biotransformation remains a key process in both natural and applied biological systems. Ongoing research into its mechanisms and applications could significantly advance environmental remediation and sustainable manufacturing practices.

3. Applications of Bioremediation

Bioremediation has a wide range of applications, including the clean-up of oil spills, heavy metal contamination, pesticide residues, and industrial waste.

3.1. Oil Spill Clean-up

Bioremediation has been effectively applied to clean-up marine oil spills, such as the Exxon Valdez and Deep water Horizon incidents, where bacteria capable of degrading hydrocarbons were utilized to mitigate the environmental damage (Das and Chandran, 2011). This process, known as bio spill clean-up, involves using biological agents—such as microorganisms, enzymes, or plants—to degrade, transform, or immobilize pollutants into less harmful substances (Atlas and Hazen, 2011).

The degradation of hydrocarbons, which are the main components of crude oil, is central to bioremediation efforts. Microorganisms such as *Pseudomonas*, *Alcanivorax*, and *Rhodococcus* utilize hydrocarbons as their carbon and energy sources. Key enzymes like monooxygenases and dioxygenases play a vital role in breaking down complex hydrocarbons into simpler, more bioavailable compounds (Prince, 2015).

There are two primary approaches to bio spill clean-up: in-situ and ex-situ bioremediation. In situ bioremediation treats the contaminated site directly, often through biostimulation or bioaugmentation. Biostimulation adds nutrients (such as nitrogen and phosphorus) or electron acceptors to promote the growth and activity of indigenous microorganisms. Bioaugmentation, on the other hand, involves introducing specialized microbial strains capable of breaking down specific pollutants (Hazen *et al.*, 2010). Ex-situ bioremediation involves removing the contaminated material from the site and treating it off-site, usually in controlled environments like bioreactors or composting systems.

Environmental conditions significantly affect the success of bioremediation. Factors like temperature, pH, oxygen availability, and the chemical nature of the pollutants influence microbial activity and degradation rates. For example, cold marine environments, such as those in the Arctic, present challenges due to reduced microbial activity and lower solubility of hydrocarbons (Margesin and Schinner, 2001).

Recent advancements in biotechnology have improved the effectiveness and versatility of bio spill clean-up. Genetic modifications have led to the development of microbial strains with enhanced degradation capabilities and tolerance to harsh environmental conditions. Additionally, biosurfactants produced by certain microbes, such as *Bacillus* and *Pseudomonas* species, can enhance the bioavailability of hydrophobic pollutants, speeding up their degradation (Desai and Banat, 1997).

Despite its potential, bio spill clean-up faces some limitations. These include incomplete degradation of complex hydrocarbons, the possibility of harmful by-products, and variability in site-specific conditions. Moreover, regulatory concerns and public acceptance of genetically engineered organisms can impact their use in the field (Singh *et al.*, 2006). Nonetheless, bio spill clean-up presents a sustainable and cost-effective alternative to traditional physical and chemical remediation methods. Continued research focused on optimizing bioremediation strategies, especially for extreme and sensitive environments, is crucial to improving its effectiveness and applicability

3.2. Heavy Metal Remediation

Techniques like phytoremediation and microbial bioremediation are being used to extract heavy metals from contaminated soils and aquatic environments. Advances in genetic modification of plants and microbes have improved the efficiency of metal uptake and detoxification (Pilon-Smith, 2005). Heavy metal contamination poses a significant environmental threat due to its toxicity, persistence, and ability to accumulate in living organisms. Metals such as lead (Pb), mercury (Hg), cadmium (Cd), chromium (Cr), and arsenic (As) can have severe negative effects on ecosystems and human health. The goal of remediation strategies is to reduce the mobility, bioavailability, or concentration of these metals in polluted areas like soils, water, and sediments (Gadd, 2010).

3.2.1. Biological Remediation

Bioremediation, using either microorganisms or plants, presents an environmentally friendly and sustainable method for remediating heavy metal contamination. Key biological techniques include:

1. **Microbial Bioremediation**: Microorganisms, including species like *Pseudomonas*, *Bacillus*, and *Aspergillus*, can either transform or immobilize heavy metals through processes such as biosorption, bioaccumulation, and bioprecipitation. For instance, sulphate-reducing bacteria can convert metals into insoluble sulphides, rendering them less bioavailable (Lloyd and Lovley, 2001).
2. **Phytoremediation**: Certain plants, such as *Brassica juncea* (Indian mustard) and *Populus deltoides* (poplar), are capable of accumulating or stabilizing heavy metals within their tissues or the surrounding soil. Techniques such as phytoextraction (uptake and storage of metals in plant tissues) and phytostabilization (binding metals in the soil) are actively being studied (Ali *et al.*, 2013).

3.2.2. Chemical and Physical Remediation

Though biological methods are increasingly preferred, chemical and physical techniques remain essential for large-scale, rapid remediation:

1. **Chemical Stabilization**: Additives like lime, biochar, and phosphates can reduce the solubility and mobility of metals in soils (Ahmad *et al.*, 2014).

2. **Soil Washing and Electrokinetic Remediation**: Soil washing utilizes chemical solutions to extract metals from contaminated soil, while electrokinetic remediation uses electric fields to move and remove metals from soils and sediments (Acar and Alshawabkeh, 1993).

3.2.3. Emerging Technologies

Advances in nanotechnology and genetic engineering are offering new possibilities for heavy metal remediation:

1. **Nano-remediation**: Nanomaterials, such as zero-valent iron (nZVI) and carbon-based nanoparticles, have proven highly effective in removing heavy metals through adsorption and reduction processes (Karn *et al.*, 2009).
2. **Engineered Microbes**: Genetic modification has enabled the creation of microorganisms with enhanced tolerance to heavy metals and greater bioaccumulation capacity. For example, recombinant bacteria expressing metallothioneins or phytochelatins are effective at sequestering heavy metals (Rafeeq *et al.*, 2023).

Despite the promising nature of these techniques, challenges persist in scaling up remediation processes, managing complex mixtures of contaminants, and addressing secondary pollution. Site-specific factors like pH, organic content, and metal speciation complicate remediation efforts (Alloway, 2012). To ensure sustainable practices, integrating biological, chemical, and physical strategies, along with robust monitoring and risk assessment frameworks, is vital. Future research should focus on developing cost-effective, energy-efficient, and community-accepted solutions.

3.3. Pesticide Degradation

Microbial degradation of pesticides has proven to be an effective means of reducing their toxicity and persistence in agricultural soils. Various bacteria and fungi are capable of breaking down pesticides such as organophosphates, carbamates, and other pesticide groups (Cycon *et al.*, 2009). Pesticide degradation refers to the breakdown of pesticide compounds into less harmful or bioavailable forms through biological, physical, or chemical processes. Effective pesticide degradation is essential to mitigate environmental impacts such as soil and water contamination, biodiversity loss, and bioaccumulation (Fenner *et al.*, 2013).

3.3.1. Mechanisms of Pesticide Degradation

1. ***Abiotic Degradation***: Abiotic processes like photolysis, hydrolysis, and oxidation occur independently of biological processes. Photolysis,

driven by sunlight, is particularly important for the degradation of organophosphates and pyrethroids in both aquatic and terrestrial settings. Hydrolysis, which involves breaking chemical bonds in water, is effective for pesticides like carbamates and triazines. Chemical oxidation, often facilitated by reactive oxygen species, also contributes to pesticide breakdown.

2. **Biotic Degradation**: Microbial degradation is a dominant pathway for pesticide breakdown. Bacteria, fungi, and other microorganisms metabolize pesticides as carbon or energy sources, often converting them into less toxic compounds. For example, *Pseudomonas putida* and *Bacillus subtilis* degrade organophosphates, while white-rot fungi like *Phanerochaete chrysosporium* break down persistent chemicals like DDT and atrazine (Singh and Walker, 2006).

3.3.2. *Factors Affecting Pesticide Degradation*

The rate of pesticide degradation depends on factors like the chemical structure of the pesticide, environmental conditions, and the presence of microorganisms capable of degradation. Persistent pesticides, such as organochlorines, are more resistant to breakdown due to their stable chemical structures and lipophilicity. Environmental factors, such as temperature, pH, moisture content, and organic matter, also influence degradation rates by affecting microbial activity and chemical reactivity (Fenner *et al.*, 2013).

3.3.3. Emerging Technologies for Enhanced Degradation

1. **Bioremediation and Bioaugmentation**: Bioremediation uses native or introduced microorganisms to break down pesticides in contaminated areas. Bioaugmentation, the process of introducing specific microbial strains with enhanced degradation capabilities, has been successfully used to accelerate the breakdown of pesticides like lindane and chlorpyrifos (Rasool *et al.*, 2022). A review of interactions of pesticides within various interfaces of intrinsic and organic residue amended soil environment. *Chemical Engineering Journal Advances*, *11*, 100301.).
2. **Enzyme Engineering**: Advances in enzyme engineering have led to the development of enzymes with improved efficiency and specificity for degrading pesticides. For example, organophosphorus hydrolases and carbamate hydrolases have been engineered to target pesticide residues effectively (Singh *et al.*, 2009).
3. **Nanotechnology**: Nanoparticles, like titanium dioxide (TiO_2), are being explored for enhancing photocatalytic degradation of pesticides under sunlight. This method shows promise for removing pesticide

residues from both soil and water in an environmentally friendly manner (Abdennouri *et al.*, 2016).

3.3.4. *Challenges and Environmental Implications*

While pesticide degradation reduces the persistence of these chemicals in the environment, incomplete degradation can lead to the formation of toxic by-products. For instance, the breakdown of organophosphates can yield intermediates like Chlorpyrifos-oxon, which are more toxic than the original pesticide. Therefore, continuous monitoring of degradation pathways and their by-products is necessary to minimize unintended ecological risks (Sidhu *et al.*, 2029).

Pesticide degradation involves a combination of abiotic and biotic processes, each playing a critical role in reducing the environmental impact of pesticides. The integration of biotechnological and chemical innovations with natural degradation pathways holds significant potential for enhancing pesticide remediation strategies while minimizing ecological risks.

3.4. Treatment of Industrial Waste

Industrial effluents, which often contain a mix of organic and inorganic pollutants, can be treated effectively using bioreactors that employ specially designed microbial consortia to break down complex waste streams. This approach helps minimize the environmental impact of industrial operations and encourages the principles of a circular economy (Cao *et al.*, 2022). Industrial waste management includes the processes of waste collection, treatment, and disposal, and it is essential for reducing environmental pollution, conserving resources, and safeguarding public health. Industrial waste is generally categorized into hazardous and non-hazardous types, encompassing solid, liquid, and gaseous forms, with each requiring specific methods for proper management and treatment (Slack *et al.*, 2007).

3.4.1. Key Principles of Industrial Waste Management

1. **Waste Reduction**: The first step in handling industrial waste is to reduce its generation at the source. Techniques like process optimization, material substitution, and cleaner production technologies are used to minimize waste. In the chemical industry, for instance, green chemistry principles are employed to decrease the production of harmful by-products (Anastas and Eghbali, 2010).
2. **Recycling and Recovery**: Many industries adopt practices in line with the circular economy by recycling waste into valuable products. For example, fly ash from coal-fired power plants is repurposed in the

creation of building materials like cement and concrete (Ahmaruzzaman, 2010). Furthermore, waste heat recovery systems are employed in energy-intensive sectors to improve efficiency and reduce emissions (Fallis, 2013).

3. **Treatment Methods**: Waste treatment technologies are selected based on the waste type and its potential environmental consequences:
   a) **Physical Treatment**: Techniques such as filtration, sedimentation, and centrifugation are used to separate and concentrate pollutants.
   b) **Chemical Treatment**: Processes like neutralization, oxidation, and precipitation are used to transform harmful chemicals into less toxic forms. For example, chemical precipitation is commonly applied to treat wastewater contaminated with heavy metals.
   c) **Biological Treatment**: Microorganisms are employed in bioremediation and anaerobic digestion to break down organic waste and generate biogas (Sevillano *et al.*, 2021).
4. **Secure Disposal**: In cases where recycling or treatment is not feasible, waste must be safely disposed of, typically in engineered landfills or through incineration. Modern incineration systems, especially those with energy recovery capabilities, such as waste-to-energy plants, are increasingly used to minimize reliancet on landfills while producing electricity (Arena, 2012).

3.4.2 Challenges in Industrial Waste Management

1. **Handling Hazardous Waste**: Hazardous materials like chemicals, heavy metals, and toxic organic compounds present significant risks to human health and the environment. Improper handling can lead to contamination of soil and water, as seen in major incidents like the Bhopal disaster (Jasanoff, 1994).
2. **Regulatory Compliance**: Industries must adhere to strict environmental regulations such as the Resource Conservation and Recovery Act (RCRA) in the U.S. or the European Union's Waste Framework Directive to ensure responsible waste management.
3. **Economic Constraints**: The high costs associated with advanced waste treatment technologies and disposal methods may limit their widespread adoption, especially in developing countries. Financial solutions such as incentives, subsidies, and public-private partnerships can help address these challenges.

3.4.2. Innovations and Future Outlook

Several emerging technologies and practices are redefining industrial waste management:

1. **Digital Monitoring**: The Internet of Things (IoT) and data analytics are being increasingly used to track waste generation and optimize treatment processes in real time.
2. **Biotechnological Approaches**: Advances in synthetic biology have led to the creation of engineered microbes capable of degrading complex pollutants, including persistent organic pollutants (POPs) and plastics (Singh *et al.,* 2009).
3. **Circular Economy Models**: Industrial symbiosis, where the waste from one industry becomes a resource for another, is gaining momentum as a strategy to reduce waste production and enhance resource efficiency (Chertow, 2000).

The field of industrial waste management is rapidly evolving, requiring multidisciplinary approaches to tackle environmental, regulatory, and economic challenges. By adopting cutting-edge technologies and sustainable practices, industries can reduce their environmental impact while supporting a circular economy.

4. Advantages of Bioremediation

Bioremediation offers numerous advantages over traditional remediation techniques, making it an ideal choice for sustainable development:

1. **Eco-friendly**: Bioremediation utilizes natural processes, reducing the need for chemical treatments that may have secondary environmental consequences.
2. **Cost-effective**: Bioremediation is generally more affordable than physical or chemical remediation methods, particularly for large-scale applications.
3. ***In Situ* Application**: Bioremediation can often be performed directly on-site, which eliminates the need for excavating and transporting contaminated materials.
4. **Versatility**: A wide range of pollutants can be addressed using various microorganisms, making bioremediation adaptable to different contamination scenarios.

5. Challenges and Future Directions

Although bioremediation holds great promise, several challenges must be addressed to enhance its effectiveness:

1. **Environmental Variability**: Factors like temperature, pH, and nutrient availability can impact bioremediation processes. Developing resilient microbial strains and optimizing environmental conditions are key to success (Singh and Ward, 2004).
2. **Concerns about Genetic Modification**: The use of genetically modified organisms (GMOs) in bioremediation raises regulatory and public acceptance issues. It is essential to ensure the safety and environmental monitoring of GMOs (Timmis *et al.,* 2014).
3. **Scale-up Challenges**: Translating successful laboratory outcomes to field applications requires careful consideration of scale-up factors, such as the distribution and delivery of microorganisms at contaminated sites (Steffan and Sperry, 1999).

Future research should focus on advancing bioremediation methods like bioaugmentation, biostimulation, and synthetic biology to create more efficient microbial strains. Additionally, combining bioremediation with other sustainable approaches, such as phytoremediation and eco-engineering, can further strengthen environmental management strategies.

6. Modern Applications and Future Outlook

Bioremediation has found diverse applications, from oil spill clean-up and heavy metal removal to treating agricultural and industrial waste. The integration of "omics" technologies (such as genomics, proteomics, and metabolomics) has provided deeper insights into microbial communities, enabling more targeted and effective bioremediation strategies (Deutschbauer *et al*., 2006).

Additionally, advances in bioaugmentation and biostimulation have made bioremediation processes more efficient. Bioaugmentation involves adding specific microorganisms to contaminated sites, while biostimulation optimizes environmental conditions to enhance the activity of indigenous microbes (Tyagi *et al*., 2011).

7. Conclusion

Bioremediation is a crucial tool for achieving sustainable development by addressing environmental pollution in a cost-effective and natural manner. Ongoing research and technological innovations hold the potential to make bioremediation a mainstream solution to various environmental challenges, leading to a cleaner and healthier world for future generations.

References

Abdennouri, M., Baâlala, M., Galadi, A., El Makhfouk, M., Bensitel, M., Nohair, K., & Barka, N. (2016). Photocatalytic degradation of pesticides by titanium dioxide and titanium pillared purified clays. Arabian Journal of Chemistry, 9, S313-S318.

Acar, Y. B., & Alshawabkeh, A. N. (1993). Principles of electrokinetic remediation. Environmental science & technology, 27(13), 2638-2647.
Ahmad, M., Rajapaksha, A. U., Lim, J. E., Zhang, M., Bolan, N., Mohan, D., ... & Ok, Y. S. (2014). Biochar as a sorbent for contaminant management in soil and water: a review. Chemosphere, 99, 19-33.
Ahmaruzzaman, M. (2010). A review on the utilization of fly ash. Progress in energy and combustion science, 36(3), 327-363.
Ahrens, L., & Bundschuh, M. (2014). Fate and effects of poly-and perfluoroalkyl substances in the aquatic environment: A review. Environmental toxicology and chemistry, 33(9), 1921-1929.
Alexander, M. (1999). Biodegradation and Bioremediation. Academic Press San Diego CA. Biodegradation and bioremediation.
Ali, H., Khan, E., & Sajad, M. A. (2013). Phytoremediation of heavy metals—concepts and applications. Chemosphere, 91(7), 869-881.
Alloway, B. J. (Ed.). (2012). Heavy metals in soils: trace metals and metalloids in soils and their bioavailability (Vol. 22). Springer Science & Business Media.
Amran, R. H., Jamal, M. T., Pugazhendi, A., Al-Harbi, M., Ghandourah, M., Al-Otaibi, A., & Haque, M. F. (2022). Biodegradation and Bioremediation of Petroleum Hydrocarbons in Marine Ecosystems by Microorganisms: A Review. Nature Environment & Pollution Technology, 21(3).
Anastas, P., & Eghbali, N. (2010). Green chemistry: principles and practice. Chemical Society Reviews, 39(1), 301-312.
Ardern, E., & Lockett, W. T. (1914). Experiments on the oxidation of sewage without the aid of filters. Journal of the society of chemical industry, 33(10), 523-539.
Arena, U. (2012). Process and technological aspects of municipal solid waste gasification. A review. Waste management, 32(4), 625-639.
Atlas, R. M., & Hazen, T. C. (2011). Oil biodegradation and bioremediation: a tale of the two worst spills in US history.
Boening, D. W. (2000). Ecological effects, transport, and fate of mercury: a general review. Chemosphere, 40(12), 1335-1351.
Bornscheuer, U. T., Huisman, G. W., Kazlauskas, R. J., Lutz, S., Moore, J. C., & Robins, K. (2012). Engineering the third wave of biocatalysis. Nature, 485(7397), 185-194.
Cao, Z., Yan, W., Ding, M., & Yuan, Y. (2022). Construction of microbial consortia for microbial degradation of complex compounds. Frontiers in Bioengineering and Biotechnology, 10, 1051233.
Cases, I., & Lorenzo, V. D. (2005). Genetically modified organisms for the environment: stories of success and failure and what we have learned from them.
Chertow, M. R. (2000). Industrial symbiosis: literature and taxonomy. Annual review of energy and the environment, 25(1), 313-337.
Connolly, J. P., & Pedersen, C. J. (1988). A thermodynamic-based evaluation of organic chemical accumulation in aquatic organisms. Environmental science & technology, 22(1), 99-103.
Cycoń, M., Piotrowska-Seget, Z., Kaczyńska, A., & Kozdrój, J. (2006). Microbiological characteristics of a sandy loam soil exposed to tebuconazole and λ-cyhalothrin under laboratory conditions. Ecotoxicology, 15, 639-646.
Das, N., & Chandran, P. (2011). Microbial degradation of petroleum hydrocarbon contaminants: an overview. Biotechnology research international, 2011(1), 941810.
Desai, J. D., & Banat, I. M. (1997). Microbial production of surfactants and their commercial potential. Microbiology and Molecular biology reviews, 61(1), 47-64.

Deutschbauer, A. M., Chivian, D., & Arkin, A. P. (2006). Genomics for environmental microbiology. Current opinion in biotechnology, 17(3), 229-235.

Eisler, R. (2000). Handbook of chemical risk assessment: Health hazards to humans, plants, and animals-Vol. 2: Organics.

Fallis, A. (2013). Waste Heat recovery: technology and opportunities in US industry. J. Chem. Inf. Model, 53(9), 1689-1699.

Fenner, K., Canonica, S., Wackett, L. P., & Elsner, M. (2013). Evaluating pesticide degradation in the environment: blind spots and emerging opportunities. science, 341(6147), 752-758.

Gadd, G. M. (2004). Microbial influence on metal mobility and application for bioremediation. Geoderma, 122(2-4), 109-119.

Gadd, G. M. (2010). Metals, minerals and microbes: geomicrobiology and bioremediation. Microbiology, 156(3), 609-643.

Guengerich, F. P. (2008). Cytochrome p450 and chemical toxicology. Chemical research in toxicology, 21(1), 70-83.

Hazen, T. C., Dubinsky, E. A., DeSantis, T. Z., Andersen, G. L., Piceno, Y. M., Singh, N., ... & Mason, O. U. (2010). Deep-sea oil plume enriches indigenous oil-degrading bacteria. Science, 330(6001), 204-208.

Head, I. M., Jones, D. M., & Röling, W. F. (2006). Marine microorganisms make a meal of oil. Nature Reviews Microbiology, 4(3), 173-182.

Jasanoff, S. (Ed.). (1994). Learning from disaster: Risk management after Bhopal. University of Pennsylvania press.

Karn, B., Kuiken, T., & Otto, M. (2009). Nanotechnology and in situ remediation: a review of the benefits and potential risks. Environmental health perspectives, 117(12), 1813-1831.

Keasling, J. D. (2010). Manufacturing molecules through metabolic engineering. Science, 330(6009), 1355-1358.

Koch, R. (1881). Zur untersuchung von pathogenen organismen. Norddeutschen Buchdruckerei und Verlagsanstalt.

Lavoie, R. A., Jardine, T. D., Chumchal, M. M., Kidd, K. A., & Campbell, L. M. (2013). Biomagnification of mercury in aquatic food webs: a worldwide meta-analysis. Environmental science & technology, 47(23), 13385-13394.

Leahy, J. G., & Colwell, R. R. (1990). Microbial degradation of hydrocarbons in the environment. Microbiological reviews, 54(3), 305-315.

Lloyd, J. R., & Lovley, D. R. (2001). Microbial detoxification of metals and radionuclides. Current opinion in biotechnology, 12(3), 248-253.

Löffler, F. E., Tiedje, J. M., & Sanford, R. A. (1999). Fraction of electrons consumed in electron acceptor reduction and hydrogen thresholds as indicators of halorespiratory physiology. Applied and environmental microbiology, 65(9), 4049-4056.

Mackay, D., & Fraser, A. (2000). Bioaccumulation of persistent organic chemicals: mechanisms and models. Environmental pollution, 110(3), 375-391.

Margesin, R., & Schinner, F. (2001). Biodegradation and bioremediation of hydrocarbons in extreme environments. Applied Microbiology and biotechnology, 56, 650-663.

Pasteur, L. (1879). Studies on fermentation: the diseases of beer, their causes, and the means of preventing them. Macmillan & Company.

Pignatello, J. J. (2013). Adsorption of dissolved organic compounds by black carbon. Molecular environmental soil science, 359-385.

Pilon-Smith, E. (2005). Phytoremediation: Annual Review Plant Biology.

Prince, R. C. (2015). Oil spill dispersants: boon or bane?. Environmental science & technology, 49(11), 6376-6384.

Rafeeq, H., Afsheen, N., Rafique, S., Arshad, A., Intisar, M., Hussain, A., ... & Iqbal, H. M. (2023). Genetically engineered microorganisms for environmental remediation. Chemosphere, 310, 136751.

Rasool, S., Rasool, T., & Gani, K. M. (2022). A review of interactions of pesticides within various interfaces of intrinsic and organic residue amended soil environment. Chemical Engineering Journal Advances, 11, 100301.

Sevillano, C. A., Pesantes, A. A., Peña Carpio, E., Martínez, E. J., & Gómez, X. (2021). Anaerobic digestion for producing renewable energy—the evolution of this technology in a new uncertain scenario. Entropy, 23(2), 145.

Shimada, T., & Fujii-Kuriyama, Y. (2004). Metabolic activation of polycyclic aromatic hydrocarbons to carcinogens by cytochromes P450 1A1 and1B1. Cancer science, 95(1), 1-6.

Sidhu, G. K., Singh, S., Kumar, V., Dhanjal, D. S., Datta, S., & Singh, J. (2019). Toxicity, monitoring and biodegradation of organophosphate pesticides: a review. Critical reviews in environmental science and technology, 49(13), 1135-1187.

Singh, A., & Ward, O. P. (Eds.). (2004). Biodegradation and bioremediation (Vol. 2). Springer Science & Business Media.

Singh, A., Kuhad, R. C., & Ward, O. P. (Eds.). (2009). Advances in applied bioremediation (pp. 1-19). Berlin: Springer-Verlag.

Singh, B. K., & Walker, A. (2006). Microbial degradation of organophosphorus compounds. FEMS microbiology reviews, 30(3), 428-471.

Slack, R. J., Gronow, J. R., & Voulvoulis, N. (2005). Household hazardous waste in municipal landfills: contaminants in leachate. Science of the total environment, 337(1-3), 119-137.

Steffan, R. J., Sperry, K. L., Walsh, M. T., Vainberg, S., & Condee, C. W. (1999). Field-scale evaluation of in situ bioaugmentation for remediation of chlorinated solvents in groundwater. Environmental Science & Technology, 33(16), 2771-2781.

Testa, B., & Krämer, S. D. (2007). The biochemistry of drug metabolism–an introduction: Part 2. Redox reactions and their enzymes. Chemistry & biodiversity, 4(3), 257-405.

Tyagi, M., da Fonseca, M. M. R., & de Carvalho, C. C. (2011). Bioaugmentation and biostimulation strategies to improve the effectiveness of bioremediation processes. Biodegradation, 22, 231-241.

UNEP (2013). Global Mercury Assessment 2013: Sources, Emissions, Releases, and Environmental Transport. United Nations Environment Programme.

Van Beilen, J. B., & Funhoff, E. G. (2007). Alkane hydroxylases involved in microbial alkane degradation. Applied microbiology and biotechnology, 74, 13-21.

Wackett, L. P. (2004). Evolution of enzymes for the metabolism of new chemical inputs into the environment. Journal of Biological Chemistry, 279(5), 3465-3474.

3

Exploring the Concept of Bioremediation in Air Pollution Mitigation

Soniya Santwani, Vikrant Pachouri* and Prafull Kothari

Institute of Engineering and Technology, Mohanlal Sukhadia University, Udaipur - 3130001, Rajasthan, India

**Corresponding Author Email: vikrantpachouri121@gmail.com*

Abstract

Anthropogenic activities have risen over the years, and this has led to a more serious deterioration of the environment, especially by polluting the air more than before. A novel, efficient and eco-friendly method of eradicating air pollutants from the environment is bioremediation, which employs the utilisation of microorganisms to degrade and neutralise pollutants. This chapter specifically looks at the topic of bioremediation and particularly the viability of using it in combating air pollution. The general approaches to bioremediation include bio-stimulation, bio-augmentation, bio-venting, bio-piles and bio-attenuation through which hazardous pollutants are converted into less damaging forms. Bioremediation, when used in air pollution control, is complemented by bio-systems, for instance, bio-filters, bio-trickling filters and bio-scrubbers that are used in the purification of polluted gases from industries.

This chapter also seeks to explain the processes of bioremediation, different bioremediation methods and the biostructures in combating air pollution. Thus, the effectiveness and possibilities of bioremediation are demonstrated through case studies and real-world examples of air pollutant reduction. The chapter also describes the issues and opportunities of bioremediation concerning air pollution control, normalizing processes, and stressing the idea of clean and healthy air.

Keywords: *Bioremediation, Air Pollution Mitigation, Bio-filters, Industrial Emissions, Bio-scrubbers, Pollutant Degradation*

1. Introduction

One of the foremost environmental challenges the world faces today is air pollution. As industrialisation, urbanisation, and transportation continue unabated, harmful substance levels in the atmosphere have become alarmingly high (Thanvisitthpon *et al.*, 2024). This general contamination in the air is not just a serious risk to humanity but also harms ecosystems while contributing to global climate change. Controlling air pollution has become a critical necessity at this stage, with the need for innovative solutions in this respect. Bioremediation is one of the innovative solutions, where certain microorganisms break down or neutralise pollutants in the air.

1.1. Overview of Environmental Contamination and Air Pollution

Recently, air pollution has emerged as a new environmental issue not only in cities but also in the countryside. Simply stated, it is defined as the emission of harmful chemicals, particulate matter, and biological material into the atmosphere, which affects the quality of air (Kumar *et al.*, 2023). These include industrial emissions, exhaust gases from vehicles, agricultural practices, and poor waste disposal mechanisms. It also contains gases such as sulfur dioxide (SO_2), NO_x, carbon monoxide (CO), particulate matter (PM), etc, and volatile organic compounds (VOCs) (Gurav *et al.*, 2024). All these are harmful both to human health and the environment.

This concentration, therefore, brings about these pollutants into the atmosphere, resulting in massive health problems, such as respiratory and cardiovascular diseases, and may even cause preterm deaths. The contribution of these pollutants also goes to the degradation of ecosystems, affecting plants, fauna, and soils. Again, carbon dioxide (CO_2) and methane (CH_4) are strong contributors to the greenhouse effect among the many causing the alarming environmental crisis (Ramirez-Corredores *et al.*, 2023).

Efforts to control air pollution have been ongoing for decades, but the problem remains, especially in highly industrialised areas. Traditional methods for air pollution control, like filtration and chemical scrubbers, do not provide effective alternatives that are sustainable and cost-effective (Elehinafe *et al.*, 2024). Because of this, there have been alternative nature-based solutions, bioremediation, as an approach to controlling this escalating problem of pollution.

1.2. Importance of Addressing Air Pollution

Air pollution has turned out to be a very sensitive issue with far-reaching implications in the world at large, both concerning public health and environmental issues (Ofremu *et al.*, 2024). According to the World Health

Organisation, millions of people die prematurely every year. These include vulnerable groups, mainly including children, the elderly, and individuals suffering from previous medical conditions. Exposure to polluted air for an extended period is asserted to be hazardous since it results in increased susceptibility to suffering from chronic respiratory diseases, like asthma and COPD. Cardiovascular diseases are included in this list along with stroke and heart attacks (Bălă *et al.*, 2024). Besides the effects on health, air pollution is one of the main causes of environmental degradation. It impairs visibility by causing smog, acid rain, and haze that disturb ecosystems and lower agricultural productivity.

1.3. Fundamentals of Bioremediation and Its Significance

Bioremediation is also a promising and rapidly advancing strategy for addressing and dealing with pollution. It comes with advantages over conventional approaches to air pollution control, which are less environmentally friendly and costlier. Bioremediation is a form of biodegradation which involves utilising the abilities of tiny microorganisms to store, move, reduce or neutralise pollutants dangerous to human health or inhibitive to life (Ahmed *et al.*, 2022). To prevent air pollution, the approach employed bacterial, fungal and other microbial interventions to break down or immobilise the pollutant airborne particles.

Several reasons show that the potential offered by bioremediation is important because it may provide an effective solution to air pollution that would be viable, sustainable, and conform to the principles of sustainable use of resources and prevention of harm to the environment. They are relatively autochthonous systems and can sometimes operate in fairly minimalistic manners as regards energy input, the major distinction being that bioremediation systems are not known to procreate secondary pollutants. Bioremediation is the targeted removal of pollutants, and it is possible to remediate specific contaminants, including volatile organic compounds (VOCs), nitrogen oxides, and sulfur compounds, which are otherwise hard to remove by conventional methods (Hussain *et al.*, 2022).

The growing concern for bioremediation is due to its multifold advantages cost, low energy, and, as a byproduct, being harmless. Bioremediation has been, therefore, successfully practised in wastewater treatments, soil remediation, and increasingly air pollution controls (Bala *et al.*, 2022). A promising solution through techniques, such as bio-filtration and bio-trickling, harnessing microorganisms to clean the polluted air, is offered with the perspective of reducing toxic emissions through industries, transport, or urban environments.

There cannot be an overemphasis on the importance of bioremediation in controlling air pollution. With increasing momentum towards global climate change and pollution mitigation efforts, bioremediation has become one of the sustainable natural approaches to one of the most significant environmental challenges that face the world today.

2. Mechanisms of Bioremediation

Bioremediation is based on the inherent ability of microorganisms to degrade or transform harmful pollutants into less toxic or harmless substances. Bioremediation is effective in degrading a large number of air pollutants through biological processes (Bala *et al.*, 2022). The mechanisms that are involved in bioremediation provide an appreciation of its potential as a feasible approach to the remediation of environmental contamination.

2.1. Biological Process Involved in Bioremediation

Bioremediation involves a major biological mechanism called microbial degradation, where complex pollutants are broken down by microorganisms, such as bacteria, fungi, and algae, into simpler and less harmful substances (Sharma *et al.*, 2021). It occurs naturally but can be enhanced in controlled conditions for better removal of pollutants.

- **Microbial Degradation:** These microbes degrade toxic compounds as an energy source and a carbon source. Some also happen inorganic pollutants like sulfur, where bacteria have to convert the toxic gaseous forms of sulfur compounds to safe, non-toxic states.
- **Assimilation:** It is a process in which microorganisms incorporate pollutants within the cellular structure (Fouad *et al.*, 2022). It simply means that NO_x or ammonia would get absorbed by the microorganism and then used in microbial growth and reproduction, thereby removing it from the environment.
- **Detoxification:** In some instances, microbes metabolise harmful substances into harmless end products through enzymatic actions (Kumari & Das, 2023). This is most applicable to the biodegradation of complex organic pollutants, in which enzymes catalyse transformations that make the compounds harmless.

2.2. Types of Micro-organisms Used

Different types of microbes are involved in bioremediation, targeting different classes of pollutants. The three most commonly used organisms in bioremediation are bacteria, fungi, and algae. These organisms are naturally adapted to metabolise different types of pollutants in the air.

- **Bacteria:** Bacteria represent one of the most commonly applied microbial agents in bioremediation, primarily owing to their diversity and variability (Kour *et al.*, 2021). The microorganism displays the capability to transform various pollutant categories that extend from volatile organic compounds to hydrocarbons, as well as inorganic sulfur compounds. Some bacterial strains demonstrate superior air contaminant degrading capabilities through bio-filters while also functioning in bio trickling filters of Pseudomonas, Bacillus and Rhodococcus.
- **Fungi:** The white-rot fungi stand out as the most renowned among specialised fungi, which degrade complex organic compounds. These microorganisms degrade organic compounds which remain indigestible to bacterial breakdown methods. Such microorganisms enact enzymatic detoxification processes on pollutants, which include both polycyclic aromatic hydrocarbons and VOCs.
- **Algae:** The bio-scrubbing process utilises algae along which perform bioremediation duties. Algal systems are quite useful in removing CO_2 and absorbing other pollutants, among others, like ammonia and heavy metals (Pachouri *et al.*, 2024). Photosynthesis helps improve the absorption and transduction of pollutants into biomass further treatable.

These microorganisms can be used in a variety of bioremediation systems such as bio-filters, bio-trickling filters and bio-scrubbers. Suits specific kinds of pollutants or treatment needs.

2.3. Transformation of Pollutants into Less Harmful Forms

Bioremediation has a principal benefit, its capability to turn bad pollutants into non-toxic compounds with the action of microbes. Microorganisms degrade harmful substances using a myriad of biochemical pathways and convert those substances into harmless byproducts like water, carbon dioxide, and simple metabolites.

- **Volatile Organic Compounds (VOCs):** These are generally the most common industrial process pollutants. Bacteria usually degrade these through aerobic respiration (Kumari & Das, 2023). In this case, VOCs are oxidised to carbon dioxide and water, thus removing them from the air.
- **Carbon Monoxide (CO):** Carbon monoxide is one of the toxic gases generally formed by incomplete combustion. Carbon dioxide is non-toxic and can, therefore, be emitted freely into the atmosphere.
- **Sulfur Dioxide (SO_2):** SO_4 is the major gaseous precursor to acid rain. SO_2 can be biologically oxidised by sulfur-oxidising bacteria, which can

be assimilated by plants or harmlessly removed from the environment as sulfate (LaCount *et al.*, 2021).

- **Nitrogen Oxides (NO_x):** These nitrogen oxides, some of which are emitted as a result of vehicle exhausts and industrial processes, are, by some bacteria, changed into either nitrogen gas (N_2) or nitrate (NO_3^-), which are not harmful.

This is not only detoxification but also results in low secondary pollutant production in the bioremediation process. Therefore, this is the most sustainable method for the control of air pollution.

By using natural microorganisms, bioremediation provides a future prospect of reducing air pollution. The ability of these organisms to break down or transform pollutants into harmless substances is central to their role in improving air quality and reducing the impact of industrial emissions.

3. Bioremediation Techniques

Bioremediation refers to the use of various techniques that exploit the natural action of microorganisms to clean up pollutants. There are mainly two types: in situ and ex-situ (Bala *et al.*, 2022). Both have their merits and demerits depending on the type of pollutants, environment, and result. Thus, these techniques have to be known to optimize their use in controlling air pollution. Fig. 1. Displays the techniques involved in Bioremediation.

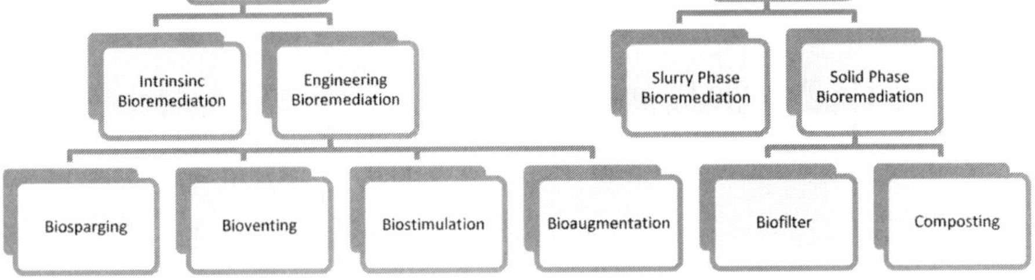

Fig. 1: Bioremediation Techniques

3.1. *In-situ* Bioremediation

It involves the treatment of contaminated air at the source of pollution with or without the removal of the pollutants from the environment. This approach makes use of the microbial population in the contaminated area while elevating their activity to catalyse the degradation of pollutants in this in-situ bioremediation approach (Bhatt *et al.*, 2021). Some of the advantages of in-situ bioremediation include low cost, little disruption to the environment, and little infrastructure.

- **Application to Air Pollution**: In the air, in-situ bioremediation is most applied to bio-filtration or bio-trickling filters, whereby contaminated air is passed over a bioreactor containing microorganisms capable of degrading particular pollutants. Usually, these bio-filters come in the form of some media, for example, compost, activated carbon, or soil on which microbial life grows. The microorganisms will thus consume all those harmful components through the air passing in these bio-filters and break them into non-harmful elements, such as carbon dioxide and water.
- **Advantages**: In-situ techniques do not require the removal and transportation of contaminated air and, hence, are relatively cheap (Hussain *et al.*, 2022). Also, they aid in maintaining environmental stability because treatment takes place where pollution is produced to prevent the dispersion of pollutants in the atmosphere.

3.2. Ex-situ Bioremediation

Ex-situ bioremediation is the capturing and treatment of contaminated air at a remote location or system, distinct from the pollution source. Unlike in-situ methods, ex-situ bioremediation requires the collection and transportation of the pollutant-laden air into some controlled setting for processing. Here, microorganisms can be applied to degrade the pollutant.

- **Application to Air Pollution**: For ex-situ bioremediation, bio-trickling filters or bio-scrubbers are quite popular where air pollutants can be captured and purified after drawing the contaminated air from any source (Hussain *et al.*, 2022). The extracted pollutants are then processed in a system that applies a concentration of microbial populations to improve the breakdown of contaminants.
- **Advantages**: Ex-situ methods offer better controls on the environment, including temperature, humidity, and pH, for the growth of micro-organisms. This method is also very effective when pollutants are too concentrated or too toxic for in-situ treatment.

3.3. Detailed Explanation of Bio-stimulation, Bio-augmentation, Bio-venting, Bio-piles, and Bio-attenuation

Several sub-techniques include bioremediation, which facilitates the activity of microbes for the degradation of pollutants. These sub-techniques include bio-stimulation, bio-augmentation, bio-venting, bio-piles, and bio-attenuation. All have different roles in improving the bioremediation processes.

- **Bio-stimulation**: Bio-stimulation modifies the environmental conditions to stimulate the growth and activity of indigenous microorganisms. This may include adding nutrients, such as nitrogen or phosphorus, modification of the pH levels, or oxygenation of the system (Omenna *et al.*, 2024). Bio-stimulation enhances the natural microbial populations to degrade the pollutants more efficiently.
- **Bioaugmentation**: The technique involves the introduction of special microbial flora into the contaminated medium (Nwankwegu *et al.*, 2022). The organisms selected must possess specific capabilities for degrading one form of pollutant that would prove very difficult to degrade otherwise, with the available indigenous flora.
- **Bio-venting**: Bio-venting increases the ventilation of contaminated soil or air to enhance the microbial activity of the microorganisms that degrade organic pollutants. Bio-venting is generally applied for soil remediation. In a few cases, bio-venting can be used for air treatment. In bio-venting systems, air is passed through a microbial filter or directly above contaminated sites, and then necessary oxygen feeds into the microbial degradation of the pollutants. This is very effective in removing VOCs from industrial emissions.
- **Bio-piles**: The bio-piles are the process by which contaminated soil or material is piled up to activate microbial activities. In case of atmospheric pollution, the bio-piles can be applied as an integral component of the bio-filters in which piles of organic material like compost, wood chips, etc., support microbial life and will help remove pollutants (Farid *et al.*, 2023). These piles will be aerated generally for the growth of microbes, along with the speedier removal of pollutants.
- **Bio-attenuation**: It is the reduction of levels of a pollutant through microbial activity without human involvement. It occurs naturally as time passes and microorganisms that live in the environment learn to degrade the pollutant. Although bio-attenuation is slower than bio-stimulation or bio-augmentation, it could be integrated into a long-term strategy, especially for bioremediation if pollutant concentrations are relatively low and natural microbial populations are already degrading them.

There are many bioremediation techniques, from in-situ treatments such as bio-filtration to ex-situ systems such as bio-trickling filters. All of these offer versatile and environmentally friendly ways of controlling air pollution.

Each has its specific advantages, depending on the type of pollutant and environmental conditions. Bio-stimulation, bio-augmentation, and other specialised techniques increase the efficiency of these systems, which makes bioremediation a promising approach for sustainable air pollution control.

4. Bioremediation in Air Pollution Mitigation

Bioremediation is one of the promising means of reducing air pollution that utilises the natural abilities of microbes to degrade or transform harmful pollutants into harmless substances. Industrialisation and urbanisation, two factors contributing to poor air quality, are increasing further (Bala *et al.*, 2022). In this regard, bioremediation is also a sustainable means of reducing hazardous emissions in different sectors. This section discusses how bioremediation is applied to air pollution control, points out specific bioremediation systems, and looks into their mechanisms and effectiveness in treating polluted off-gases.

4.1. Application of Bioremediation to Air Pollution Control

Different sorts of industries, agricultural industries and various sections within city economies have become increasingly successful in solving the severe pollution problem of air due to bioremediation (Sharma *et al.*, 2021). Advantage can thus be taken to clean up many forms of pollution, such as VOC, NO_x, SO_2, and PM, taking advantage of microorganisms while having bioremediation applied.

- **Industrial Applications**: Huge quantities of harmful pollutants from petrochemical plants, refineries, and chemical manufacturing facilities include VOCs and other dangerous gases. Bioremediation is used through bio-filters and bio-trickling filters to capture the emitted amount of these pollutants and subsequently degrade them before their emissions into the atmosphere.
- **Agricultural Applications**: Agriculture is the greatest source of air pollution, which comprises ammonia (NH_3) from animal waste and fertiliser. Bio-scrubbers have been used for cleaning poultry farms and other agricultural settings (Wyer *et al.*, 2022). The systems absorb and neutralise ammonia and other pollutants by passing them through a liquid medium containing microorganisms that can absorb and break down the contaminants.
- **Urban Applications**: Bioremediation technologies fit the bill of cities having heavy loads of traffic and heavy-intensity industrial activities. In recent years, there has been extensive use of cities having large-scale ventilation systems wherein the polluted air emanated from highways, factories, and waste management plants using bio-filters and bio-

scrubbers that reduce the amount of these harmful gases and particulates such as nitrogen oxides, sulfur compounds in the environment that improve public health due to quality air.

4.2. Overview of Bio-filters, Bio-trickling Filters, and Bio-scrubbers

Bioremediation systems such as bio-filters, bio-trickling filters, and bio-scrubbers have become a very common method in controlling air pollution. The application of microorganisms for degrading pollutants in bioremediation systems makes it very effective and appropriate for certain types of pollutants and environments.

- **Bio-filters**: Bio-filters are the most common method of abating air pollution. This system contains a porous medium that can be composed of compost, wood chips, or activated carbon, and the medium is the residence place of the microorganisms (Gandu *et al.*, 2021). When this polluted air passes through the filter, the polluted air travels through the porous medium and gets degraded by microorganisms. Bio-filters are most effective in VOCs, ammonia, and other organic compounds. It has many common applications in such industrial installations for the existence of VOC in sites for paint booths or chemical plants. *Fig.* 2 showcases the bio-filter for the air treatment process with the separation of Leachate.

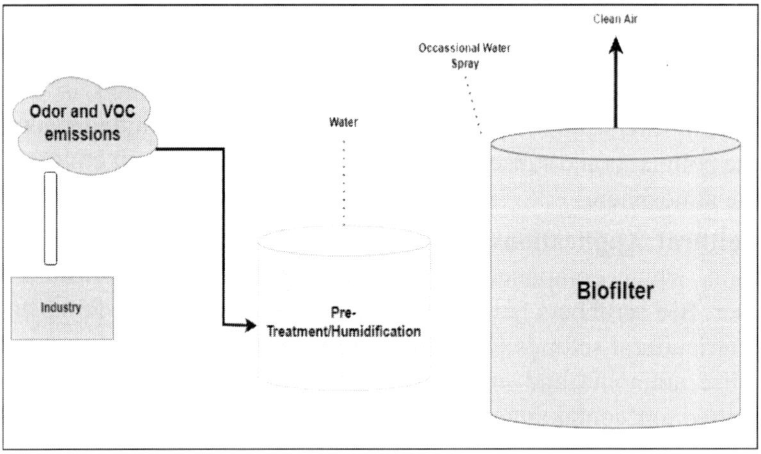

Fig. 2: Bio-Filter for Air Treatment

- **Bio-trickling Filters**: It is like bio-filters in that they are also biological filters based on trickling contaminated air through an impregnated medium by microorganisms. The trick is that the design is different: the moist medium of bio-trickling filters is continuously irrigated with a nutrient solution for maximal activation of microorganisms. This continuous supply of moisture enables the degradation ability of microorganisms to more pollutants, hence a wider working range. Bio-trickling filters find very wide usage for the treatment of hydrogen sulfide gas, sulfur dioxide, and ammonia gas. Fig. 3 depicts the bio-trickling filter used to purify the air.

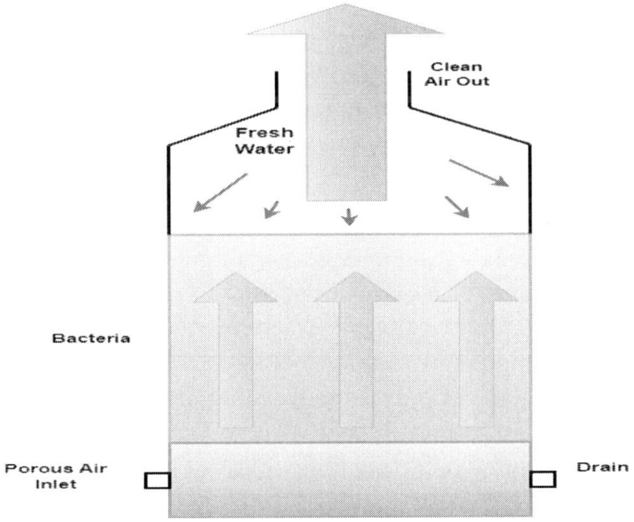

Fig. 3: Bio-trickling Filter

Bio-scrubbers: In bio-scrubbers, there is a liquid phase, usually water or nutrient solution, which absorbs and dissolves the incoming gases and pollutants in it. In this system, contaminated air passes through a liquid medium in which the microorganisms, such as bacteria or algae, degrade the pollutants (Arora *et al.*, 2023). Bio-scrubbers are especially effective for stripping hydrogen sulfide (H_2S), VOCs, and ammonia. Bio-scrubbers may be very versatile, in that the liquid medium could be formulated to address those particular pollutants of interest for the conditions of the application.

4.3. Mechanisms and Effectiveness of These Bio-systems in Treating Polluted Off-gas

Bio-filters, bio-trickling filters, and bio-scrubbers work on microbial processes for treating harmful off-gases. These noxious pollutants are converted into relatively non-nocuous products, resulting in less harmfulness of the air.

- **Bio-filters**: Bio-filters have the potential to break down organic pollutants such as VOCs by supplying air with microorganisms such as bacteria and fungi, where the aerobic degradation takes place because the microorganisms, including bacteria and fungi, break down the organic pollutants as a source of carbon and energy in the breakdown process into harmless products, including carbon dioxide and water, during the passing of contaminated air through the filter medium (Naha *et al.*, 2022). Factors of effectiveness for bio-filters depend upon factors such as pollutant types, temperature, humidity, and filter medium composition. Bio-filters are normally very effective for the removal of low to moderate concentration levels of VOCs and other organic compounds, removal efficiency often goes over 90%.

- **Bio-trickling Filters**: Bio-trickling filters improve the microbial degradation process by constantly providing moisture and nutrients to the microorganisms. This provides for the possibility of higher pollutant degradation rates and the capacity to treat higher concentrations of pollutants (Xie *et al.*, 2021). Continuous irrigation of the medium provides optimal conditions for the growth of microorganisms and the breakdown of pollutants. Bio-trickling filters have excellent removal efficiencies for sulfur compounds, such as hydrogen sulfide and sulfur dioxide, as well as ammonia. Generally, bio-trickling filters have higher removal efficiencies for gaseous pollutants as compared to bio-filters, especially in industries having relatively high pollutant concentrations.
- **Bio-scrubbers**: This technology requires using microbes to oxidise the soluble gases in liquid solutions to oxidize. This is among the efficient treatment methods, hence used in removing acidic or alkaline gases such as ammonia (NH_3), carbon dioxide (CO_2), and hydrogen sulfide (H_2S). The products of the contaminants in bio-scrubbers are absorbed into the solution. They get degraded by microorganisms. These systems are usually applied for high-volume air streams, especially in agricultural facilities or waste treatment plants (Gandu *et al.*, 2021). The removal efficiency of pollutants within the gas phase is very effective, especially in bio-scrubbers, and their removal efficiency can reach as high as 95%, depending on specific pollutants and system design.

5. Case Studies and Applications

Bioremediation has been introduced to show high potential in the stride toward discovering a sustainable approach to mitigate air pollution. This section presents real-life success stories illustrating the use of bioremediation technology in industries, along with the discovery and elaboration of the actual types of applications in which bioremediation was used to remove air pollution. Challenges, solutions, and benefits are discussed.

5.1. Successful Bioremediation Projects

Bioremediation has been successfully used in various industries. There are some examples of successful bioremediation projects. It has successfully been used to reduce the level of air pollution as well as to raise the level of air quality.

Case Study 1: Petrochemical Industry – Bio-filtration at a Refinery

Perhaps the most significant application of bioremediation at an industrial scale is at one of the biggest petrochemical refineries in the United States. This plant had huge emissions of VOCs as a result of its operation through chemicals such as benzene, toluene, and xylene. To abate this, the refinery set up a bio-filter system, which used compost as a growth medium for microorganisms (Jha *et al.*, 2021). With continuous nutrients and oxygen supply, the microorganisms in this bio-filter degraded the VOCs to a significant degree when the air passed through. It took them months of operating before shooting the removal efficiency up to 95%, thereby reducing the harmful emissions in the refinery become less. It made this project a sure success regarding the compliance by the refinery with the strict environment that goes along with improving the quality of the air.

Case Study 2: Municipal Waste Treatment - Bio-scrubbing for Landfill Gas

The main application of the bio-scrubbers involves the treatment of landfill gas, mostly made up of methane and H_2S, at the municipal waste facility in Europe. A nutrient solution that had directly targeted microorganisms to consume methane and sulfur compounds, for instance, was applied at the bio-scrubber system in this specific study (Tan *et al.*, 2023). The polluted air was bypassed through the scrubber, and with the microorganisms present, toxic gases were degraded in such a way that greatly lowered the odour and harmful emissions. It indeed recorded impressive results with more than 90% efficiency of removal of methane and almost total removal of sulfur, and, hence, the environmental footprint was considerably reduced. The local air quality improved, but this had the effect of reducing the emission of greenhouse gases into the atmosphere. This made the facility operate sustainably.

Case Study 3: Agricultural Industry - Ammonia Removal in Livestock Facilities

NH_3 emissions from animal waste might considerably contribute to the pollution of the air in agricultural areas, especially in livestock farming. A large U.S. poultry farm installed a bio-scrubber system to control this emission source. Air from the farm was mostly removed of ammonia by passing this air through a solution containing active microorganisms that absorbed and degraded the gas (Wang *et al.*, 2021). The removal efficiencies reached about 80% to significantly reduce the ammonia concentrations in the air. Besides addressing foul odour and air pollution, the bioremediation approach also enhanced the air quality in the operation for the farm workers and made the farm more sustainable.

5.2. Analysis of Specific Case Studies in Industrial Settings

An analysis of bioremediation applications in industrial settings is based upon detailed reviews of such problems and their feasibility as such systems.

a) **Oil Refining Industry – Bio-trickling Filter for Hydrogen Sulfide Removal**: This led to problems with H_2S emissions from crude oils via the main body oil refining in California. H_2S is bad for health, but when it forms by it translates to acid rain, which falls in acidic rainfall. To solve this, the refinery built a bio trickling filter system. A wet medium with sulfur-oxidising bacteria was packed into the filter and decomposed hydrogen sulfide into environmentally friendly sulfate compounds (Chan *et al.*, 2021). The bio-trickling filter was maintained through continuous observation to ascertain if the microbial populations were viable and active. The reduction level of H_2S emissions reached up to 95%, resulting in clean air in the locality of the refinery. The operation and maintenance costs of the system were less than those of chemical scrubbers, and the system also turned out to be very efficient while dealing with fluctuating levels of pollutants.

b) **Wastewater Treatment Plant – Bio-filters for VOC Removal**: One major metropolitan municipal water treatment facility in Asia experienced the release of VOCs from aeration tanks. The compost-based bio-filter supported diverse species of microbes capable of biodegradation VOCs. This resulted in the average removal efficiency for the VOCs being at about 90%, thus ensuring the plant met its standards of emissions. Besides that, the bio-filter system operates at a lower cost of operation than traditional carbon filtration.

c) **Chemical Manufacturing Plant - Bio-scrubbing for Ammonia and Nitrogen Oxides**: Regulatory problems of this company related to high ammonia and nitrogen oxide (NO_x) emission output occurred in its North American chemical manufacturing plant. However, to do that, we have installed a system to feed microorganisms that will break down these gases to remove ammonia and NO_x. An efficiency of up to 85% has been recorded in the bio-scrubber for the removal of NO_x and ammonia. They integrated the system with existing plant infrastructure, and since it was successful, the environmental impact and operation cost for the plant were reduced (Izah *et al.*, 2024). It is a case study on how effective the bio-scrubbing systems are when dealing with industrial emissions, so that the standards of the environment are met.

5.3. Benefits and Outcomes of Bioremediation in Air Pollution Mitigation

Discovering bioremediation as an abatement technology is an attractive option that provides many benefits to abate air pollution. These include:

- **Cost Savings**: The best advantage of the bioremediation is to save the costs (Kuppan *et al.*, 2024). Bioremediation systems have usually lower capital and operating costs than chemical scrubbers or activated carbon filters.
- **Environmental Sustainability**: Bioremediation is not environmentally friendly. Unlike chemical treatment systems, bioremediation does not produce secondary pollutants (Bala *et al.*, 2024). It applies natural processes that do not cause harmful by-products. Using microorganisms to degrade pollutants, these systems contribute to a cleaner environment with no additional chemical contaminants.
- **Improved Air Quality**: Bioremediation methods, including bio-scrubbers, bio-trickling filters, and bio-filters, have shown incredible efficiency in decreasing aerial pollutants that include VOCs, ammonia, sulfur compounds, and nitrogen oxides (James *et al.*, 2024). It helps to remove the gaseous pollutants from the atmosphere, hence improving the quality of air, which will help mitigate the risk of toxicity when exposed to the pollutants. This is crucial in urban and industrial centres where the levels of pollution always surpass acceptable limits.
- **Health Benefits**: Bioremediation that decreases harmful pollutants improves not only the environmental quality but also directly improves human health. It decreases hazardous gases such as ammonia, sulfur compounds, and VOCs, which helps to minimise respiratory problems, cardiovascular diseases, and other pollution-related health issues (Kumar *et al.*, 2021). Bioremediation systems help decrease respiratory, cardiovascular, and other problems resulting from pollution in industrial and agricultural fields. Workers are mostly exposed to dangerous levels of these pollutants while working in industries and farms, and bioremediation helps to reduce those levels.
- **Regulatory Compliance**: The majority of industries come under strict environmental regulations concerning control of air pollution. Using case analysis, it appears that bioremediation can offer a way to attain

compliance without a large disruption in operations. In this chapter, it is considered that different operating companies logically applied the bioremediation systems into their practices, to meet the standards on emission levels and avoid penalties for negligence, in building up their image as environmental corporate citizens.

6. Challenges and Future Prospects

Although bioremediation holds many promises towards air pollution mitigation, it's not a problem-free field of endeavour. Technical, operational, economic, and regulatory hurdles characterise the challenges facing such issues. However, with increasing technological advances in bioremediation, it is possible to overcome all the obstacles and improve the applicability of bioremediation for the control of air pollution. The subsequent part will explain several technical, economic, and regulatory challenges in bioremediation and sketch several trends that might be incorporated to determine the future advancement and promotion of the process.

6.1. Technical and Operational Challenges in Bioremediation

Despite the great expectations in bioremediation, several technical and operational challenges lie ahead that affect the applicability of the technique in the particular context.

- **Efficiency Concerns**: The most critical problems with bioremediation are the removal efficiencies, which can be quite high if the pollutant is at high concentrations or chemically resistant. In a system such as a bio-filter or a bio-scrubber, most of the pollutants are removed (Bala *et al.*, 2022). However, very high concentrations are beyond the efficiency of bioremediation systems or even for those recalcitrant compounds where, for the optimisation and the microbial process for the degradation of the pollutants, there lies a major challenge.
- **Environmental Conditions**: Microbes tend to increase under some favourable conditions, including temperature, humidity, pH and oxygen. For these reasons, bioremediation systems should provide optimized conditions for the growth of microorganisms as well as the degradation of pollutants. Such demands in turn raise the design and operating complexity of the systems in question to even greater levels.
- **Scaling Up**: Where new structures are developed on a large scale, this would require the scale-up of those systems, especially where the site is an industrial complex or major city. This is because the bioremediation processes need to be just as efficient at very large scales. More extensive

monitoring and optimisation of system parameters such as airflow, microbial density, and pollutant concentrations have to be achieved. A larger volume of bio-reactor materials (e.g., compost, activated carbon) is needed for large-scale systems, thereby complicating operations and increasing costs.

- **Microbial Adaptation and Stability**: It is another challenge because microbes have to adapt to various pollutants (Wani *et al.*, 2022). The microbial populations take time to acclimate to new pollutants. In some cases, specific microorganisms might have to be added to enhance the biodegradation of pollutants, and maintaining the stability of populations over time can be hard, especially for systems whose pollutant concentrations or environment are fluctuating.

6.2. Economic and Regulatory Considerations

Economic viability and regulatory considerations are two major factors affecting the wide-scale application of bioremediation. While it offers potential cost advantages, several factors must be considered before its implementation.

- **Cost of Implementation**: It is very expensive to install a bioremediation system. Bio-filters, bio-trickling filters, and bio-scrubbers are costly and consume much material, structure, and microbial inoculants. Maintenance costs will involve nutrient addition, monitoring the microbes, and general operation management. However, in comparison to traditional chemical and physical pollution control methods, bioremediation systems generally have lower long-term operating costs related to energy consumption and chemical usage.
- **Return on Investment (ROI)**: Economic benefits are derived in bioremediation from cost-effectiveness due to reduced operational expenses, efficient use of energy, and possibly conserving penalty environmental or regulatory fines (Baskaran & Byun, 2024). Again, return on investment would largely depend on the scope of application, character of the pollutants for remediation, and geospatial location. There will be no interest in investments by the industries if there is no real profit motive towards the end. Some governments can provide incentives through subsidies or tax credits for environmental technologies.
- **Regulatory Landscape**: Bioremediation remains an emerging regulatory landscape, which may influence its adoption. There are a few issues of regulatory hurdles for adoption, such as standardised procedures for monitoring and validation of these systems. Another challenge is to show that bioremediation technologies meet the required

emission standards to achieve the ambient air quality requirements, which, of course, vary between countries.

- **Incentives for Adoption**: The most significant help that governments can make to speed up bioremediation adoption comes through financial incentives, funding for R and D, and clear regulation leading to sustainable technologies (Bala *et al.*, 2022). Industries might then adopt bioremediation for environmentally set targets and reduce their dependence on chemical treatments. Policies in this regard may also increase demand for bioremediation systems.

6.3. Future Trends and Potential Advancements in Bioremediation Techniques

Several promising developments and trends might help conquer the current challenges and advance the cause of bioremediation in air pollution mitigation, looking ahead.

- **Genetic Engineering of Microorganisms**: The most recent developments in genetic engineering and synthetic biology opened new horizons for improving the efficiency of bioremediation. Scientists can more and more modify microorganisms to increase their degradation capacity for specific pollutants (Maqsood *et al.*, 2024). These biologic systems would be integrated into a bio-scrubber or a bio-trickling filter to enhance the rate of pollutant removal, hence also opening a larger spectrum of possible treatment options for these pollutants.
- **Novel Filtration Technologies**: New filtration technologies can contribute to enhancing the performance of the bioremediation systems. Investigation into biofilm reactors, electro-kinetic systems, or hybrid systems integrating both bioremediation with physical and/or chemical processes could be done to raise removal efficiency in bioreactors (Naha *et al.*, 2022). In addition, smart sensors and control systems are integrated into bioremediation setups to enhance operational conditions and monitor performance in real-time.
- **Increased Adoption of Integrated Bioremediation Systems**: In the coming decades, integrated systems that combine bioremediation with other pollution control technologies might become more commonly adopted (Kumar *et al.*, 2025). These would allow a combination of bioremediation's merits with conventional methods to obtain an optimal removal of all kinds of pollutants at low operation cost and reduced negative impacts on the environment.

- **Wider Use of Bioremediation in Urban Air Quality Management**: Bioremediation might turn out to be the only viable technology to assume an overall dominating position within the urban space in this area of concern, specifically through bio-filters and bio-scrubbers installed into ventilation systems, public transportation hubs, and along highways as more people move to the city (Isha *et al.*, 2024). These would target common pollutants in urban air, such as nitrogen oxides (NO_x), VOCs, and particulate matter. Integration of bioremediation into smart city initiatives can be a very innovative approach to the management of urban air quality and the reduction of health impacts from air pollution.

7. Conclusion

One of the most promising and sustainable means of abating air pollution has become bioremediation. This is a low-cost, natural alternative to conventional approaches to pollution control. This process works on the microorganisms' abilities to degrade such harmful air pollutants as VOCs, SO_2, NO_x, and NH_3 into harmless by-products such as carbon dioxide and water. This would reduce environmental pollution and, at the same time, contribute to progress in public health through reduced degradation from air pollution.

Bioremediation can be proven to be an effective application in the specific area of air pollution among the sectors of industry, agriculture, and municipal waste. Real case studies of employing bio-filters in petrochemical refineries, bio-scrubbers in agricultural facilities, and bio-trickling filters in chemical plants are significant examples that have proven the practicality of bioremediation systems to considerably decrease emissions and, at the same time, provide a sustainable and cost-effective option.

Like all new technology, however, bioremediation is not problem-free. Its main problems are the removal efficiency of pollutants, the scalability of systems, and environmental conditions affecting microbial activity. In addition, research has to be carried out to improve microbial processes. Besides, economic concerns and regulatory issues have to be overcome so that bioremediation can be feasible at larger scales of application.

The future of bioremediation will only be realised when the currently faced challenges are overcome by relentless innovation and research. Other breakthroughs in genetic engineering, novel filtration technologies, or integration of bioremediation with other methods of pollution will make these systems even more practical and efficient. Increased desire for sustainable environmental management, supplemented by government policies and incentives that encourage the adoption of the technology, will be key in ensuring the broader application of bioremediation in both industrial and urban areas.

References

Ahmed, S. F., Mofijur, M., Rafa, N., Chowdhury, A. T., Chowdhury, S., Nahrin, M., ... & Ong, H. C. (2022). Green approaches in synthesising nanomaterials for environmental nanobioremediation: Technological advancements, applications, benefits and challenges. Environmental Research, 204, 111967.

Arora, K., Kumar, V., Nayak, S. J., Surya, A., & Sugumar, S. (2023). Recent Advances in Biological Remediation of Volatile Organic Compounds (VOCs) and Heavy Metals. Bio-Nano Filtration in Industrial Effluent Treatment, 49-68.

Bălă, G. P., Râjnoveanu, R. M., Tudorache, E., Motișan, R., & Oancea, C. (2021). Air pollution exposure—the (in) visible risk factor for respiratory diseases. Environmental Science and Pollution Research, 28(16), 19615-19628.

Bala, S., Garg, D., Thirumalesh, B. V., Sharma, M., Sridhar, K., Inbaraj, B. S., & Tripathi, M. (2022). Recent strategies for bioremediation of emerging pollutants: a review for a green and sustainable environment. Toxics, 10(8), 484.

Baskaran, D., & Byun, H. S. (2024). Current trend of polycyclic aromatic hydrocarbon bioremediation: Mechanism, artificial mixed microbial strategy, machine learning, ground application, cost and policy implications. Chemical Engineering Journal, 155334.

Bhatt, P., Gangola, S., Bhandari, G., Zhang, W., Maithani, D., Mishra, S., & Chen, S. (2021). New insights into the degradation of synthetic pollutants in contaminated environments. Chemosphere, 268, 128827.

Chan, Y. H., Loy, A. C. M., Cheah, K. W., Chai, S. Y. W., Ngu, L. H., How, B. S., ... & Lam, S. S. (2023). Hydrogen sulfide (H2S) conversion to hydrogen (H2) and value-added chemicals: Progress, challenges and outlook. Chemical Engineering Journal, 458, 141398.

Elehinafe, F. B., Aondoakaa, E. A., Akinyemi, A. F., Agboola, O., &Okedere, O. B. (2024). Separation processes for the treatment of industrial flue gases–Effective methods for global industrial air pollution control. Heliyon.

Farid, M., Zubair, M., Numra, Shakoor, M. B., Fatima, M., Farid, S., ... & Liaqat, N. (2023). Bioremediation of Sites Contaminated with Heavy Metals, Techniques, and Their Application. In Climate-Resilient Agriculture, Vol 2: Agro-Biotechnological Advancement for Crop Production (pp. 733-750). Cham: Springer International Publishing.

Fouad, F. A., Youssef, D. G., Shahat, F. M., & Abd El-Ghany, M. N. (2022). Role of microorganisms in the biodegradation of pollutants. In Handbook of biodegradable materials (pp. 1-40). Cham: Springer International Publishing.

Gandu, B., Palanivel, S., Juntupally, S., Arelli, V., Begum, S., &Anupoju, G. R. (2021). Removal of NH3 and H2S from odor causing tannery emissions using biological filters: Impact of operational strategy on the performance of a pilot-scale bio-filter. Journal of Environmental Science and Health, Part A, 56(6), 625-634.

Gurav, M. T., Shinde, S. P., Nimbalkar, N. J., Jadhav, P. S., Zambare, D. N., &Wategaonkar, S. B. (2024). Air Pollutants from Sugar Industry Chimneys: Evaluating Environmental Hazards and Global Mitigation Strategies.

Hussain, A., Rehman, F., Rafeeq, H., Waqas, M., Asghar, A., Afsheen, N., ... & Iqbal, H. M. (2022). In-situ, Ex-situ, and nano-remediation strategies to treat polluted soil, water, and air–A review. Chemosphere, 289, 133252.

Isha, Ali, S., Khalid, A., Naseer, I. A., Raza, H., & Chang, Y. C. (2024). Bioremediation of Smog: Current Trends and Future Perspectives. Processes, 12(10), 2266.

Izah, S. C., Ogwu, M. C., Etim, N. G., Shahsavani, A., & Namvar, Z. (2024). Short-term health effects of air pollution.

James, A., Rene, E. R., Bilyaminu, A. M., & Chellam, P. V. (2024). Advances in amelioration of air pollution using plants and associated microbes: An outlook on phytoremediation and other plant-based technologies. Chemosphere, 142182.

Jha, A., Narasimhaiah, N. K., Sreekumar, N., Babu, P. A. M., Umashankar, P., Mohan, S., ... & Bharti, R. K. (2022). Biofiltration techniques in the remediation of hazardous inorganic and organic contaminants. In An Innovative Role of Biofiltration in Wastewater Treatment Plants (WWTPs) (pp. 137-154). Elsevier.

Kour, D., Kaur, T., Devi, R., Yadav, A., Singh, M., Joshi, D., ... & Saxena, A. K. (2021). Beneficial microbiomes for bioremediation of diverse contaminated environments for environmental sustainability: present status and future challenges. Environmental Science and Pollution Research, 28, 24917-24939.

Kumar, A., Kumar, R., Parnian, A., Mahbod, M., & AbdelRahman, M. A. (2025). Bioremediation: a better technique for wastewater treatment and resource recovery. In Biotechnologies for Wastewater Treatment and Resource Recovery (pp. 17-32). Elsevier.

Kumar, P., Singh, A. B., Arora, T., Singh, S., & Singh, R. (2023). Critical review on emerging health effects associated with indoor air quality and its sustainable management. Science of The Total Environment, 872, 162163.

Kumar, V., Roy, S., Behera, B. K., Swain, H. S., & Das, B. K. (2021). Biofloc microbiome with bioremediation and health benefits. Frontiers in Microbiology, 12, 741164.

Kumari, S., & Das, S. (2023). Bacterial enzymatic degradation of recalcitrant organic pollutants: catabolic pathways and genetic regulations. Environmental Science and Pollution Research, 30(33), 79676-79705.

Kuppan, N., Padman, M., Mahadeva, M., Srinivasan, S., & Devarajan, R. (2024). A comprehensive review of sustainable bioremediation techniques: Eco friendly solutions for waste and pollution management. Waste Management Bulletin.

LaCount, M. D., Haeuber, R. A., Macy, T. R., & Murray, B. A. (2021). Reducing power sector emissions under the 1990 Clean Air Act Amendments: A retrospective on 30 years of program development and implementation. Atmospheric Environment, 245, 118012.

Maqsood, Q., Waseem, R., Sumrin, A., Wajid, A., Tariq, M. R., Ali, S. W., & Mahnoor, M. (2024). Recent trends in bioremediation and bioaugmentation strategies for mitigation of marine based pollutants: current perspectives and future outlook. Discover Sustainability, 5(1), 524.

Naha, A., Saha, S., Singh, H. R., Shukla, S. K., Tripathi, V. K., & Jha, S. K. (2022). Recent trends and future perspectives in applications of biofiltration. In An Innovative Role of Biofiltration in Wastewater Treatment Plants (WWTPs) (pp. 113-136). Elsevier.

Nwankwegu, A. S., Zhang, L., Xie, D., Onwosi, C. O., Muhammad, W. I., Odoh, C. K., ... &Idenyi, J. N. (2022). Bioaugmentation as a green technology for hydrocarbon pollution remediation. Problems and prospects. Journal of Environmental Management, 304, 114313.

Ofremu, G. O., Raimi, B. Y., Yusuf, S. O., Dziwornu, B. A., Nnabuife, S. G., Eze, A. M., &Nnajiofor, C. A. (2024). Exploring the Relationship between Climate Change, Air Pollutants and Human Health: Impacts, Adaptation, and Mitigation Strategies. Green Energy and Resources, 100074.

Omenna, E. C., Omage, K., Ezaka, E., &Azeke, M. A. (2024). Bio-augmentation and bio-stimulation with kenaf core enhanced bacterial enzyme activities during bio-degradation of petroleum hydrocarbon in polluted soil. Scientific Reports, 14(1), 8.

Pachouri, V., Chandramauli, A., Singh, R., Gehlot, A., Priyadarshi, N., & Twala, B. (2024). Removal of contaminants by chlorella species: an effort towards sustainable remediation. Discover Sustainability, 5(1), 19.

Ramirez-Corredores, M. M., Goldwasser, M. R., & Falabella de Sousa Aguiar, E. (2023). Carbon dioxide and climate change. In Decarbonization as a Route Towards Sustainable Circularity (pp. 1-14). Cham: Springer International Publishing.

Shareefdeen, Z. (2020). High-performance biofilters for air treatment applications. In FromBiofiltration to Promising Options in Gaseous Fluxes Biotreatment (pp. 111-127). Elsevier.

Sharma, J., Goutam, J., Dhuriya, Y. K., & Sharma, D. (2021). Bioremediation of Industrial pollutants. Microbial Rejuvenation of Polluted Environment: Volume 2, 1-31.

Tan, V. W. G., Chan, Y. J., Arumugasamy, S. K., & Lim, J. W. (2023). Optimizing biogas production from palm oil mill effluent utilizing integrated machine learning and response surface methodology framework. Journal of Cleaner Production, 414, 137575.

Thanvisitthpon, N., Kallawicha, K., & Chao, H. J. (2024). Effects of urbanization and industrialization on air quality. In Health and Environmental Effects of Ambient Air Pollution (pp. 231-255). Academic Press.

Wang, H., Zhao, Z., Winiwarter, W., Bai, Z., Wang, X., Fan, X., ... & Ma, L. (2021). Strategies to reduce ammonia emissions from livestock and their cost-benefit analysis: A case study of Sheyang county. Environmental Pollution, 290, 118045.

Wani, A. K., Akhtar, N., Sher, F., Navarrete, A. A., & Américo-Pinheiro, J. H. P. (2022). Microbial adaptation to different environmental conditions: molecular perspective of evolved genetic and cellular systems. Archives of Microbiology, 204(2), 144.

Wyer, K. E., Kelleghan, D. B., Blanes-Vidal, V., Schauberger, G., & Curran, T. P. (2022). Ammonia emissions from agriculture and their contribution to fine particulate matter: A review of implications for human health. Journal of Environmental Management, 323, 116285.

Xie, D., Gao, M., Yang, M., Wu, C., Meng, J., Xu, M., ... & Sun, X. (2021). Nitrate-rich wastewater discharged from a bio-trickling filter can be reused as a moisture conditioning agent for organic waste composting. Environmental Technology & Innovation, 24, 101932.

4

Bioremediation of Agroindustrial Waste: An Ecofriendly Synergistic Approach

Manishita Das Mukherji[1*], Rana Mukherji[2], Mridul Jain[1] and Ravneet Chug[1*]

[1] *Amity Institute of Biotechnology, Amity University Rajasthan, Jaipur, India*

[2] *ICFAI Tech, The ICFAI University, Jaipur*

**Corresponding Authors Email: mdmukherji@jpr.amity.edu; rchug@jpr.amity.edu*

Abstract

Over the past decade, recycled agro-industrial waste has gathered significant attention among researchers and innovators as a cost-effective and sustainable substrate for fermentation processes and obtaining useful products. The types of products, yields, and potential applications largely depend on the composition of the waste, which varies in terms of composition of biomolecules like proteins, carbohydrates, lipids, etc. Commonly reported microbial products include enzymes, biosurfactants, antibiotics, and each serving various industrial purposes. Advancements in research focuses on exploring waste sources as potential nutrient reservoirs and then the optimization of scaling techniques, in transitioning this agro-industrial waste from laboratory settings to industrial scales. This review provides insights into the potential of agro-industrial waste as possible substrates and the high-value products derived from microbial conversion, aiming to guide future research efforts toward commercializing sustainable microbial products effectively.

***Keywords**: Agro-industrial waste; Bioremediation, Microorganisms, Microbial Products, Sustainability*

1. Introduction

Agro-industrial waste plays a crucial role in bioremediation, the process of using biological organisms to remove or neutralize pollutants from contaminated environments (Garg, 2020). With the rise of industrialization and agricultural activities, there's a corresponding increase in waste production. However, these waste materials, often rich in organic compounds, can be repurposed as valuable resources as mentioned in scientific studies (Sahota and Sharma, 2020; Yadav *et al.*, 2022).

One key aspect is their ability to serve as substrates for microbial communities. Many agro-industrial wastes, such as crop residues, animal manure, and food processing byproducts, are rich in nutrients that support the growth of diverse microbial populations. These microbes can metabolize a wide range of agricultural waste, waste coming from industries, pollutants including heavy metals, pesticides, and hydrocarbons, through various enzymatic pathways, ultimately transforming them into less harmful substances (Bala *et al.*, 2022)

Moreover, the use of agro-industrial waste in bioremediation is cost-effective and environmentally sustainable. By utilizing these waste materials, we not only reduce the burden on landfills but also mitigate pollution in contaminated sites, contributing to the restoration and preservation of ecosystems (Sharma *et al.*, 2022). The extensive exploitation of non-renewable energy resources has sparked significant environmental, social, and economic apprehensions. Consequently, policymakers and research institutions are increasingly focusing on the sustainable utilization of agro-industrial food and crop wastes as an alternative. Agro-industrial waste serves a pivotal role in the realm of bioremediation, offering sustainable solutions to environmental contamination issues.

Agroindustrial waste, if not managed properly, poses significant environmental concerns. The large-scale generation of organic residues such as fruit peels, bagasse, and animal by-products can lead to the emission of foul odors and attract pests, creating unhygienic conditions that affect both human health and the surrounding ecosystem. When these wastes are dumped in open areas or near water bodies, they often leach harmful substances into the soil and groundwater, leading to contamination and loss of soil fertility. Moreover, the uncontrolled discharge of liquid waste from industries like dairies and slaughterhouses can severely degrade water quality, harming aquatic life. Burning agro-industrial waste, which is still a common practice in some areas, releases harmful gases and particulate matter into the atmosphere, contributing to air pollution and climate change. In the absence of sustainable disposal and recycling methods, agro-industrial waste becomes a major environmental burden rather than a resource with potential for reuse in energy production, composting, or as animal feed.

Another major concern is the contribution of agro industrial waste to greenhouse gas emissions. When organic waste decomposes anaerobically in landfills or open dumps, it produces methane a potent greenhouse gas that significantly contributes to global warming. Additionally, the leachate from improperly stored waste can seep into nearby water sources, endangering both human and animal populations who rely on these waters for drinking and irrigation. The accumulation of solid waste also reduces land availability for other productive uses and increases the risk of fires, especially in dry seasons. Without efficient waste management strategies like segregation at source, recycling, and conversion into value-added products, the environmental footprint of agro-industrial activities will continue to expand, threatening ecological balance and sustainable development goals.

2. Agro-Industrial Waste: Classification and Composition

Agro-industrial waste is composed of two parts, i.e., in-farm waste (agricultural waste) and out-farm waste (industrial waste) (Yaashikaa *et al.*, 2022). Stalks, leaves, husks, roots and pods encompasses in-farm residues, while peels, pomace and molasses, which comes from food processing, is included under out-farm waste (Singh *et al.*, 2021). These residues generally contain compounds such as cellulose, hemicellulose, and lignin, which require pretreatment methods to breakdown its complex structure for further processing. Methods including acid/alkali hydrolysis, autohydrolysis, enzymatic hydrolysis, microwaving, and ultrasound are some of the widely used methods for pretreatment (Prado-Acebo *et al.*, 2024). Agricultural waste of some crops such as corn, wheat, and sugarcane provides a rich carbon content which can be used to produce succinic acid, PHB, ethanol, and furfural, while lignin can be used as an energy source (Singh *et al.*, 2021). On the other hand, industrial wastes from grape, potato, barley, which are available in smaller quantities, can be processed to produce xylitol, lactic acid, PHAs, and biogas (Prado-Acebo *et al.*, 2024).

India produces a large number of agro resources which includes, 296 MT fruits and vegetables, 187.7 MT milk, 532 MT poultry (Yaashikaa *et al.*, 2022). Agro-industrial sectors including, plant-based (grain mills, sugar units), animal-based (meat, poultry), and non-food (textile, leather), generates a substantial amount of waste in the form of husks, shells, rinds, wastewater, and slaughterhouse waste. While crop residues are generally reused as animal feed or mulch, animal and fish processing byproducts are highly persistent and can cause environmental pollution. Improper disposal (e.g., open dumping or burning) leads to emission of green house gases (CH_4, N_2O), toxic chemicals (SO_2, NOx), and carcinogens (dioxins, furans), polluting air, water, and soil quality. Waste from textile and pulp industries can adversely affect the aquatic ecosystems as it contains mutagens and endocrine disruptors (Singh *et al.*, 2021). Unreasonable usage of untreated agro-industrial waste can lead to pesticide/herbicide buildup, soil degradation, and eutrophication, affecting biodiversity and human health. However, residues like sugarcane bagasse, fruit peels, animal slurry, and poultry litter can be used as a substrate for the production of bioenergy, compost, and biocontrol agents, decreasing the dependency on synthetic chemicals (Yaashikaa *et al.*, 2022). Fig.ure 1 describes classification of agro-industrial waste, its sources and types.

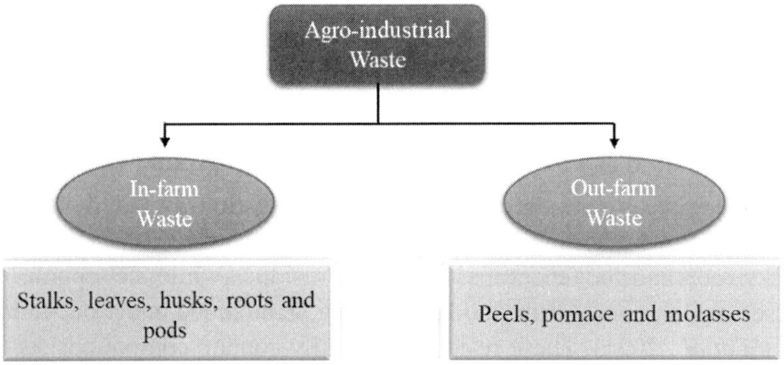

Fig. 1: Classification of Agro-Industrial Waste

3. Principles of Bioremediation

Bioremediation refers to the controlled degradation of organic wastes through biological processes. By providing essential nutrients and chemicals, harmful substances can be broken down or detoxified by organisms. These organisms utilize pollutants as energy source and nutrients, degrading them into harmless and environmentally friendly end products such as carbon dioxide (CO_2), water (H_2O), microbial biomass, and different metabolites (Najam *et al.*, 2021). Enzymes, including oxidoreductases, lyases, transferases, and hydrolases, are pivotal in this metabolic process, enabling the degradation of a wide range of substrates. Successful bioremediation relies on enzymatic action to transform pollutants effectively (Chen *et al.*, 2016; Malik *et al.*, 2022).

Microorganisms responsible for bioremediation can either be indigenous to the contaminated site or introduced from external sources. When external microbes are employed, the process is referred to as bioaugmentation. In this method, specific microbial strains which are often genetically modified or naturally capable of superior pollutant degradation, are introduced to the polluted environment to enhance the detoxification process. Bioaugmentation is particularly useful when the native microbial population lacks the capability to efficiently degrade complex contaminants such as petroleum hydrocarbons. However, the effectiveness of this strategy depends heavily on the ability of the introduced microbes to survive, adapt to the local environmental conditions, and compete with native microorganisms (Najam *et al.*, 2021). Additionally, biostimulation is another vital strategy in bioremediation. It focuses on enhancing the activity of the native microbial community already present at the contaminated site. This is achieved by modifying environmental conditions and adding essential nutrients such as nitrogen, phosphorus, oxygen, and carbon, which are crucial for microbial growth and metabolism. By improving these factors, the native microbes are stimulated to break down pollutants more efficiently. Parameters like temperature, pH, and oxygen levels must be carefully controlled as they significantly influence the microbial degradation rate and overall bioremediation success (Sanjana *et al.*, 2024).

Bioremediation works best when environmental parameters such as temperature, pH, oxygen levels, and nutrient availability are adjusted to support optimal microbial activity. This is because microbes, being the main agents of biodegradation, need specific conditions to thrive and efficiently break down pollutants. When these conditions are met, the microorganisms can convert harmful organic pollutants into non-toxic and environmentally safe substances like carbon dioxide and water, which do not pose risks to humans or ecosystems (Ren *et al.*, 2018; Malik *et al.*, 2022).

Bioremediation is valuable particularly due to is its eco-friendly and cost-effective nature, along with its high level of public acceptance. Since it relies on natural processes, it minimizes environmental disruption while promoting sustainability. For more challenging pollutants such as azo dyes or petroleum-based compounds, scientists often use specially isolated or genetically engineered microbial strains that have enhanced degradation abilities. These modified microbes play an essential role in industrial-scale applications, where pollutant loads are heavier and more complex (Najam *et al.*, 2021).

Ultimately, the core principle of bioremediation is to utilize the metabolic functions of microbes, under carefully controlled environmental conditions, to transform pollutants into harmless end-products. The technique combines biostimulation (adjusting environmental conditions to boost native microbial activity) and bioaugmentation (introducing powerful external microbes) to ensure efficient and sustainable remediation of polluted environments (Sanjana *et al.*, 2024).

4. Role of Bioremediation in Treatment of Agro-Industrial Waste

Bioremediation, the process of using biological organisms to mitigate pollutants in the environment, relies heavily on the utilization of various organic materials, including agro-industrial waste, due to their rich nutrient content and potential for microbial activity (Garg, 2020; Conde Molina *et al.*, 2023).

Bioremediation addresses both environmental concerns and resource management challenges. Agro-industrial waste, stemming from agricultural processes like cereal industries producing husk, food processing like coffee pulp from coffee industries, and related industries, presents significant environmental risks if not managed properly (Cusenza *et al.*, 2021; Yaashikaa *et al.*, 2022). However, through bioremediation, these waste materials can be effectively treated and converted into valuable products, contributing to environmental sustainability. Microorganisms including bacteria, fungi, and algae, can metabolize these organic compounds through enzymatic processes, leading to the degradation or conversion of pollutants into less harmful forms (Bala *et al.*, 2022)

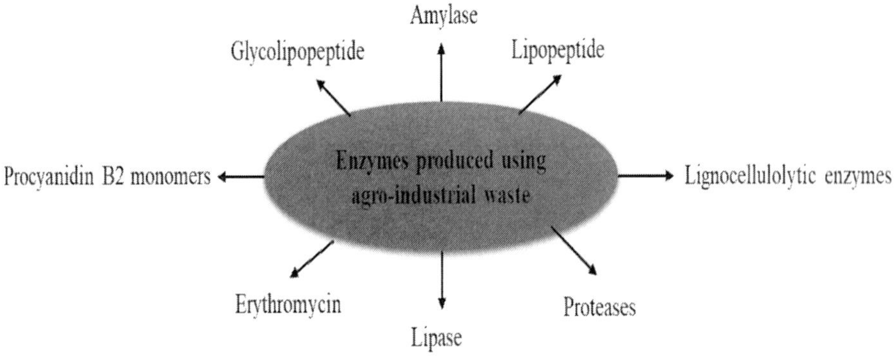

Fig. 2: Enzyme Production using Agro-industrial Waste

In the context of agro-industrial waste treatment, bioremediation can be implemented through various approaches, including composting, anaerobic digestion, and microbial fermentation (Sadh *et al.*, 2018; Balasubramani *et al.*, 2023). Composting, for instance, involves the controlled decomposition of organic materials by microbial communities under aerobic conditions. This process not only reduces the volume of waste but also produces compost, a nutrient-rich soil amendment beneficial for agriculture and landscaping purposes.

Anaerobic digestion is another bioremediation method suitable for treating agro-industrial waste, particularly organic residues rich in carbohydrates and fats. Microorganisms in anaerobic environments break down complex organic compounds into methane and carbon dioxide through a series of biochemical reactions, yielding biogas as a renewable energy source.

Microbial fermentation, on the other hand, utilizes specific microorganisms to metabolize organic substrates and produce value-added products such as biofuels, organic acids, enzymes, and biopolymers. Agro-industrial waste materials can serve as feedstocks for microbial fermentation processes, offering a sustainable alternative to conventional fossil-based fuels and chemical production methods.

In recent times, scientists and researchers have also been trying to focus on correlation between bioeconomy and bioremediation. Bioeconomy encompasses economic activities that utilize renewable biological resources to produce goods, energy, and services while minimizing environmental impact. Bioremediation, a key aspect of the bioeconomy, leverages biological processes to clean up polluted environments (D'Adamo *et al.*, 2022; Akinsemolu *et al.*, 2023). By employing microorganisms, plants, or enzymes, bioremediation can degrade, immobilize, or transform contaminants into less harmful substances, effectively restoring ecosystems and safeguarding human health. This approach offers a sustainable alternative to conventional remediation methods, such as chemical treatment or excavation, which may be costly and disruptive. Bioremediation not only addresses environmental challenges but also creates economic opportunities by repurposing waste materials and promoting circular economy principles. As the bioeconomy continues to expand, integrating bioremediation practices into various sectors can enhance resource efficiency, reduce pollution, and contribute to a more sustainable and resilient future.

The application of bioremediation for agro-industrial waste treatment not only mitigates environmental pollution but also offers economic and social benefits. By converting waste materials into valuable products, bioremediation creates opportunities for revenue generation, job creation, and local economic development, particularly in rural and agricultural communities. Moreover, bioremediation reduces the dependence on landfilling and incineration, minimizing greenhouse gas emissions and conserving natural resources (Onyeaka, and Ekwebelem, 2022; Thran, 2022; Akinsemolu *et al.*, 2023).

Growth of agro-industries has resulted in an increased wide range of complex and hazardous compounds to the environment. Excess growth of hazardous waste has led to reduced availability of clean water and disturbances of soil thus limiting crop production. Waste generated from above mentioned sources have different kinds of chemical compounds i.e. organic or inorganic. Traditional methods are not able to deal with some of these chemical compounds. Bioremediation process is a good option in such environmental problems as it provides a solution for cleaning up pollution by enhancing the natural biodegradation processes. It treats such waste with the help of microorganisms. Number of microbes including aerobes, anaerobic and fungi are involved in bioremediation process. Specific types of microbes are used to treat specific type of chemical contaminant.

5. Production of Bioactive Microbial Compounds/Products from Agro-Industrial Waste Using Bioremediation

Metabolic capabilities and remarkable adaptability of microorganisms ensure more cost-effective and eco-friendly approaches for producing valuable bioactive microbial compounds and accomplishing bioremediation goals. While it's possible to extract various compounds directly from agro-industrial waste, repurposing and transforming this nutrient-rich waste via fermentation processes offers a wide range of opportunities to generate valuable and innovative microbial products. Various useful products like enzymes, biosurfactants, phenolic compounds can be obtained from agro-industrial waste by exploiting potential of microorganisms which vary in terms of their activities, as well as the complexity. Few of these have been listed in Table 1.

Agro-industrial waste presents both a challenge and an opportunity in sustainable resource management. Microbial products derived from agro-industrial waste offer a promising avenue for addressing environmental concerns while providing valuable resources for various industries. The utilization of microbial products from agro-industrial waste not only adds value to otherwise underutilized resources but also contributes to waste reduction, environmental sustainability, and the circular economy.

Table 1. Various microorganisms utilizing agro-industrial waste for production of enzymes

Name of Microorganism	Enzyme/ Product/	Agro-industrial Waste used	Reference
Bacillus sp.	α-amylase	Potato peels, mango peels and lemon peels	(Saleh *et al.*, 2020)
Pseudomona-saaeruginosa	Lipase	Palm oil mill effluent	(Hermansyah *et al.*, 2018)
Aspergillus flavipes	Proteases	Wheat bran	(Zanutto-Elgui *et al.*, 2019)
Aspergillus niger	Lipase	Rice bran with Jatropha seed cake	(Putri *et al.*, 2020)
Bacillus subtilis	Lipopeptide	Molasses	(Rane *et al.*, 2017)
Lactococcus	Glycolipopeptide	Vinasse	(Vera, *et al.*, 2018)
Aspergillus niger	Procyanidin B_2 monomers	Hass avocado seeds	(Yepes-Betancur *et al.*, 2021)
Saccharopolyspora erythraea	Erythromycin	Sugarcane bagasse, beet sugar root and oatmeal	(Shata, 2018)
Aspergillus niger *Aspergillus ibericus* SSF	Ergosterol Lignocellulolytic enzymes	Olive mill waste with winery waste	(Filipe *et al.*, 2019)

6. Conclusion and Future Prospects

In conclusion, agro-industrial waste represents a valuable resource in bioremediation strategies, harnessing the power of nature to combat environmental pollution effectively. Embracing these practices can lead to cleaner environments and more sustainable agricultural and industrial practices.

Reducing environmental issues associated with agro-industrial waste can be achieved by bio converting this waste into bioactive compounds in an environmentally friendly manner, thus recycling nutrients with the help of microorganisms. Utilizing agro-industrial waste in the production of microbial products not only decreases production costs but also enhances the sustainability of industrial processes. Recent studies have concentrated on optimizing fermentation conditions to maximize bioactive compound production and suggesting novel types of agro-industrial waste as suitable substrates. However, scaling up fermentation processes presents significant challenges, prompting further research avenues in the valorisation of agro-industrial waste. Therefore, it's imperative to direct research efforts towards pretreatment alternatives and downstream processes for efficient product recovery, aiming to develop industrial-scale technologies for sustainable microbial product production that prioritize human and environmental safety.

References

Akinsemolu, A., Onyeaka, H., Fagunwa, O., & Adenuga, A. H. (2023). Toward a resilient future: the promise of microbial bioeconomy. Sustainability, 15(9), 7251.

Bala, S., Garg, D., Thirumalesh, B. V., Sharma, M., Sridhar, K., Inbaraj, B. S., & Tripathi, M. (2022). Recent strategies for bioremediation of emerging pollutants: a review for a green and sustainable environment. Toxics, 10(8), 484.

Balasubramani, R., Awasthi, M. K., Varjani, S., & Karmegam, N. (2023). Aerobic and anaerobic digestion of agro-industrial and livestock wastes: A green and sustainable way toward the future. Agronomy, 13(10), 2607.

Chen, B. Y., Ma, C. M., Han, K., Yueh, P. L., Qin, L. J., & Hsueh, C. C. (2016). Influence of textile dye and decolorized metabolites on microbial fuel cell-assisted bioremediation. Bioresource technology, 200, 1033-1038.

Conde Molina, D., Liporace, F., & Quevedo, C. V. (2023). Bioremediation of an industrial soil contaminated by hydrocarbons in microcosm system, involving bioprocesses utilizing co-products and agro-industrial wastes. World Journal of Microbiology and Biotechnology, 39(12), 323.

Cusenza, M. A., Longo, S., Cellura, M., Guarino, F., Messineo, A., Mistretta, M., & Volpe, M. (2021). Environmental assessment of a waste-to-energy practice: The pyrolysis of agro-industrial biomass residues. Sustainable Production and Consumption, 28, 866-876.

D'Adamo, I., Gastaldi, M., Morone, P., Rosa, P., Sassanelli, C., Settembre-Blundo, D., & Shen, Y. (2022). Bioeconomy of sustainability: Drivers, opportunities and policy implications. Sustainability, 14(1), 200.

Filipe, D., Fernandes, H., Castro, C., Peres, H., Oliva-Teles, A., Belo, I., & Salgado, J. M. (2020). Improved lignocellulolytic enzyme production and antioxidant extraction using solid-state fermentation of olive pomace mixed with winery waste. Biofuels, Bioproducts and Biorefining, 14(1), 78-91.

Garg, S. (2017). Bioremediation of agricultural, municipal, and industrial wastes. In Handbook of research on inventive bioremediation techniques (pp. 341-363). IGI Global.

Hermansyah, H., Maresya, A., Putri, D. N., Sahlan, M., & Meyer, M. (2018). Production of dry extract lipase from Pseudomonas aeruginosa by the submerged fermentation method in palm oil mill effluent. International Journal of Technology, 9(2), 325-334.

Malik, S., Dhasmana, A., Kishore, S., & Kumari, M. (2022). Microbes and microbial enzymes for degradation of pesticides. In Bioremediation and Phytoremediation Technologies in Sustainable Soil Management (pp. 95-127). Apple Academic Press.

Najam, T., Shah, S. S. A., & Rahman, M. M. (2021). Water-stable metal–organic framework for environmental remediation. In Bioremediation for environmental sustainability (pp. 585-621). Elsevier.

Onyeaka, H., & Ekwebelem, O. C. (2023). A review of recent advances in engineering bacteria for enhanced CO2 capture and utilization. International Journal of Environmental Science and Technology, 20(4), 4635-4648.

Prado-Acebo, I., Cubero-Cardoso, J., Lu-Chau, T. A., & Eibes, G. (2024). Integral multi-valorization of agro-industrial wastes: A review. Waste Management, 183, 42-52.

Putri, D. N., Khootama, A., Perdani, M. S., Utami, T. S., & Hermansyah, H. (2020). Optimization of Aspergillus niger lipase production by solid state fermentation of agro-industrial waste. Energy Reports, 6, 331-335..

Rane, A. N., Baikar, V. V., Ravi Kumar, V., & Deopurkar, R. L. (2017). Agro-industrial wastes for production of biosurfactant by Bacillus subtilis ANR 88 and its application in synthesis of silver and gold nanoparticles. Frontiers in Microbiology, 8, 492

Ren, X., Zeng, G., Tang, L., Wang, J., Wan, J., Wang, J., ... & Peng, B. (2018). The potential impact on the biodegradation of organic pollutants from composting technology for soil remediation. Waste management, 72, 138-149.

Sadh, P. K., Duhan, S., & Duhan, J. S. (2018). Agro-industrial wastes and their utilization using solid state fermentation: a review. Bioresources and bioprocessing, 5(1), 1-15.

Sahota, N. K., & Sharma, R. (2020). Bioremediation: harnessing natural forces for solid waste management. Handbook of Solid Waste Management: Sustainability through Circular Economy, 1-32.

Saleh, F., Hussain, A., Younis, T., Ali, S., Rashid, M., Ali, A., ... & Qamer, S. (2020). Comparative growth potential of thermophilic amylolytic Bacillus sp. on unconventional media food wastes and its industrial application. Saudi Journal of Biological Sciences, 27(12), 3499-3504.

Sanjana, M., Prajna, R., Urvi, S. K., & Kavitha, R. V. (2024). Bioremediation–the recent drift towards a sustainable environment. Environmental Science: Advances, 3(8), 1097-1110.

Sharma, V., Tsai, M. L., Nargotra, P., Chen, C. W., Kuo, C. H., Sun, P. P., & Dong, C. D. (2022). Agro-industrial food waste as a low-cost substrate for sustainable production of industrial enzymes: a critical review. Catalysts, 12(11), 1373.

Shata, H. M., El Deen, A. M. N., Abdelwahed, N. A., & Farid, M. A. (2018). Statistical optimization of erythromycin production by Saccharopolyspora erythraea under solid state fermentation of agro-industrial materials using response surface methodology. J. Microbiol. Biotechnol. Food Sci, 8, 692-697.

Singh, R., Das, R., Sangwan, S., Rohatgi, B., Khanam, R., Peera, S. P. G., ... & Misra, S. (2021). Utilisation of agro-industrial waste for sustainable green production: a review. Environmental Sustainability, 4(4), 619-636.

Thrän, D. (2022). Introduction to the Bioeconomy System. In The bioeconomy system (pp. 1-19). Berlin, Heidelberg: Springer Berlin Heidelberg.

Vera, E. C. S., de Azevedo, P. O. D. S., Domínguez, J. M., & de Souza Oliveira, R. P. (2018). Optimization of biosurfactant and bacteriocin-like inhibitory substance (BLIS) production by Lactococcus lactis CECT-4434 from agroindustrial waste. Biochemical Engineering Journal, 133, 168-178.

Yaashikaa, P. R., Kumar, P. S., & Varjani, S. (2022). Valorization of agro-industrial wastes for biorefinery process and circular bioeconomy: A critical review. Bioresource Technology, 343, 126126.

Yadav, A. N., Suyal, D. C., Kour, D., Rajput, V. D., Rastegari, A. A., & Singh, J. (2022). Bioremediation and waste management for environmental sustainability. Journal of Applied Biology and Biotechnology, 10(2), 1-5.

Yepes-Betancur, D. P., Márquez-Cardozo, C. J., Cadena-Chamorro, E. M., Martinez-Saldarriaga, J., Torres-León, C., Ascacio-Valdes, A., & Aguilar, C. N. (2021). Solid-state fermentation–assisted extraction of bioactive compounds from hass avocado seeds. Food and Bioproducts Processing, 126, 155-163..

Zanutto-Elgui, M. R., Vieira, J. C. S., do Prado, D. Z., Buzalaf, M. A. R., de Magalhães Padilha, P., de Oliveira, D. E., & Fleuri, L. F. (2019). Production of milk peptides with antimicrobial and antioxidant properties through fungal proteases. Food chemistry, 278, 823-831.

5

Cyanobacteria-Based Bioremediation Strategies for Pesticide Minimization: Mechanisms, Applications and Future Directions

Premlata Vikal, Vishal Sharma* and Manisha Nehra

Department of Botany, S.M.B. Govt. P.G. College, Nathdwara, Rajasthan, India

**Corresponding Author Email: vishalsharmabotany@gmail.com*

Abstract

Pesticides widely used in agriculture has resulted in environmental contamination and health risks for humans, including cancer, neurological disorders, endocrine disruption and ecosystem damage. They also harm ecosystems by eliminating beneficial insects, contaminating soil and water, and harming non-target species. Bioremediation is an affordable, eco-friendly, capable and efficient technology for pesticide detoxification. Bioremediation, which ideally aims to completely break down pesticides into harmless molecules, involves a diverse of techniques. These include bioabsorption, bioadsorption and biotransformation utilizing enzymes and nano enzymes, as well as comprehensive degradation facilitated by microorganisms like bacteria, fungi, macro and microalgae, or phytoremediation. Among these microorganisms, cyanobacteria have emerged as promising agents for the bioremediation of pesticide-contaminated environments. This chapter explores the potential of cyanobacteria in minimizing pesticide pollution through biodegradation and transformation processes. We provide an overview of the biological characteristics of cyanobacteria that make them suitable for pesticide degradation, including their metabolic pathways and tolerance mechanisms. Key enzymatic and genetic factors involved in the biodegradation of various pesticides are discussed, along with real-world applications of cyanobacteria-based bioremediation in agricultural and industrial settings. Despite their potential, the application of cyanobacteria in bioremediation faces challenges such as environmental constraints and economic viability. This chapter also addresses these challenges and presents future prospects, including advancements in genetic engineering and the integration of cyanobacteria with other bioremediation strategies. By examining both the present status and future directions of cyanobacteria-based bioremediation, this chapter highlights the significant role that these microorganisms can play in reducing pesticide pollution and promoting environmental sustainability.

Keywords: *Cyanobacteria, Bioremediation, Pesticide degradation, Environmental sustainability, Enzymatic factors, Ecosystem protection.*

1. Introduction

1.1. The Growing Problem of Pesticide Pollution

Pesticides are used in very great amounts in agricultural production for the protection of crops against pests, diseases and weeds. They are also very instrumental in boosting agricultural productivity by ensuring food security and preventing losses that may be caused by outbreaks of pests (Popp *et al.*, 2013). However, the extensive use of pesticides has generally resulted in huge environmental concerns. Improperly managed pesticide residues may remain in soil, water and even air for many years and create severe ecological and health hazards in the long run (Zhang *et al.*, 2024). This widespread existence of pesticides in the environment has raised a serious alarm concerning their adverse impacts on biodiversity, soil fertility and quality of water, further bioaccumulation in the food chain that affects wildlife and human populations (Aktar *et al.*, 2009).

Research evidence is increasingly showing that a large number of pesticides in common use, such as organochlorines, organophosphates and pyrethroids, may result in the contamination of ecosystems, reducing biodiversity and further polluting freshwater systems. For example, pesticide residues in water bodies can affect aquatic organisms such as fish, amphibians and aquatic plants, thus disturbing the natural ecological balance of such ecosystems (Stehle and Schulz, 2015). Moreover, pesticides may leach into the soil and affect soil microbial communities, essential nutrients and plant growth. These are but a few of the many concerns that have risen to date and called for sustainable solutions with regard to managing and reducing pesticide contamination, putting more emphasis on methods of removing, neutralizing, or even restoring ecological balance to pesticides (Zhang *et al.*, 2024).

Traditional methods to abate pesticide pollution involved either chemical treatment or physical processes like activated carbon filtration, adsorption and incineration. These have a short-term effect, yet they are usually quite expensive, energy-intensive and often lead to hazardous by-products (Geissen *et al.*, 2015). On the other hand, bioremediation, or the use of living organisms to detoxify or remove pollutants from the environment, is relatively more eco-friendly, cost-effective and sustainable. Recently, bioremediation has received considerable attention as one of the most promising ways of mitigating pesticide pollution, since it is capable of degrading, transforming or sequestering pesticide molecules in such a way that their impact on the environment is minimized (Jaiswal *et al.*, 2019).

One of the most promising bioremediation techniques employs the use of cyanobacteria, an extremely diverse group of photosynthetic microbes, generally known as blue-green algae. They have shown immense tolerance to a wide range of abiotic stresses and are capable of growing in different ecosystems: freshwater lakes and rivers, soil and even extreme environments (Singh *et al.*, 2016). These microorganisms have been attracting much attention for detoxifying a wide array of pollutants, including pesticides, through enzymatic degradation, absorption and microbial transformation. Being primary producers in aquatic ecosystems and contributors to biogeochemical cycles such as nitrogen fixation and carbon sequestration, cyanobacteria are well suited to play an important role in the cleaning up of the environment (Carrillo Campos *et al.*, 2023).

1.2. Cyanobacteria as Bioremediators

Cyanobacteria are believed to be among the most ancient forms of life known to have existed on the Earth; their evolutionary history goes back more than 3 billion years. These prokaryotic microorganisms carry out oxygenic photosynthesis, wherein sunlight energy is converted to chemical energy with the release of oxygen (Whitton and Potts, 2012). Cyanobacteria contain chlorophyll, as well as other pigments, which enable them to capture sunlight for metabolic processes. Originally mistaken for simple algae due to their relatively simple structure, they are quite dissimilar to eukaryotic algae. Colonial forms, filamentous colonies and biofilms are all built up by cyanobacteria, which has made possible survival in several different systems where they have adapted: aquatic and fresh water, soils, hot springs and arid desserts (Singh *et al.*, 2016).

An important aspect that makes cyanobacteria one of the best candidates for bioremediation involves being able to survive and grow under extreme conditions in which various microorganisms would not. They possess the ability for nitrogen fixation, thus facilitating its assimilation from atmospheric air and conversion into an active, accessible biologically functional form for all higher plants and organisms (Jaiswal *et al.*, 2019). The cyanobacteria can also live symbiotically with other microorganisms to enhance resilience and metabolic versatility (Carrillo Campos *et al.*, 2023). Apart from their ecological functions, cyanobacteria were found efficient in detoxification of various environmental pollutants. In literature, there are reports of degradation or transformation of toxicants like heavy metals, organic compounds and pesticides by cyanobacteria. The bioremediation modes mediated by cyanobacteria are quite versatile, comprising enzymatic degradation, absorption, bioaccumulation and microbial transformation. For instance, cyanobacteria have been noted to detoxify pesticides by actual enzymatic degradation; pesticide molecules are broken down by these microorganisms into usually less toxic, more biodegradable parts. Another thing is that they assimilate pesticide residues intracellularly, hence sequestrating them from biological availability to other organisms present in the environment (Whitton and Potts, 2012; Singh *et al.*, 2016).

Although several species of cyanobacteria synthesize toxic compounds, including microcystins, most of the species are not harmful to human health and hence, can safely be used for bioremediation. This is because of the great diversity in the species of cyanobacteria and their different environmental adaptability, rendering them versatile and highly efficient in the bioremediation of pesticide contamination within different ecosystems (Jaiswal *et al.*, 2019).

2. Cyanobacteria: An Overview

2.1. Introduction to Cyanobacteria

Cyanobacteria or blue-green algae represent a diverse group of photosynthetic prokaryotic microorganisms. For more than 3 billion years, cyanobacteria represent one of the oldest sets of organisms. Unlike most eukaryotic algae, cyanobacteria are single-celled organisms that can carry on oxygenic photosynthesis or the process in which carbon dioxide (CO_2) and water (H_2O) are converted, using sunlight, into a simple sugar called glucose ($C_6H_{12}O_6$) and oxygen (O_2). This ability to produce oxygen through photosynthesis has contributed a lot to the development of the Earth's atmosphere and was absolutely necessary for the evolution of aerobic forms of life (Whitton and Potts, 2012).

They are very versatile, from freshwater and marine ecosystems to terrestrial environments, extreme hot springs, polar regions and even desert soils. It has the ability to be flexible in environments with varied light intensities, temperature extremes and fluctuating nutrient levels. Cyanobacteria play many crucial ecological roles in numerous ecosystems, including their contribution to nutrient cycling, primary production and as a major part of food webs (Singh *et al.*, 2016).

2.2. Taxonomy of Cyanobacteria

Cyanobacteria are prokaryotic microorganisms of the kingdom of bacteria, while true algae are eukaryotic. Although lacking a membrane-bound nucleus and most other organelles apply to all prokaryotes, cyanobacteria never the less show considerable diversity in cellular morphology, pigmentation and ecological niches that they occupy. Up to this time, more than 1,500 species have been described, with many more yet to be discovered (Rastogi and Sinha, 2009).

Cyanobacteria are divided into several morphological types based on cell structure and cellular arrangement:

- **Unicellular cyanobacteria**: Single-celled organisms, which may form small colonies. Examples include *Synechococcus* and *Microcystis.*
- **Filamentous cyanobacteria**: Elongated chains or filaments of cells that can be unbranched or form more complex structures. Examples include *Anabaena*, *Nostoc* and *Oscillatoria.*
- **Colonial cyanobacteria**: Large multicellular colonies that are usually spherical or irregular. Examples: *Gloeocapsa.*
- **Heterocyst-forming cyanobacteria**: Certain filamentous cyanobacteria, such as *Anabaena* and *Nostoc*, have heterocysts, which are specialized cells where nitrogen fixation can occur. These heterocysts will convert the atmospheric nitrogen, (N_2) to ammonia (NH_3), to be used by them and also by plants in general, while this process plays an important role in nutrient cycling in ecosystems.

Such structural diversity underlines the ability of cyanobacteria to colonize a wide variety of ecosystems, being highly contributory to biogeochemical cycles, including carbon and nitrogen cycling (Whitton and Potts, 2012).

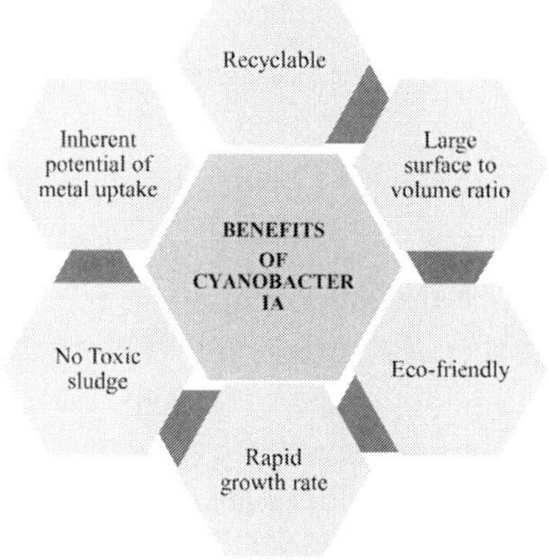

Fig. 1: Benefits of Cyanobacteria

2.3. Photosynthesis and Metabolism

Among all prokaryotes, only cyanobacteria possess the property of oxygen-producing photosynthesis process that ultimately fixes light energy into chemical energy. Variations in different pigments, such as chlorophyll a, phycocyanin, phycoerythrin and carotenoids, form the basis of photosynthetic activities in them. Of these, phycocyanin imparts rather blue pigmentation to Cyanobacteria and efficient absorption of light of red and orange parts of electromagnetic spectrum is another important traits of poor nutrient conditions (Rastogi and Sinha, 2009).

In the case of cyanobacteria, photosynthesis is the process by which carbon dioxide (CO_2) and water (H_2O) are converted into products, namely glucose ($C_6H_{12}O_6$) and oxygen (O_2), using light energy. These processes take place in specialized regions of the cytoplasm, thylakoid membranes, where the photosynthetic apparatus containing Photosystems I and II (PSI and PSII), is responsible for the light-dependent reactions. Products produced through this process can be stored either in the form of cyanophycin, which is a nitrogen-rich polymer, or glycogen itself, which can then be utilized metabolically (Whitton and Potts 2012).

While plants are able to accomplish the process of photosynthesis within their chloroplasts, cyanobacteria do not have any membrane-bound organelles, relying instead on their thylakoid membranes to carry out photosynthesis. Besides, cyanobacteria exhibit nitrogen fixation as part of their metabolic activities. They can convert atmospheric nitrogen (N_2) to ammonia (NH_3) in specialized cells called heterocysts. Such a capacity to execute nitrogen fixation provides these organisms with an ecological advantage for survival under nitrogen-limiting conditions. In this sense, it supports metabolic versatility for cyanobacteria by way of enabling them to synthesize various key biomolecules such as amino acids, proteins and nucleotides (Cassier *et al.*, 2020).

2.4. Ecological Roles of Cyanobacteria

The ecological roles played by cyanobacteria may be enlisted broadly as under:

1. **Primary Producers**: The aquatic food webs have a big foundation due to the cyanobacteria. It has a tremendous domination in environments poor in nutrients. They prepare chemical energy from sunlight with the help of photosynthesis, thus building organic matter. Organic matters provide a foundation to zooplankton and herbivores. As shown by Falkowski *et al.* (2008), the usual case is the dominance in primary productivity in oligotrophic waters.

2. **Nitrogen Fixation**: Several cyanobacteria are known to, like *Anabaena* and *Nostoc*, fix atmospheric nitrogen (N_2) to ammonia (NH_3) in a form accessible to most living things. It is mainly crucial in systems with deficiencies in nitrogen since it makes the soils, freshwater ecosystems and plant life fertile, thereby raising the production levels in ecosystems as a whole (Capone *et al.*, 2008).
3. **Biofilm Formation**: Most of the Cyanobacteria develop biofilms that are thick attached colonies of microorganisms to the surfaces. These biofilms play an important role in nutrient cycling, sediment stabilization and pollutant sequestration. Besides protecting the cyanobacterium by providing a shield, the biofilms thus formed may help them to overcome extreme conditions of pH and temperature fluctuations (Battin *et al.*, 2016).
4. **Symbiotic Relationships**: Cyanobacteria form symbiosis with a wide array of organisms, which contributes a tremendous nutrient source in the environment. Besides aquatic examples, such as *Anabaena* live within leaf cavities of a water fern *Azolla* fix and excrete nitrogen to that fern; cyanobacteria generally play an important role for both nutrient economies of lichen complexes, including some root systems that enable their host plants to spread upon very poor soils (Ekman *et al.*, 2008).
5. **Aquatic Bloom Formation**: In eutrophic (nutrient-rich) waters, cyanobacteria like *Microcystis*, *Anabaena* and *Aphanizomenon* are capable of forming huge blooms. Though these serve ecological functions in natural conditions, they may be problematic since they provide hypoxia and even release cyanotoxins like microcystins. These cyanotoxins threaten the life cycle of aquatic organisms and include health risks for humans or animals when they contaminate drinking water or recreational areas (Filatova *et al.*, 2021).

3. Mechanisms of Pesticide Biodegradation by Cyanobacteria

Cyanobacteria utilize multiple mechanisms to degrade pesticides, playing a pivotal role in mitigating pesticide contamination in the environment. These mechanisms are:

3.1. Enzymatic Degradation

Enzymatic degradation is one of the most extensively studied mechanisms of pesticide biodegradation by cyanobacteria. Various cyanobacteria have the following enzymes capable of certain degradative actions on pesticides. Hydrolases, such as esterases or phosphatases - capable of hydrolyzing an ester bond in such organophosphate pesticides and furthermore, oxidase and dehydrogenases are known to mediate the oxidized degradation of such organochlorine pesticides. Others are laccases and peroxidases, responsible for the catalysis involved in the breakdown of more common phenolic compounds broadly used in herbicides. These enzymatic transformations result in the conversion of toxic pesticides into nontoxic or less toxic metabolites by reducing environmental harm (Jaiswal *et al.*, 2019).

3.1.1. Hydrolases

Hydrolases catalyze the breaking of chemical bonds using water, which is generally referred to as hydrolysis. Included among the enzymes of this class are many that serve to detoxify the hazardous organophosphate pesticides, malathion and parathion.

Phosphatases and esterases are other specific types of enzymes that cleave the ester bonds of various organophosphate pesticides to mostly harmless byproducts like alcohols and carboxylic acids, thus detoxifying these toxins. This enzymatic action might detomidine pesticides in ways that involve metabolites either taken up or released by the concerned cyanobacteria. Several other classes of cyanobacteria are reported to produce hydrolyzing enzymes and phosphatases, which act on the detoxification of malathion from environmental toxicity (Lin *et al.*, 2020).

3.1.2. Oxidoreductases (Oxidases and Dehydrogenases):

Oxidases and dehydrogenases, a subclass of oxidoreductases catalyzing redox reactions by adding or removing oxygen atoms from pesticide molecules. These enzymes play an important role in the breakdown of more complex pesticides, which include organohalides and organophosphorus compounds, of which DDT (dichlorodiphenyltrichloroethane), is representative. The oxidoreductases of the cyanobacteria oxidize the chlorinated pesticides, which ultimately causes the cleavage of either the aromatic ring or a halogen substitution, thus making the molecules less toxic and more biodegradable. This has been achieved in the degradation of DDT and its related compounds by cyanobacteria such as *Oscillatoria* and *Anabaena* (Singh *et al.*, 2016).

3.1.3. Peroxidases and Laccases

Biodegradation also depends on the activity of peroxidases, laccases-oxidation reactions, catalytic degradation of many common herbicides and other phenolic pesticides. One type, the laccases catalyze the oxidation of phenolic groups into quinones, those being further degraded to smaller less toxic compounds. The other type, peroxidases, using hydrogen peroxide as substrate, catalyze the reactions that degrade the aromatic ring of the pesticide glyphosate and various phenolic fungicides. Such enzymes have higher degradation capability for resistant pollutants in agricultural runoff (Carrillo Campos *et al.*, 2023).

The enzymatic degradation of pesticides by cyanobacteria acts specifically on the cleavage of particular functional groups and structures, thereby allowing detoxification through reduction of persistence, which is helping in the recovery of the environment as a whole.

3.2. Absorption and Bioaccumulation

Besides enzymatic degradation, the other major strategies developed in the process of pesticide bioremediation by cyanobacteria include the absorption and bioaccumulation of pesticides present in the immediate surroundings. This includes active absorption of water-soluble pesticides across the cellular membrane of cyanobacteria, whereby pesticides get concentrated within cyanobacterial cells.

3.2.1. Absorption

Because of their large surface area relative to their volume, Cyanobacteria could passively absorb the dissolved pesticides and other chemicals from contaminated water or soil. Actually, a number of pesticides, mostly herbicides, but also water-soluble insecticides are taken up through general cell membrane transport systems across the cell membrane (Lin *et al.*, 2020).

3.2.2. Bioaccumulation

Once taken up, cyanobacteria can express the bioaccumulation of pesticides. These pesticide molecules shall either be stored in the vacuoles and lipid membranes or other membrane-bound structures, effectively thereby sequestering the pesticides away from other subcellular components. This limits the bioavailability of bioaccumulated pesticides to interacting organisms like fish or plants and even soil microbial populations (Carrillo Campos *et al.*, 2023).

It is very useful in aquatic environments, where the cyanobacteria act as a "sink" for pesticide contaminants, thus lessening their effect on the surrounding ecosystem. However, long-term effects due to bioaccumulation depend on the type of pesticide and its potential for toxicity. Sometimes, bioaccumulated pesticides are released back into the environment once the cyanobacterial population dies or is disrupted (Lin *et al.*, 2020).

3.3. Microbial Transformation

The microbial transformation of a pesticide molecule involves a chemical structural change in the induced metabolic activity. Unlike enzymatic degradation, microbial transformation of pesticide molecules, in general, refers to changes in pesticide structure, possibly without degradation. Microbial transformation of pesticide molecules might result in either a decrease or increase in the toxicity depending on the type of modification the pesticide molecule undergoes (Lin *et al.*, 2020).

3.3.1. Hydroxylation

Hydroxylation, meaning the addition of a hydroxyl group to the pesticide molecule, is one of the most frequent types of microbial transformation taking place in cyanobacteria. Thus, this increases their hydrophilicity and hence enhances the degradation rate or removal process. Such modifications, normally representing the initial step in these lipophilic compounds-which otherwise might not be available for microbial attack due to poor water solubility-increase the chemical aqueous solubility, hence increasing the possibility of its further microbial catabolism or physical abatement from surroundings (Carrillo Campos *et al.*, 2023). Hydroxylation of certain insecticides or herbicides, for example, may decrease their persistence in soil or water by promoting breakdown by other organisms in the environment.

3.3.2. Demethylation and Ring Cleavage

Some cyanobacteria possess enzymes capable of removing methyl groups from the pesticide molecule through demethylation. Such a transformation is of great importance in the degradation of alkylated pesticides, such as organophosphates and carbamates. By removing methyl groups, cyanobacteria make the pesticide molecule less toxic or easily metabolized by other microorganisms.

The microbial transformation includes ring cleavage, whereby the cyanobacteria cleave open the aromatic ring of several persistent pesticides such as PCBs and DDT. This can then result in simpler, nontoxic molecules that become much easier to further degrade or mineralize (Lin *et al.*, 2020).

3.4. Synergistic Interactions with Other Microorganisms

Generally, cyanobacteria, in natural environments, may live in complex microbial communities in which they interact with other microorganisms, such as bacteria, fungi and algae. Such interactions will improve the biodegradation of pesticides because different microbial populations may have complementary enzymatic activities or metabolic pathways responsible for pesticide degradation (Lin *et al.*, 2020).

For example, bacterial species in consortia with cyanobacteria may provide supplementary addition of degradation enzymes that complement the functions of the enzymes produced by the cyanobacteria. Through such metabolic interactions, some bacteria metabolize intermediate products derived from the degradation of xenobiotics by cyanobacteria, thereby completing its degradation. Some species show greater capabilities in the complete degradation of aromatic or halo-substituted pesticides, which otherwise normally resist degradation (Carrillo Campos *et al.*, 2023).

These interactions may prove to be beneficial for cyanobacteria themselves by making nutrients or by-products of other microorganisms more available. Certain bacteria may degrade pesticides into simpler compounds, more easily utilized by cyanobacteria, thus facilitating the growth of the cyanobacterial population (Maslova *et al.*, 2024).

It means that, collectively, the metabolic activities of cyanobacteria with other microorganisms degrade the pesticide more efficiently and completely; thus, microbial consortia will prove more effective in the field of bioremediation. These interactions may benefit the cyanobacteria themselves either by accessing nutrients or some metabolic by-products produced by other microorganisms. For example, some bacteria degrade pesticides into simpler compounds which can be easily utilized by cyanobacteria, thus increasing the growth of the cyanobacterial population (Maslova *et al.*, 2024).

In this way, the integrated metabolic activities of cyanobacteria and other microorganisms can lead to more effective and complete degradation of pesticides, making microbial consortia more efficient in bioremediation. These various mechanisms followed by cyanobacteria prove that they can act as efficient agents in pesticide bioremediation. The ability of adaptation and interaction with other microorganisms shows the promising application of cyanobacteria in the environmental pesticide contamination remediation.

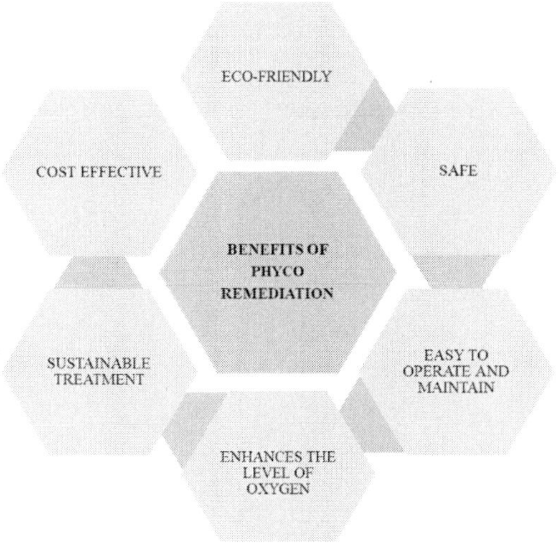

Fig. 3: Benefits of Phycoremediation

4. Applications of Cyanobacteria in Pesticide Bioremediation

The growing concern for pesticide contamination in the environment has paid much attention to effective and sustainable eco-friendly manners aimed at mitigating pesticide pollution. Common treatments against pesticides through methods like washing the soil or by chemical oxidation generally bear too much cost and environmental risks besides bearing poor effectiveness. On the other hand, bioremediation is one of the promising natural methods involving living organisms to degrade, transform, or sequester contaminants and hence, reduce their toxicity and ecological impact (Pathak *et al.*, 2022).

The metabolic potential, resilience and wide adaptability of cyanobacteria have evolved them as an important tool for pesticide bioremediation due to their unique features. They are known to detoxify pesticides through numerous mechanisms: enzymatic degradation, bioaccumulation and transformation. Because of the presence of various metabolic pathways, they are capable of degrading a wide range of pesticides such as herbicides, insecticides, fungicides and organophosphates in aquatic and terrestrial environments (Carrillo Campos *et al.*, 2023).

Applications of cyanobacteria in bioremediation of pesticides include the current role, advantages and bottlenecks pertinent to deploying these organisms in the remediation of various groups of pesticide contaminants.

4.1. Degradation of Herbicides

Herbicides are among the most widely applied chemicals in agriculture to control unwanted plants and weeds. However, overuse of herbicides has resulted in persistence in the environment and thus is potentially toxic to non-target organisms, including aquatic species and useful microorganisms in the soil. Species belonging to several cyanobacteria have been reported capable of degrading some common herbicides through enzymatic or microbial transformations (Lin *et al.*, 2020).

4.1.1. Glyphosate Degradation

Glyphosate is also a member of chemical entities belonging to a category of widely spread herbicides throughout the whole world, possessing high-degree level and exhibiting broad-spectrum actions that result in potential control of weeds. However, considerable concern has arisen about their toxic properties towards microorganisms and real-time persistence of the former compounds naturally leads to interest in this bioremediation involving some cyanobacterial, species of *Anabaena*, *Nostoc* being major ones, able to degradate glyphosate, a process which includes oxidative de-phosphorylation. This enzymatic process entails the removal of the phosphate group from glyphosate, hence turning it into a less toxic compound, aminomethylphosphonic acid (AMPA). Through this metabolic breakdown, AMPA from microbial transformation by cyanobacteria can be degraded to yield much smaller, nontoxic metabolites (Gibb *et al.*, 2022).

It has been noted that in aquatic environments, cyanobacteria may cause significant reductions in the concentration of glyphosate, preventing the contamination of water bodies and hence, protecting aquatic ecosystems from the toxic effects of this herbicide. Besides, the accumulation of glyphosate by cyanobacteria may also contribute to its removal from the environment, though further research is needed regarding the process of this accumulation (Wang *et al.*, 2021).

4.1.2. Atrazine Biodegradation

Among them, atrazine may be considered one of the most popular herbicides in agriculture, widely used in the control against broadleaf and grassy weeds on a variety of crops, mostly on corn and sorghum. However, its diffuse application made it also one of the most important pollutants because of its persistence on soil and water and acute toxic effect on aquatic organisms. Dechlorination and dealkylation reactions have been found to degrade atrazine by some strains of cyanobacteria, such as *Synechococcus* and *Microcystis*. These transformations break down the toxic atrazine molecule into nontoxic metabolites, including hydroxyatrazine and desethylatrazine (Hong *et al.*, 2022).

Various studies have evidenced that the application of cyanobacteria would result in effective atrazine removal from contaminated water to mitigate further damage to the aquatic organism. Furthermore, using cyanobacteria in this process might help minimize financial expenses or chemical treatments that have their own harmful effects for solving atrazine contamination in agricultural runoff (Hong *et al.*, 2022).

4.2. Degradation of Insecticides

Insecticides are the chemicals used in the management of insect pests, though their indiscriminate use developed into certain concerns about their effect on non-target organisms, such as humans, wildlife and useful insects. Cyanobacteria can exhibit a role in the bioremediation of insecticides via different enzymatic and metabolic pathways in organophosphates, carbamates and pyrethroids (Lin *et al.*, 2020).

4.2.1. Organophosphate Degradation

Organophosphates are one common class of insecticides, the key representatives of which include malathion, parathion and chlorpyrifos. These major classes of insecticides do provide effectiveness against pests, while to humans, animals and advantageous organisms alike, the potential for toxicity exists because such an interference could cause critical malfunctioning of the nerves; all act via an inhibition of acetylcholinesterase. *Anabaena* and *Nostoc* represent the species which have been found to harbor enzymes capable of degrading organophosphates due to hydrolysis of the ester bond of phosphate with minor toxicity to insecticides (Hong *et al.*, 2022; Mahmoud *et al.*, 2022).

While a number of cyanobacteria may produce phosphatases or esterases that hydrolyze organophosphate molecules by an ester bond into metabolites, the produced metabolic byproducts will be alcohol and acids, apart from this ability for detoxification of the organophosphates through such enzymatic activity. Also, this enzymatic capability may help degrade organophosphate persistence in the natural environment. Degradation through cyanobacteria, thus, is a type of gentle and eco-friendly remediation process which will avoid hazardous chemicals probably causing more toxic residues and affects the species out of targets (Zhang *et al.*, 2024).

4.2.2. Carbamate and Pyrethroid Degradation

Carbamates and pyrethroids are the two classes of synthetic insecticides used most frequently in agriculture. These are also found to have toxic effects on aquatic life, pollinators and other beneficial insects. Certain species of *Oscillatoria* and *Microcystis* have the potential for carbamate and pyrethroid insecticide degradation through oxidative transformation and hydrolysis (Lin *et al.*, 2020).

For example, methomyl can be degraded to less toxic by-products via enzymatic dehydrogenation and hydrolysis under metabolic activities of cyanobacteria. The degradative enzymes from cyanobacteria also initiate the hydrolysis of an ester group and the oxidation of the aromatic part in the chemical structure of pyrethroids. Such processes have a role not only in decreasing the toxicity of insecticides but also in preventing accumulation within the ecosystem for health and environmental sustainability (Lin *et al.*, 2020).

4.3. Bioremediation of Pesticide-Contaminated Soils & Water Bodies

Various conventional remediation techniques have different drawbacks that the application of cyanobacteria in the bioremediation of pesticide-contaminated soil and water bodies overcomes. Cyanobacteria, which can be employed on pesticide-laden effluent or contaminated soil in a number of systems such as constructed wetlands, ponds and bioreactors, present advantages in treating pesticide-laden effluent or contaminated soil (Lin *et al.*, 2020).

4.3.1. Pesticide Degradation in Aquatic Systems

The cyanobacteria are more effective in the aquatic system where they have the ability for the degradation of pesticide residues of ponds, lakes and rivers. In an aquatic system, cyanobacteria are capable of metabolizing pesticides or acting as a substrate for some other microorganisms, for example bacteria or fungi, for degradation of pollutants. Cyanobacteria can accumulate the pesticide in water and degrade it, reducing its contamination level in water and thus helping in the restoration process of water quality (Lin *et al.*, 2020; Pathak *et al.*, 2022).

Microalgae-bacteria consortia, especially cyanobacteria combined with specific bacteria with the ability to degrade pesticides, have been promising in improving the process. The consortium can exhibit a synergistic interaction and degrade pesticide and its metabolites, thus minimizing persistence in the ecosystem (Lin *et al.*, 2019).

4.3.2. Soil Bioremediation

In agricultural soils, for example, the application of cyanobacteria to the environment could help degrade pesticides. Application is probably possible for crops to improve their growth. In soil enrichment using nitrogen fixation, it provides better soil health, enhancing its remediation effectiveness. In return, they would help degrading a wider range from pesticides to organophosphate insecticides such as glyphosate by enabling the amendment in the fertility and even structure of soil (Singh *et al.*, 2016; Zhang *et al.*, 2024).

Sometimes, the cyanobacteria may form biofilms on soil particles and interact directly with contaminants. In situ degradation of pesticides by cyanobacteria can prevent the spreading of these pesticides to deeper layers of soil and groundwater, hence reducing the environmental impact of pesticide contamination (Lin *et al.*, 2020; Te *et al.*, 2022).

5. Challenges and Limitations

Although the application of cyanobacteria in bioremediation of pesticides is very promising, several challenges and limitations do exist:

- **Specificity of pesticide degradation**: Not all cyanobacteria are capable of degrading all types of pesticides. Efficiency might relate to the species of cyanobacteria, type of pesticide and other environmental factors. Some show effectiveness in degrading certain classes of pesticides, while against other classes of compounds, it may have limited or no activity at all (Lin *et al.*, 2020; Zhang *et al.*, 2024).
- **Environmental factors:** Temperature, light intensity, nutrient availability and other contaminant interplay may influence performance

in the bioremediation of pesticides by cyanobacteria. For example, according to Singh *et al*., 2016, too much nutrient availability could trigger algal blooms that would negatively affect the efficiency of pesticide degradation. Additionally, very extreme temperatures or significant changes in light may have deep effects on the metabolism and pollutant processing potential of this cyanobacterium (Te *et al*., 2022).

- **Toxin production**: Several cyanobacterial species, including the genus *Microcystis*, are known to produce potent toxins like microcystins that could be toxic to aquatic ecosystems and possibly humans. Large-scale bioremediation applications may also heighten such risks. While developing and operating a cyanobacterial bioremediation system, especially around places of possible human or animal contact, any toxin production risk involved in it must be considered (Zhang *et al*., 2024).

6. Future Prospects and Innovations in Cyanobacteria-Based Pesticide Bioremediation

The use of cyanobacteria for the bioremediation of pesticides has been one of the most promising approaches toward the environmental and health hazards resulting from chemical pesticides. A few more novel approaches, besides, go on to further increase efficiency, scalability and applicability of bioremediation assisted by cyanobacteria. The next section outlines some of the highlights of future prospects and innovations regarding genetic engineering for monitoring the quality of the environment and carrying integrated bioremediation systems.

6.1.1. Enhanced Degradation Pathways

Genetic engineering can extend the range of pesticides degradable by cyanobacteria or enhance the rate of their degradation. Using genetic engineering, various workers have prepared genetically modified strains of cyanobacteria capable of enhanced degradation of pesticides by introducing or over-expressing specific biodegradative enzymes involved in pesticide degradation, such as phosphatases, esterases and oxidoreductases. For example, genes responsible for the degradation of specific pesticides, like glyphosate, chlorpyrifos, or atrazine, may be introduced into cyanobacteria so that they can efficiently detoxify such chemicals in polluted environments. Besides, metabolic engineering of cyanobacteria may be carried out to produce variant enzymes capable of degrading new classes of pesticides or pesticide metabolites which are more toxic than their parent compounds. This may greatly improve the potential and efficiency of cyanobacteria in remediating pesticide pollution in soil and water (Kusakabe *et al*., 2013).

6.1.2. Improved Stress Tolerance

The pesticide, in most environments, is usually not the only stressor that is impacting cyanobacteria. High levels of salinity, extreme temperatures, high light intensity and toxic by-products interfere with growth and performance in the use of the cyanobacteria for bioremediation applications. Such stresses could be combated through genetic engineering to further enable the cyanobacteria to perform under less-than-ideal conditions in pesticide-contaminated environments.

For example, engineering cyanobacteria for osmotic tolerance will enable them to grow with high concentrations of pesticides or in salty conditions, such as coasts, because of runoff. The introduction of antioxidant-producing genes will be useful in enhancing the resistance of cyanobacteria against oxidative stress due to pesticide degradation and associated toxic by-products (Bhardwaj *et al.*, 2022).

6.2. Bioreactor and Phytoremediation Systems

Methods in bioremediation techniques using cyanobacteria will further evolve with large-scale applications requiring the engineering of efficient, scalable and cost-effective delivery systems. One of the bright directions in scaling up is the use of cyanobacteria with bioreactors or in phytoremediation systems.

6.2.1. Bioreactor Systems for Pesticide Remediation

Bioreactors are controlled environments whereby microorganisms can grow in a closed system for bioremediation. In the future, genetic engineering could allow the construction of bioreactors that, by means of cyanobacteria-based mechanisms, would clean large volumes of water or soil contaminated with pesticides. Immobilized cyanobacteria can, under such systems, be constantly grown to degrade pesticides on a flow-through basis in continuous cultures, hence offering a fitting and efficient treatment solution against contaminated effluents generated either by agricultural or industrial activities.

The bioreactor designs may further include cocultures or consortia of cyanobacteria with other microorganisms, such as bacteria or fungi, in the optimization of pesticide degradation because of their synergistic metabolic processes. It would thus imply integrated systems for better biodegradation, probably at reduced environmental toxicity, by ensuring that the by-products from pesticide degradation are further degraded into nontoxic compounds (Aslanli *et al.*, 2024).

6.2.2. Phytoremediation Integration

Cyanobacteria may also contribute to plant growth through nitrogen fixation and improving soil health, which would result in better survival of plants and enhanced phytoremediation efficiency. A combination of cyanobacteria and plants in constructed wetlands or vegetated filter strips could make those systems not only for pesticide removal but also other valuable ecosystem services such as habitat creation, nutrient removal and water filtration. The integrated systems could have a holistic approach in cleaning up the environment by capitalizing on the advantages of plant-based and microbial remediation strategies (Nawaz *et al.*, 2024).

6.3. Multi-Functional Bioremediation and Circular Economy Applications

Further research and development in pesticide bioremediation using cyanobacteria may lead to advanced systems of bioremediation with multiple functionalities and wider environmental issues. This can also help in the abatement of other pollutants such as heavy metals, nitrogen and phosphorus, while aiding in the improvement of soil health, water purification and biodiversity restoration, apart from the removal of pesticides.

Such engineering strategies of waste into resources might include, during treatment, the production of useful co-products from cyanobacteria, possibly including biofuels, bioplastics, or biofertilizers. For instance, a cleanup scenario using cyanobacteria against pesticides in water could include biofuels or other highadded-value metabolites from specific processes in an economically viable treatment against agricultural runoff and factory effluents.

Such integrated bioremediation systems would be a holistic approach toward management of pollution whereby biodegradation of pesticides is coupled with the recovery of some valuable resources. This would minimize the environmental impact and hence, enhance the scope of sustainability (Chakraborty *et al.*, 2024).

7. Conclusion

Cyanobacteria-based bioremediation has been identified as an effective and environmentally friendly method for cleaning up environmental pesticide contamination. Through processes like biosorption, bioaccumulation and enzymatic degradation, cyanobacteria can break down or detoxify a variety of pesticides, including carbamates, organophosphates and organochlorines, due to their well-known metabolic adaptability and photosynthetic capability. Such complex pesticide molecules are degraded into less toxic or nontoxic metabolites, for which the production of certain enzymes like hydrolases, monooxygenases and dehalogenases is critical. Thus, bioremediation stands out as a better alternative to conventional chemical and physical means of pesticide removal, which tend to be more costly, energy-consuming and prone to secondary pollution.

Applications of cyanobacteria in bioremediation have come to include a whole array of environmental matrices from soil-water systems. In addition to breaking down the residual pesticide, Cyanobacteria enhances fertility and contributes to nutrient cycling in agricultural soils, thus improving their microbial function. In aquatic ecosystems, adsorption and metabolism make them adequate candidates to be used in a number of water purification techniques, either in constructed wetlands or in algal biofilms. Further, the incorporation of cyanobacteria in biofertilizers presents dual advantages due to their role in promoting plant growth by nitrogen fixation and at the same time degrading pesticide residues in support of sustainable agriculture.

Despite their potential, several drawbacks mean that large-scale application on cyanobacteria-based bioremediation is impossible. For example, environmental factors in all aspects, such as pH or temperature, may affect their abilities to bioremediate pesticide contents. Besides, the complete decomposition of some synthetic pesticides proves impossible due to their recalcitrant nature. Possible harmful algal bloom under certain conditions does awake environmental safety concerns. Further, the cost, efficiency and operational complexity in scaling up laboratory findings into real applications have to be considered. This, therefore, calls for future research efforts toward the optimization of the strains, using genetic engineering for better degradative capabilities. The development of consortia combining cyanobacteria with other complementary organisms might achieve this broadened spectrum of pesticides being degraded. Thirdly, high-value-added bioreactor systems should be developed that are able to provide a controlled environment for the maintenance of optimum activity of cyanobacteria. Long-term field applications should be done to validate the lab data by testing these technologies regarding environmental and economic viability.

In conclusion, cyanobacteria thus represent a cost-effective and environment friendly response to pesticide pollution. Further improvements, together with strategic development, have so far enabled the use of cyanobacteria-based bioremediation strategies with full potential to revolutionize environmental practices in the name of healthier ecosystems and truly sustainable agriculture.

References

Aktar, M. W., Sengupta, D., Chowdhury, A. (2009). Impact of pesticides use in agriculture: Their benefits and hazards. Interdisciplinary Toxicology, 2(1), 1–12. https://doi.org/10.2478/v10102-009-0001-7

Aslanli, A., Maslova, O. (2024). Using fungi in artificial microbial consortia to solve bioremediation problems. Microorganisms, 12(3), 470. https://doi.org/10.3390/microorganisms12030470

Battin, T. J., Besemer, K., Bengtsson, M. M., Romani, A. M., Packmann, A. I. (2016). The ecology and biogeochemistry of stream biofilms. Nature Reviews Microbiology, 14(4), 251–263. https://doi.org/10.1038/nrmicro.2016.15

Bhardwaj, N., Gupta, R. K., Kumar, A. (2022). Mechanisms of stress tolerance in cyanobacteria under extreme conditions. Stresses, 2(4), 531–549. https://doi.org/10.3390/stresses2040036

Capone, D. G., Zehr, J. P., Paerl, H. W., Bergman, B., Carpenter, E. J. (2008). Trichodesmium, a globally significant marine cyanobacterium. Science, 276(5316), 1221–1229. https://doi.org/10.1126/science.276.5316.1221

Cassier-Chauvat, C., Chauvat, F. (2020). Recent Advances in the Photoautotrophic Metabolism of Cyanobacteria. Life, 10(5), 71. https://doi.org/10.3390/life10050071

Carrillo Campos, J., Garcia-Sanchez, A., Díaz-Raviña, M. (2023). Microbiology and biochemistry of pesticide biodegradation. International Journal of Molecular Sciences, 24(21), 15969. https://doi.org/10.3390/ijms242115969

Chakraborty, S., Singh, R., Kumar, R. (2024). Multi-functional cyanobacteria-based systems for pollutant removal and resource recovery: Towards a circular economy. Environmental Science and Technology, 58(4), 2435-2444. https://doi.org/10.1021/es5034568

Ekman, M., Tollbäck, P., Bergman, B. (2008). Proteomic analysis of the cyanobacterium of the Azolla symbiosis: Identity, adaptation, and NifH modification. Journal of Experimental Botany, 59(5), 1023–1034. https://doi.org/10.1093/jxb/erm282

Falkowski, P. G., Barber, R. T., Smetacek, V. (2008). Biogeochemical controls and feedbacks on ocean primary production. Science, 281(5374), 200–206. https://doi.org/10.1126/science.281.5374.200

Filatova, D., Jones, M.R., Haley, J.A. (2021). Cyanobacteria and their secondary metabolites in three freshwater reservoirs in the United Kingdom. Environ Sci Eur, 33, 29. https://doi.org/10.1186/s12302-021-00472-4

Geissen, V., Mol, H., Klumpp, E., Umlauf, G., Nadal, M., van der Ploeg, M., van de Zee, S. E. A. T. M. (2015). Emerging pollutants in the environment: A challenge for water resource management. International Soil and Water Conservation Research, 3(1), 57-65. https://doi.org/10.1016/j.iswcr.2015.03.002

Gibb, S. W., Ekosse, G.-I. (2022). Review of glyphosate-based herbicide and aminomethylphosphonic acid (AMPA): Environmental and health impacts. Applied Sciences, 12(17), 8789. https://doi.org/10.3390/app12178789

Hong, J., Boussetta, N., Enderlin, G., Merlier, F., Grimi, N. (2022). Degradation of residual herbicide atrazine in agri-food and washing water. Foods, 11(16), 2416. https://doi.org/10.3390/foods11162416

Jaiswal, S., Singh, D.K., & Shukla, P. (2019). Gene editing and systems biology tools for pesticide bioremediation: A review. Frontiers in Microbiology, 10, 87. https://doi.org/10.3389/fmicb.2019.00087

Kusakabe, T., Tatsuke, T., & Tsuruno, K. (2013). Engineering a synthetic pathway in cyanobacteria for isopropanol production directly from carbon dioxide and light. Metabolic Engineering, 20, 101-109. https://doi.org/10.1016/j.ymben.2013.09.007

Lin, Z., Zhang, W., Pang, S., Huang, Y., Mishra, S., Bhatt, P., Chen, S. (2020). Current approaches to and future perspectives on methomyl degradation in contaminated soil/ water environments. Molecules, 25(3), 738. https://doi.org/10.3390/molecules25030738.

Mahmoud, H. M., Mostafa, S. S. M., Husein, D. Z., El-Sayed, M. A. (2022). Growth optimization and secondary metabolites evaluation of Anabaena variabilis and its potential for organophosphate biodegradation. Plants, 11(6), 735. https://doi.org/10.3390/ plants11060735

Maslova, O., Lyagin, I. (2024). Using fungi in artificial microbial consortia to solve bioremediation problems. Microorganisms, 12(3), 470. https://doi.org/10.3390/ microorganisms12030470

Nawaz, T., Gu, L., Fahad, S., Saud, S., Bleakley, B., Zhou, R. (2024). Exploring sustainable agriculture with nitrogen-fixing cyanobacteria and nanotechnology. Molecules, 29(11), 2534. https://doi.org/10.3390/molecules29112534

Popp, J., Pető, K., Nagy, J. (2013). Pesticide productivity and food security: A review. Agronomy for Sustainable Development, 33(1), 243–255. https://doi.org/10.1007/ s13593-012-0105-x

Pathak, V. M., Verma, V. k., Rawat, B. S., Dewali, S., Yadav, M. (2022). Current status of pesticide effects on environment, human health and it's eco-friendly management as bioremediation: A comprehensive review. Front. Microbiol., 13. https://doi.org/10.3389/ fmicb.2022.962619

Rastogi, R. P., Sinha, R. P. (2009). Biotechnological and industrial significance of cyanobacterial secondary metabolites. Biotechnology Advances, 27(4), 521-539. https://doi. org/10.1016/j.biotechadv.2009.04.009

Stehle, S., Schulz, R. (2015). Agricultural insecticides threaten surface waters at the global scale. Proceedings of the National Academy of Sciences, 112(18), 5750–5755. https:// doi.org/10.1073/pnas.1500232112

Singh, J. S., Kumar, A., Rai, A. N., Singh, D. P. (2016). Cyanobacteria: A precious bio-resource in agriculture, ecosystem and environmental sustainability. Frontiers in Microbiology, 7, 529. https://doi.org/10.3389/fmicb.2016.00529

Te, S. H., Boo, C. Y., Gin, K. Y.-H., Swarup, S. (2022). Effects of light and temperature on the metabolic profiling of two habitat-dependent bloom-forming cyanobacteria. Metabolites, 12(5), 406. https://doi.org/10.3390/metabo12050406

Whitton, B. A., Potts, M. (2012). The ecology of cyanobacteria: Their diversity in time and space (2nd ed.). Springer Science and Business Media. https://doi.org/10.1007/978-94-007-3855-3

Wang, Y., Zhang, M., Huang, Q. (2021). The role of cyanobacteria in the bioremediation of glyphosate in aquatic systems. Environmental Pollution, 285, 117428. https://doi. org/10.1016/j.envpol.2021.117428

Zhang, Q., Jia, L., Chen, Y., Yan, H., Chen, Q., Zhang, J., Sun, H. (2024). Molecular mechanisms of the cyanobacterial response to different phosphorus sources. Sustainability, 16(13), 5642. https://doi.org/10.3390/su16135642

6

Mycoremediation to Combat Heavy Metal Pollutants

***Harshita Saini*[1]*, Sheetal Sharma*[1]*, Anuya Verma*[2] *and Rohini Trivedi*[1]**

[1]*Department of Botany, MLS University, Udaipur - 313001*

[2]*Department of Environment Science, MLS University, Udaipur - 313001*

[*]*Corresponding Author Email: rohinitrivedi08@gmail.com*

Abstract

Bioremediation employs living units and biological processes to remediate or restore the environment to its original state. The techniques of bioremediation can be in situ and ex situ. Microorganism play a crucial role in bioremediation. Fungi and their enzymes help in bringing the environment back to its original state, eliminating the toxic contaminant or pollutant. Mycoremediation is a promising clean-up technique within the field of bioremediation. Several fungi can degrade a wide range of pollutants. Industrial and mining processes are resulting in contamination of soil and water by heavy metals. Mycoremediation can take a future lead in the bioremediation of heavy metal (HMs) polluted water and soil. The high adsorption and accumulation capacity of fungi serve as a tool for the removal of HMs. Various physical and chemical parameters like time, pH, temperature, concentration, and nature of HMs, along with the type and biomass of the fungi involved, are the most influential factors that affect the process of mycoremediation of HMs.

Keywords: *Bioremediation, mycoremediation, in situ remediation, ex situ remediation, heavy metals.*

1. Introduction

Bioremediation is an environmentally friendly process that uses numerous different microbes (e.g., fungi & bacteria), green plants, or combinations of them to weaken and detoxify dangerous pollutants in a sequential or successional manner. The pollutant composites are converted by biological organisms through a response that takes place as part of their metabolic processes. They convert poisonous pollutants into carbon dioxide (CO_2) and water (H_2O), microbial biomass, and other products that are less poisonous. The organic wastes are biologically degraded under controlled, harmless, and safe condition into inoffensive products below the admissible limits established by supervisory authorities.

Bioremediation utilizing microorganisms is primarily the most effective and efficient way to degrade, detoxify, and break down environmental pollutants into less poisonous forms. The microorganisms may be indigenous to a polluted area or transported from away and brought to the defiled spots. For bioremediation to be effective, microorganisms must enzymatically attack the pollutants and convert them to inoffensive products.

The environment is facing challenges from different pollutants including halogenated solvents, petroleum hydrocarbons, explosives, agrochemicals, heavy metals/ions, and radionuclides, heavy metals (HMs) (Cd, Cu, Hg, Pb, Mn, As, Ni, Zn, Fe, etc.), which are extensively used in different types of industries and released in high amounts with their disposing effluents. These contaminants directly or indirectly come into the environment. The wide-scale occurrence of these heavy metals leads to adverse health effects such as multiple organ failure and cancer.

Mycoremediation is a strong bioremediation technique that utilizes fungi to eliminate poisons from environmental components (Jeyakumar *et al.,* 2023). Fungi bear an indestructible morphology with a specific mechanism that captures heavy metals on the surface, which is, accumulated inside the fungal cell. Fungi absorb the metal ions into the cell and also chelate them, so that they can tolerate heavy metal compounds (Gadd & White 1993; Anahid *et al.* 2011). The cell wall of fungi comes into contact with the metal ion in a solution where the metal can be deposited on the surface or within the cell wall structure before interacting with cytoplasmic materials (Gadd 1990).

2. Bioremediation Techniques

2.1. In-situ

In situ term refers to when defiled waste is treated right at its point of origin or actuality. The benefit of *in situ* treatment is that it prevents the spread of impurity during the relocation and transport of the polluted material.

2.1.1. Bioventing

Bioventing is a technique that requires a considerable amount of air to be blown into the soil to increase the rate of oxygen in the contaminated soil. This method is suitable especially for degrading chemicals or metal contaminants that are sensitive to oxygen. It is a technique used to degrade any degradable contaminant that is susceptible to aerobic conditions (Rockne and Reddy, 2003).

This process uses nutrients and water to improve bioremediation with the sole aim of transforming pollutants to a less harmful form, employing microbial activities (Philip and Atlas, 2005). This technique is particularly suitable, especially for light spills of hydrocarbons like gasoline, oil, petroleum, and so on (Hohener and Ponsin, 2014; Eyvazi and Zytner, 2010).

2.1.2. Bioaugmentation

Bioaugmentation is the introduction of microbes isolated from the polluted site. Bioaugmentation uses indigenous microorganism and their increment at the site with contaminated soil and water (Ueno *et al.* 2007). Bioaugmentation is useful for polluted soil sites having a low number of contaminant-degrading microorganisms, for small-scale land on which bioaugmentation is economically suitable as compared to other non-biological methods, and for soils polluted with persistent toxic compounds (Frosyth *et al.* 1995).

2.1.3. Biostimulation

Biostimulation refers to the addition of nutrients, oxygen, or other electron donors and acceptors to the contaminated soil to increase the microbial activity or their population (Perfumo *et al.,* 2007). It is also defined as a remediation process that occurs naturally, such that the degradation of pollutants is improved by maximizing the conducive physical conditions, such as aeration, nutrient addition, changes in pH, and temperature (Margesin and Schinner, 2001). Thus, biostimulation involves the alteration of the environment of the soil so that indigenous microorganisms capable of bioremediation can be stimulated.

2.2. Ex-situ

Ex-situ term refers to treatment that occurs after the polluted waste has been removed to a treatment area. The main advantage is that it helps to contain and control the bioremediation products, as well as making the area that was defiled available for use.

2.3. Phytoremediation

The term 'Phytoremediation' consists of the Greek prefix 'Phyto' that connotes plant and the suffix 'Remediation', which is obtained from the Latin word remedies, which means to correct, restore, or remove an evil (Erakhrumen and Agbontalor, 2007). Phytoremediation simply refers to using plants to remove contaminants from either water or soil. The plants are capable of drawing and holding contaminants in their structures, which leads to effective removal of them from soil or water. There are several uptake mechanisms employed by plant for remediation technology such as phytoextraction(plants remove dangerous elements or compounds from soil, water or heavy metals), phytovolatilization(process in which plants absorb contaminants, such as heavy metals and organic pollutants, through their roots and release them into the atmosphere as volatile compounds), phytostabilization (the reduction of the mobility of heavy metals in soil), and rhizofiltration (it involves filtering contaminated water through roots to remove toxic substances) (Tangahu *et al.*, 2011).

2.4. Mycoremediation

The term mycoremediation can be broken down as myco (fungus) and remediation (to clean), and indeed, mycoremediation is the use of fungi for creating simple yet effective biomass capable of breaking down environmental and industrial pollutants effectively and economically. Mycoremediation is one of the biotechniques that recruits fungi to remove toxic contaminants from the environment. Mycoremediation is a form of bioremediation that decontaminate the environment from contaminants. Mycoremediation relies on the efficient enzymes produced by fungi for the degradation of various types of substrates and pollutants.

3. Factors Affecting Bioremediation Processes

The control and optimization of bioremediation processes is a complex system, governed by numerous factors like humidity, pH, oxygen content, nutrients, temperature, types of soil, etc.

3.1. Humidity

Humidity (water vapour) affects the rate of degradation of contaminants because biological units need water to survive. It also affects the type and volume of remaining substances, leading to answerable and accountability. The quantum of water in the pore spaces of soil also affects the exchange of oxygen. The humidity content of the soil should be between 25% and 85% of the water holding capacity, and between 50% – 80% results in enhanced biodegradation. The water holding capacity of soil, also known as the "field capacity," is the amount of water left in an impregnated soil where gravitational drainage has stopped (Eskander & Saleh, 2017).

3.2. pH

Soil pH helps in detecting the degree of acidity or alkalinity. The soil's pH can affect the metabolic rate of microbes. The rate of degradation of soil pollutants can either increase or decrease due to the metabolic activity of microbes. In general, optimum biodegradation takes place at a pH value ranging from 6.5 to 8.5 for most submarine and terrestrial systems (Abatenh *et al.*, 2017). The pH of the soil also affects the availability of nutrients.

3.3. Oxygen content

The bioremediation process occurs both in aerobic and anaerobic conditions. Aerobic conditions are promotive as oxygen is available to increase the metabolic rate of microorganisms, as the case of petroleum adulterants oxygenase (Macaulay, 2015). Oxygen is needed in the respiration process as well as in the exertion of oxygenases, crucial in the degradation of crude oil painting, and adulterants.

3.4. Nutrients

For microorganisms to thrive, indeed, continue their adaptation and conditioning to the new environment. They require to bear some quantum of essential nutrients like nitrogen, phosphorus, and carbon. The C: N: P ratio is crucial in the rate and efficiency of biodegradation as the ratio is very small in the decline of hydrocarbon. Thus, the absence or low presence of essential nutrients affects microbial growth and metabolism. An optimal quantity of nutrients is needed for the biodegradation of organic matter, but too much of these nutrients can negatively impact the biodegradation of hydrocarbons due to the low metabolic rate of the microorganisms (Koshlaf & Ball, 2017).

3.5. Temperature

Among all the factors affecting the rate of biodegradation, temperature is the most important because it ascertains or ensures which microorganisms can flourish. It can also affect the composition of the pollutants (Das & Chandran, 2011). Temperature can either increase or decrease the rate of biodegradation. Each pollutant requires a particular temperature for degradation to take place. The rate of microbial action increases with temperature, and reaches its maximum exertion at optimum temperature, after which any further increase or drop in temperature results in a reduction in the metabolic rate of the microbes, leading to low biodegradation (Abatenh *et al.*, 2017).

3.6. Types of Soils

The structure and condition of the soil determine the movement of the pollutants, affecting the biodegradation rate. An advanced amount of organic matter in the soil yields high biomass as well as better microbial growth, bringing about an increased biodegradation rate of hydrocarbons (Scherr *et al.*, 2007; Chaerun & Tazaki, 2005).

According to Kensa (2011) soil moisture (50% - 85% of water holding capacity), soil pH 6.5 – 8.5, and oxygen content of at least 10% air-filled pore space, along with nutrient content- specifically N and P for microbial growth- and temperature (0 °c) ranging from 15 to 45, plus a low clay or silt content in the soil, influence the bioremediation.

4. Mycoremediation of Heavy Metals (HMs)

Mycoremediation is a bioremediation process that utilizes fungi to eliminate harmful compounds, especially effective in removing poisonous heavy metals from various environmental components. Mycoremediation is dependent on the selection and use of an appropriate fungal species for the target or heavy metal pollutant. Mycoremediation can be used as a biological instrument for degradation, transformation, or immobilization of pollutants (Bosco & Mollea, 2019).

4.1. Fungi used in HMs Remediation

The interaction between fungi and HMs may be positive (promotive) or negative (inhibitory) to growth and metabolic rate, it may result in the death of fungi in the presence of HMs. The positive interaction of fungi with HMs, which is due to higher fungal resistance towards HMs, leads to a new approach of using fungi in HM removal. A few fungal members have multi-metal removal capacity in viable or in dead biomass form are promising tools (Chakraborty *et al.,* 2014; Kan *et al.,*2015).

Table 1. Heavy metal tolerant candidate fungi for HMs removal

S. No.	Fungi	HMs pollutants	Reference
1	*Aspergillus flavus*	Cu, Pb, Cr, and Cd	Iram & Sumera, (2015), Qayyum *et al.,* (2016)
2	*Aspergillus fumigatus*	Cr	Panda *et al.*, (2014), Sakthivel *et al.*, (2016)
3	*Aspergillus* sp.	Cu, Zn, and Cr	Pundir & Dastidar (2015), Sathvika *et al.*, (2015), Pundir *et al.,* (2018)
4	*Fusarium oxysporum*	Cr and Pb	Migahed *et al.,* 2017
5	*Fusarium solani*	Cd	Kumar *et al.*, (2019)
6	*Neurospora* sp.	Pb, Ni, Co, Cr, Cu, and Zn	Joshi (2014), Desai *et al.,* (2016)
7	*Penicillium chrysogenum*	Zn and Cu	Tahir *et al.,* (2017)
8	*Penicillium simplicissimum*	Cd, Pb, Cu, Zn, and Cr	Xu *et al.,* (2015), Mohammadian *et al.*, (2017)
9	*Alternaria* sp.	Ni and Cd	Netpae *et al.*, (2015), Costa & Tavares (2017)
10	*Pleurotus eryngii*	Cu, Zn, and Hg	Li *et al.* (2018)
11	*Trichoderma viride*	Cr and Pb	Sugasini & Rajagopal (2015), Migahed *et al.,* (2017)

4.2. Mechanism Adopted by Fungi in HM Remediation

Fungi are a class of microorganism that are veritably opportunistic, have a wide range of rigidity, but are quick to respond to stressful terrain. They can break down poisonous substances and complex hydrocarbon chains into less dangerous, biodegradable forms to be disposed of in environmental remediation. Mycoremediation involves several crucial response processes and mechanisms of action adapted to Fig.ht stressful conditions. Various mechanisms of mycoremediation are depicted in Fig. 1.

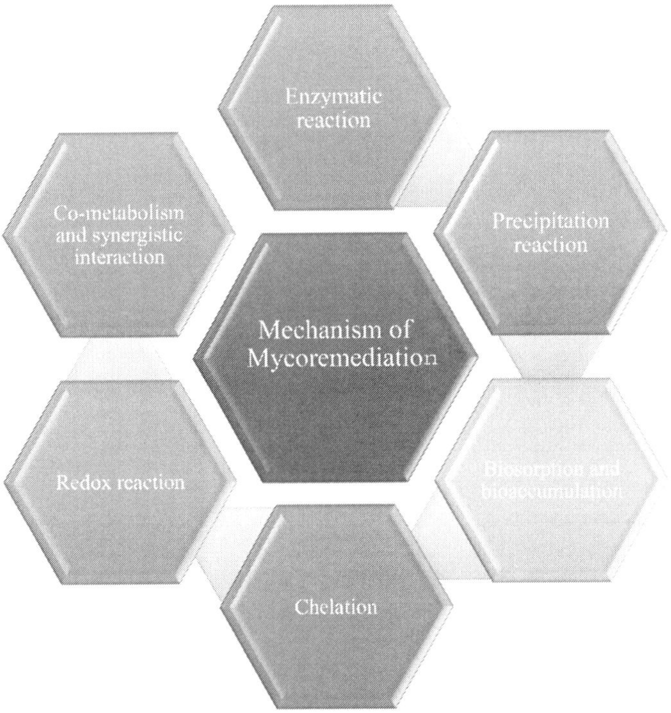

Fig. 1: Various mechanisms of mycoremediation

4.2.1. Enzymatic Response

HM induces a high production of ROS in fungi, one of the major factors for damaging the cell components due to oxidative stress. Fungi generally Fig. ht oxidative stress by multiple antioxidative mechanisms. The antioxidant mechanisms are driven by numerous antioxidative enzyme like catalases (CAT), peroxidase, and superoxide dismutase (SOD). Fungi produce these major enzymatic antioxidants under HMs stress. *Tricoderma lixii* and *Aspergillus flavus* produce significantly high CAT, peroxidase, and SOD under Cr (VI) exposure to Fig.ht the ROS product (Kumar & Dwivedi 2019). *Pleurotus ostreatus* produces CAT, peroxidase, and SOD under the presence of Pb stress (Li *et al.*, 2017). Jacob *et al.*, (2004) reported that SOD production increases with an increase in the incubation period in Cd- treated fungus

Paxillus involutus which is essential in detoxification mechanisms of ROS. Teng *et al.,* (2017) noted a significant increase in exertion of SOD, CAT, peroxidase, glutathione reductase (GR), and glutathione- S- transferase (GST) in *Penicillium janthinellum* under Zn stress. The increase in the production of CAT, peroxidase, SOD, was also reported in *Trichoderma lixii, A. flavus,* and *Fusarium solani* (Kumar & Dwivedi 2019.; Kumar *et al.,* 2019). Solanki *et al.,* (2019) reported increase in the lipase enzyme in *Trichoderma* sp., *Curvularia* sp., *Drechslera* sp., *Fusarium* sp., *Aspergillus* sp., *Rhizopus* sp., in H_2O_2 oxidation on aromatic compounds.

Table 2. Mechanism of Action of Fungal Enzymes

S. No.	Enzymes	Fungi	Mechanisms of action	References
1	Lipase	*Trichoderma* sp., *Curvularia* sp., *Drechslera* sp., *Fusarium* sp., *Aspergillus* sp., *Rhizopus* sp.	Single-electron oxidation using H_2O_2 on aromatic compounds	Solanki, *et al.,* (2019)
2	Catalase	*Curvularia* sp., *Acremonium* sp., *Aspergillus* sp.,	Single-electron oxidation using H_2O_2 on aromatic compounds	Bosco, (2019)
3	Laccase	*Bjerkandera adusta, Phanerochaete chrysosporium, Trametes versicolor, Pleurotus* sp. *Doratomyces nanus, Thermoascus*	Single-electron oxidation using O_2 on organic compounds.	Mouhamadou *et al.*, (2013)
4	Peroxidase	*Aspergillus niger, Aspergillus foetidus, Trichoderma viride, Aspergillus sojae,*	Single-electron oxidation using H_2O_2 on aromatic compounds	Jakovljevi *et al.*, (2018)

4.2.2. Rush responses

Rush response refers to the rapid cellular and physiological reaction fungi show, when facing stress condition. Fungi can induce rush responses by converting usable ions into unusable forms. Fungi can store organic acids, which then react with ions to produce phosphates or oxalates that precipitate out of the mixture. *Aspergillus niger* has been proven to grease the rush of lead as pre-eminent oxalate, effectively, reducing supereminent attention to polluted media (Niehaus *et al.,* 2018, Alam *et al.,* 2018; Yousaf *et al.*, 2018).

4.2.3. Biosorption and bioaccumulation

The primary mechanism by which fungi remove heavy metals from contaminated surroundings is biosorption. Fungi have cell walls rich in functional groups similar to carboxyl, hydroxyl, and amine groups, which can bind ions through ionic exchange, complexation, and chelation. *Aspergillus niger* and *Penicillium chrysogenum* are known for their high biosorption capacity for Cd, Pb, and Hg (Yang *et al.*, 2019). *Trichoderma harzianum* can bioaccumulate significant quantities of Pb and Zn, effectively removing these ions from polluted soil (Smith *et al.*, 2018).

4.2.4. Chelation

Aspergillus niger and other phosphate-solubilizing fungus store organic acids such as citric and oxalic acids, which are essential for the chelation and rush of heavy metals. Citric acid can reduce the bioavailability and toxicity of Pb, Cd, and Al by forming stable complexes. (Yang *et al.*, 2019).

4.2.5. Redox Responses

Fungi can also effectively induce redox responses that play a critical part in the metamorphosis and detoxification of pollutants. These responses involve the transfer of electrons between the contaminant and the fungal cells, leading to the reduction or oxidation of the pollutants. For illustration, *Aspergillus niger* has been reported to reduce Cr (VI) to Cr (III) under both aerobic and anaerobic conditions, making it a potent agent for the bioremediation of Cr-defiled spots (Yang *et al.*, 2019).

4.3.6. Co-metabolism and Synergistic Relations

Fungi can also engage in co-metabolism, where the declination of one emulsion facilitates the breakdown of another. This process is particularly useful in surroundings defiled with complex fusions of pollutants. For example, fungi can degrade hydrocarbons while contemporaneously enhancing the breakdown of heavy oils by altering the pH and redox conditions of the soil (Burton *et al.*, 2019). Fungi can interact synergistically with other microorganisms, similar to bacteria, to enhance the overall bioremediation process. These relations can lead to the successional degradation of pollutants, where fungi break down complex molecules into simpler intercedes that are further degraded by another microorganism.

5. Conclusion

Bioremediation, especially mycoremediation, offers an innovative and environmentally sustainable system of cleaning up the terrain. Using the special metabolic styles of some fungi, mycoremediation gives an affordable, environmentally friendly, and adaptable way to remove a variety of pollutants, such as fungicides, heavy metals, hydrocarbons, and more. The effectiveness of this strategy is demonstrated by fungi's capability to convert complex pollutants into simpler, less dangerous forms or into accoutrements that are more readily broken down by other organisms. These pollutants also contain numerous types of contaminants other than HMs, including nitrate, phosphate, sulphites, fluoride, polyaromatic hydrocarbons, fungicides, etc., and studies have reported the capability of numerous fungal species to remove these pollutants. So, mycoremediation could affect the contemporaneous treatment of HMs with other adulterants, making it a new approach having implicit operation in bioremediation.

References

Abatenh, E., Gizaw, B., & Tsegaye, Z. (2017). Application of microorganisms in bioremediation-review Journal of Environmental Microbiology,1(1),02-09.

Alam, M. Z., Tan, I. K., Amor, R., & Khan, N. A. (2018). Role of oxalic acid in heavy metal precipitation by Aspergillus niger. Environ. Technol. Innov. 11, 63–71.

Anahid S, Yaghmaei S, & Ghobadine A. D. (2011). Heavy metal tolerance of fungi. Scintica Iranica 18(3):502–508.

Atlas, R. M., & Philp, J. (2005). Bioremediation. Applied microbial solutions for real-world environmental cleanup (pp. xi+-366).

Bosco, F. & Mollea, C. (2019). Mycoremediation in soil, in: Environmental Chemistry and Recent Pollution Control Approaches, p. 173.

Burton, J.L., Smith, E.A., Peterson, S. M., & Brown, M.T., (2019). Degradation of polycyclic aromatic hydrocarbons by Phanerochaete chrysosporium. J. Environ. Chem. Eng. 7 (5), 4262–4271.

Chaerun, S., & Tazak, K. (2005). How kaolinite plays an essential role in remediating oil-polluted seawater. Clay Miner, 40,481–491.

Chakraborty, S., Mukherjee, A., Khuda-Bukhsh, A. R., & Das, T. K. (2014). Cadmium-induced oxidative stress tolerance in cadmium-resistant Aspergillus foetidus: its possible role in cadmium bioremediation. Ecotoxicol Environ Saf 106,47–53.

Costa, F., & Tavares, T. (2017). Bioremoval of Ni and Cd in the presence of diethyl ketone by fungi and by bacteria, A comparative study. Int Bio deterior Bio degrad 120,115–123.

Das, N., & Chandran, P. (2011). Microbial degradation of petroleum hydrocarbon contaminants: An overview. Biotechnology Research International, 1-13.

Desai, H., Patel, D., & Joshi, B. (2016). Screening and Characterisation of Heavy Metal Resistant Fungi for their Prospects in Bioremediation of Contaminated Soil. Int J Curr Microbiol App Sci 5(4),652–658.

Erakhrumen, A., & Agbontalor, A. (2007). Review Phytoremediation: an environmentally sound technology for pollution prevention, control, and remediation in developing countries. Educational Research and Review, 2(7), 151–156.

Eskander, S. B., & Saleh, H. M. (2017). Biodegradation: process mechanism. Environmental Science and Engineering, Vol. 8: 8: Biodegradation and Bioremediation.

Eyvazi, M. J., & Zytner, R. G. (2010). A correlation to estimate the bioventing rate constant. Bioremediation Journal, 13,141-153.

Forsyth, J. V., Tsao, Y. M., & Bleam, R. D. (1995). Bioremediation: when is augmentation needed? (No. CONF-950483-). Battelle Press, Columbus, OH (United States).

Gadd GM (1990). Fungi and yeasts for metal accumulation. In: CL EHLB (ed) Microbiol mineral recovery. McGraw-Hill, New York, pp 249–276.

Gadd GM, & White C (1993) Microbial treatment of metals pollution- a working biotechnology? Trends Biotechnol 11(8):353–359.

Hohener, P., & Ponsin, V. (2014). In situ vadose zone bioremediation. Current Opinion in Biotechnology 27,1–7.

Iram, S., & Sumera, A. (2015). Biosorption of Copper and Lead by Heavy Metal Resistant Fungal Isolates. Int J Sci Res Publ 5(1),1–5.

Jacob, C., Courbot, M., Martin, F., Brun, A., & Chalot, M. (2004). Transcriptomic responses to cadmium in the ectomycorrhizal fungus Paxillus involutus. FEBS Lett 576,423–427.

Jakovljevi, V. D., C, M.M., & Vrvi, c. (2018). Potential of pure and mixed cultures of Cladosporium cladosporioides and Geotrichum candidum for application in bioremediation and detergent industry, Saudi J. Biol, Sci. 25 (3), 529–536.

Jeyakumar, P., Deb Nath, C., Vijayaraghavan, R., & Muthuraj, M. (2023). Trends in bioremediation of heavy metal contaminations. Environmental Engineering Research, 28(4).

Joshi, B. H. (2014). Evaluation and Characterisation of Heavy Metal Resistant Fungi for their Prospects in Bioremediation. J Environ Res Dev 8(4),876–882.

Kan, S.H., Sun, B. Y., Xu, F., Song, Q. X., & Zhang, S. F. (2015). Biosorption of aquatic copper (II) by mushroom biomass Pleurotus eryngii: kinetic and isotherm studies. Water Sci Technol 71(2),383–288.

Kensa, V. M. (2011). Bioremediation-an overview. Journal of Industrial Pollution Control, 27(2), 161-168.

Koshlaf, E., & Ball, A. S. (2017). Soil bioremediation approaches for petroleum hydrocarbon-polluted environments. AIMS Microbiology, 3(1),25-49.

Kumar, V., & Dwivedi, S. K. (2019). Hexavalent chromium reduction ability and bioremediation potential of Aspergillus flavus CR500 isolated from electroplating wastewater. Chemosphere 237,124-567.

Kumar, V., Singh, S., Singh, G., & Dwivedi, S.K. (2019). Exploring the Cadmium Tolerance and Removal Capability of a Filamentous Fungus, Fusarium solani. Gemicrobiol J 36(9),782–791.

Li, J., Sun, Y., Jiang, X., Chen, B., & Zhang, X. (2018). Arbuscular mycorrhizal fungi alleviate arsenic toxicity to Medicago sativa by influencing arsenic speciation and partitioning. Ecotoxicol Environ Saf 157,235 243.

Li, X., Wang, Y., Pan, Y., Yu, H., & Zhang, X. (2017). Mechanisms of Cd and Cr removal and tolerance by the macro fungus Pleurotus ostreatus HAU-2. J Hazard Mater.

Macaulay, B. M. (2015). Understanding the behaviour of oil-degrading microorganisms to enhance the microbial remediation of spilt petroleum. Applied Ecology and Environmental Research, 13,247–262.

Margesin, R., & Schinner, F. (2001). Bioremediation (Natural Attenuation and Bio stimulation) of Diesel-Oil-Contaminated Soil in an Alpine Glacier Skiing Area. Applied Environmental Microbiology, 67,3127-3133.

Migahed, F., Abdelrazak, A., & Fawzy, G. (2017). Batch and continuous removal of heavy metals from industrial effluents using microbial consortia. Int J Environ Sci Technol 14(6),1169–1180.

Mohammadian, E., Ahari, A. B., Arzanlou, M., Oustan, S., & Khazaei, S. H. (2017). Tolerance to heavy metals in filamentous fungi isolated from contaminated mining soils in the Zanjan Province, Iran. Chemosphere 185,290–296.

Mouhamadou, B., Faure, M., Sage, L., Marçais, J., Souard, F., & Geremia, R. A. (2013). Potential of autochthonous fungal strains isolated from contaminated soils for degradation of polychlorinated biphenyls, Fungal Biol. 117 (4),268–274.

Neu, T. R., & Lawrence, J. R. (2010). Extracellular polymeric substances in microbial biofilms. In Microbial glycobiology (pp. 733-758). Academic Press.

Niehaus, J.W., Lim, J.M., & Kim, H.S. (2018). Fungal mediated precipitation of Lead from contaminated water: a case study with Aspergillus niger. Chemosphere 211, 1–9.

Panda, S. S., Sahoo, K., Muduli, S. D., Sahoo, G., Ahemad, M. D. J., Nayak, B. B., & Dhal, N. K. (2014). Chromium-tolerant indigenous fungal strains from Industrial Effluents of Anugul District, Odisha, India. Biolife 2(2),634–640.

Perfumo, A., Banat, I. M., Marchant, R., & Vezzulli, L. (2007). Thermally enhanced approaches for bioremediation of hydrocarbon-contaminated soils. Chemosphere, 66(1),179-184.

Pundir, R., & Dastidar, M G. (2015). Feasibility of cheese whey as an energy source for the growth of Aspergillus sp. and for the removal of heavy metals in a batch reactor. Desalin Water Treat 57(33),15467-15474.

Pundir, R., Chary, G. H. V. C., & Dastdar, M. G. (2018). Application of the Taguchi method for optimising the process parameters for the removal of copper and nickel by growing Aspergillus sp. Water Resour. Ind.

Qayyum, S., Khan, I., Meng, K., Zang, X., Zhoa, Y., Gu, Q., & Peng, C. (2016). Bioaccumulation of heavy metals from aqueous solution using indigenous fungal isolates. Indian J Geo-Mar Sci 45(4),499–507.

Rockne, K., & Reddy, K. (2003). Bioremediation of contaminated sites. Proceedings international e-conference on modern trends in foundation engineering: geotechnical challenges and solutions, IIT Madras, India, Pp 22.

Rockne, K., &Reddy, K. (2003). Bioremediation of contaminated sites, University of Illinois at Chicago.

Sakthivel, M., Ayyasamy, P. M., & Hemalatha, N. (2016). Biotransformation of chromium (Cr-VI to Cr III) from tannery effluent using bacteria and fungi. Int J Adv Res Biol Sci 3(11),68–79.

Sathvika, T., Manasi, Rajesh, V., & Rajesh, N. (2015). Prospective application of Aspergillus species immobilised in sodium montmorillonite to remove toxic hexavalent chromium from wastewater. RSC Adv 5,107031–107044.

Scherr, K., Aichberger, H., & Braun, R. (2007). Influence of soil fractions on microbial degradation behaviour of mineral hydrocarbons. European Journal of Soil Biology, 43,341–350.

Smith, H. E., Tiedje, J. E., & Brown, M. T. (2018). Fungal phosphate solubilization and its role in heavy metal bioremediation. Fungal Ecol. 36,103–112.

Solanki, M. K., Kashyap, B. K., Solanki, A. C., Malviya, M. K., & Surapathrudu, K. (2019). Helpful linkages of Trichoderma sp. in the process of Mycoremediation and Mycorestoration, in: Plant Health Under Biotic Stress: Volume 2: Microbial Interactions, pp. 51–64.

Sugasini, A., & Rajagopal, K. (2015). Hexavalent chromium removal from aqueous solution using Trichoderma viride. Int J Pharm Bio Sci 6(1)(B),485–495.

Tahir, A., Lateef, Z., Abdel, Megeed, A., Sholkamy, E. N., & Mostafa, A. A. (2017). In vitro compatibility of fungi for the biosorption of zinc (II) and copper (II) from electroplating effluent. Curr Sci 112(4),839–844.

Tangahu, B. V., Abdullah, S. R. S., Basri, H., Idris, M., Anuar, N., & Mukhlisin, M. A. (2011). Review on heavy metals (As, Pb, and Hg) uptake by plants through phytoremediation. International Journal of Chemical Engineering, Volume, Article ID 939161, 31 pages.

Teng, Y., Du, X., Wang, T., Mi, C., Yu, H., & Zou, L. (2017). Isolation of a fungus, Pencicillium sp., with zinc tolerance and its mechanism of resistance. Arch Microbiol 200(1),159–169.

Ueno, A., Ito, Y., Yumoto, I., &Okuyama, H. (2007). Isolation and characterisation of bacteria from soil contaminated with diesel oil and the possible use of these in autochthonous bioaugmentation. World Journal of Microbiology and Biotechnology, 23,1739–1745.

Xu, W., Jian, H., Liu, Y., Zeng, G., Li, X., Gu, Y., &Tan, X. (2015). Removal of Chromium (VI) from Aqueous Solution Using Mycelial Pellets of Penicillium simplicissimum Impregnated with Powdered Biochar. Biorem J 19(4),259–268.

Yang, L., Li, T., Li, J., & Chen, J. (2019). Chelation of heavy metals by phosphate-solubilising Fungi for bioremediation, Ecotoxicol. Environ. Saf. 175, 1–8.

Yousaf, S. A., Khan, N. A., &Ahmed, Z. (2018). Precipitation of heavy metals by Aspergillus niger: a mechanistic study.J. Hazard. Mater. 349 (2018), 15–22.

7

Algal Bioremediation: Sustainable Strategies for Toxic Metal Removal from Wastewater

Prafull Kothari, Vikrant Pachouri* and Soniya Santwani

Institute of Engineering and Technology, Mohanlal Sukhadia University, Udaipur - 3130001, Rajasthan, India

**Corresponding Author Email: vikrantpachouri121@gmail.com*

Abstract

The concentration of metals has become a major factor in streams and rivers around the world because of the growing industrialization and developmental processes. This chapter focuses on the possible application of algae in bioremediation, which entails the use of algae to remove toxic metal ions from polluted water. The work goes deeper into the processes of bio-sorption, bio-accumulation, bio-assimilation, and bio-remediation and the comparative analysis of the representatives of various algal genera. This chapter introduces several advantages of the algal bioremediation including the efficiency of resources, diversification, and sustainability of the method, and points out the existing issues linked to the method like scaling up problems and proper management of biomass. Real-life examples range from business organizations to healthcare to showcase successful strategies; technological innovations are also described. Finally, this chapter provides concise information about the future research prospects and policy recommendations for the uptake and implementation of algal bioremediation in WWTPs and also other sustainable practices to improve the overall environmental quality around the world.

Keywords: *Bioremediation, Bio-accumulation, Bio-sorption, Bio-assimilation, Algae, Toxic metals, Sustainability*

1. Introduction

Metals pollution in water particularly in wastewater has become a center of concern due to social changes such as industrialization and urban development (Blaszczak-Boxe *et al.*, 2023). It is in this context that most of the conventional wastewater treatment fails to remove the heavy metals in question satisfactorily and that there is a pressing need for new ideas that would be sustainable. This chapter aims to discuss algal bioremediation as an environmentally suitable process for the decontamination of wastewater containing toxic metals.

1.1. Context and Importance of Wastewater Treatment

The pollutants in wastewater are numerous and cut across domestic, industrial, and agricultural uses and some of the heavy metals include lead, mercury, cadmium, and arsenic. These metals are toxic to aquatic organisms and have detrimental effects on aquatic ecosystems and humans even at very low concentrations (Jeong *et al.,* 2023). Processes such as chemical precipitation, ion exchange, and membrane filtration have high operating costs, complicated operation processes, and secondary pollution problems.

The need for sophisticated treatment techniques is made apparent by the ever-tightening environmental measures and policies focused on water quality preservation (Iqbal *et al.,* 2024). Effective and environmentally friendly recovery of heavy metals from wastewater is a milestone that should be achieved to address the effects of heavy metals pollution and have accessible clean water for different purposes.

1.2. Role of Algae in Bioremediation

Algae are photosynthetic aquatic organisms that in the last years, have been studied intensively for bioremediation technologies because of their ability to grow almost in any water type and the presence of inherent features that enable them to concentrate heavy metals (Dubey *et al.,* 2023). Edible algal bioremediation is environmentally friendly and has been known to be cheaper when compared to conventional methods.

Heavy metal removal is possible using three methods bio-sorption, bio-accumulations, and bio-assimilations all of which are achieved by algae. These processes include the chelation of metal ions to the external surface of algal cells, internalization of these ions into structure compartments, and reduction of the ions to forms that are less toxic to the cells. The Rate of metal removal also depends on the species of algae, growth conditions, concentrations of metals, and the level of competing ions.

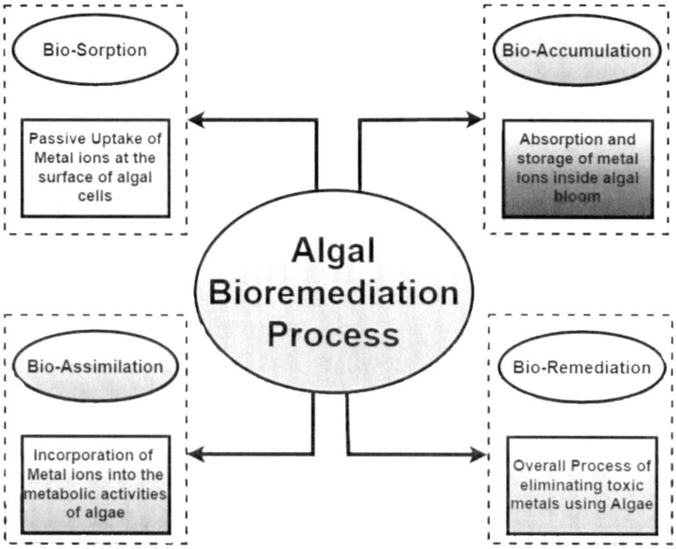

Fig. 1: Algal Bioremediation Process

Fig. 1 shows algal bioremediation process which include bio-sorption, bio-accumulation, bio-assimilation, and bio-remediation. The incorporation of algal bioremediation into wastewater treatment has several advantages. In particular, it increases the efficiency of extracting toxic metals, decreases the use of chemicals, and contributes to the general recovery processes' sustainability (Pachouri *et al.,*2024). Also, biomass formed during algal bioremediation can be used for bioenergy generation which increases the biosphere's economic value and supports the circular economy principle.

This chapter provides detailed information regarding the working principle of algal bioremediation, the prominence, and limitations of the bioprocess, as well as case studies and further research regarding the use of algae for wastewater management.

2. Mechanisms of Algal Bioremediation

The biological treatment of algae involves different procedures that can facilitate the elimination of toxic metals from water. Various activities that involve the use of living organisms in eradicating pollutants and toxic compounds are known as bio-sorption, bio-accumulation, bio-assimilation, and bio-remediation. Knowledge of these processes is important for the algae-based wastewater treatment and enhancement of metal removal efficiency.

2.1. Bio-sorption and Bio-accumulation

Bio-sorption

Bio-sorption is known as the process of passive uptake of metal ions at the surface of the algal cells. This happens through adsorption that involves physical and chemical changes in the metal ions with functional groups in the cell envelopes, mainly carboxyl, hydroxyl, and amino groups (Mal and Rangabhashiyam, 2021). These functional groups and the large surface area due to the small size of the algal cells make algae very suitable for metal ions adsorption from the aqueous medium.

Key factors influencing bio-sorption include

- **pH:** This is because the solution's pH affects the charges on functional groups at the surface of the adsorbent and the form of metal ions in the solution.
- **Metal concentration:** An increase in the concentration of the metal ions increases the adsorption capacity up to the point where saturation is achieved.
- **Contact time:** Sufficient contact time provides for the balance in the interactions between the metal ions and the algal biomass.

Bio-sorption is relatively fast and also is known to be a reversible process, therefore; it is appropriate for the first step of metal elimination from the wastewater (Razzak *et al.,* 2022).

Bio-accumulation

The bioaccumulation process includes the absorption of metal ions through the uptake mechanisms of different algal cells. Ions of metals once inside the cell can be stored in a vacuole, incorporated in intracellular proteins, or even in the structural components of the cell (Chugh *et al.,* 2022). This process is dependent on such factors as the physiological status of the algae and the surroundings.

Key aspects of bio-accumulation include

- **Active transport:** Metals complexes move in the cell through the cell membranes by carrier proteins channel.
- **Intracellular binding:** This prevents the metallic ions from getting in contact with other biomolecules, and some of the metals can be sequestered within vacuoles.
- **Metabolic transformation:** This implies that while there is some metal which after getting metabolized in the body changes to forms that are not as toxic to the body.

Thus, bio-accumulation is quite slower than bio-sorption, though the latter is quite effective and considered to be a more efficient technique for removing metals from the solution as these metals get trapped in the cells of the bio-mass (Breil *et al.,* 2022).

2.2. Bio-assimilation and Bio-remediation

Bio-assimilation

Bio-assimilation is a process whereby a metal ion is incorporated into the body of an alga in the process of the alga's metabolic activities. Some metals when present in low concentrations are very vital for algae to grow and to carry out their normal processes (Kumar *et al.,* 2024). Such a process minimizes the degree of solubility of toxic metals in the environment and therefore its bioavailability is decreased.

Key elements of bio-assimilation include

- **Nutrient uptake:** Zinc, copper, and iron are some of the metals that are crucial for enzymatic processes that happen in the algae.
- **Metabolicintegration:** Some metals are incorporated into the metabolic process which decreases the free concentration of the metals in the water.

Bio-accumulation plays a role in the removal of metals from the effluent through biotransformation where the metals formed in less toxic states are adsorbed on the bio-mass of the algae (Deb and Nag, 2022).

Bio-remediation

Bio-remediation is defined as a general process of employing algae in the elimination or even fixing of toxic metals from wastewater (Srivastava and Bhargava, 2022). This entails bio-sorption, bio-accumulation, and also bio-assimilation all at the same time. Bio-remediation efficiency can be affected by algae kinds, types of wastewater, and the conditions that are used in the treatment system.

Key factors in bio-remediation include

- **Algal species:** This entry of metals can either be active or passive and organisms have diverse abilities and susceptibilities to such metals.
- **Environmental conditions:** Some of the factors are light intensity, temperature, and concentration of the nutrients as they determine the pace at which the algae grow and its capability to remove metal.
- **System design:** Algae, bioreactors, or treatment systems determine the contact point between algae and the wastewater from where the process of metal removal occurs (Bhatt *et al.,* 2022).

2.3. Algal Strains and Their Specification

It was also found that different strains of algae also have differences in the efficiency of the absorption of the metal depending on its morphological and physiological features (Chan *et al.,* 2022). Commonly used algal strains in bioremediation include:

- ***Chlorella:*** It is a well-known organism due to its very high tolerance to heavy metals and high bio-sorption capacities.
- ***Scenedesmus:*** High efficiency in bioaccumulation and bio-assimilation and applicable for wastewater with more than one metal ion (Goswami *et al.,* 2022).
- ***Spirogyra*:** But, can exhibit good bio-sorption properties and can survive different environmental conditions.
- ***Dunaliella*:** Famous for bio-accumulating heavy metals and whose survival it has been established is equally efficient under saline conditions.

Table1: Comparative Analysis of Algal Strains

Algal Strains	***Chlorella***	***Scenedesmus***	***Spirogyra***	***Dunaliella***
Metal Uptake Efficiency	High	Moderate	Low	High
Growth Condition	Versatile	Specific	General	Saline
Tolerance to Pollutant	High	Moderate	High	High
References	Ciurli *et al.,* 2021	Waqar *et al.,*2023	Almeida *et al.,* 2023	Chen *et al.,* 2024

In the Table 1, one can find a brief comparison of various algal strains applied for bioremediation with the references to their efficiency and growth conditions. The choice of the given algal strains depends on the characteristics of the wastewater being treated, the type and concentration of metals, as well as the conditions within the wetland system (Abdelfattah *et al.,*2023). Therefore, algal bioremediation which includes bio-sorption, bio-accumulation, bio-assimilation, and bio-remediation is a complete strategy in the removal of toxic metals in wastewater. Due to the potential to vary the properties of different algal strains, this cleaner method improves the intensiveness and effectiveness of wastewater treatment operations.

3. Advantages of Algal Bioremediation

Algal bioremediation has several advantages over conventional wastewater treatment techniques making the method sustainable for the removal of toxic metals (Priya *et al.,* 2022). The following area deals with the advantages such as sustainability, the possibility to use the method cost-effectively, and its applicability to various types of wastewater.

3.1. Sustainability and Environment Benefits

a) **Reduced Chemical Use:** The use of algal bioremediation reduces the inclusion of chemicals that are usually used in conventional treatment procedures (Almomani and Omar, 2022). It lowers the chance of secondary pollution and the creation of undesirable chemicals enhancing an environmentally sound disposition.

b) **Resource Efficiency:** Algae can adapt to many conditions of the environment as it can capture sunlight and the nutrients from the water. This efficiency in the usage of the available resources makes algal bioremediation to be environmentally friendly and sustainable (Roy *et al.,* 2022). Algae can also derive carbon dioxide from industrial emissions, thus helping in carbon capture and thus, reducing greenhouse gases.

c) **Biodiversity Enhancement:** When algae for bioremediation is grown, it could be beneficial to the local wildlife since it creates a home for microorganisms, thereby achieving an ecological balance. The optimization of this biodiversity is beneficial to the health of the water bodies and will help boost the ecosystem.

d) **Renewable Biomass Production:** Algal bioremediation therefore yields biomass as a secondary product which can thereafter be used in more related processes such as bio-energy production, the production of bio-fertilizers, and animal feed (Rasool *et al.,* 2023). This sustainable biomass enhances the value of the bioremediation process and the best practice of the circular economy.

Compared to the traditional processes of wastewater treatment focused on toxic metals elimination, bioremediation by means of algae is advantageous in several ways, which makes it a more suitable approach to the problem. Some of the above advantages and some of the limitations are shown in the tabular form given below:

Table 2. Advantages and Limitation of Algal Bioremediation

References	**Advantages**	**Limitations**
Mishra *et al.,*2023	Sustainability	Scalability Issues
Kumar and Shukla, 2023	Cost – Effectiveness	Contamination Risks
Upadhyay *et al.,*2024	Resource Efficiency	Harvesting and Processing
Wicker *et al.,*2023	Biodiversity Enhancement	Metal Recovery Challenges

3.2. Economic Feasibility

a) **Cost-Effectiveness:** Algal bioremediation is more efficient than conventional practices primarily because operational and maintenance expenses are comparatively cheaper. Algae, which are cultivated at a rather low cost, show a high ability to take metals from water, and thus the general expenses of cleaning the wastewater are considerably decreased.

b) **Resource Recovery:** Algal biomass generated during algal bioremediation can be used for valuable products such as fuel, biopolymers, and chemicals (Kumar *et al.,* 2022). This recovery of resources affords extra revenues to contribute to the corporation's income as it leads to the acclamation of the bioremediation expense.

c) **Scalability:** Incorporating algal bioremediation into wastewater treatment plants can be done depending on the size of specific wastewater treatment plants, from small-scale decentralized treatment plants to mega large-scale treatment plants (Tazkiaturrizki *et al.,* 2021). In this respect, the system's implementation can easily be scaled within industrial and several municipal environments, ensuring availability to a large number of users.

d) **Employment Opportunities:** Therefore, jobs are produced in the formation and operations of algal bioremediation centers in areas of biotechnology, engineering, and the environmental and waste sectors. These employment opportunities are beneficial to the nation's economy as well as foster the development of communities.

3.3. Adaptability to Different Wastewater Sources

a) **Versatility:** Algae are versatile organisms and are capable of growing in almost all forms of wastewater industrial effluents, municipal sewage, and agricultural drainages (Zainith *et al.,* 2021). Due to this aspect, Algal Bioremediation can be applied to various sources of wastewater and it performs a dual job of removing the contaminants from water.

b) **Tolerance to Contaminants:** Some algal species can even take on a high affinity to toxic metals and other pollutants, thus, they are efficient in environments that are polluted. This enables bioremediation systems to work effectively and withstand different conditions with high reliability and resilience.

c) **Integration with Existing Systems:** There is always a possibility to combine algal bioremediation with other processes of wastewater treatment, for instance, the primary and the secondary treatment steps.

This integration offers the best of the two worlds, by adding on to the conventional methods while adding on to the algal-based systems.

d) **Flexible Operation:** It is though possible to design algal bioremediation systems for batch and continuous use which makes them very effective in controlling the flow of wastewater and their treatment. This flexibility enables the systems to be customized according to the nature of the treatment and the restrictions of operations.

The benefits depicted by algal bioremediation such as; sustainability, economics, and versatility indicate that acts as a good solution to the issues of metal toxicity in wastewater (Krishnamoorthy and Manickam, 2021). Given the features of algae, this method is considered one of the most effective for the improvement of wastewater treatment efficiency, protection of the environment, and intensification of the development of new and effective technologies.

4. Challenges and Limitations

Despite the enormous benefits of algal bioremediation, there are some constraints and issues that need to be met and solved to obtain the maximum result from algal bioremediation processes. The following section presents the limitations and problems faced during operation, Metal Recovery and Biomass disposal, and Environmental impacts of Algal bioremediation.

4.1. Operational Challenges

- **Growth Conditions:** The prevention of any changes in the growth pattern of algae is important if bioremediation has to produce the desired results. Environmental conditions like light intensity, temperature, pH of the medium, and the availability of nutrients at the place shall be regulated effectively (Taghavijeloudar *et al.,* 2021). It also implies that the efficiency of the algal uptake of metals and growth is affected by variation in the environmental conditions a factor that poses some challenges in the treatment process.
- **Scalability:** The transition of algal bioremediation systems, from Small scale or semi-industrial to large-scale has some difficulties. ESG always must be larger in scale to handle a massive quantity of specimens and therefore becomes critically dependent on infrastructure elaborated conditions and constant monitoring of processes.
- **Contamination and Competition:** In an open pond, the growth of algae is vulnerable to competitors like bacteria and fungi besides other algae species. Such contaminants are capable of competing with the wanted species of algae, therefore, diminishing the efficiency of the

metal removal (Lin *et al.*, 2023). Another significant operation problem is the management of contamination and guaranteeing the prevalence of the bioremediation-specific algae species.

- **Harvesting and Processing:** Proper techniques in the collection and transformation of the algal mass are inevitable in the outcomes of bioremediation (Yan *et al.*, 2022). Harvesting has to be cheap, this means that harvesting technologies have to be easily scalable so that they can be used in a large number of applications and this should be done while consuming as little energy as possible. Moreover, the method of processing biomass for the extraction of metals and other usages involves the use of complicated machinery and apparatus since the extraction is not as easy as that of the metals found in ores.

4.2. Metal Recovery and Biomass Disposal

- **Metal Recovery:** The attempts to recover metals from algal biomass are essential to integrate bioremediation and make it cyclic. However, in this process, the retention stunning might be a problem as many metals tie strongly to the cellular components. The techniques of metal recovery including chemical extraction, pyrolysis, and electrochemical deposition are the aspects of current interest.
- **Biomass Disposal:** The most significant problem with algal biomass is the method of its disposal, as it can be considered hazardous because of its contamination (Znad *et al.*, 2022). It is pointed out that if metals are not properly reclaimed, Disposal of biomass aggravated secondary pollution. Methods like biochemical recycling into biochar or utilization in constructed wetlands for disposal need to be designed and put into use concerning this problem.

4.3. Potential Environmental Risks

- **Release of Secondary Pollutants:** Flaring is also one of the drawbacks of bioremediation wherein some secondary pollutants like organic compounds or products of the metabolism of algae are released into the environment. The following pollutants could be destructive to aquatic life as well as to human beings. One risk associated with the bioremediation process is the chances of the pollutant re-emerging into the environment once the microorganism is dead.
- **Impact on Local Ecosystems:** The occurrence of massive amounts of algae in open waters could be potentially damaging to the habitats of the affected bodies of water. Algal blooms are characterized by excessive growth of algae in water; they can reduce the oxygen levels in the water

and affect all aquatic life (Kazmi *et al.*, 2022). Also, some species of algae can release toxic compounds into water that affect living organisms and human beings.

- **Regulatory and Compliance Issues:** Algal bioremediation regulation is still in its developing stage and the possibility of meeting the standard and regulatory procedures set is still much of a question mark. The systems of algal bioremediation must conform with all the legal and regulatory measures.

However, the mentioned challenges and limitations are essential for algal bioremediation to be more effective and implemented on a large scale. Due to these barriers, further study and investment as well as policy backing are needed to remove all these barriers to harness the full potential of this environmentally friendly wastewater treatment method.

5. Case Studies and Applications

The problem starts by looking at the real-world practices of algal bioremediation to determine the feasibility and efficacy of the procedure. It features good practices, strategies and methods, technologies, and cultivation and utilization of algae for the removal of metal toxicity from wastewater at regional and global levels.

5.1. Successful Implementation

Fig. 2: Case Studies of Algal Bioremediation

Fig. 2 demonstrates the geographical spread of the case studies to illustrate the worldwide incorporation of algal bioremediation in municipal, industrial, and agricultural activities.

Case Study 1: Algal Bioremediation in Industrial Wastewater Treatment

In the textile industry of India, algal bioremediation was used to treat wastewater containing many heavy metals including chromium and cadmium. The industry arranged an algal pond system where *Chlorella* and *Scenedesmus* were brought up (Arutselvan *et al.*,2022). In six months, the system proved that metal concentrations had been reduced and the removal efficiency was higher than 90%. This project proved that algae for industrial wastewater treatment was cheap and that it could be useful in other fields.

Case Study 2: Municipal Wastewater Treatment in the United States

An example is when a wastewater treatment plant in a specific municipality in California used algal bioremediation in the secondary treatment process. The plant has applied the photo bio-media/photo-bioreactor system with Spirulina and *Chlorella* species for the reduction of any remaining heavy metals like lead and zinc (Adams *et al.*, 2022). The specific outcomes were realized in the modification of the water quality, where the metal concentrations were brought to the revised regulatory thresholds. Further, the produced algal biomass was utilized and transformed into biofuel, which further enhances the energy sources plant of the project.

Case Study 3: Agricultural Runoff Management in Europe

There is an agricultural runoff treatment in Spain whereby metal of fertilizers and pesticide contaminant was managed by a process called algal bioremediation. The application of the open Pond system with *Spirogyra* and *Dunaliella* was used for the treatment of runoff waters before discharge to natural water basins (Saber *et al.*, 2022). In ARA, the project indicated an impressive decrease in metal concentrations and nutrients as proof of how the algal bioremediation project applies to agricultural scenarios and is expressive of the protection of aquatic life forms.

5.2. Innovative Techniques and Technologies

a) **Photo-bioreactors:** Photo-bioreactor systems are phototrophic systems that assist in enhancing better conditions for microalgae which in turn aids in enhancing efficiency in the removal of metals (Dange *et al.*, 2022). The systems provide accommodation that allows good amounts of light to penetrate them, and nutrients and gases are also appropriately balanced. Innovations in methods of photo-bioreactor include but are not limited to the use of light-emitting diode (LED) and automation in monitoring devices making it easy to upscale algal bioremediation systems.

b) **Genetically Engineered Algae:** Recent studies have emerged that indicated that it is feasible to conduct an enhancement of algae regarding the metals absorption efficiency. There are some speciated and successful attempts to alter the strains of *Chlorella* and *Scenedesmus* that can increase the bio-sorption and bio-accumulation of the microalgae (Mohanty and Patra, 2023). These genetically modified algae can be used in bioremediation and the genetic modifications can lead to enhanced efficiency of the bioremediation processes.

c) **Hybrid Systems:** Integrating algal bioremediation with the constructed wetlands or bio-filters and many other treatments has proved to be highly effective. These systems take advantage of the effectiveness of various methods for total pollutant elimination of metals, organic compounds, and nutrients. The mentioned hybrid systems can give specific approaches to individual complicated problems in wastewater treatment processes.

5.3. Regional and Global Initiatives

a) **European Union's Horizon 2020 Program:** Horizon 2020 has provided support to several projects dealing with the assessment and advancement of algal bioremediation solutions (Samoraj *et al.,* 2024). Another major collaborative project included the ALGADISK project which sought to produce efficient and sustainable algal-based technology for wastewater treatment and secondary resource reuse. From the project, it was shown that metal extraction by using algae is possible and showed the economical prospect of such systems for metal removal as well as the environmental advantages.

b) **Asia-Pacific Bioremediation Network:** Various groups, including some of the leading countries in the Asia-Pacific region such as the People's Republic of China, India, and Japan among others have bolstered the use of bioremediation technologies by forming a network. Today, it helps disseminate the latest findings in the field, collaborate on common research, and advocate for the use of algal bioremediation worldwide. In general, regional initiatives have been able to successfully carry out pilot projects in implementing algae, illustrating the potential of algae to solve regional water pollution problems.

c) **United Nations Environment Programme (UNEP):** The UNEP has acknowledged algal bioremediation as one of the bounce technologies for sustainable wastewater management. The capacity of developing nations and other regions has been built through specific programs, as well as UNEP has offered technical support (Udom *et al.,* 2023). As part of the UNEP, global endeavors have created awareness of algal bioremediation and other similar methodologies and their inclusion into the national water management schemes.

The examples provided in this section including the case studies and applications reflect the practical possibilities of algal bioremediation and the advantages of this process. Efficacious programs and practices, new approaches to the use of algae, and the successes of regional and global algae programs prove that algae may become a sustainable solution to metal toxicity in the treatment of wastewater. The above cases form the basis for more studies, innovation, and implementation of algal bioremediation systems.

6. Future Directions and Recommendations

Thus, for the algal bioremediation of toxic metal from the wastewater further studies and functioning policies have to overcome its present drawbacks and discover new horizons. In this section, major research directions are presented, prerequisites for public policies and legislation are identified, and possibilities of combining algal bio-remedial processes with other forms of sustainability.

6.1. Research and Development Need

1. **Advanced Genetic Engineering:** Therefore, future research should be aimed at improving the existing processes of genetic engineering to create new algal strains that possess a higher ability to uptake and tolerate metals. Therefore, according to the given genetic influences on the metal-binding and stress response genes, one can produce algae that ramp up on the bio-sorption, the bio-accumulation, and the bio-assimilation. To realize these gains consistently, the work between the biotechnologists and environmental scientists will bear a lot of fruits.
2. **Optimization of Growth Conditions:** More investigations will be required concerning the growth conditions of algae in large-scale bioremediation processes. Further research should be conducted to consider the impact of the environmental conditions including light intensity, temperature, pH, and nutrient status on the growth of algae and the efficiency of metal uptake. Also, an implementation of automatic monitoring and control systems assists in maintaining appropriate processing conditions all the time making the algal bioremediation processes more reliable.
3. **Innovative Bioreactor Designs:** The most important aim is to design new bioreactor conditions that will improve the productivity of algae and the efficiency of metal absorption. As a result of that it is suggested that future studies search for new conFig.urations of more than one-stage

reactors, rotating photo-bioreactors, combined systems, which involve algae and other technologies. It is on this premise that these designs should be made in a bid to enhance scalability, cost, and operation.

4. **Metal Recovery and Biomass Utilization:** Better ways to extract metals from algal biomass and more importantly to use biomass in other useful applications have to be invented. There is much need to develop purification techniques on different extraction methods that include chemical leaching and electrochemical processing as well as bioleaching for recovery of metals in a more efficient, cheaper, and sustainable manner with very high purity and yield. Also, specific research on the possible applications of algal biomass, for example, biochar, biofertilizers, and bioplastics will improve the economic feasibility of algal bioremediation.

To apply algal bioremediation of toxic metals from wastewater, more research and effective polices should be carried out. The important areas for the promotion of R&D are presented in Table 3 below.

Table 3. Research and Development Needs

References	Key Areas	Description
Abhirami *et al.*, 2024	Genetic Engineering	The developed algal strains need to be higher in uptake to warrant the creation of new algal strains
Upadhyay *et al.*, 2023	Growth Condition Optimization	Studying environmental impacts
Narciso-Ortiz *et al.*, 2023	Bioreactor Designs	This of course involves the creation of effective and cost efficient designs
Satya *et al.*, 2023	Metal Recovery	Innovating sustainable extraction methods

6.2. Policy and Regulatory Support

1. **Development of Standards and Guidelines:** It is therefore very essential that standards and guidelines are set and followed when implementing algal bioremediation systems. Policy-making bodies should engage researchers and players in the biomass yields and retail business to come up with some rules of thumb related to algal strain, systems designs, monitoring, and lastly the way biomass is addressed. These standards should be reviewed and revised over time depending on the new findings in science and new experiences in the field.
2. **Incentives for Adoption:** The governments and the regulatory authorities should offer incentives for the use of algal bioremediation technologies. Price finesse includes grants, tax credits, and subsidies which can assist

the various organizations to offset the cost of establishing bioremediation systems. Thus, measures and incentives for encouraging companies and municipalities to implement sustainable solutions for wastewater treatment can be effective in advancing the domain and provoking the development of better practices.

3. **Integration into National Water Management Policies:** Algal bioremediation needs to be incorporated into the national and regional guidelines of water resource management as one of the possible treatments for wastewater. Thus, the environmental, economic, and social outcomes of algal bio-remedial must be taken into consideration by policymakers in the course of the formation of water quality standards and treatment regulations. Efforts in the future should involve the cooperation of the government, research organizations, and industrial players to integrate algal bioremediation into the present management strategies of water.

6.3. Integration with Other Sustainable Practices

1. **Combining with Renewable Energy Production:** Technological coupling of bioenergy from algal bioremediation with the generation of bioenergy like biofuel production is possible. Algal biomass that is formed during bioremediation can be further used for biofuel generation restoring the case of wastewater management and generation of energy in the same step. The areas of these studies should include the possibility of operating bioremediation and biofuel production plants simultaneously to save more resources and money.
2. **Circular Economy Approaches:** Using algae for wastewater treatment bioremediation is also a good strategy that can be adopted when implementing circular economy systems (Alazaiza *et al.*, 2022). It integrates waste streams as substrates for algal growth and recycles valuable products from algal biomass hence minimizing the effects on the environment. For future work, stakeholders should aim to design multifaceted circular economy strategies in which algal bioremediation might be an integrated element.
3. **Collaboration with Agriculture:** There is a need for synergy between the wastewater treatment plants and the sectors that partake in agriculture in the utilization of algal bioremediation. From algal biomass feedstocks, bio-fertilizer generation is possible which can replace chemical fertilizers and enhance the quality of soil. Also, algae offer the potential to mitigate agricultural discharge effluents, cutting down nutrient input and metals in water basins. The best examples of such integrated approaches are

partner studies and pilot interventions which can prove the efficiency of such types of activities and facilitate their extension.

Discussions in this section provide a snapshot of directions for the future and recommendations that need to be met that are imperative to take algal bioremediation to the next level toward sustainable wastewater treatment solutions. Hence, through research and development in algal bioremediation, policy and regulatory support, and synergy with other sustainable technologies, stakeholders can work around existing barriers to achieving large-scale application of algal bioremediation technologies.

7. Conclusion

Algal bioremediation has been deemed as a sustainable source of toxic metals elimination from wastewater. Thus, this approach is environmentally friendly and economically feasible due to the natural properties of algae like bio-sorption, bio-accumulation, and bio-assimilation. In situations ranging from crisis to optimized processes, positive examples and modern developments illustrate that algal bioremediation is efficient and feasible.

The method, however, possesses several drawbacks, which include achieving the best growth conditions, effective metal recovery, and management of environmental impacts. Therefore, future work should be devoted to the enhancement of the genetic engineering of microorganisms, the creation of new bioreactor structures, and the improvement of methods for metals recovery. It was noted that policy support in the form of standards and incentives is very important for enabling large-scale implementation of the solutions. Application of algal bioremediation with other related concepts like renewable energy production and circular economy solutions can strengthen the results. Thus, with the further development of knowledge, carefully planned investments, and heuristic policies, algal bioremediation has the potential to become a substantial solution to metal toxicity in WWTPs. Their use can help enhance water quality, protect the environment, and increase the knowledge in the field of environmental wastewater treatment hence could go a long way in supporting the quest for environmental sustainability in the world.

References

Abdelfattah, A., Ali, S. S., Ramadan, H., El-Aswar, E. I., Eltawab, R., Ho, S. H., ... & Sun, J. (2023). Microalgae-based wastewater treatment: Mechanisms, challenges, recent advances, and future prospects. Environmental science and ecotechnology, 13, 100205.

Abhirami, S., Kumaran, A., Sahadevan, S., Manjunatha, K. G., Akshay, S. D., Aithal, P., ... & Akhila, D. S. (2024). Role of algae and its enzymes in the bioremediation of pollutants. In Enzyme Biotechnology for Environmental Sustainability (pp. 439-463). Academic Press.

Adams, W., Blust, R., Dwyer, R., Mount, D., Nordheim, E., Rodriguez, P. H., & Spry, D. (2020). Bioavailability assessment of metals in freshwater environments: A historical review. Environmental Toxicology and Chemistry, 39(1), 48-59.

Alazaiza, M. Y., Albahnasawi, A., Ahmad, Z., Bashir, M. J., Al-Wahaibi, T., Abujazar, M. S. S., ... & Nassani, D. E. (2022). Potential use of algae for the bioremediation of different types of wastewater and contaminants: Production of bioproducts and biofuel for green circular economy. Journal of Environmental Management, 324, 116415.

Almeida, Â., Cotas, J., Pereira, L., & Carvalho, P. (2023). Potential Role of Spirogyra sp. and Chlorella sp. in bioremediation of mine drainage: a review. Phycology, 3(1), 186-201.

Almomani, F., & Omar, A. (2022). Algal bioremediation versus conventional wastewater treatment. In Petroleum Industry Wastewater (pp. 71-86). Elsevier.

Arutselvan, C., kumar Seenivasan, H., Oscar, F. L., Ramya, G., Chi, N. T. L., Pugazhendhi, A., & Thajuddin, N. (2022). Review on wastewater treatment by microalgae in different cultivation systems and its importance in biodiesel production. Fuel, 324, 124623.

Bhatt, P., Bhandari, G., Turco, R. F., Aminikhoei, Z., Bhatt, K., & Simsek, H. (2022). Algae in wastewater treatment, mechanism, and application of biomass for production of value-added product. Environmental Pollution, 309, 119688.

Blaszczak-Boxe, C. S., Karle, N. N., Wang, S., Yu, M., Golosov, N., Riyad, M., ... & Sen, P. (2023). Environmental Assessment and Monitoring of Heavy Metals in New York City Potable Water Systems: Case Study at Medgar Evers College, Correlation Analysis, and Public Health Impacts. Water, 15(24), 4233.

Breil, P., Pons, M. N., Armani, G., Amer, R., Pienaar, H., Oberholster, P., & Namour, P. (2022). Natural-Based Solutions for Bioremediation in Water Environment. Sustainable solutions for environmental pollution: Air, water and soil reclamation, 1-93.

Chan, S. S., Low, S. S., Chew, K. W., Ling, T. C., Rinklebe, J., Juan, J. C., ... & Show, P. L. (2022). Prospects and environmental sustainability of phyconanotechnology: A review on algae-mediated metal nanoparticles synthesis and mechanism. Environmental research, 212, 113140.

Chen, H., Yu, S., Yu, Z., Ma, M., Liu, M., & Pei, H. (2024). Phycoremediation Potential of Salt-Tolerant Microalgal Species: Motion, Metabolic Characteristics, and Their Application for Saline–Alkali Soil Improvement in Eco-Farms. Microorganisms, 12(4), 676.

Chugh, M., Kumar, L., Bhardwaj, D., & Bharadvaja, N. (2022). Bioaccumulation and detoxification of heavy metals: an insight into the mechanism. In Development in wastewater treatment research and processes (pp. 243-264). Elsevier.

Ciurli, A., Di Baccio, D., Scartazza, A., Grifoni, M., Pezzarossa, B., Chiellini, C., ... & Pardossi, A. (2021). Influence of zinc and manganese enrichments on growth, biosorption and photosynthetic efficiency of Chlorella sp. Environmental Science and Pollution Research, 28, 8539-8555.

Dange, P., Gawas, S., Pandit, S., Mekuto, L., Gupta, P. K., Shanmugam, P., ... & Banerjee, S. (2022). Trends in photobioreactor technology for microalgal biomass production along with wastewater treatment: Bottlenecks and breakthroughs. In An Integration of Phycoremediation Processes in Wastewater Treatment (pp. 135-154). Elsevier.

Deb, S., & Nag, S. (2022). Use of microalgae for the removal of emerging contaminants from wastewater. In Biodegradation and Detoxification of Micropollutants in Industrial Wastewater (pp. 193-210). Elsevier.

Dubey, S., Chen, C. W., Haldar, D., Tambat, V. S., Kumar, P., Tiwari, A., ... & Patel, A. K. (2023). Advancement in algal bioremediation for organic, inorganic, and emerging pollutants. Environmental Pollution, 317, 120840.

Goswami, R. K., Agrawal, K., Mehariya, S., & Verma, P. (2022). Current perspective on wastewater treatment using photobioreactor for Tetraselmis sp.: an emerging and foreseeable sustainable approach. Environmental Science and Pollution Research, 29(41), 61905-61937.

Iqbal, A., Cevik, E., Mustafa, A., Qahtan, T. F., Zeeshan, M., & Bozkurt, A. (2024). Emerging developments in polymeric nanocomposite membrane-based filtration for water purification: A concise overview of toxic metal removal. Chemical Engineering Journal, 148760.

Jeong, H., Byeon, E., Kim, D. H., Maszczyk, P., & Lee, J. S. (2023). Heavy metals and metalloid in aquatic invertebrates: a review of single/mixed forms, combination with other pollutants, and environmental factors. Marine Pollution Bulletin, 191, 114959.

Kazmi, S. S. U. H., Yapa, N., Karunarathna, S. C., & Suwannarach, N. (2022). Perceived intensification in harmful algal blooms is a wave of cumulative threat to the aquatic ecosystems. Biology, 11(6), 852.

Krishnamoorthy, S., & Manickam, P. (2021). Phycoremediation of industrial wastewater: challenges and prospects. Bioremediation for Environmental Sustainability, 99-123.

Kumar, A. N., Yoon, J. J., Kumar, G., & Kim, S. H. (2021). Biotechnological valorization of algal biomass: an overview. Systems Microbiology and Biomanufacturing, 1, 131-141.

Kumar, N., & Shukla, P. (2023). Microalgal-based bioremediation of emerging contaminants: Mechanisms and challenges. Environmental Pollution, 122591.

Kumar, R., Kundu, D., Kormoker, T., Joshi, S., Rose, P. K., Kumar, S., ... & Lamba, J. (2024). Phycoremediation of potentially toxic elements for agricultural and industrial wastewater treatment: Recent advances, challenges, and future prospects. Desalination and Water Treatment, 100505.

Lin, M., Li, F., Li, X., Rong, X., & Oh, K. (2023). Biochar-clay, biochar-microorganism and biochar-enzyme composites for environmental remediation: a review. Environmental Chemistry Letters, 21(3), 1837-1862.

Mal, J., & Rangabhashiyam, S. (2021). Bioremediation and phytoremediation as the environmentally sustainable approach for the elimination of toxic heavy metals. In Sustainable Technologies for Water and Wastewater Treatment (pp. 227-258). CRC Press.

Mishra, V., Mudgal, N., Rawat, D., Poria, P., Mukherjee, P., Sharma, U., ... & Sharma, R. S. (2023). Integrating microalgae into textile wastewater treatment processes: Advancements and opportunities. Journal of Water Process Engineering, 55, 104128.

Mohanty, M., & Patra, H. K. (2023). Phytoremediation of Chromium from Soil and Water. In Chromium in Plants and Environment (pp. 253-291). Cham: Springer Nature Switzerland.

Narciso-Ortiz, L., Tec-Caamal, E. N., Aguirre-García, G. J., & Lizardi-Jiménez, M. A. (2023). Bioreactors for bioremediation of polluted water. In Current Status of Marine Water Microbiology (pp. 345-364). Singapore: Springer Nature Singapore.

Pachouri, V., Chandramauli, A., Singh, R., Gehlot, A., Priyadarshi, N., & Twala, B. (2024). Removal of contaminants by chlorella species: an effort towards sustainable remediation. Discover Sustainability, 5(1), 19.

Priya, A. K., Jalil, A. A., Vadivel, S., Dutta, K., Rajendran, S., Fujii, M., & Soto-Moscoso, M. (2022). Heavy metal remediation from wastewater using microalgae: Recent advances and future trends. Chemosphere, 305, 135375.

Rasool, S., Rasool, T., & Gani, K. M. (2023). Unlocking the potential of wetland biomass: Treatment approaches and sustainable resource management for enhanced utilization. Bioresource Technology Reports, 101553.

Razzak, S. A., Faruque, M. O., Alsheikh, Z., Alsheikhmohamad, L., Alkuroud, D., Alfayez, A., ... & Hossain, M. M. (2022). A comprehensive review on conventional and biological-driven heavy metals removal from industrial wastewater. Environmental Advances, 7, 100168.

Roy, A., Gogoi, N., Yasmin, F., & Farooq, M. (2022). The use of algae for environmental sustainability: trends and future prospects. Environmental Science and Pollution Research, 29(27), 40373-40383.

Saber, A. A., El-Refaey, A. A., Saber, H., Singh, P., van Vuuren, S. J., & Cantonati, M. (2022). Cyanoprokaryotes and algae: classification and habitats. In Handbook of Algal Biofuels (pp. 1-38). Elsevier.

Samoraj, M., Çalış, D., Trzaska, K., Mironiuk, M., & Chojnacka, K. (2024). Advancements in algal biorefineries for sustainable agriculture: Biofuels, high-value products, and environmental solutions. Biocatalysis and Agricultural Biotechnology, 58, 103224.

Satya, A. D. M., Cheah, W. Y., Yazdi, S. K., Cheng, Y. S., Khoo, K. S., Vo, D. V. N., ... & Show, P. L. (2023). Progress on microalgae cultivation in wastewater for bioremediation and circular bioeconomy. Environmental Research, 218, 114948.

Srivastava, S., & Bhargava, A. (2022). Heavy Metal Remediation: The Microbial Approach. In Microbes in Agri-Forestry Biotechnology (pp. 197-220). CRC Press.

Taghavijeloudar, M., Yaqoubnejad, P., Amini-Rad, H., & Park, J. (2021). Optimization of cultivation condition of newly isolated strain Chlorella sorokiniana pa. 91 for CO 2 bio-fixation and nutrients removal from wastewater: Impact of temperature and light intensity. Clean Technologies and Environmental Policy, 1-13.

Tazkiaturrizki, Rinanti, A., Fachrul, M. F., Hendrawan, D. I., Aphirta, S., Marendra, S. M. P., & Sipahutar, N. O. L. (2024). Implementation of Algal Approach in Techno-socio-economical Aspect of Wastewater Treatment. In Algae as a Natural Solution for Challenges in Water-Food-Energy Nexus: Toward Carbon Neutrality (pp. 199-260). Singapore: Springer Nature Singapore.

Udom, G. J., Frazzoli, C., Ekhator, O. C., Onyena, A. P., Bocca, B., & Orisakwe, O. E. (2023). Pervasiveness, bioaccumulation and subduing environmental health challenges posed by polycyclic aromatic hydrocarbons (PAHs): A systematic review to settle a one health strategy in Niger Delta, Nigeria. Environmental Research, 226, 115620.

Upadhyay, M., Hansa, A., Devi, A., Lajayer, B. A., Price, G. W., & Sharma, R. (2024). Status and future scope of algal biomass-based remediation for various environmental contaminants. Journal of Water Process Engineering, 65, 105809.

Waqar, R., Kaleem, M., Iqbal, J., Minhas, L. A., Haris, M., Chalgham, W., ... & Mumtaz, A. S. (2023). Kinetic and Equilibrium Studies on the Adsorption of Lead and Cadmium from Aqueous Solution Using Scenedesmus sp. Sustainability, 15(7), 6024.

Wicker, R. J., Kwon, E., Khan, E., Kumar, V., & Bhatnagar, A. (2023). The potential of mixed-species biofilms to address remaining challenges for economically-feasible microalgal biorefineries: a review. Chemical Engineering Journal, 451, 138481.

Yan, C., Qu, Z., Wang, J., Cao, L., & Han, Q. (2022). Microalgal bioremediation of heavy metal pollution in water: Recent advances, challenges, and prospects. Chemosphere, 286, 131870.

Zainith, S., Saxena, G., Kishor, R., & Bharagava, R. N. (2021). Application of microalgae in industrial effluent treatment, contaminants removal, and biodiesel production: Opportunities, challenges, and future prospects. Bioremediation for environmental sustainability, 481-517.

Znad, H., Awual, M. R., & Martini, S. (2022). The utilization of algae and seaweed biomass for bioremediation of heavy metal-contaminated wastewater. Molecules, 27(4), 1275.

8

Role of Microorganism in Heavy Metal Bioremediation

Khushbu Singhal and Kavita Paliwal

Department of Botany, U.S. Ostwal P.G. College, Mangalwad, Chittorgarh, Rajasthan

Corresponding author's mail ID: khushbusinghal021@gmail.com

Abstract

The world is currently dealing with the issue of many forms of environmental pollution. Unsafe agricultural methods, industrialization, and urbanization all contribute to the environment's ongoing heavy metal buildup, which contaminates soil and water. Microorganisms have long been known to play a unique and distinct role in the detoxification of contaminated environments; in more recent years, this process has been referred to as bioremediation or bioreclamation. Microorganisms survive in all places in the biosphere. They can be found in a wide variety of environmental settings. The process of bioremediation involves transforming waste materials into a form that may be used and recycled by other living beings. The use of microorganisms for the bioremediation of metals is an economical, effective, and environmentally benign approach to metal detoxification. Metal pollutants are converted into less hazardous forms by microbes, which use them as a source of energy. This research aims to explore the many roles that microorganisms play in the process of heavy metal bioremediation. This becomes important for devising strategies for metal bioremediation.

***Keywords**: Bioremediation, Heavy metals, pollution, Microorganism, environment friendly, Metal-microbe interaction*

1. Introduction

The world's ecosystem is facing significant challenges because of rising populations, more urbanization, and industrialization (Juwarkar *et al.*, 2014). Significant environmental contamination has been caused by modern civilizations, industrialization, population increase, intensive agricultural practices, human activities, and technological advancement (Lalevic *et al.*, 2016).

Although heavy metals are naturally occurring in the environment, certain activities such as mining and smelting metals, using fertilizers and pesticides in agricultural sewage sludge, disposing of municipal waste, and producing batteries and other metal products in industries raise their concentration to levels that are detrimental to both plants and animals (Chibuike and Obiora, 2014).

Thus, heavy metal pollution has arisen as a main pressing issue in the present environment. Various conventional physical and chemical techniques, including electrochemical treatment, evaporation, reverse osmosis, and chemical precipitation, were employed for the remediation of soil and groundwater. However, due to technological complexity, inefficiency, public disapproval, expense, environmental unfriendliness, invasiveness, and laboriousness, new approaches were developed that focused on biologically driven contaminant removal through mineralization or sequestration (Lalevic *et al.*, 2016; Mustapha and Halimoon, 2015).

Currently, biotechnology is viewed as a developing area of research focused on environmental conservation (Singh *et al.*, 2014). To utilize organism, their cells, or parts of them, along with molecular elements for products and services, biotechnology is defined as " the integration of natural sciences and engineering" (Gavrilescu, 2010).

This method uses microorganisms to biologically remediate air, water, and soil contaminants. Cost-effective methods of lowering wastewater contamination potential have been discovered through biotechnology advances for hazardous waste under controlled environmental circumstances, which has improved public acceptability and adherence to environmental regulations (Singh *et al.*, 2014).To eliminate or lessen contamination from air, soil, and water, bioremediation uses biological systems, such as bacteria, fungi, and yeast (Luka *et al.*, 2018).

This method relies on fostering the growth of certain microbial communities that are either already present at the polluted area or may be introduced from other ecosystems in the hopes that they will perform beneficial functions (Agarwal,1998). Changing pollutants into less dangerous or non-hazardous forms entails biotransformation and biodegradation (Singh *et al.*, 2014). An interdisciplinary strategy combining microbiology, ecology, engineering, geology, and chemistry is essential for successful bioremediation (Luka *et al.*, 2018). Therefore, bioremediation is a cutting-edge method that could reduce the harmful contamination (Singh *et al.*, 2014).

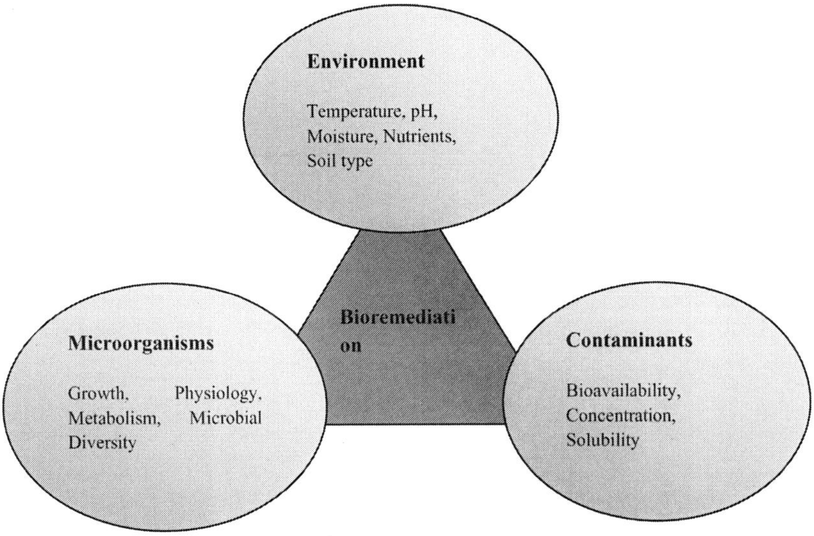

Fig. 1: Factors affecting bioremediation

2. Heavy metals and their Toxicity

Heavy metals are a group of metals and metalloids that have relatively high density, atomic weight, or atomic number and are toxic even at low concentrations. The indestructibility and inherent persistence of heavy metals make them a permanent environmental threat, despite their origins as crustal elements of the Earth (Duruibe *et al.* 2007). They negatively impact the environment and living things, and their specific density is greater than 5g/cm3. Their compositions differ in various locations, which causes spatial differences in the concentrations around them (Jaishankar *et al.*, 2014). There are about 40 elements that fall into this category. The anthropogenic contribution of heavy metals to the biosphere has grown due to urbanization and industrialization (Sharma and Agarwal, 2005). The most common heavy metals found in wastewater are arsenic, cadmium, chromium, copper, lead, nickel, and zinc; all of these are detrimental to ecosystems and human health (Lambert *et al.*, 2000).

2.1. Arsenic (As)

Arsenic is considered one of the most significant heavy metals. It is semi-metallic, highly carcinogenic and poisonous, and can be found as a salt of iron, sodium, calcium, copper, among other elements, or as oxides and sulfides. Arsenite and arsenate complexes are examples of their inorganic forms; they are toxic to both plants and animals (Jaishankar *et al.*, 2014). It is also referred to as the "king of poisons" (Balali-Mood *et al.*, 2021). Trivalent As (III) oxide intake of 70–80 mg is lethal to humans (WHO, 1987). Although small quantities are essential for maintaining good health, lower exposure levels can lead to symptoms such as nausea, vomiting, leukocyte changes, irregular heartbeat, and vascular damage. Prolonged exposure may result in neurological disorders, lung diseases, internal cancers, hypertension, heart disease, Alzheimer's disease, multiple sclerosis, and Parkinson's disease (Jaishankar *et al.*, 2014). Inorganic As can cause various toxic effects in humans, including acute, sub-acute, and chronic conditions like abdominal cramps, abnormal electrocardiograms, and numbness in the extremities (Sharma and Agarwal, 2005). As decreases the germination of seeds, the generation of dry matter and leaf area, and the height of seedlings in rice (Asati *et al.*, 2016).

2.2. Zinc (Zn)

Zn is one of the major metals found in the effluents released by the metallurgical, electroplating, and galvanization sectors (Plum *et al.*, 2010). Although Zn in its metallic form is environmentally benign and has a low bioavailability, it can combine with other substances, such as oxygen and acids, to generate a possibly perilous atom that can genuinely disturb organic frameworks (Fosmire, 1990). It is a shiny, bluish-white metal that is highly malleable and somewhat reactive. Zn is a necessary trace element for microbes, plants, and animals. Excessive exposure results in weight loss, immune system impairment, and obstruction of vital cell reactions in people, while lower exposure levels can cause hair, skin lesions, and diarrhea (Usman *et al.*, 2020).

Zn aids in the production of chlorophyll in plants; however, a Zn-deficient soil slows down plant growth. Chlorosis is caused by Zn deficiency. High levels of Zn in soil limit root and shoot growth and impede numerous metabolic processes, leading to senescence (Asati *et al.*, 2016).

2.3. Lead (Pb)

Lead is a shiny, silver-colored metal that is slightly bluish in dry conditions. It is a very hazardous metal that has led to widespread environmental contamination and health issues. Human exposure to Pb is either due to contaminated food or drinking water (Jaishankar *et al.*, 2014). It impacts nearly all the body's organs and systems in humans (Usman *et al.*, 2020). It results in hemoglobin synthesis inhibition, renal, cardiovascular, and central nervous system dysfunction (Duruibe *et al.*, 2007).

Plant roots can absorb Pb. Lead is known to inhibit *Allium* species, *barley*, and *Raphanus sativas* from elongating their roots and stems and expanding their leaves. Reduced plant biomass, decreased germination rate, and decreased plant protein content have all been observed in maize (Asati *et al.*, 2016).

2.4. Cadmium (Cd)

During World War I, this metal was used as a pigment in the paint industry and as a tin replacement (Jaishankar *et al.*, 2014). To make Zn, it is necessary to produce this byproduct. The element Cd is found in a variety of products, including semiconductors, Ni-Cd batteries, polyvinyl chloride, pigment, and control rods for nuclear reactors (Usman *et al.*, 2020).

The main ways that humans can be exposed to this metal are by ingestion and inhalation (Jaishankar *et al.*, 2014). High blood pressure, damage to the kidneys, lungs, liver, and testes, bone discomfort, and indications of renal impairments are just a few of the harmful effects of cadmium exposure in humans (Usman *et al.*, 2020). The toxicity of Cd disrupts plasma membrane permeability, decreases nitrate absorption and transport from root to shoot by inhibiting nitrate reductase activity, and disturbs chloroplast function by hindering chlorophyll synthesis. In wheat, an excess of Cd hampers seed germination, lowers plant nutrient levels, and shortens both shoot and root length. In maize, it limits shoot growth and inhibits root development (Asati *et al.*, 2016).

2.5. Chromium (Cr)

Most of the Earth's crust is composed of Cr (Jaishankar *et al.*,2014). Many stable forms have been found, including trivalent and hexavalent forms, with oxidation states ranging from -2 to +6. The 2018 evaluation by the International Agency for Research on Cancer (IARC) classified hexavalent chromium as a category I occupational carcinogen. A significant amount of Cr is released into the atmosphere by metallurgical, refractory, and oil well drilling. Cr causes a variety of human diseases due to its bioaccumulation, including neurological illnesses, gastrointestinal disorders, skin disorders, renal disorders, and the development of various cancers in the thyroid, lungs, larynx, kidneys, and bladder (Balali-Mood *et al.*, 2021).

The phytotoxic effect of Cr includes reduced root development, chlorosis of leaves, decreased biomass, inhibition of seed germination, and necrosis. It significantly impacts the biological functions of various plants, including wheat, maize, barley, cauliflower, and several vegetables (Jaishankar *et al.*, 2014).

2.6. Mercury (Hg)

As a naturally occurring metal, metallic mercury is a white, odorless, glossy liquid that, when heated, changes into a gas that has neither color nor odor (Jaishankar *et al.*, 2014). Compared to the liquid form, Hg vapor is more dangerous (Balali-Mood *et al.*, 2021). Hg in inorganic forms causes GI problems, congenital malformations, and spontaneous abortion. An organic form causes pink disease, gingivitis, stomatitis, neurological disorders, and damage to the brain and CNS (Duruibe *et al.*, 2007). Anthropogenic activities such as mining, incineration, industrial wastewater discharge, municipal wastewater discharge, and agriculture are major causes of Hg pollution (Jaishankar *et al.*, 2014).

Plants exposed to high levels of Hg develop physiological disorders and visual damage. Excess Hg in *Oryza sativa* reduces plant height, diminishes the development of tillers and panicles, lowers yield, and increases the bioaccumulation of mercury in seedling shoots and roots (Asati *et al.*, 2016).

Table 2. Effect of Heavy Metals on Plants

S. No.	Heavy metal	Plant	Toxic effects	References
1.	As	Rice (*Oryza sativa*)	Decreased seed germination, lower seedling height, diminished leaf area, & less dry matter production.	(Marin *et al.*, 1993; Abedin *et al.*, 2002)
2.	Cd	Wheat (*Triticum* sp.)	Decrease in seed germination, lower nutrient levels in plants, & shortened shoot and root lengths.	Yourtchi and Bayat, 2013; Ahmad *et al.*., 2012
3.	Co	Tomato (*Lycopersicon esculentum*)	A decrease in the nutrient levels of plants	(Jayakumar *et al.*, 2013)
4.	Cu	Bean (*Phaseolus vulgaris*)	Build-up of Cu in plant roots, deformities in roots, & decreased accumulation in both shoots and roots, along with a reduction in shoot and root length.	(Cook *et al.*, 1998)
5.	Mn	Broad bean (*Vicia faba*)	Build-up in both shoot & root, decrease in shoot & root length.	(Arya and Roy, 2011)
6.	Zn	*Phaseolus vulgaris*	Decrease in growth and development, metabolism, and induction of oxidative damage	Fernandes and Henriques, 1991
7.	Hg	*Oryza sativa*	Decrease plant height, reduces tiller and panicle formation	Du *et al.*, 2005 Kibra, 2008
8.	Cr	*Allium cepa*	Inhibition of the germination process & reduction of plant biomass	Nematshahi *et al.*, 2012
9.	Pb	*Zea mays*	Reduction in germination percentage, reduced plant biomass, and suppressed growth	Hussain *et al.*, 2013
10.	Ni	*Cajanus cajan*	Decrease chlorophyll content and stomatal conductance, decrease enzyme activity	Sheoran *et al.*, 1990

3. Bioremediation Strategies

The biological approach of remediation is a distinct technique that uses biological activity to break down or remove heavy metals (Sayqal and Ahmed, 2021). According to Khalid *et al.*, (2017), "bioremediation" encompasses a wide range of practices that rely on plants and microbes to clean up polluted areas. There are two subfields within bioremediation: both I*n* situ and *Ex*-situ.

3.1. *In-situ* Bioremediation

According to Sayqal and Ahmed (2021), *In*-situ bioremediation procedures eliminate pollution on the spot without removing wastewater or soil. It includes bioventing, biostimulation, bioaugmentation, biosparging, bioattenuation, aquatic phytoremediation, and aquatic zooremediation.

3.1.1. Biostimulation

The first recyclers in nature were microorganisms like fungi and bacteria. To start the bioremediation process, it entails increasing the development and activity of resident bacteria. The ability of natural microbes to convert contaminants into energy sources and valuable commodities. The technique is inexpensive and eco-friendly (Ali *et al.*, 2023; Sayqal and Ahmed, 2021).

3.1.2. Bioaugmentation

In this method, to remove contaminants from a contaminated place, a collection of microbes or genetically modified microorganisms is introduced (MUTTALEB and Ali, 2022). Because of their unique DNA alterations and varied metabolic profiles, genetically engineered microorganisms are more efficient than normally occurring bacteria in breaking down contaminants (Ali *et al.*, 2023).

3.1.3. Bioattenuation

Microorganisms use a variety of metabolic pathways, including diffusion, transformation, solubilization, volatilization, ion exchange, and chemical processes, to break down environmental contaminants in this biological process (Ali *et al.*, 2023).

3.1.4. Biosparging

Biosparging is the process of injecting low-pressure air into a saturated zone to promote the growth and activity of microorganisms for the bioremediation of pollutants (Bala *et al.*, 2022; Vidali, 2001; Sayqal and Ahmed, 2021).

3.1.5. Bioventing

To increase microbial activity for the breakdown of pollutants, nutrients, and oxygen are delivered to the contaminated site through wells in this bioremediation procedure (MUTTALEB and Ali, 2022). To minimize pollutant volatilization and release to the environment, bioventing necessitates low oxygen rates and restricted air flow for microbial biodegradation (Sayqal and Ahmed, 2021).

3.1.6. Bioslurping

This system removes pollutants from soil and groundwater by combining several processes, including vacuum-enhanced pumping and soil vapor extraction. Volatile and semi-volatile organic pollutants are eliminated using these techniques (Vidali, 2001; MUTTALEB and Ali, 2022).

3.1.7. Phytoremediation

Phytoremediation employs plants to clean up water and soil of contaminants (Sayqal and Ahmed, 2021). Depending on the type and amount of pollution, phytoremediation uses a range of techniques (Bala *et al*., 2022). Common reed (*Phragmites communis*), pondweed (*Potamogeton spp.*), whorl-leaf watermilfoil (*Myriophyllum verticillatum*), cattail (*Typha latifolia*), water lettuce (*Pistia stratiotes*), and duckweed (*Lemnagibba*) are among the plants used to clean wastewater (Ali *et al*., 2023).

3.2. Ex-situ bioremediation strategies

Pollutants are pumped or dug to an appropriate treatment site in ex-situ bioremediation (MUTTALEB and Ali, 2022). It uses a variety of techniques, including landfarming, composting, windrows, biopile, bioreactors, and biofilter.

3.2.1. Biopile

Biopiling is a combination of land farming and composting methods. It is also known as the heaping method. It uses a special engineered technology designed for various microorganisms to bioremediate polluted environments. For contaminated soils, Biopiling is a practical and affordable method (MUTTALEB and Ali, 2022; Sayqal and Ahmed, 2021).

3.2.2. Landfarming

Due to its low cost and absence of specialized equipment, landfarming is the most straightforward and important bioremediation technique (Bala *et al*., 2022). It entails excavating polluted soil over a prepared area and tilling it periodically until microorganisms break down the toxins.

3.2.3. Composting

To promote the aerobic decomposition of complex organic waste, including plant and garden waste, coffee grounds, and food scraps, into substances like humus, this method combines contaminated soil with organic, non-hazardous agricultural waste at a temperature of 40 to 65 ^{0}C (Ali *et al.*, 2023; Sayqal and Ahmed, 2021).

3.2.4. Windrows

The piled contaminated soil is rotated frequently in this method to promote the biodegradation process of both native and transient hydrocarbonoclastic bacteria (Ali *et al.*, 2023; Bala *et al.*, 2022).

3.2.5. Bioreactor

A bioreactor is a container that uses a series of biological processes to transform raw materials into a specific product (Ali *et al.*, 2023). This procedure involves controlled treatment of contaminated water and solid waste (sludge, soil, and sediment) (Vidali, 2001; MUTTALEB and Ali, 2022).

4. Metal-microbe Interaction

Different heavy metals interact with microorganisms in different ways. The most crucial heavy metal remediation method is resistance to the metal (Kushwaha and Kashyap, 2021). To break down or detoxify heavy metals, microorganisms have developed specialized resistance mechanisms (Jeyakumar *et al.*,2023). Numerous processes, including biosorption, bioaccumulation, bioreduction, bioprecipitation, bioleaching, and biovolatilization, are used to bioremediate heavy metals.

Metal ions attach non-specifically to proteins and polysaccharides on the cell surface through an interaction process called biosorption. A biosorbent may be made from a variety of natural resources, such as biomass from plants, animals, fungi, algae, and even industrial waste. The process by which pollutants are absorbed by living biomass or cells is called bioaccumulation. This includes heavy metal metabolism that relies on active uptake. Bioaccumulation is a metabolically active process that involves passive diffusion, ion pumps, protein channels, and carrier-mediated transport. Bioleaching and bio-oxidation are both included in the wide phrase "Biomining." Bioleaching is the process by which organic acids generated by acidophilic bacteria dissolve and complex heavy metals from sediment ores. In bioleaching, microorganisms, including bacteria and fungi, dissolve metal sulfides and oxides from secondary wastes and ore deposits (Tayang and Songachan, 2021; Jeyakumar *et al.*, 2023).

Microbes may change metalloids and soluble metal species into insoluble structures such as phosphates, carbonates, sulfides, and hydroxides Heavy metal ions can be readily retrieved by making minor pH adjustments, and the precipitated metals stay affixed to the microbial cells. Organic acids and electron donors precipitate the heavy metal into the extracellular medium that microorganisms release (Jeyakumar *et al.*, 2023; Kapahi and Sachdeva, 2019).

Mineral formation is a natural process. Biomineralization comprises the natural synthesis of minerals such as carbonates, oxides, silicates, sulphates, and phosphates by the actions of living organisms. Metal-microbes interaction transforms both organic and metal compounds from harmful to less toxic forms. Microbes have evolved this method to adapt to environmental changes (Tayang and Songachan, 2021). Through oxidation, reduction, demethylation, methylation, and complexation, microorganisms can affect the biotransformation of heavy metals (Jeyakumar *et al.*, 2023).

5. Microbial Population for the Bioremediation Process

Microorganisms are widely distributed and naturally occurring in the environment, making them a key component of bioremediation (Singh *et al.*, 2014). Microbial bioremediation techniques are economically viable and show encouraging outcomes in the breakdown of contaminants (Usman *et al.*, 2020). The idea that all organisms can extract materials from their surroundings for their development and metabolism is the foundation for using organisms to eliminate contaminants. Numerous microorganisms, including bacteria and fungi, are effective in breaking down complex molecules, while plants, algae, and protozoa can absorb a variety of minerals and metals from their surroundings (Gavrilescu, 2010).

5.1. Bacteria

The three metals Pb, Cu, and Fe were bioremediated from polluted soil at the mining site in Zamfara city by Audu *et al.*, in 2020, using the *Pantoea agglomerans* bacteria. 96% Pb, 60% Cu, and 99.6% Fe were absorbed by *Pantoea agglomerans*. Oziegbe *et al.*, (2021) evaluated the bioremediation potential of the native bacteria that were screened and the impact of pH and carbon source on the improvement of bioremediation processes.

Table 2. Bacterial species for bioremediation

S. No.	Bacteria	Heavy Metal	Efficiency	References
1.	Algal-bacterial aerobic granular sludge	Pb	Adsorption capacity of 72.4 mg g^{-1}	Wang *et al.*, 2023
2.	*Bacillus cereus* (ID1)	Pb	80%	Idrees *et al.*, 2023
3.	*Bacillus* sp. (ID3)	Pb	88%	Idrees *et al.*, 2023
4.	*Sporosarcina pasteurii* & urease-producing isolate	Zn	70.36% 71.46%	Ghorbanzadeh *et al.*, 2022
5.	*Acinetobacter junii* strain b2w	Cr	83.06%	Sevak *et al.*, 2023
6.	*Stenotrophomonas maltophilia* (TJ-1)	Cr	99.77%	Chen *et al.*, 2022
7.	*Brucella intermedius*	Cr	93.86%	Chen *et al.*, 2022
8.	*Lactobacillus delbrueckii*	Pb	42-50%	Hasan *et al.*, 2022
9.	Subsp. *bulgaricus* LDMB02	Cd	40-58%	Hasan *et al.*, 2022
10.	*Lysinibacillus fusiformis, Bacillus licheniformis*	Cd	23.72-43.79%	Ma *et al.*, 2023

5.2. Fungi

The safest and most environmentally friendly way to break down contaminants is through fungi bioremediation. Fungi can reduce pollutants in a variety of ways, including by secreting a potent fungal enzyme (Tomar *et al.*, 2021). *Aspergillus niger* has been shown to produce copper oxalates, which can be utilized to advance copper bioremediation from soil (Frank-Kamenetskaya *et al.*, 2021).

Table 3. Fungal species for bioremediation

S.No.	Fungi	Heavy Metal	Efficiency	References
1.	*Cladosporium sp. Didymella glomerata*	Hg	47.975	Vacar *et al.*, 2021
2.	*Trichoderma harzianum*	Cd Mn Zn Fe	98.63% 94.35% 39.77% 47.43%	Hussain and Mutlag, 2021
3.	*Aspergillus penicillioides* (F12) (MN210327)	Pb (II)	73.145	Paria *et al.*, 2022
4.	White rot fungi	Ni	99-98%	Sharma *et al.*, 2023
5.	*Phlebia floridensis*	Cd	98-97%	Sharma *et al.*, 2023
6.	*Phlebia brevispora*	Pb	12-98%	Sharma *et al.*, 2023
7.	*Geotricum* sp. CS-67	Cu Zn Ni	66.67% 85.15% 79.55%	Hu *et al.*, 2022
8.	*Trichoderma brevicompactum*	Pb(II) Cu(II)	97.50% 64.50%	Zhang *et al.*, 2020

5.3. Algae

Heavy metals like As, Cd, Pb, Cr, and Hg harm algae, whereas other heavy metals, including B, Cu, Fe, Mn, Zn, and Mo, serve as essential trace elements for their metabolic functions and enzymatic activities. Because of their resistance to heavy metal stress, several cyanobacterial species, including *Oscillatoria*, *Phormidiumm*, *Spirogyra*, and *Anabaena*, naturally flourish in water contaminated with heavy metals (Jeyakumar *et al.*, 2023).

Table 4. Algal species for bioremediation

S. No.	Algae	Heavy Metal	Efficiency	References
1.	*Zygnema pectiantum*	Zn	82.19%	Farid *et al.*, 2023
2.	*Spyrogyra* species	Cu	85.13%	Farid *et al.*, 2023
3.	*Chlorella sorokiniana*	Cd	65%	Leon-Vaz *et al.*, 2021
4.	*Chlorella sorokiniana*	Cu	$2.2 g/kg^{-1}$	Leon-Vaz *et al.*, 2021
5.	*Chlorella sorokiniana*	As (V)	$0.145 g/kg^{-1}$	Leon-Vaz *et al.*, 2021
6.	*Chlorella kessleri*	Pb	97.1%	Sultana *et al.*, 2020
7.	*Arthrospira maxima*	Fe	97.9%	Blanco-Vieites *et al.*, 2022
8.	*Chlamydomonas renihardtii*	Cd	90.2%	Pina-Olavide *et al.*, 2020
9.	*Aphanothess* sp.	Pb	99.9%	Keryanti & Mulyono, 2021
10.	*Neochloris oleoabundans*	Pb	93%	Gu *et al.*, 2021

5.4. Genetically engineered microorganism

A range of complex pollutants created by industrial activities and human actions remains ineffective, addressed by naturally occurring microorganisms because of their chemical structure and strong bond energy. Heavy metals can change the surface properties and enzyme activity of microorganisms (Sharma *et al.*, 2022). By altering their genetic makeup, natural microorganisms—known as genetically modified organisms, or GMOs—can be made more capable of cleaning up pollutants through the multidisciplinary application of molecular methods, bioinformatics, genomics, and microbiology (Saravana *et al.*, 2022)

Table 5. Genetically engineered microorganism for bioremediation

S.No.	Genetically moified microorganism	Modified gene	Mode of expression	References
1.	*Deinococcus radiodurans*	MerH; Ion transporter	Produce resistance against mercury and degrade *marinum strain*	Meruvu, 2021
2.	*Acidithiobacillus ferrooxidans*	Mer C; Ion transporter	Mercury degradation	Arshadi and Yaghmaei, 2020
3.	*Mesorhizobium huakuii*	Incorporated genes from *Arabidopsis thaliana* encoding PCs	Degradation and accumulation of Cd2+	Luo *et al.* (2020)
4.	*E. coli* SE5000 strain	nixA: express nickel transporting system	Degradation of nickel from an aqueous system	Tsyganov *et al.* (2020)
5.	*Pseudomonas* K-62	Exhibit expression of organomercurial lyase and Hg degradation	Degradation of mercury	(Sharma, 2020a)
6.	*Ralstonia eutropha* AUM-01	Recombinant rhizobacteria bound to Chambliss	Removes heavy metals	(Sharma, 2020b)
7.	*Rhodopseu-domonas palustris*	Express mercury transporting system	Degradation of Mercury	Brown *et al.*, 2020
8.	*Pseudomonas putida* X3 strain	Incorporation of EGFP and methyl parathion cadmium degradation gene	Degradation of cadmium and methyl parathion	Ayogu *et al.*, 2020

6. Conclusion

Bioremediation is an environmentally friendly method that uses microorganisms to clean up contaminated environments by increasing natural biodegradation processes. When contrasted with other technologies, such as physical and chemical procedures, which might potentially produce more pollutants in the treated region, this one is advantageous in terms of cost-effectiveness, environmental friendliness, efficiency, and promise. New bioremediation methods need to undergo extensive field testing before they may be considered economically viable, despite their obvious attractiveness. When one method fails in the field, it's usually because of environmental factors or the toxicity of the targeted chemical. Therefore, hybrid solutions are necessary that can adjust to the ever-changing toxicity of compounds and environmental circumstances.

There is a tremendous need for these technologies to include scientific and technical inputs for use in the field.

References

Agarwal, S. K. (2009). Heavy metal pollution. APH Publishing Corporation, New Delhi, India

Agarwal, S. K. (1998). Environment Biotechnology (1st Ed.). APH Publishing Corporation, New Delhi, India

Ahmad, I., Akhtar, M. J., Zahir, Z. A., & Jamil, A. (2012). Effect of cadmium on seed germination and seedling growth of four wheat (Triticum aestivum L.) cultivars. Pak. J. Bot, 44(5), 1569-1574.

A. Hussain, N. Abbas, F. Arshad (2013) "Effects of diverse doses of lead (Pb) on different growth attributes of Zea mays L.," Agricultural Sciences, 4(5), pp. 262–265

Abedin, M. J., Cotter-Howells, J., & Meharg, A. A. (2002). Arsenic uptake and accumulation in rice (Oryza sativa L.) irrigated with contaminated water. Plant and soil, 240, 311-319.

Arya, S. K., & Roy, B. K. (2011). Manganese induced changes in growth, chlorophyll content and antioxidants activity in seedlings of broad bean (Vicia faba L.). Journal of Environmental Biology, 32(6), 707-711.

Arshadi, M., & Yaghmaei, S. (2020). Advances in bioleaching of copper and nickel from electronic waste using Acidithiobacillus ferrooxidans: evaluating daily pH adjustment. Chemical Papers, 74, 2211-2227.

Ayogu, C. V., Ifeanyi, V. O., & Obasi, N. P. (2020). Monitoring of metabolic compounds from degradation of petrochemicals using indigenous consortium of Pseudomonas strains. Nigerian Journal of Microbiology, 34(2), 5221-5238.

Asati, A., Pichhode, M., & Nikhil, K. (2016). Effect of heavy metals on plants: an overview. International Journal of Application or Innovation in Engineering & Management, 5(3), 56-66. DOI: 10.13140/RG.2.2.27583.87204

Ali, S., Sharma, B., Deoli, K., Saini, D., & Bisht, M. (2023). An Overview on Bioremediation Strategies for Waste Water Treatment and Environmental Sustainability. Appl. Ecol. Environ. Sci, 11, 64-70.

Audu, K. E., Adeniji, S. E., & Obidah, J. S. (2020). Bioremediation of toxic metals in mining site of Zamfara metropolis using resident bacteria (Pantoea agglomerans): A optimization approach. Heliyon, 6(8).

Blanco-Vieites, M., Suárez-Montes, D., Delgado, F., Álvarez-Gil, M., Battez, A. H., & Rodríguez, E. (2022). Removal of heavy metals and hydrocarbons by microalgae from wastewater in the steel industry. Algal Research, 64, 102700. https://doi.org/10.1016/j.algal.2022.102700.

Brown, B., Immethun, C., Wilkins, M., & Saha, R. (2020). Rhodopseudomonas palustris CGA009 polyhydroxybutyrate production from a lignin aromatic and quantification via flow cytometry. Bioresource Technology Reports, 11, 100474.

Bala, S., Garg, D., Thirumalesh, B. V., Sharma, M., Sridhar, K., Inbaraj, B. S., & Tripathi, M. (2022). Recent strategies for bioremediation of emerging pollutants: a review for a green and sustainable environment. Toxics, 10(8), 484.

Balali-Mood, M., Naseri, K., Tahergorabi, Z., Khazdair, M. R., & Sadeghi, M. (2021). Toxic mechanisms of five heavy metals: mercury, lead, chromium, cadmium, and arsenic. Frontiers in pharmacology, 12, 643972.

Cook, C. M., Kostidou, A., Vardaka, E., & Lanaras, T. (1998). Effects of copper on the growth, photosynthesis and nutrient concentrations of Phaseolus plants. Photosynthetica, 34, 179-193.

Chibuike, G. U., & Obiora, S. C. (2014). Heavy metal polluted soils: effect on plants and bioremediation methods. Applied and environmental soil science, 2014(1), 752708. http://dx.doi.org/10.1155/2014/752708.

Chen, W., Li, W., Wang, T., Wen, Y., Shi, W., Zhang, W., ... & Yang, Y. (2022). Isolation of functional bacterial strains from chromium-contaminated site and bioremediation potentials. Journal of Environmental Management, 307, 114557.

Duruibe, J. O., Ogwuegbu, M. O. C., & Egwurugwu, J. N. (2007). Heavy metal pollution and human biotoxic effects. International Journal of physical sciences, 2(5), 112-118.

Fernandes JC, Henriques FS (1991) Biochemical, physiological and structural effects of excess copper in plants. Bot Rev 57:247–273.

Farid, N., Ullah, A., Khan, S., Butt, S., Khan, A. Z., Afsheen, Z., ... & Ali, Q. (2023). Algae and hydrophytes as potential plants for bioremediation of heavy metals from industrial wastewater. Water, 15(12), 2142.

Fosmire, G. J. (1990). Zinc toxicity. The American journal of clinical nutrition, 51(2), 225-227. https://doi.org/10.1093/ajcn/51.2.225

Frank-Kamenetskaya, O. V., Zelenskaya, M. S., Izatulina, A. R., Vereshchagin, O. S., Vlasov, D. Y., Himelbrant, D. E., & Pankin, D. V. (2021). Copper oxalate formation by lichens and fungi. Scientific Reports, 11(1), 24239.

Gu, S., Boase, E. M., & Lan, C. Q. (2021). Enhanced Pb (II) removal by green alga Neochloris oleoabundans cultivated in high dissolved inorganic carbon cultures. Chemical Engineering Journal, 416, 128983.

Gavrilescu, M. (2010). Environmental biotechnology: achievements, opportunities and challenges. Dynamic biochemistry, process biotechnology and molecular biology, 4(1), 1-36.

Ghorbanzadeh, N., Ghanbari, Z., Farhangi, M. B., & Rad, M. K. (2022). Zinc bioremediation in soil by two isolated L-asparaginase and urease producing bacteria strains. Applied Geochemistry, 140, 105271.

Hu, T., Chen, A., Jiang, Y., Sun, C., Luo, S., & Shao, J. (2022). Application of a newly recorded diazotrophic cyanobacterium in acidified and Cd contaminated paddy soil: Promotes rice yield and decreases Cd accumulation. Science of the Total Environment, 814, 152630.

Hussain, D. F., & Mutlag, N. H. (2021, June). Assessment the ability of Trichoderma harzianum Fungi in Bioremediation of some of Heavy Metals in Waste Water. In IOP conference series: Earth and environmental science (Vol. 790, No. 1, p. 012087). IOP Publishing.

Hasan, M. S., Islam, M. Z., Liza, R. I., Sarker, M. A. H., Islam, M. A., & Harun-ur-Rashid, M. (2022). Novel probiotic lactic acid bacteria with in vitro bioremediation potential of toxic lead and cadmium. Current Microbiology, 79(12), 387.

Idrees, M., Ali, S., Rehman, A., Hussain, S. Z., & Bukhari, D. A. (2023). Uptake of lead by bacteria isolated from industrial effluents and their potential use in bioremediation of wastewater. Saudi Journal of Biological Sciences, 30(8), 103740.

I. S. Sheoran, H. R. Singal, and R. Singh (1990) "Effect of cadmium and nickel on photosynthesis and the enzymes of the photosynthetic carbon reduction cycle in pigeon pea (Cajanus cajan L.)," Photosynthesis Research, 23(3), pp. 345–351.

Jaishankar, M., Tseten, T., Anbalagan, N., Mathew, B. B., & Beeregowda, K. N. (2014). Toxicity, mechanism and health effects of some heavy metals. Interdisciplinary toxicology, 7(2), 60-72.

Jayakumar, K., Rajesh, M., Baskaran, L., & Vijayarengan, P. (2013). Changes in nutritional metabolism of tomato (Lycopersicon esculantum Mill.) plants exposed to increasing concentration of cobalt chloride. Int J Food Nutr Saf, 4(2), 62-69.

Jeyakumar, P., Debnath, C., Vijayaraghavan, R., & Muthuraj, M. (2023). Trends in bioremediation of heavy metal contaminations. Environmental Engineering Research, 28(4). https://doi.org/10.4491/eer.2021.631

Juwarkar, A. A., Misra, R. R., & Sharma, J. K. (2014). Recent trends in bioremediation. Geomicrobiology and biogeochemistry, 81-100.

Khalid, S., Shahid, M., Niazi, N. K., Murtaza, B., Bibi, I., & Dumat, C. (2017). A comparison of technologies for remediation of heavy metal contaminated soils. Journal of geochemical exploration, 182, 247-268.

Keryanti, K., & Mulyono, E. W. S. (2021). Determination of Optimum Condition of Lead (Pb) Biosorption Using Dried Biomass Microalgae Aphanothece sp. Periodica Polytechnica. Chemical Engineering, 65(1), 116-123.

Kapahi, M., & Sachdeva, S. (2019). Bioremediation options for heavy metal pollution. Journal of health and pollution, 9(24), 191203.

Kushwaha, P., & Kashyap, P. L. (2021). A review of advances in bioremediation of heavy metals by microbes and plants. Journal of Natural Resource Conservation and Management, 2(1), 65-80.

Lambert, M., Leven, B. A., & Green, R. M. (2000). New methods of cleaning up heavy metal in soils and water. Environmental science and technology briefs for citizens, 7(4), 133-163.

León-Vaz, A., León, R., Giráldez, I., Vega, J. M., & Vigara, J. (2021). Impact of heavy metals in the microalga Chlorella sorokiniana and assessment of its potential use in cadmium bioremediation. Aquatic Toxicology, 239, 105941. https://doi.org/10.1016/j.aquatox.2021.105941.

Lalevic, B., Raicevic, V., Kikovic, D., Talaiekhozani, A., Hamidovic, S., Gkorezis, P., & Van Hamme, J. (2016, December). Exploiting microorganisms for the removal of organic pollutants: an environmental perspective. In International conference of applications (Vol. 29).

Luka, Y., Highina, B. K., & Zubairu, A. (2018). Bioremediation: A solution to environmental pollution-a review. Am J Eng Res, 7(2), 101-109.

Luo, S., Yin, J., Peng, Y., Xie, J., Wu, H., He, D., ... & Cheng, G. (2020). Glutathione is involved in detoxification of peroxide and root nodule symbiosis of Mesorhizobium huakuii. Current microbiology, 77, 1-10.

M. G. Kibra (2008) “Effects of mercury on some growth parameters of rice (Oryza sativa L.),” Soil & Environment, 27(1), pp. 23–28.

Marin, A. R., Pezeshki, S. R., Masschelen, P. H., & Choi, H. S. (1993). Effect of dimethylarsenic acid (DMAA) on growth, tissue arsenic, and photosynthesis of rice plants. Journal of Plant Nutrition, 16(5), 865-880.

Muttaleb, W. H., & ALI, Z. H. (2022). Bioremediation an eco-friendly method for administration of environmental contaminants. International Journal of Applied Sciences and Technology, 4(6), 22-32.

Ma, B., Song, W., Zhang, X., Chen, M., Li, J., Yang, X., & Zhang, L. (2023). Potential application of novel cadmium-tolerant bacteria in bioremediation of Cd-contaminated soil. Ecotoxicology and Environmental Safety, 255, 114766.

Mustapha, M. U., & Halimoon, N. (2015). Screening and isolation of heavy metal tolerant bacteria in industrial effluent. Procedia Environmental Sciences, 30, 33-37.

Meruvu, H. (2021). Bacterial bioremediation of heavy metals from polluted wastewaters. In New trends in removal of heavy metals from industrial wastewater (pp. 105-114). Elsevier.

N. Nematshahi, M. Lahouti, and A. Ganjeali (2012) “Accumulation of chromium and its effect on growth of (Allium cepa cv. Hybrid),” European Journal of Experimental Biology, 2(4), pp. 969– 974

Oziegbe, O., Oluduro, A. O., Oziegbe, E. J., Ahuekwe, E. F., & Olorunsola, S. J. (2021). Assessment of heavy metal bioremediation potential of bacterial isolates from landfill soils. Saudi Journal of Biological Sciences, 28(7), 3948-3956.

Plum, L. M., Rink, L., & Haase, H. (2010). The essential toxin: impact of zinc on human health. International journal of environmental research and public health, 7(4), 1342-1365. https://doi.org/10.3390/ijerph7041342

Paria, K., Pyne, S., & Chakraborty, S. K. (2022). Optimization of heavy metal (lead) remedial activities of fungi Aspergillus penicillioides (F12) through extra cellular polymeric substances. Chemosphere, 286, 131874.

Piña-Olavide, R., Paz-Maldonado, L. M., Alfaro-De La Torre, M. C., García-Soto, M. J., Ramírez-Rodríguez, A. E., Rosales-Mendoza, S., ... & García De la-Cruz, R. F. (2020). Increased removal of cadmium by Chlamydomonas reinhardtii modified with a synthetic gene for γ-glutamylcysteine synthetase. International Journal of Phytoremediation, 22(12), 1269-1277. https://doi.org/10.1080/15226514.2020.1765138.

Sayqal, A., & Ahmed, O. B. (2021). [Retracted] Advances in Heavy Metal Bioremediation: An Overview. Applied Bionics and Biomechanics, 2021(1), 1609149.

Sharma, I. (2020). Bioremediation techniques for polluted environment: concept, advantages, limitations, and prospects. In Trace metals in the environment-new approaches and recent advances. IntechOpen.

Sultana, N., Hossain, S. Z., Mohammed, M. E., Irfan, M. F., Haq, B., Faruque, M. O., ... & Hossain, M. M. (2020). Experimental study and parameters optimization of microalgae based heavy metals removal process using a hybrid response surface methodology-crow search algorithm. Scientific reports, 10(1), 15068. https://doi.org/10.1038/s41598-020-72236-8.

Saravanan, A., Kumar, P. S., Ramesh, B., & Srinivasan, S. (2022). Removal of toxic heavy metals using genetically engineered microbes: Molecular tools, risk assessment and management strategies. Chemosphere, 298, 134341.

Sharma, K. R., Giri, R., & Sharma, R. K. (2023). Efficient bioremediation of metal containing industrial wastewater using white rot fungi. International Journal of Environmental Science and Technology, 20(1), 943-950.

Sevak, P., Pushkar, B., & Mazumdar, S. (2023). Mechanistic evaluation of chromium bioremediation in Acinetobacter junii strain b2w: a proteomic approach. Journal of Environmental Management, 328, 116978.

Sharma, I. (2020). Bioremediation techniques for polluted environment: concept, advantages, limitations, and prospects. In Trace metals in the environment-new approaches and recent advances. IntechOpen.

Sharma, P., Bano, A., Singh, S. P., Sharma, S., Xia, C., Nadda, A. K., ... & Tong, Y. W. (2022). Engineered microbes as effective tools for the remediation of polyaromatic aromatic hydrocarbons and heavy metals. Chemosphere, 306, 135538.

Sharma, R. K., & Agrawal, M. (2005). Biological effects of heavy metals: an overview. Journal of environmental Biology, 26(2), 301-313.

Singh, R. (2014). Microorganism as a tool of bioremediation technology for cleaning environment: a review. Proceedings of the International Academy of Ecology and Environmental Sciences, 4(1), 1.

Tsyganov, V. E., Tsyganova, A. V., Gorshkov, A. P., Seliverstova, E. V., Kim, V. E., Chizhevskaya, E. P., ... & Tikhonovich, I. A. (2020). Efficacy of a plant-microbe system: Pisum sativum (L.) cadmium-tolerant mutant and Rhizobium leguminosarum strains, expressing pea metallothionein genes PsMT1 and PsMT2, for cadmium phytoremediation. Frontiers in Microbiology, 11, 15.

Tomer, A., Singh, R., Singh, S. K., Dwivedi, S. A., Reddy, C. U., Keloth, M. R. A., & Rachel, R. (2021). Role of fungi in bioremediation and environmental sustainability. Mycoremediation and Environmental Sustainability: Volume 3, 187-200.

Tayang, A., & Songachan, L. S. (2021). Microbial bioremediation of heavy metals. Current Science, 120(6), 00113891.

Văcar, C. L., Covaci, E., Chakraborty, S., Li, B., Weindorf, D. C., Frențiu, T., ... & Podar, D. (2021). Heavy metal-resistant filamentous fungi as potential mercury bioremediators. Journal of Fungi, 7(5), 386.

Vidali, M. (2001). Bioremediation. an overview. Pure and applied chemistry, 73(7), 1163-1172.

WHO. (1987). Air quality guidelines for Europe. WHO Regional Publications, European Series No. 23. World Health Organization officer for Europe, Copenhagen.

Wang, Z., Wang, H., Nie, Q., Ding, Y., Lei, Z., Zhang, Z., ... & Yuan, T. (2023). Pb (II) bioremediation using fresh algal-bacterial aerobic granular sludge and its underlying mechanisms highlighting the role of extracellular polymeric substances. Journal of Hazardous Materials, 444, 130452.

X. Du, Y.-G. Zhu, W.-J. Liu, and X.-S. Zhao (2005) "Uptake of mercury (Hg) by seedlings of rice (Oryza sativa L.) grown in solution culture and interactions with arsenate uptake," Environmental and Experimental Botany, 54(1), pp. 1–7.

Yourtchi, M. S., & Bayat, H. R. (2013). Effect of cadmium toxicity on growth, cadmium accumulation and macronutrient content of durum wheat (Dena CV.). 6(15), pp. 1099–1103.

Zhang, D., Yin, C., Abbas, N., Mao, Z., & Zhang, Y. (2020). Multiple heavy metal tolerance and removal by an earthworm gut fungus Trichoderma brevicompactum QYCD-6. Scientific Reports, 10(1), 6940.

Zango Usman, U., Mukesh, Y., Vandana, S., Sharma, J. K., Sanjay, P., Sidhartha, D., & Sharma Anil, K. (2020). Microbial bioremediation of heavy metals: Emerging trends and recent advances. Research Journal of Biotechnology, 15(1).

9

Emerging Technologies and Sustainable Solutions in Bioremediation Strategies for Textile Dyes and Effluents

Riddhi Kathrotiya[1], Shivani V. Raval[1, 2] and Vikram H. Raval[1]*

[1]Department of Microbiology and Biotechnology, University School of Sciences, Gujarat University, Navrangpura - 380009, Ahmedabad

[2]Department of Biotechnology, M G Science Institute (Autonomous), Navrangpura - 380009, Ahmedabad

**Corresponding Author Email: vhraval@gujaratuniversity.ac.in; vikramhraval@gmail.com*

Abstract

Textile effluent is among the key sources leading to environmental pollution and causing potential health hazards due to synthetic azo dyes and heavy metals. The remediation of such effluents poses multi-dimensional challenges due to their varied classification, allied toxicity, and complex physico-chemical characteristics. Here we attempt an inclusive synopsis of the bioremediation strategies employed effectively for the treatment of textile dye effluents, proposing a greener solution compared to the preexisting and traditional ones. The review explores the downsides of physicochemical effluent treatments with the advantages of biological methods. Advancements in bioremediation techniques, including emerging integrated and hybrid technologies, are focused, highlighting their effectiveness in mitigating pollution and its ill-effects. Studies suggest that combining biological processes with physio-chemical processes including electro-coagulation, adsorption, ultra-filtration, membrane, and advanced oxidation, is effective for textile wastewater, considering the specific pollutants and discharge regulations. This advancement highlights the growing importance of industrial wastewater management, offering a sustainable alternative to traditional methods. Furthermore, current trends and prospects in textile dye bioremediation are evaluated, emphasizing innovative approaches and various technologies. Genetically engineered bacterial systems show potential for enhanced pollutant degradation, though concerns persist regarding the release of transgenic bacteria and their impact on biodiversity.

Efforts are underway to address these issues and detect transgenic bacteria in the environment. Despite the controversy, genetically engineered bacteria offer a promising solution for managing contaminated sites. Additionally, Microbial electrochemical systems/bio electrochemical systems (MES/BES) utilize microbes to convert waste into electricity and clean toxic compounds from wastewater. Nanotechnology and nanomaterials are effective in dye decolorization due to their high surface area and stability, with an increasing focus on eco-friendly biological synthesis methods. Overall, this review provides valuable insights into the state-of-the-art methods and technologies employed in textile dye bioremediation, offering a roadmap for researchers toward achieving more effective and sustainable solutions for mitigating hazardous pollution menace.

Keywords: *Bioremediation of dye, Azo dyes, Biological methods, Microbial Electrochemical Technologies, Textile wastewater, Integrated/Hybrid Technologies*

1. Introduction

In recent decades, environmental contamination has emerged as a serious threat to our modern society. The relentless growth of urbanization and industrialization leads to the ever-increasing production of hazardous chemicals. Safe disposal of these hazardous materials is crucial, as releasing them into an already strained environment has severe consequences. The exponential rise in human activity is directly related to the alarming rise in local and global pollution levels, which severely disturbs the balance of ecosystems. Industrial activities, in particular, are a major source of highly hazardous chemicals such as dyes, pigments, and aromatic compounds, which pose a serious threat to the environment. In India, the textile industry, a major force driving the country's economic growth, also ranks as the second largest polluter. The printing, finishing, and dyeing processes used in textile production generate significant waste (Dubey *et al.*, 2018). Dyes, a major source of pollution in this industry, are synthetic and complex in structure. This complexity makes them poorly soluble in water, hindering biodegradation and posing a long-term environmental concern. The global textile industry utilizes a wide range of dyes, with estimates indicating over 100,000 commercially available options. The worldwide production of these dyes is estimated to be 700,000 tons per year (Al-Tohamy *et al.*, 2020). However, dye contamination is a major challenge in this industry. Studies indicate that during synthesis and dyeing processes, a significant portion, ranging from 10% to 15% of the total dyestuffs used ends up released into global water bodies (Leal *et al.*, 2020). A significant portion of textile dyes, ranging from 2% to 20%, are directly released into water reservoirs through wastewater effluents. This is concerning because dyes are highly concentrated, and even small amounts (less than 1 milligram per liter) can produce significant color changes in water. Even worse, this contamination renders the water unsuitable for consumption (Garg *et al.*, 2020). The textile industry is a major consumer of water and chemicals. Estimates suggest that around 50-90 grams of dye and 200 liters of water are needed to produce just one kilogram of textiles (Bera & Tank, 2021). On a global scale, this translates to roughly 280,000 tons of textile dyes being released into the environment each year through industrial wastewater (Berradi *et al.*, 2019). Improper treatment of this wastewater allows the dyes to persist in the environment for extended periods. For example, a commonly used dye, hydrolyzed reactive blue 19, can remain in the environment for 46 years under normal conditions (Awasthi & Datta, 2019). Untreated discharges of textile wastewater pose serious threats. Dyes can enter the food chain, accumulate in organisms (bioaccumulate), disrupt photosynthesis, and potentially cause ecotoxic, mutagenic, and carcinogenic effects (Desai *et al.*, 2021). Thus, in the modern era, efficiently treating wastewater containing textile dyes is a crucial environmental concern (Goswami *et al.*, 2020; Kiran *et al.*, 2022). Industries often use physical and chemical methods to remove synthetic dyes from effluent. These techniques include ozonation, oxidation, membrane filtration, ion exchange, adsorption, and coagulation (L. Liu *et al.*, 2018). Conventional methods of treating textile dye wastewater have developed, but bioremediation offers significant advantages. Numerous research has shown that bioremediation is more effective than other approaches (Desai *et al.*, 2021; Rafique *et al.*, 2022). A recent study explores micro-biotechnology methods utilizing algae, bacteria, yeast, and fungi for bioremediation, with promising results in ecologically friendly wastewater treatment (Rafique *et al.*, 2022). The substantial increase in articles on ScienceDirect using the search term "bioremediation of dyes" (over 1400 publications by 2022) shows the rising corpus of research on this topic. This review paper addresses the gap in a comprehensive analysis of biological methods for treating textile wastewater. It focuses on bacterial approaches and their effectiveness in dye removal. Additionally, the paper summarizes the impact of textile dyes on various organisms and the environment. Notably, it presents recent findings on diverse bioreactors used for dye remediation and explores next-generation techniques for textile wastewater treatment, offering insights into the future of bioremediation in this field (Sudarshan *et al.*, 2023).

1.1. Characterization of effluent

Textile wastewater ranks as the most contaminated of all industrial effluents. High production volumes and the unique composition of their effluent make their wastewater some of the most problematic industrial waste. Two factors significantly contribute to this pollution: excess dye from the dyeing process itself, and improper disposal practices that lead to direct dye release into the environment. The chemicals used in textile manufacturing further contribute to the issue. Salts and sodium hydroxide, employed for color fixation during wet processing, elevate salinity and alkalinity levels (pH reaching 11.0 to 11.5) in the wastewater. Additionally, these effluents typically contain dye concentrations between 16 and 20 milligrams per liter (Guadie *et al.*, 2017). Textile industry wastewater is a complex mixture with several concerning properties. These include intense color, high chemical oxygen demand and biological oxygen demand, salinity, and extreme pH levels (both high and low). Additionally, the presence of salts, hazardous metals (Cd^{+2}, Cr^{+2}, Cu^{+2}, Pb^{+2}, Ni^{+2}, Zn^{+2}, etc.), and surfactants is a major concern. This contaminated water is unfit for human consumption due to the presence of synthetic dyes. Furthermore, overall pollution disrupts aquatic ecosystems and threatens water resources (Hernández-Zamora & Martínez-Jerónimo, 2019; Kiran *et al.*, 2022). Multiple studies (Behl *et al.*, 2020; Gowri *et al.*, 2020) have shown that various pollutants in dyeing industry effluent can surpass regulatory limits. These pollutants include color, odor, pH, total solids (TS), as well as breakdowns like total dissolved solids (TDS) and total suspended solids (TSS). Additionally, chloride and chemical oxygen demand (COD) are often found to exceed allowable levels.

2. Synthetic Dyes

Dyes are divided into natural and synthetic categories (as shown in Fig.ure 1). These categories are based on the dye's origin and chemical structure. Dyes can also be classed by their ionic properties (ionic or non-ionic), such as dispersion dyes. Furthermore, dyes can be classed according to their affinity for fabrics (anionic, cationic, direct, acidic, and reactive) (Vikrant *et al.*, 2018). This classification system considers both the dye's and the fibers' physical and chemical qualities. Ultimately, the presence of chromophores in the dye's structure allows it to give color to materials (Mashkoor & Nasar, 2020).

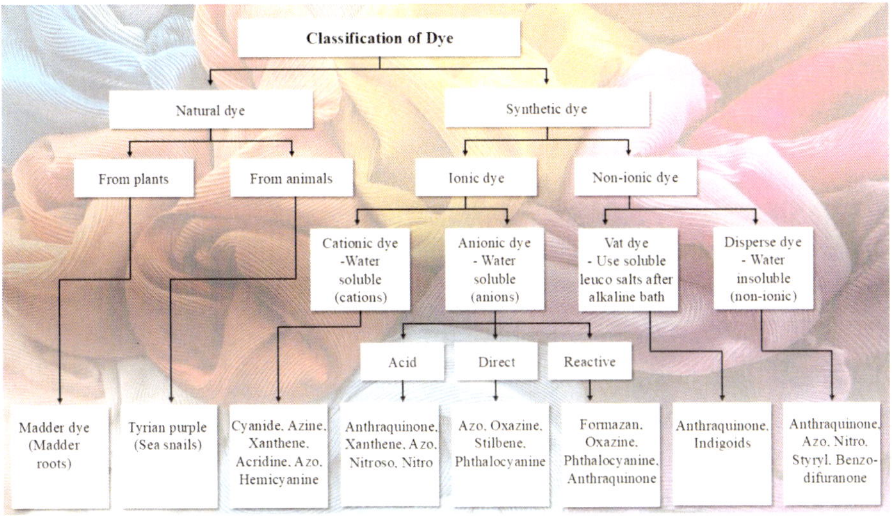

Fig. 1: Classification of Dye

The color of a molecule is determined mostly by chromophores. These groups are made up of certain atomic conFig.urations that allow them to absorb certain wavelengths of light. This absorption is caused by a system of delocalized electrons, often associated with the presence of double bonds. Common chromophores include functional groups such as azo groups (-N=N-), nitro groups ($-NO_2$), carbonyl groups (-CO), and conjugated aromatic systems ($O=C_6H_4=O$) (Kalogerakis *et al.*, 2006). Auxochromes, on the other hand, do not directly determine color but rather can influence the intensity or shade of a color provided by a chromophore. These functional groups, including carboxylic acids (-COOH), amines ($-NH_3$), sulfonic acids ($-SO_3H$), and hydroxyls (-OH), can interact with chromophores, altering their light-absorbing capabilities (Pinheiro *et al.*, 2022). Azo dyes reign supreme in the realm of synthetic colorants, constituting an impressive 60-70% of all known dye structures. This dominance can be attributed to their exceptional performance during the dyeing process, user-friendly application methods, and cost-effective production costs (Prabhakar *et al.*, 2022; Sathishkumar *et al.*, 2019; Shetty & Krishnakumar, 2020). Anthraquinone dyes follow closely behind as the second most prevalent class, favored for similar reasons. The vast palette of colors achievable in dyes is further enhanced by the existence of an estimated 25 distinct chromophore-based dye groups. These chromophores, functional groups responsible for light absorption and color manifestation, offer a diverse array of coloring possibilities (Varjani *et al.*, 2020). Textile wastewater discharge is a significant cause of pollution. According to studies, these effluents contain sulfonate dyes, which are extremely soluble in water.

This high solubility can lead to widespread contamination. The characteristics of textile wastewater often exceed the standards established by organizations such as the World Health Organization (WHO). These characteristics are most likely related to the well-documented environmental and health effects associated with textile effluent, as highlighted in various scientific studies (Donkadokula *et al.*, 2020). The high visibility of synthetic dyes in wastewater can be aesthetically displeasing and potentially disrupt aquatic ecosystems. Furthermore, their resistance to degradation means they can persist in the environment, potentially accumulating in the food chain and causing harm to wildlife and humans (Srivastava *et al.*, 2022).

3. Conventional Approaches

Several physicochemical processes have emerged as promising methods for dye removal from wastewater. These methods include coagulation-flocculation, adsorption, membrane separation, ion exchange, photodegradation, and oxidation. However, these methods face several challenges hindering their widespread application. These limitations include high operational costs, the generation of significant sludge volumes requiring disposal, the formation of harmful byproducts, and limited commercial scalability (Yang *et al.*, 2020). Coagulation and flocculation are well-established physical methods for dye removal from effluent. These methods utilize coagulants such as ferric chloride and ferrous sulfate, which destabilize dye molecules in the effluent and cause them to clump together. Subsequent flocculation facilitates the settling or removal of these dye aggregates. Research has also shown success with alternative coagulants like magnesium chloride, polyaluminum chloride, chitosan, and aluminum chloride. However, coagulation and flocculation have limitations. These methods may not achieve complete decolorization, are expensive due to chemical costs, and produce significant sludge that requires additional treatment or disposal (Vikrant *et al.*, 2018). Adsorption is a widely adopted water treatment method for removing various pollutants because it is simple to use, inexpensive, and efficient. This process has received special attention for its remarkable ability to decolorize dyes present in textile wastewater. A wide range of adsorbents have been investigated for dye removal, including readily available and inexpensive options such as activated carbon, wood, fly ash, and rice husk. Furthermore, more advanced materials such as carbon nanotubes, metal oxides, zeolites, and metal-organic frameworks (MOFs) have demonstrated promise (Khalid *et al.*, 2018). An ideal adsorbent for water treatment should have several key characteristics, including high affinity for the target dye molecules, efficient uptake capacity, low cost, and the ability to regenerate for reuse. However, a notable drawback of adsorption is that it doesn't degrade the dye molecules; they simply accumulate on the adsorbent surface (K. Ali *et al.*, 2021). Another promising method for treating textile industry wastewater is membrane filtration, which includes techniques such as reverse osmosis (RO), nanofiltration (NF), and ultrafiltration (UF) (Tay *et al.*, 2018). These membrane-based approaches offer many advantages. They require minimal space for installation, making them suitable for a variety of applications (low footprint requirement). They are also easier to maintain and operate than other treatment methods. Researchers developed advanced membranes specifically designed to selectively remove various dye molecules from wastewater (Van der Bruggen, 2018). However, membrane filtration also has limitations. It can be challenging to handle a large volume of dye-containing wastewater. Furthermore, the process generates concentrated waste streams that require further treatment or disposal. Finally, dye molecules can clog the membranes, hindering their long-term effectiveness. Textile wastewater treatment also explores oxidation techniques for dye degradation. These techniques fall into two main categories: chemical oxidation and advanced oxidation processes (AOPs). Chemical oxidation utilizes oxidizing agents like hydrogen peroxide (H_2O_2) or ozone (O_3) to generate hydroxyl radicals. These highly reactive radicals break down the complex dye structures into smaller, non-colored molecules through a process called mineralization. This approach is particularly effective for dyes containing double bonds. AOPs, on the other hand, involve the generation of even stronger oxidants, often through combinations of chemicals, light, or ultrasound. These AOPs can target a wider range of dye structures but may require more complex or expensive setups (Ihsanullah *et al.*, 2020). While chemical oxidation offers a viable solution, its effectiveness is limited compared to AOPs. This is because chemical oxidants generate fewer hydroxyl radicals, leading to slower degradation rates. Ozonation, a popular method for water purification, utilizes ozone (O_3) to degrade pollutants. It boasts advantages like minimal sludge production and no wastewater volume increase (Wang & Chen, 2020). However, drawbacks include the formation of potentially toxic byproducts, high operational costs, and ozone's sensitivity to temperature, pH, and salinity. Ion exchange resins represent another approach for dye removal from textile effluents. However, this method is often less efficient and more expensive compared to other techniques (Ali *et al.*, 2019; Ihsanullah, 2020).

4. Biological Methods for Degradation of Dyes

Current methods for dye removal like physical and chemical treatments are often expensive. Researchers are focusing on developing more cost-effective and efficient biological techniques for dye decolorization. Typically, biological techniques used alone or in conjunction with chemical and physical methods, are gaining favor due to their lower cost and environmental benefits. However, there's a growing emphasis on advanced biological approaches that are more cost-effective, and efficient, and produce less sludge as a byproduct compared to traditional approaches (Bala *et al.*, 2022). Microbial techniques are particularly effective for treating various organic and inorganic pollutants (Das *et al.*, 2023; Patel *et al.*, 2022). Microorganisms can break down, detoxify, and mineralize various synthetic dyes under specific environmental conditions (Goud *et al.*, 2020). Bacterial treatment has been highlighted as particularly efficient for dye decolorization based on several research studies.

4.1. Bacterial Remediation

Bacteria stand out amongst microbial decolorization agents due to their exceptional capabilities. Biotechnologically, they offer several advantages. One key strength is their vast range of degradative enzyme production capacity, enabling them to tackle a wide spectrum of dye molecules. This enzymatic versatility allows them to adapt to various dye compositions, making them highly effectively break down many different types of dyes, making them great at remediating dye pollution. Another significant advantage is their rapid growth rate. Bacteria reproduce quickly, leading to faster dye removal in bioremediation applications. Additionally, their ease of cultivation makes them a manageable solution for dye treatment. Bacteria can be readily cultured and maintained in controlled environments, facilitating large-scale implementation. Additionally, certain bacteria can target specific dye structures, such as those containing sulfur groups, and break them down into simpler molecules like sulfuric acid (H_2SO_4) (Nguyen *et al.*, 2016). There are two main approaches to dye decomposition using bacteria: pure cultures with a single bacterial strain, and mixed cultures known as consortia. Several studies have demonstrated efficient bio-decolorization of dyes using pure bacterial cultures. Pioneering research in the 1970s identified pure bacterial cultures of *Bacillus cereus*, *Aeromonas hydrophila*, and *B. subtilis* capable of degrading azo dyes (Wuhrmann *et al.*, 1980). Since then, diverse bacterial strains capable of dye degradation have been identified across various habitats. Wang *et al.* (2022) studied the decolorization ability of *Enterococcus faecalis* R1107 against several azo dyes. Their study included diazo dyes (Congo red, Reactive Black 5) and a triazo dye (Direct Black 38). The results indicated a negative correlation between decolorization efficiency and both the number of azo bonds within the dye molecule and the concentration of azo dyes. These data support the hypothesis that dye structure, initial dye concentration, and type of bacteria used have a significant impact on biodegradation (Wang *et al.*, 2022).

Louati *et al.* achieved complete decolorization of dyes at pH 8 with *Pseudomonas aeruginosa* strain Gb30 (Louati *et al.*, 2020). Montanez-Barragan *et al.* observed over 90% decolorization across pH 6–11 using *Halomonas* species (Montañez-Barragán *et al.*, 2020). Shi *et al.* reported 100% removal of Brilliant Crocein dye with *Providencia rettgeri* (Shi *et al.*, 2021), and Fareed *et al.* achieved 80–100% decolorization using *Bacillus cereus* cells at temperatures of 32°C, 37°C, and 45°C (Fareed *et al.*, 2022). Tripathi *et al.* demonstrated significant dye decolorization (98% of crystal violet) with a native isolate, *Aeromonas caviae* MT-1 in a recent study. Srinivasan *et al.* achieved excellent decolorization (88-96%) of various azo dyes (A. Srinivasan & Viraraghavan, 2010), and Du *et al.* reported complete decolorization of specific dyes (malachite green and crystal violet) under a wide range of pH and temperature conditions using a newly identified *Serratia* strain (Du *et al.*, 2013). Microbial communities can exhibit synergistic metabolic interactions, leading to more efficient dye degradation compared to single-strain cultures. This synergy can involve the sequential utilization of breakdown products. One member of the consortium may initiate dye degradation, producing simpler intermediates that other members can then further metabolize. Consequently, consortia often possess a broader spectrum of degradative activities compared to pure cultures. For example, Saratale *et al.* (2010) constructed a consortium (GR) composed of *Proteus vulgaris* and *Micrococcus glutamicus*. This consortium effectively decolorized six different sulfonated reactive dyes (Green HE4BD, Golden yellow HE4R, Orange 3R, Violet 5R, Red ME4BL, and Red M2BN). Notably, the consortium achieved faster decolorization of Green HE4BD compared to either bacterial strain alone (Saratale *et al.*, 2011). Abou-El-Souod and El-Sheekh (2016) examined the degradation of basic fuchsin and methyl red by the cyanobacterial strains *Hydrocoleum oligotrichum* and *Oscillatoria limnetica*. Both strains effectively degraded basic fuchsin (>90% after 7 days), but exhibited lower efficiency with methyl red (53.23% and 50.18% for *H. oligotrichum* and *O. limnetica*, respectively) (Abou-El-Souod & El-Sheekh, 2016). In their study, Wang *et al.* (2009a) highlighted the superior efficiency of biodegradation over biosorption using *Citrobacter* CK3, a bacterium isolated from textile mill-activated sludge. The living cells of *Citrobacter* CK3 were able to degrade more than 95% of reactive red 180 within 48 hours. In contrast, the decolorization achieved by heat-killed cells was only 29.7% after 120 hours of incubation. Similarly, Wang *et al.* (2013) demonstrated that living *Bacillus* sp. YZU1 could degrade the azo dye reactive black 5 significantly within 120 hours, whereas heat-killed cells managed to decolor only 3.1% of the dye (H. Wang *et al.*, 2009). The enhanced efficiency observed in living cells is primarily attributed to the activity of azoreductase. Other enzymes such as laccase and NADH-DCIP reductase also play significant roles in biodegradation processes. For instance, *Micrococcus* sp. BD15 utilizes these enzymes to effectively degrade malachite green (Du *et al.*, 2013). *Pseudomonas desmolyticum* NCIM 2112 employs oxidative enzymes, including lignin peroxidase, laccase, and tyrosinase, to completely biodegrade the diazo dye direct blue-6. Numerous enzymes are involved in the decolorization of dyes, such as laccase, azoreductase, NADH-DCIP reductase, peroxidase, manganese peroxidase, lignin peroxidase (Chen *et al.*, 2018), tyrosinase, catalase (Sudarshan *et al.*, 2023), and veratryl alcohol oxidase (S. Srinivasan & Sadasivam, 2021). Several degradation pathways have been identified. For example, the degradation of direct black 38 by *Enterococcus gallinarum* involves the elimination of the azo linkage through the action of azoreductase, resulting in the production of benzidine and 4-aminobiphenyl (4-ABP). The resulting secondary metabolites are significantly less hazardous and mutagenic compared to the original dye. For instance, The cyanobacterium *Phormidium autumnale* UTEX 1580 exhibits a high capacity for indigo dye degradation, ultimately transforming it into less hazardous and mutagenic secondary metabolites. Dellamatrice *et al.* (2017) identified these metabolites as anthranilic acid (m/z 399) and isatin (indole-2,3-dione, m/z 421). Their study suggests a multi-step degradation pathway for indigo by *P. autumnale*. The initial stage likely involves oxidative cleavage of the indigo molecule's central double bond, leading to the formation of isatin as an intermediate. Subsequent steps presumably involve hydrolysis, reduction, and decarboxylation of isatin, ultimately resulting in the formation of the non-toxic anthranilic acid (Dellamatrice *et al.*, 2017). Several studies have demonstrated the ability of single microbial strains to remove multiple dyes. For example, the marine cyanobacterium *Phormidium valderianum* has been shown to efficiently decolorize over 90% of three textile dyes—Acid Red, Acid Red 119, and Direct Black 155—primarily through adsorption (V. Shah *et al.*, 2001). Bheemaraddi *et al.* (2014) identified *Paracoccus* sp. GSM2 as ideal for treating textile wastewater containing a mixture of reactive azo dyes. This bacterium exhibited efficient decolorization not only for the single dye, reactive violet 5 but also for a mixture of six different azo dyes, achieving over 70% removal within 38 hours (Bheemaraddi *et al.*, 2014). Shah (2014) investigated the potential of *Bacillus megaterium* strain PMS82 for bioremediation of textile wastewater. This isolated strain exhibited remarkable biodegradation capabilities. Under static conditions, *B. megaterium* PMS82 completely utilized acid orange dye as its sole carbon source. Furthermore, the strain demonstrated broad-spectrum activity, effectively decolorizing over 70% of five structurally diverse azo dyes within 38 hours. The utilization of dyes by microorganisms for biodegradation varies depending on the specific strain and dye structure. Some bacteria, like the previously mentioned *Bacillus megaterium* PMS82 (Shah, 2014), can utilize dyes as their primary source of carbon, nitrogen, or phosphorus. This metabolic versatility allows them to thrive solely on the dye as an energy source. However, other microbial strains lack the necessary enzymatic machinery to break down dyes efficiently as sole carbon sources. In such cases, co-substrates, readily available organic compounds, are required to facilitate dye degradation. Karim *et al.* (2018) demonstrated this dependence on co-substrates for optimal dye removal by certain microbial strains and also investigated the biodegradation potential of indigenous bacterial isolates for the treatment of textile wastewater. The study employed a range of bacteria commonly found in such environments, including *Neisseria*, *Vibrio*, *Bacillus* sp., and *Aeromonas* sp. These isolates were evaluated for their ability to decolorize five commercially available textile dyes (Novacron orange FN, Novacron brilliant blue, Bezema yellow, and Bezema red) and a mixture of these dyes. Interestingly, The study revealed that in the lack of co-substrates like glucose and yeast extract, these bacterial isolates were unable to initiate the decolorization process. This suggests that the tested dyes were not readily utilized as the primary energy source by these bacterial strains (Karim *et al.*, 2018).

4.2. Synthetic Dye Degradation by Fungi

Due to their unique physiological and enzymatic capabilities, fungi have emerged as potent bioremediation agents for textile wastewater treatment. Their adaptable metabolism allows them to thrive under fluctuating environmental conditions often encountered in wastewater streams (Singh *et al.*, 2019). Additionally, fungi possess a robust cell morphology that enhances their resilience in harsh environments (Pietro-Souza *et al.*, 2020). Most importantly, they are equipped with efficient extracellular enzymatic machinery, particularly the ligninolytic enzyme system, which plays a crucial role in dye degradation. These enzymes and other intra and extracellular enzymes contribute to the fungi's metabolic activity, enabling them to utilize various dye components as a source of carbon and energy for growth and survival. Fungal biodegradation of textile dyes relies on the action of extracellular enzymes produced within the hyphae. These enzymes initiate a two-step degradation process: adsorption and enzymatic cleavage. First, dye molecules are adsorbed onto the fungal cell wall through various interactions like electrostatic attraction or hydrophobic binding (Rajhans *et al.*, 2021). Following adsorption, the enzymes break down the complex structures of dye molecules via enzymatic cleavage. Among fungi, white-rot fungi are particularly well-suited for dye removal due to their production of non-specific ligninolytic enzymes. These enzymes exhibit broad activity against various aromatic compounds, including dyes. However, despite their effectiveness, several limitations hinder the widespread application of white-rot fungi in dye bioremediation (Dao *et al.*, 2019; Mir-Tutusaus *et al.*, 2018). One challenge is the inherent variability in enzyme production by these fungi, which can significantly impact dye degradation efficiency. Additionally, the complex nature of certain dyes may necessitate large-scale reactors to ensure complete degradation (Das *et al.*, 2016). Furthermore, white-rot fungi often require extended growth phases before reaching their peak degradation activity. Finally, these fungi may thrive in specific nutrient conditions, such as environments with limited nitrogen availability, which can pose challenges for practical wastewater treatment applications. Employing single-strain fungal cultures for dye bioremediation holds promise but may not be optimal for long-term applications. Extended treatment periods (20-30 days) can lead to the introduction and proliferation of bacteria within the system. These bacterial populations can compete with the fungi for nutrients and space, disrupting the fungal dominance and potentially hindering efficient dye degradation. Fungal biodegradation of textile dyes is a complex process influenced by various environmental and nutritional factors. Key parameters include initial growth conditions, nutrient concentration (dose), temperature, pH, oxygen availability, characteristics of the dye wastewater, any pretreatment steps employed, and the presence of specific carbon and nitrogen sources (Singh *et al.*, 2019). The interplay between these factors can significantly impact fungal metabolic activity and enzyme production, ultimately affecting dye degradation efficiency. For instance, Neoh *et al.* (2015) investigated the impact of carbon and nitrogen sources on the degradation of Congo Red, Acid Orange 7, and Reactive Black 5 dyes by the fungus *Curvularia clavata* NZ2. Their study revealed that while most dyes were degraded in both the presence and absence of additional carbon and nitrogen sources, Acid Orange 7 decolorization was inhibited by the presence of a nitrogen source. The inhibition of decolorization in the presence of a nitrogen source suggests that the metabolic pathways involved in decolorization might be influenced or even dependent on the availability of nitrogen (Neoh *et al.*, 2015b). Akar *et al.* investigated the biosorption of Acid Red 57 by *Neurospora crassa*, analyzing the impact of various factors (Akar *et al.*, 2006). The study revealed rapid dye uptake, reaching equilibrium within 40 minutes. Interestingly, increasing temperature resulted in a decrease in the adsorption capacity of the fungus for AR57. At higher pH values, biosorption efficiency declined due to competition between the negatively charged AR57 ions and the abundance of hydroxyl ions for binding sites on the fungal cell wall. This competition phenomenon aligns with observations reported for the biosorption of another textile dye, Acid Blue

40, by *Thuja orientalis* cone biomass (Akar *et al.*, 2008). Studies by Rajesh *et al.* (2023) investigated the impact of the carbon-to-nitrogen ratio (C/N) and initial Brilliant Green dye concentration on its decolorization by *Aspergillus* sp. strain CB-TKL-1. Their findings revealed a positive correlation between the C/N ratio and decolorization percentage. Conversely, decolorization efficiency decreased when the initial dye concentration was high (Rajesh *et al.*, 2023). Similarly, research by Arunprasath *et al.* (2019) examined the influence of pH, temperature, and Malachite Green concentration on its biodegradation by *Lasiodiplodia* sp. The isolated fungal strain exhibited the ability to degrade MG within an acidic pH range, with degradation efficiency declining significantly at neutral and alkaline conditions. Similarly, the biodegradation process increased with rising temperature, reaching a maximum of 30°C. Beyond this point, degradation efficiency gradually decreased. Furthermore, increasing the initial dye concentration led to a reduction in the percentage of MG degradation (Arunprasath *et al.*, 2019). While fungal growth and reproduction typically involve spore formation, their potential for pollutant removal through biosorption has received limited research attention. Li *et al.* 2019 investigated the use of pre-formed pellets of *Aspergillus niger* for the simultaneous removal of acidic anionic dyes and fungal growth. Notably, these pellets demonstrated a high removal efficiency, exceeding 98% for a dye concentration of 200 mg/L. Additionally, the study explored the growth cycle of *Aspergillus niger*, demonstrating the formation of new spores or pellets within approximately three days (Li *et al.*, 2019). The study observed progressive dye adsorption throughout the fungal pellets, taking approximately six days for completion. The decolorization rate exhibited a two-stage pattern: slow initially for the first three days, followed by a significant increase in the rate of decolorization. This suggests a potential mechanism with initial binding followed by a more efficient pollutant sequestration process. Electrostatic interactions and complexation are likely the primary mechanisms for dye uptake by the fungal biomass. The extensive network of fungal hyphae and spores provides a large surface area, facilitating efficient pollutant mobilization and binding, further enhancing the overall removal efficiency (Deshmukh *et al.*, 2016). Hoque *et al.* studied that various fungal components from aquatic *Mucor hiemalis* strains (microbiomes, germinated spores, cell walls) effectively remove metals (>81% to 99%) via biosorption and precipitation on the spore surface. Electrostatic interactions between negatively charged spore sites and positively charged metal cations likely facilitate this process (Hoque & Fritscher, 2019). Spore formation in *Aspergillus carbonarius* membranes for wastewater treatment can compromise membrane permeability, indicating a potential trade-off between metal bioremediation and membrane function (Isik *et al.*, 2019).

4.3. Synthetic Dye Degradation by Algae

Algae are gaining significant interest for their potential as sustainable biosorbents in wastewater treatment due to their widespread presence in both saltwater and freshwater environments. This abundant biomass offers several advantages, including high biosorption capacity and biocoagulant properties, which are attributed to their large surface area and high binding affinity for pollutants. Furthermore, specific functional groups like phosphate, hydroxyl, amino, and carboxylate on the algal cell wall play a crucial role in pollutant removal through complexation and electrostatic interactions. A comprehensive analysis of the existing literature suggests that algae utilize three primary mechanisms for dye degradation in wastewater treatment (Holkar *et al.*, 2016). These mechanisms include 1. By transformation: dyes are transformed into colorless intermediates through enzymatic or photolytic processes within algal cells 2. Nutrient source: some algae can directly utilize dyes as a source of carbon and energy for their growth and proliferation, or 3. By biosorption: dyes can be adsorbed onto the algal cell wall through physical interactions such as chromophore adsorption, potentially involving electrostatic interactions and complexation with functional groups on the cell surface (Kulkarni *et al.*, 2018). Several factors, including pH, temperature, and the presence of specific functional groups such as phosphate, hydroxyl, amino, and carboxylate groups influence the efficiency of algal remediation in treating textile wastewater. Algal biomass is particularly effective in removing common textile dyes, including Tartrazine, Navy Blue HE22, Congo Red, Rhodamine B, Remazol Black B, Amido Black 10B, Malachite Green, Methyl Red, and Basic Fuchsin. The equilibrium data for the biosorption of Methylene Blue (MB) on *Scenedesmus* at various temperatures and pH values are represented in this study (Afshariani & Roosta, 2019). Increasing the temperature from 22°C to 30°C positively affected the adsorption capacity, but a further increase to 37°C led to a decrease in capacity. Additionally, the adsorption capacity improved with an increase in pH from 5 to 7. This enhancement at higher pH values is likely due to the reduced concentration of hydrogen ions, resulting in less competition between MB molecules and hydrogen ions for the adsorption sites on the microalgae. The experimental adsorption data for MB closely aligned with the Langmuir isotherm model at 30°C. Other researchers have reported similar findings. Aksu and tezer *et al.* 2005 found that the adsorption capacity of various reactive dyes on the green alga *Chlorella vulgaris* decreased with increasing temperature but increased with pH up to a value of 2, after which it declined with further pH increase (Aksu & Tezer, 2005). Conversely, Baldev *et al.* 2013 observed that the decolorization percentage of Rhodamine B by *Coelastrella* sp. increased with rising pH (Baldev *et al.*, 2013).

4.4. Synthetic Dye Degradation by Plants

The potential of plants for degrading azo dyes has emerged as a promising avenue for bioremediation efforts. Untreated industrial wastewater containing azo dyes poses a significant threat to aquatic ecosystems due to their toxicity and persistence. Therefore, exploring phytoremediation alongside established bioremediation techniques is crucial for effective wastewater treatment. Plants possess the capability to degrade azo dyes through the production of specific enzymes, including lignin peroxidases and manganese peroxidases. These enzymes play a key role in breaking down the complex azo bonds present in the dye molecules. Furthermore, plant roots can facilitate the creation of both aerobic and anaerobic environments within the rhizosphere (zone of influence around the roots). This dual environment is advantageous for efficient dye degradation, as some azo dye breakdown processes occur under aerobic conditions while others require anaerobic conditions. Additionally, plant roots can directly absorb azo dyes from wastewater, contributing to decolorization and reducing the overall dye load in the treated water (Goud *et al.*, 2020). Azo dye degradation often involves an initial anaerobic step that produces aromatic amines as intermediates. These aromatic amines are well-documented to possess mutagenic and carcinogenic properties, posing a potential environmental threat. However, subsequent aerobic processes can effectively mineralize these harmful intermediates. Plant-microbe consortia can be particularly efficient in facilitating this sequential degradation. Plants can take up azo dyes and provide an anoxic environment within their rhizosphere for initial anaerobic breakdown. The resulting aromatic amines can then be mineralized by aerobic microbes present in the same rhizosphere, leading to complete degradation of the dye molecule. This collaborative approach has been demonstrated with azo dye Methyl Red, achieving up to 92% decolorization through plant-microbe association (Sudarshan *et al.*, 2022).

4.5. Pros and Cons of Bioremediation

Bioremediation offers several advantages over conventional wastewater treatment strategies, making it an attractive and environmentally friendly option. Here's a breakdown of its key benefits:

- Natural Process: Bioremediation utilizes naturally occurring microorganisms to degrade pollutants. This eliminates the need for harsh chemicals and aligns with natural ecological processes.
- Cost-Effectiveness: Compared to conventional methods requiring complex infrastructure, bioremediation can be significantly less expensive, particularly for on-site applications.

- *In-Situ* Treatment: Bioremediation can be implemented directly at the source of contamination, minimizing transportation costs and environmental risks associated with transporting wastewater.
- Complete Degradation: Bioremediation facilitates the complete breakdown of pollutants like dyes into harmless byproducts, such as carbon dioxide and water, eliminating the need for further treatment or disposal of hazardous residues.
- Simple and Sustainable: Bioremediation processes are often simpler to operate and maintain compared to conventional methods. Additionally, they rely on naturally occurring microbial populations, promoting long-term sustainability.
- Reduced Environmental Impact: Bioremediation avoids the transfer of contaminants from one environmental medium to another, minimizing the risk of secondary pollution events.

Overall, bioremediation offers a promising and sustainable approach to wastewater treatment, leveraging natural processes to achieve effective and environmentally responsible pollutant removal. Despite the numerous advantages of bioremediation, several factors limit its universal application:

- Substrate Specificity: Bioremediation is primarily effective for biodegradable contaminants. Complex pollutants or recalcitrant compounds (those resistant to degradation) may not be readily broken down by microorganisms.
- Time Constraints: Bioremediation processes can be slower than some conventional methods. Factors like microbial growth rate, nutrient availability, and contaminant complexity influence the rate of degradation.
- Environmental Dependence: Optimal bioremediation relies on specific environmental conditions, including pH, temperature, and nutrient levels. Adjusting these parameters in situ can be challenging, and natural fluctuations can impact treatment efficiency.
- Microbial Community: The success of bioremediation hinges on the presence of metabolically capable microbial populations within the contaminated environment. Introducing or enhancing desired microbial communities may be necessary in some cases.

By overcoming these challenges, bioremediation has the potential to become an even more powerful tool for sustainable wastewater treatment and environmental remediation. Additional research is required to develop improved bioremediation methods capable of treating wastewater containing various contaminants (Ghosh *et al.*, 2016; Ihsanullah *et al.*, 2020a; Mishra *et al.*, 2021)

5. Next-Generation Techniques for Textile Wastewater Treatment

Conventional biological processes offer a cost-effective and ecologically sound approach to sludge reduction in wastewater treatment. These methods rely on microbial activity for organic matter degradation, minimizing the need for chemical inputs and reducing overall energy consumption compared to some alternative methods. This translates to advantages in terms of both operational costs and environmental impact. However, limitations associated with conventional biological sludge reduction techniques necessitate further advancements. The primary challenge lies in the relatively slow processing times associated with complete biological degradation. Additionally, these processes often require larger land areas to accommodate the necessary infrastructure, such as aeration tanks and settling basins. Furthermore, the efficacy of biological treatment can be significantly compromised by the presence of toxic chemicals or fluctuations in influent wastewater composition (e.g., organic load, pH). In light of these limitations, ongoing research efforts are focused on developing advanced techniques or modifications to existing biological treatment processes (Vikrant *et al.*, 2018). This section delves into these recent developments, including strain improvement for enzyme immobilization, applications of nanotechnology, genetic engineering, microbial electrochemical technologies (METs), and phytoremediation with their advantages and disadvantages shown in Table 1.

Table 1. Advantages and disadvantages of various emerging Technologies for Textile Dye Wastewater Treatment

Technique	Advantages	Disadvantages	Citation
Strain Improvement	Eco-friendly, cost-effective, high efficiency	Time-consuming, requires genetic engineering expertise	(El-Sayed *et al.*, 2019)
Nanotechnology	High efficiency, selective removal, potential for reuse	Potential toxicity, cost of synthesis and application	(Zhao, 2020)
Genetic Engineering	Genetically modified microorganisms, enhanced degradation capabilities	Ethical concerns, potential for unintended consequences	(Wang *et al.*, 2021)
Microbial Electrochemical Technologies (METs)	Energy recovery, simultaneous waste water treatment and electricity generation	Complex technology, lower efficiency for some dyes	(Das, 2022)
Phytoremediation	Low-cost, environmentally friendly, aesthetic appeal	Slow process, limited to certain dyes and conditions	(Nedjimi, 2021)

5.1. Strain Improvement for Enzyme Immobilization

Microbial strain improvement is a powerful tool in bioremediation, allowing scientists to modify the characteristics of microbes to enhance their ability to break down pollutants. This technique can be achieved through two main approaches. The first involves optimizing the surrounding environment for the target microbes. This means fine-tuning factors like temperature, agitation, pH, oxygen levels, and the availability of essential nutrients (carbon and nitrogen sources) as well as specific enzymes. By creating ideal growth conditions, researchers can favor the proliferation of desired microbial strains that are most effective at degrading contaminants. The second approach utilizes more advanced genetic manipulation techniques. These methods, such as mutagenesis, protoplast fusion, and recombinant DNA technology, allow for direct modification of the microbial genome (Lou *et al.*, 2021). This targeted manipulation can introduce genetic changes that enhance a strain's ability to degrade specific pollutants. The primary goal is to boost productivity and enhance enzyme activity, enabling faster growth in a shorter time. Enzyme immobilization has emerged as a promising technique to improve the efficiency and cost-effectiveness of bioremediation processes. This approach involves physically or chemically confining enzyme molecules within a carrier material (support/matrix). The key advantage of immobilization lies in the ability to retain enzyme activity while facilitating easy separation from the treated solution (El-Sayed *et al.*, 2019). This allows for repeated use of the enzymes, minimizing production costs and enhancing overall process sustainability. Recent studies demonstrate the effectiveness of this approach for the removal of different pollutants, including dyes. For instance, Silva et al, report the successful use of immobilized fungal strains of *Trametes versicolor* on a wheat bran-filled polyethylene mesh for the decolorization of malachite green dye (reaching 90% removal within 240 minutes) (da Silva *et al.*, 2019). Another study by Antecka *et al.* investigated the use of Trametes Versicolor immobilized on different metal oxide materials (TiO_2-ZrO_2 and TiO_2-ZrO_2-SiO_2) for the removal of Reactive Black 5 up to 77%, Alizarin Red S up to 100%, and Remazol Brilliant Blue Rup to 91% at pH 5, 25 °C Temperature and 5 mg/L concentration of dye (Antecka *et al.*, 2018).

5.2. Nanomaterials in Bioremediation

Over the past few decades, nanotechnology has become a ubiquitous tool across diverse fields, including pharmaceuticals, medicine, bioremediation, and food processing industries (Liu & Dong, 2020). This is partly due to the rapid advancements in nanobiotechnology, which have enabled the creation of novel nano-structured materials specifically designed to support immobilized enzymes (Sheldon *et al.*, 2021; Zahirinejad *et al.*, 2021). More recently, there has been a growing interest in exploring the potential of nanoparticles for wastewater treatment, particularly for the removal of dyes. This interest stems from the unique properties of nanoparticles, such as their exceptionally large specific surface area, superior adsorption capacities, and reduced diffusion resistance (Homaeigohar, 2020). Current research has shifted its focus towards advanced functional materials, particularly metal-organic frameworks (MOFs), and the utilization of bio-based nanomaterials for the elimination of synthetic dyes. MOFs exhibit exceptional dye removal capabilities due to their ability to interact with both anionic and cationic dye molecules electrostatically. Numerous studies have documented the successful application of MOFs for dye removal, achieving impressive results (Karmakar *et al.*, 2019). Additionally, bio-nanomaterials, a class of materials derived from various biological sources such as peptides, nucleic acids, bacteria, plants, and fungi, hold significant potential in dye removal applications (Zhao, 2020). Despite their nanoscale dimensions, bio-nanomaterials possess remarkable functionalities due to their biological origin and inherent biocompatibility. These properties have garnered significant interest in various fields, including aerospace engineering, materials science, bioremediation, pharmacology, and biosensor development (Menon *et al.*, 2021). Recent studies have explored the potential of bio-synthesized silver nanoparticles using plant extracts from *Solanum nigrum* and *Cannabis sativa* for the degradation of synthetic textile dyes (M. Ahmad *et al.*, 2020). These studies investigated the impact of various parameters such as incubation time, nanoparticle concentration, pH, and temperature on dye removal efficiency. Notably, *Solanum nigrum*-derived nanoparticles exhibited superior decolorization activity for both yellow and red dyes (68.2% and 77.5%, respectively) compared to *Cannabis sativa* (27.02% and 45.7%, respectively) under identical conditions of pH 9 and temperature 60°C (Khatoon *et al.*, n.d.). Palm waste biochar has been investigated as a precursor for the synthesis of novel carbon dots (nCD-DBC) and nano zero-valent iron composites (nZVI-DBC) for methylthionine chloride (MB) dye removal. Studies have shown that under optimal pH conditions of 7.0, nCD-DBC exhibits a high decolorization efficiency of 83% for MB dye, while nZVI-DBC demonstrates a removal capacity of 38% (M. Ahmad *et al.*, 2020).

5.3. Genetic Engineering for the Degradation of Dyes

Microbial biodegradation can be enhanced through the targeted manipulation of microorganisms. Genetically engineered microorganisms (GEMs), also referred to as genetically modified microorganisms (GMMs), are constructed using techniques that mimic natural genetic exchange processes in microbes. These GEMs have emerged as a powerful tool for bioremediation due to their tailored capabilities for degrading specific contaminants (Kumar *et al.*, 2019). Four main engineering strategies contribute to GEM development for bioremediation applications: (1) targeted modification of enzyme specificities and affinities; (2) construction and regulation of novel metabolic pathways; (3) optimization of growth conditions, monitoring, and control systems for bioprocesses; and (4) utilization of bio-affinity for biosensing, reducing contaminant toxicity, and endpoint analysis. These strategies can involve interspecies gene transfer or direct genetic modification to achieve the desired biodegradation capabilities. Interspecies transfer of functional genes from environmentally relevant bacteria, such as *Sphingomonas desiccabilis, Escherichia coli, Bacillus idriensis, Pseudomonas putida, Mycobacterium marinum*, and *Ralstonia eutropha*, has emerged as a powerful strategy for engineering microorganisms for bioremediation applications. This approach has proven effective in degrading various pollutants, including chlorobenzene acids, toluene, halogenated pesticides, and toxic wastes (Pant *et al.*, 2021; Pimviriyakul *et al.*, 2020; Wang *et al.*, 2021). However, there is a scarcity of research exploring this technique for dye degradation (Zhu *et al.*, 2020). This highlights the potential of interspecies gene transfer as a novel approach for developing microbes capable of dye bioremediation.

5.4. Phyto-remediation for Biodegradation of Dyes

Plants growing near pollution can use their natural processes to absorb and break down the pollutants, cleaning up the environment. This is similar to a technology called phytoremediation, which uses plants to remove pollution (Nedjimi, 2021; Tiodar *et al.*, 2021). Plants possess various mechanisms for dealing with pollutants in their environment. These mechanisms include uptake and accumulation within plant tissues (phytoextraction), adsorption onto root surfaces (rhizofiltration), enzymatic transformation and release into the air (phytotransformation), volatilization from plant surfaces (phytovolatilization), degradation by root-zone microbes (rhizoremediation), and immobilization within soil particles (phytostabilization) (Dzionek *et al.*, 2016). A recent study by Shanmugam *et al.* demonstrated the effectiveness of Bacopa monnieri in degrading azo dyes (Shanmugam *et al.*, 2020). This study found that *Bacopa monnieri* could remove up to 90-100% of 14 different azo dyes (including 9 reactive and 4 direct dyes) at a concentration of 40 mg/L within a two-week incubation time. When compared to individual cultures of plant and bacteria, a consortium of *Portulaca grandiflora* and *Pseudomonas putida* exhibited a synergistic effect, achieving complete decolorization of Direct Red 5B within 72 hours. While *P. grandiflora* and *P. putida* alone achieved decolorization efficiencies of 92% and 81% for Direct Red 5B within 96 hours, respectively, the consortium demonstrated a significantly faster and more complete degradation process. Similarly, research has shown the efficacy of *Salvinia molesta Mitchell* in dye removal. Under optimal conditions of pH 6.5 and 50 mg/L dye concentration, *S. molesta* achieved a decolorization rate of 90% for Direct Red 28 (Kaushal & Mahajan, 2022).

5.5. Microbial Electrochemical Technologies (MET)

Microbial Electrochemical Technologies (METs) exploit the electron transfer capabilities of microorganisms to convert chemical energy into electrical energy. Microbial Fuel Cells (MFCs) and Biophotovoltaic systems are established examples of METs for bioenergy production and wastewater treatment. MFCs function as bio-electrochemical systems that utilize microorganisms to convert organic matter in the anode chamber into electrical energy. Electrons released during the microbial oxidation of organics at the anode are transferred to an electrode. These electrons then travel through an external circuit to the cathode, where they participate in the reduction of a terminal electron acceptor, often oxygen. The performance of MFCs is influenced by several factors, including the selection of electrode materials, external resistance, applied potentials if used, initial concentration of pollutants like azo dyes, and the presence of co-substrates that may enhance biodegradation. MFCs can be designed in two main conFig.urations: single-chambered and double-chambered. In double-chambered MFCs, a proton exchange membrane separates the anode and cathode compartments (Das, 2022). Microbial communities within the anode chamber of MFCs play a crucial role in dye removal by anaerobically oxidizing the pollutants as substrates. This process generates electrons that are transferred to the anode. Several studies have demonstrated the effectiveness of MFCs in dye degradation from wastewater. Jain and Zhen investigated a bioelectrochemical system where the interaction between electron acceptors and donors facilitated efficient dye remediation through electricity generation (Jain & Zhen, 2018). Similarly, Yang *et al.* 2009 reported dye degradation using a bioelectrical system, highlighting the role of microbes as biocatalysts for anode oxidation of acetate (Yang *et al.*, 2009). Furthermore, Yadav *et al.* 2022 observed electricity generation alongside exceptional dye removal efficiency (96-100%) within their MFC. This property holds promise as a sustainable alternative to fossil fuel-based energy generation and conventional wastewater treatment methods. METs offer the potential to generate bioelectricity and achieve bioremediation of pollutants, including those present in complex effluents like textile wastewater. Notably, recent research has demonstrated the effectiveness of MFCs in removing various textile dyes from wastewater, including Acid Orange 5, Active Red 30, Reactive Brilliant Red X-3B, and Acid Blue 29 (Fatima *et al.*, 2021; Khan *et al.*, 2021; Ma *et al.*, 2021; Qiu & Liu, 2021). While MFCs exploit microbial metabolism, Biophotovoltaic (BPV) systems leverage the photosynthetic capabilities of cyanobacteria or algae. In BPV systems, light energy is used to drive the splitting of water molecules, generating electricity (Nath *et al.*, 2021). Both MFCs and BPVs hold promise for diverse applications beyond wastewater treatment. These include the development of biosensors for environmental monitoring of toxic compounds, powering self-propelled environmental robots, and even integrated functionalities in "smart" building materials (Jadhav *et al.*, 2019). The versatility of METs highlights their potential as a robust and sustainable technology for wastewater treatment and bioenergy production (Logan *et al.*, 2019).

6. Exploring Integrated and Hybrid Technologies for Textile Dye Treatment

The accumulation of dyestuffs and dye-containing wastewater poses a significant environmental threat due to pollution and associated health and aesthetic concerns. The complex nature of these pollutants presents a challenge for wastewater treatment. Dyes are often recalcitrant, meaning they resist biodegradation. Additionally, the composition of dye wastewater can vary considerably, making a single treatment approach often inadequate for achieving regulatory compliance. Several studies have highlighted the limitations of single-treatment methods for contemporary dye wastewater treatment due to increasingly stringent environmental regulations (Dehghani *et al.*, 2021). The complex composition of industrial wastewater, often containing a mixture of dyes, heavy metals, and other contaminants, presents a significant challenge for single-treatment methods. These diverse pollutants may require different removal mechanisms, rendering a single approach ineffective (Li *et al.*, 2019). Researchers are increasingly exploring hybrid treatment processes that combine multiple techniques to address this limitation. These hybrid systems can be tailored to target specific pollutants based on the wastewater characteristics, thereby achieving more efficient and comprehensive treatment. It is crucial to differentiate between integrated and hybrid processes. Integrated processes combine physical, chemical, and biological treatment steps within a single unit, while hybrid processes involve sequential application of different treatment units (Dehghani *et al.*, 2021). Hybrid treatment processes offer several advantages over single-treatment methods, particularly for complex wastewater containing diverse contaminants. By combining multiple treatment techniques within a single system, hybrid processes can potentially reduce operational costs, minimize resource requirements, and simplify operational complexity (Patel *et al.*, 2020; Pineda Zapana *et al.*, 2022). This approach has gained significant interest not only in the treatment of textile wastewater, but also in the remediation of various other contaminated effluents, including municipal wastewater, piggery wastewater, domestic oily wastewater, pharmaceutical wastewater, tannery wastewater, hospital wastewater, and poultry slaughterhouse wastewater (Baker *et al.*, 2021; Pirsaheb *et al.*, 2020). The Fig.ure 2 illustrates the traditional methods employed for the removal of textile dyes, as well as the integration of hybrid processes.

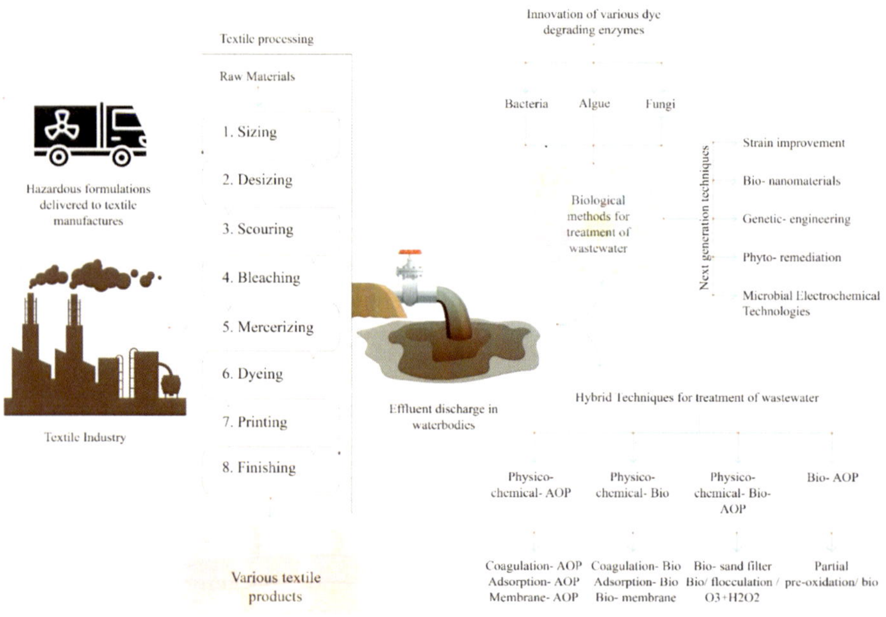

Fig. 2: Industrial Strategies for Managing Textile Processing Waste and Advanced Effluent Treatment Methods

These hybrid processes provide researchers with precise approaches for dye removal treatment. Table 2 presents various studies on dye removal, showcasing different physio-chemical techniques combined with biological processes.

Table 2. A Comparative Analysis of Hybrid Techniques for Effective Textile Effluent Management

Hybrid Technique	**Investigating Parameters/ Methods**	**Major Findings**	**Prospects**	**References**
Fenton Process + Biological Treatment	COD removal with Sequencing Batch Reactor	Effective technique with 87.1% COD removal at lower chemical dosage.	Further optimization of process parameters, scaling-up for industrial applications, and coupling with other advanced oxidation processes.	(Balcik-Canbolat *et al.*, 2019)
Ozonation + Biological Treatment	Decolorization of Acid Orange 7	98.7% decolorization at pH 7.0 and 350 mg/L ozone dosage.	Integrating with other advanced oxidation processes, exploring advanced oxidation catalysts, and optimizing operational parameters.	(Sonwani *et al.*, 2021)
Advanced Oxidation Process + Biological Treatment	Decolorization and COD, BOD removal	Complete decolorization of dyes, significant reduction in COD and BOD.	Developing more efficient microbial strains, combining with other advanced oxidation processes, and optimizing reactor conFig.urations.	(Thanavel *et al.*, 2019)
Photocatalytic Process + Biological Treatment	Decolorization, COD, and organic matter removal	97% decolorization and 63% organic matter removal.	Developing novel photocatalysts with enhanced activity, integrating with solar energy, and optimizing reactor design.	(Molinari *et al.*, 2020)
Advanced Oxidation Process (AOPs) + Biological Treatment	Decolorization of Reactive Violet 5	91.3% decolorization of Reactive Violet 5 (500 mg/L) within 60 min.	Combining with other advanced oxidation processes, developing more efficient catalysts, and optimizing operational conditions.	(Meerbergen *et al.*, 2017)
Constructed Wetlands + Biological Treatment	Removal of BOD, COD, pH, TDS, DO, and hardness	Significant reduction in pollutants, especially BOD and COD.	Combining with other advanced oxidation processes, developing more efficient catalysts, and optimizing operational conditions.	(Prasanna *et al.*, 2017)

Ozonation + Integrated Biological Treatment	Decolorization of Reactive Black 5, COD removal	94% decolorization of Reactive Black 5 and 90% COD removal.	Optimizing ozone dosage and reaction time, integrating with other biological processes, and exploring advanced oxidation catalysts.	(Venkatesh *et al.*, 2017)
Electrochemical Oxidation + Bio-oxidation	COD removal from real textile wastewater	Effective approach with 90% COD removal.	Developing more efficient electrodes, optimizing operating conditions, and integrating with other biological processes	(Aravind *et al.*, 2016)
Fenton Process + Biological Treatment	Decolorization of Remazol Red dye, COD removal	90% decolorization of Remazol Red dye and 92% COD removal.	Optimizing reaction conditions, exploring alternative catalysts, and integrating with other treatment technologies.	(Punzi *et al.*, 2015)
Oxidation Process + Biological Treatment	Decolorization, COD removal, TSS removal	Significant reduction in decolorization, COD, and TSS.	Developing more efficient oxidation processes, integrating with other treatment technologies, and optimizing operational parameters.	(Nawaz & Ahsan, 2014)
Membrane Filtration + Biological Treatment	Decolorization, COD, SS removal	High efficiency in decolorization, COD, and SS removal.	Developing more efficient membranes, optimizing membrane fouling control, and integrating with other treatment technologies.	(Dong *et al.*, 2014)

7. Current State and Future Outlook of Bioremediation in Textile Dye Treatment

While bioremediation offers a promising approach to water purification, several key challenges limit its widespread adoption in large-scale commercial applications. Addressing these challenges is crucial to fully realize the potential of bioremediation for practical water treatment (Kusumlata *et al.*, 2024).

- Scaling-up from Batch to Continuous Processes: While many studies demonstrate bioremediation's efficacy at the bench scale (batch studies), its feasibility for large-scale treatment requires investigation in continuous flow systems that mimic real-world applications.
- Microbial Strain Development: Developing genetically modified microbial strains with superior biodegradation capabilities could significantly improve treatment efficiency.
- Optimizing Microbial Fuel Cells: Although MFCs hold promise for bioremediation coupled with energy production, limitations like low power output and high installation costs need to be addressed for practical implementation.
- Bioenergy from Wastewater: The emerging field of bioremediation coupled with biofuel generation (e.g., hydrogen production) offers a sustainable approach to wastewater treatment while generating clean energy. Further research is necessary to explore the commercial viability of biohydrogen production from wastewater (Sanjana *et al.*, 2024).
- Optimizing Biodegradation Conditions: A thorough understanding of degradation mechanisms, critical operating parameters, and optimal growth conditions for pollutant-degrading microorganisms is essential for maximizing bioremediation efficacy.
- Real-World Wastewater Applicability: Most bioremediation studies focus on synthetic wastewater. Evaluating the effectiveness of bioremediation on real wastewater is essential to assess its practical applicability.
- Multidisciplinary Approach: Addressing the current hurdles and expanding bioremediation applications require a collaborative effort across scientific disciplines (Mian *et al.*, 2024).

8. Conclusion

The textile industry's heavy reliance on azo dyes poses a significant environmental threat. While conventional methods for dye removal are costly and often ineffective at low concentrations, microbial bioremediation offers a promising, eco-friendly solution. Microbial consortia, including bacteria, yeast, and fungi, along with their enzymes like lignin peroxidase and azoreductase, can effectively degrade azo dyes into non-toxic compounds under both aerobic and anaerobic conditions. This sustainable approach, combined with other advanced technologies, presents a viable and environmentally responsible strategy for mitigating the impact of azo dye pollution on ecosystems.

Acknowledgment

The authors would like to express their sincere gratitude to the Gujarat State Biotechnology Mission (GSBTM) for their generous support via the Research Support Scheme. The research funding under GSBTM/JD(R&D)/662/2022-23/00292647 is righteously acknowledged. Their support has been instrumental in advancing our understanding of these important environmental issues and the development of technological advancements. We sincerely acknowledge the Department of Microbiology and Biotechnology, Gujarat University, Ahmedabad, for providing the exceptional facilities and resources that were instrumental in conducting this research.

References

Abou-El-Souod, G. W., & El-Sheekh, M. M. (2016). Biodegradation of basic fuchsin and methyl red by the blue-green algae Hydrocoleum oligotrichum and Oscillatoria limnetica. Environmental Engineering and Management Journal, 15(2), 279–286. https://doi.org/10.30638/eemj.2016.028

Afshariani, F., & Roosta, A. (2019). Experimental study and mathematical modeling of biosorption of methylene blue from aqueous solution in a packed bed of microalgae Scenedesmus. Journal of Cleaner Production, 225, 133–142. https://doi.org/10.1016/j.jclepro.2019.03.275

Ahmad, M., Akanji, M. A., Usman, A. R. A., Al-Farraj, A. S. F., Tsang, Y. F., & Al-Wabel, M. I. (2020). Turning date palm waste into carbon nanodots and nano zerovalent iron composites for excellent removal of methylthioninium chloride from water. Scientific Reports, 10(1). https://doi.org/10.1038/s41598-020-73097-x

Akar, T., Demir, T. A., Kiran, I., Ozcan, A., Ozcan, A. S., & Tunali, S. (2006). Biosorption potential of Neurospora crassa cells for decolorization of Acid Red 57 (AR57) dye. Journal of Chemical Technology and Biotechnology, 81(7), 1100–1106. https://doi.org/10.1002/jctb.1462

Akar, T., Ozcan, A. S., Tunali, S., & Ozcan, A. (2008). Biosorption of a textile dye (Acid Blue 40) by cone biomass of Thuja orientalis: Estimation of equilibrium, thermodynamic and kinetic parameters. Bioresource Technology, 99(8), 3057–3065. https://doi.org/10.1016/j.biortech.2007.06.029

Ali, I., Burakova, I., Galunin, E., Burakov, A., Mkrtchyan, E., Melezhik, A., Kurnosov, D., Tkachev, A., & Grachev, V. (2019). High-Speed and High-Capacity Removal of Methyl Orange and Malachite Green in Water Using Newly Developed Mesoporous Carbon: Kinetic and Isotherm Studies. ACS Omega, 4(21), 19293–19306. https://doi.org/10.1021/acsomega.9b02669

Ali, K., Javaid, M. U., Ali, Z., & Zaghum, M. J. (2021). Biomass-derived adsorbents for Dye and Heavy Metal Removal from Wastewater. In Adsorption Science and Technology (Vol. 2021). Hindawi Limited. https://doi.org/10.1155/2021/9357509

Al-Tohamy, R., Sun, J., Fareed, M. F., Kenawy, E. R., & Ali, S. S. (2020). Ecofriendly biodegradation of Reactive Black 5 by newly isolated Sterigmatomyces halophilus SSA1575, valued for textile azo dye wastewater processing and detoxification. Scientific Reports, 10(1). https://doi.org/10.1038/s41598-020-69304-4

Antecka, K., Zdarta, J., Siwińska-Stefańska, K., Sztuk, G., Jankowska, E., Oleskowicz-Popiel, P., & Jesionowski, T. (2018). Synergistic degradation of dye wastewaters using binary or ternary oxide systems with immobilized laccase. Catalysts, 8(9). https://doi.org/10.3390/catal8090402

Aravind, P., Selvaraj, H., Ferro, S., & Sundaram, M. (2016). An integrated (electro- and bio-oxidation) approach for remediation of industrial wastewater containing azo-dyes: Understanding the degradation mechanism and toxicity assessment. Journal of Hazardous Materials, 318, 203–215. https://doi.org/10.1016/j.jhazmat.2016.07.028

Arunprasath, T., Sudalai, S., Meenatchi, R., Jeyavishnu, K., & Arumugam, A. (2019). Biodegradation of triphenylmethane dye malachite green by a newly isolated fungus strain. Biocatalysis and Agricultural Biotechnology, 17, 672–679. https://doi.org/10.1016/j.bcab.2019.01.030

Awasthi, A., & Datta, D. (2019). Application of Amberlite XAD-7HP resin impregnated with Aliquat 336 for the removal of Reactive Blue - 13 dye: Batch and fixed-bed column studies. Journal of Environmental Chemical Engineering, 7(6). https://doi.org/10.1016/j.jece.2019.103502

Baker, B. R., Mohamed, R., Al-Gheethi, A., & Aziz, H. A. (2021). Advanced technologies for poultry slaughterhouse wastewater treatment: A systematic review. Journal of Dispersion Science and Technology, 42(6), 880–899. https://doi.org/10.1080/01932691.2020.1721007

Bala, S., Garg, D., Thirumalesh, B. V., Sharma, M., Sridhar, K., Inbaraj, B. S., & Tripathi, M. (2022). Recent Strategies for Bioremediation of Emerging Pollutants: A Review for a Green and Sustainable Environment. In Toxics (Vol. 10, Issue 8). MDPI. https://doi.org/10.3390/toxics10080484

Balcik-Canbolat, C., Olmez-Hanci, T., Sengezer, C., Sakar, H., Karagunduz, A., & Keskinler, B. (2019). A combined treatment approach for dye and sulfate-rich textile nanofiltration membrane concentrate. Journal of Water Process Engineering, 32. https://doi.org/10.1016/j.jwpe.2019.100919

Baldev, E., MubarakAli, D., Ilavarasi, A., Pandiaraj, D., Ishack, K. A. S. S., & Thajuddin, N. (2013). Degradation of synthetic dye, Rhodamine B to environmentally non-toxic products using microalgae. Colloids and Surfaces B: Biointerfaces, 105, 207–214. https://doi.org/10.1016/j.colsurfb.2013.01.008

Behl, K., Sesha Charan, P., Joshi, M., Sharma, M., Mathur, A., Kareya, M. S., Jutur, P. P., Bhatnagar, A., & Nigam, S. (2020). Multifaceted applications of isolated microalgae Chlamydomonas sp. TRC-1 in wastewater remediation, lipid production, and bioelectricity generation. Bioresource Technology, 304. https://doi.org/10.1016/j.biortech.2020.122993

Bera, S. P., & Tank, S. K. (2021). Screening and identification of newly isolated Pseudomonas sp. for biodegrading the textile azo dye C.I. Procion Red H-3B. Journal of Applied Microbiology, 130(6), 1949–1959. https://doi.org/10.1111/jam.14920

Berradi, M., Hsissou, R., Khudhair, M., Assouag, M., Cherkaoui, O., El Bachiri, A., & El Harfi, A. (2019). Textile finishing dyes and their impact on aquatic environs. In Heliyon (Vol. 5, Issue 11). Elsevier Ltd. https://doi.org/10.1016/j.heliyon.2019.e02711

Bheemaraddi, M. C., Patil, S., Shivannavar, C. T., & Gaddad, S. M. (2014). Isolation and characterization of Paracoccus sp. GSM2 capable of degrading textile azo dye reactive violet 5. Scientific World Journal, 2014. https://doi.org/10.1155/2014/410704

Chen, Y., Feng, L., Li, H., Wang, Y., Chen, G., & Zhang, Q. (2018). Biodegradation and detoxification of Direct Black G textile dye by a newly isolated thermophilic microflora. Bioresource Technology, 250, 650–657. https://doi.org/10.1016/j.biortech.2017.11.092

da Silva, L. S., Gonçalves, M. M. M., & Raddi de Araujo, L. R. (2019). Combined photocatalytic and biological process for textile wastewater treatments. Water Environment Research, 91(11), 1490–1497. https://doi.org/10.1002/wer.1143

Dao, A. T. N., Vonck, J., Janssens, T. K. S., Dang, H. T. C., Brouwer, A., & de Boer, T. E. (2019). Screening white-rot fungi for bioremediation potential of 2,3,7,8-tetrachlorodibenzo-p-dioxin. Industrial Crops and Products, 128, 153–161. https://doi.org/10.1016/j.indcrop.2018.10.059

Das, A. (2022). Dye Degradation Using Microbial Fuel Cells: A Critical Review. International Journal for Research in Applied Science and Engineering Technology, 10(11), 1745–1754. https://doi.org/10.22214/ijraset.2022.47714

Das, S., Cherwoo, L., & Singh, R. (2023). Decoding dye degradation: Microbial remediation of textile industry effluents. In Biotechnology Notes (Vol. 4, pp. 64–76). KeAi Communications Co. https://doi.org/10.1016/j.biotno.2023.10.001

Dehghani, Y., Honarvar, B., Azdarpour, A., & Nabipour, M. (2021). Pilot-scale experiments on a hybrid membrane-electrocoagulation system to produced water treatment in a domestic oil reservoir. Water Practice and Technology, 16(1), 210–225. https://doi.org/10.2166/wpt.2020.114

Dellamatrice, P. M., Silva-Stenico, M. E., Moraes, L. A. B. de, Fiore, M. F., & Monteiro, R. T. R. (2017). Degradation of textile dyes by cyanobacteria. Brazilian Journal of Microbiology, 48(1), 25–31. https://doi.org/10.1016/j.bjm.2016.09.012

Desai, C., Jain, K., Boopathy, R., Hullebusch, E., & Madamwar, D. (2021). Eco-Sustainable Bioremediation of Textile Dye Wastewaters: Innovative Microbial Treatment Technologies and Mechanistic Insights of Textile Dye Biodegradation. In C. Desai, R. Boopathy, K. R. Jain, D. Madamwar, & E. D. van Hullebusch (Eds.), Frontiers in microbiology. Frontiers Media SA. https://doi.org/10.3389/978-2-88971-380-6

Deshmukh, R., Khardenavis, A. A., & Purohit, H. J. (2016). Diverse Metabolic Capacities of Fungi for Bioremediation. In Indian Journal of Microbiology (Vol. 56, Issue 3, pp. 247–264). Springer India. https://doi.org/10.1007/s12088-016-0584-6

Dong, B., Chen, H., Yang, Y., He, Q., & Dai, X. (2014). Treatment of printing and dyeing wastewater using MBBR followed by membrane separation process. Desalination and Water Treatment, 52(22–24), 4562–4567. https://doi.org/10.1080/19443994.2013.803780

Donkadokula, N. Y., Kola, A. K., Naz, I., & Saroj, D. (2020). A review on advanced physico-chemical and biological textile dye wastewater treatment techniques. In Reviews in Environmental Science and Biotechnology (Vol. 19, Issue 3, pp. 543–560). Springer. https://doi.org/10.1007/s11157-020-09543-z

Du, L. N., Zhao, M., Li, G., Xu, F. C., Chen, W. H., & Zhao, Y. H. (2013). Biodegradation of malachite green by Micrococcus sp. strain BD15: Biodegradation pathway and enzyme analysis. International Biodeterioration and Biodegradation, 78, 108–116. https://doi.org/10.1016/j.ibiod.2012.12.011

Dubey, S., Shri, M., Gupta, A., Rani, V., & Chakrabarty, D. (2018). Toxicity and detoxification of heavy metals during plant growth and metabolism. In Environmental Chemistry Letters (Vol. 16, Issue 4, pp. 1169–1192). Springer Verlag. https://doi.org/10.1007/s10311-018-0741-8

Dzionek, A., Wojcieszyńska, D., & Guzik, U. (2016). Natural carriers in bioremediation: A review. In Electronic Journal of Biotechnology (Vol. 23, pp. 28–36). Electronic Journal of Biotechnology. https://doi.org/10.1016/j.ejbt.2016.07.003

El-Sayed, E. S. R., Ahmed, A. S., Hassan, I. A., Ismaiel, A. A., & Karam El-Din, A. Z. A. (2019). Strain improvement and immobilization technique for enhanced production of the anticancer drug paclitaxel by Aspergillus fumigatus and Alternaria tenuissima. Applied Microbiology and Biotechnology, 103(21–22), 8923–8935. https://doi.org/10.1007/s00253-019-10129-1

Fareed, A., Zaffar, H., Bilal, M., Hussain, J., Jackson, C., & Naqvi, T. A. (2022). Decolorization of azo dyes by a novel aerobic bacterial strain Bacillus cereus strain ROC. PLoS ONE, 17(6 June). https://doi.org/10.1371/journal.pone.0269559

Fatima, M., Kiros, Y., Farooq, R., & Lindström, R. W. (2021). Low-Cost Single Chamber MFC Integrated With Novel Lignin-Based Carbon Fiber Felt Bioanode for Treatment of Recalcitrant Azo Dye. Frontiers in Energy Research, 9. https://doi.org/10.3389/fenrg.2021.672817

Garg, N., Garg, A., & Mukherji, S. (2020). Eco-friendly decolorization and degradation of reactive yellow 145 textile dye by Pseudomonas aeruginosa and Thiosphaera pantotropha. Journal of Environmental Management, 263. https://doi.org/10.1016/j.jenvman.2020.110383

Ghosh, A., Ghosh Dastidar, M., & Sreekrishnan, T. R. (2016a). Recent Advances in Bioremediation of Heavy Metals and Metal Complex Dyes: Review. Journal of Environmental Engineering, 142(9). https://doi.org/10.1061/(asce)ee.1943-7870.0000965

Goswami, M., Chaturvedi, P., Kumar Sonwani, R., Dutta Gupta, A., Rani Singhania, R., Shekher Giri, B., Nath Rai, B., Singh, H., Yadav, S., & Sharan Singh, R. (2020). Application of Arjuna (Terminalia arjuna) seed biochar in hybrid treatment system for the bioremediation of Congo red dye. Bioresource Technology, 307. https://doi.org/10.1016/j.biortech.2020.123203

Goud, B. S., Cha, H. L., Koyyada, G., & Kim, J. H. (2020). Augmented Biodegradation of Textile Azo Dye Effluents by Plant Endophytes: A Sustainable, Eco-Friendly Alternative. In Current Microbiology (Vol. 77, Issue 11, pp. 3240–3255). Springer. https://doi.org/10.1007/s00284-020-02202-0

Gowri, A. K., Karunakaran, M. J., Muthunarayanan, V., Ravindran, B., Nguyen-Tri, P., Ngo, H. H., Bui, X. T., Nguyen, X. H., Nguyen, D. D., Chang, S. W., & Chandran, T. (2020). Evaluation of bioremediation competence of indigenous bacterial strains isolated from fabric dyeing effluent. Bioresource Technology Reports, 11. https://doi.org/10.1016/j.biteb.2020.100536

Guadie, A., Tizazu, S., Melese, M., Guo, W., Ngo, H. H., & Xia, S. (2017). Biodecolorization of textile azo dye using Bacillus sp. strain CH12 isolated from alkaline lake. Biotechnology Reports, 15, 92–100. https://doi.org/10.1016/j.btre.2017.06.007

Hernández-Zamora, M., & Martínez-Jerónimo, F. (2019). Congo red dye diversely affects organisms of different trophic levels: a comparative study with microalgae, cladocerans, and zebrafish embryos. Environmental Science and Pollution Research, 26(12), 11743–11755. https://doi.org/10.1007/s11356-019-04589-1

Holkar, C. R., Jadhav, A. J., Pinjari, D. V., Mahamuni, N. M., & Pandit, A. B. (2016). A critical review on textile wastewater treatments: Possible approaches. In Journal of Environmental Management (Vol. 182, pp. 351–366). Academic Press. https://doi.org/10.1016/j.jenvman.2016.07.090

Homaeigohar, S. (2020). The nanosized dye adsorbents for water treatment. In Nanomaterials (Vol. 10, Issue 2). MDPI AG. https://doi.org/10.3390/nano10020295

Hoque, E., & Fritscher, J. (2019). Multimetal bioremediation and biomining by a combination of new aquatic strains of Mucor hiemalis. Scientific Reports, 9(1). https://doi.org/10.1038/s41598-019-46560-7

Ihsanullah, I. (2020). MXenes (two-dimensional metal carbides) as emerging nanomaterials for water purification: Progress, challenges, and prospects. In Chemical Engineering Journal (Vol. 388). Elsevier B.V. https://doi.org/10.1016/j.cej.2020.124340

Ihsanullah, I., Jamal, A., Ilyas, M., Zubair, M., Khan, G., & Atieh, M. A. (2020). Bioremediation of dyes: Current status and prospects. In Journal of Water Process Engineering (Vol. 38). Elsevier Ltd. https://doi.org/10.1016/j.jwpe.2020.101680

Isik, Z., Arikan, E. B., Bouras, H. D., & Dizge, N. (2019). Bioactive ultrafiltration membrane manufactured from Aspergillus carbonarius M333 filamentous fungi for treatment of real textile wastewater. Bioresource Technology Reports, 5, 212–219. https://doi.org/10.1016/j.biteb.2019.01.020

Jadhav, D. A., Chendake, A. D., Ashvini, M., & Chendake, D. (2019). International Journal of Alternative Fuels and Energy Advance Microbial Fuel Cell for Waste to Energy Recovery: Need of Future Era for Sustainable Development. International Journal of Alternative Fuels and Energy.

Jain, A., & He, Z. (2018). Cathode-enhanced wastewater treatment in bioelectrochemical systems. Npj Clean Water, 1(1). https://doi.org/10.1038/s41545-018-0022-x

Karim, M. E., Dhar, K., & Hossain, M. T. (2018). Decolorization of Textile Reactive Dyes by Bacterial Monoculture and Consortium Screened from Textile Dyeing Effluent. Journal of Genetic Engineering and Biotechnology, 16(2), 375–380. https://doi.org/10.1016/j.jgeb.2018.02.005

Karmakar, S., Roy, D., Janiak, C., & De, S. (2019). Insights into multi-component adsorption of reactive dyes on MIL-101-Cr metal-organic framework: Experimental and modeling approach. Separation and Purification Technology, 215, 259–275. https://doi.org/10.1016/j.seppur.2019.01.013

Kaushal, J., & Mahajan, P. (2022). Kinetic Evaluation for Removal of an Anionic Diazo Direct Red 28 by Using Phytoremediation Potential of Salvinia molesta Mitchell. Bulletin of Environmental Contamination and Toxicology, 108(3), 437–442. https://doi.org/10.1007/s00128-021-03297-2

Khalid, A., Zubair, M., & Ihsanullah. (2018). A Comparative Study on the Adsorption of Eriochrome Black T Dye from Aqueous Solution on Graphene and Acid-Modified Graphene. Arabian Journal for Science and Engineering, 43(5), 2167–2179. https://doi.org/10.1007/s13369-017-2543-x

Khan, M. D., Li, D., Tabraiz, S., Shamurad, B., Scott, K., Khan, M. Z., & Yu, E. H. (2021). Integrated air cathode microbial fuel cell-aerobic bioreactor set-up for enhanced bioelectrodegradation of azo dye Acid Blue 29. Science of the Total Environment, 756. https://doi.org/10.1016/j.scitotenv.2020.143752

Kiran, S., Ashraf, A., Afzal, G., Abrar, S., Rahmat, M., & Asif, S. (2022). Green synthesis of magnesium oxide nanoparticles using leaves of Iresine herbstii for remediation of reactive brown 9 dye. Global Nest Journal, 24(2), 291–296. https://doi.org/10.30955/gnj.003929

Kulkarni, A. N., Watharkar, A. D., Rane, N. R., Jeon, B. H., & Govindwar, S. P. (2018). Decolorization and detoxification of dye mixture and textile effluent by lichen Dermatocarpon vellereceum in fixed bed upflow bioreactor with subsequent oxidative stress study. Ecotoxicology and Environmental Safety, 148, 17–25. https://doi.org/10.1016/j.ecoenv.2017.10.001

Kumar, V., Kumar, S., Darnal, S., Patial, V., Singh, A., Thakur, V., Kumar, S., & Singh, D. (2019). Optimized chromogenic dyes-based identification and quantitative evaluation of bacterial l-asparaginase with low/no glutaminase activity bioprospected from pristine niches in Indian trans-Himalaya. 3 Biotech, 9(7). https://doi.org/10.1007/s13205-019-1810-9

Kusumlata, Ambade, B., Kumar, A., & Gautam, S. (2024). Sustainable Solutions: Reviewing the Future of Textile Dye Contaminant Removal with Emerging Biological Treatments. Limnological Review, 24(2), 126–149. https://doi.org/10.3390/limnolrev24020007

Leal, C. S., Mesquita, D. P., Amaral, A. L., Amaral, A. M., & Ferreira, E. C. (2020). Environmental impact and biological removal processes of pharmaceutically active compounds: The particular case of sulfonamides, anticonvulsants, and steroid estrogens. Critical Reviews in Environmental Science and Technology, 50(7), 698–742. https://doi.org/10.1080/10643389.2019.1642831

Li, W., Mu, B., & Yang, Y. (2019). Feasibility of industrial-scale treatment of dye wastewater via bio-adsorption technology. In Bioresource Technology (Vol. 277, pp. 157–170). Elsevier Ltd. https://doi.org/10.1016/j.biortech.2019.01.002

Liu, C., Baffoe, D. K., Zhan, Y., Zhang, M., Li, Y., & Zhang, G. (2019). Halophile, an essential platform for bioproduction. Journal of Microbiological Methods, 166, 105704. https://doi.org/10.1016/J.MIMET.2019.105704

Liu, C., Chen, L., & Zhu, L. (2018). Application of membrane distillation for the treatment of anaerobic membrane bioreactor effluent: An especial attention to the operating conditions. Chemosphere, 208, 530–540. https://doi.org/10.1016/j.chemosphere.2018.06.013

Logan, B. E., Rossi, R., Ragab, A., & Saikaly, P. E. (2019). Electroactive microorganisms in bioelectrochemical systems. In Nature Reviews Microbiology (Vol. 17, Issue 5, pp. 307–319). Nature Publishing Group. https://doi.org/10.1038/s41579-019-0173-x

Lou, W. Y., Fernández-Lucas, J., Ge, J., & Wu, C. (2021). Editorial: Enzyme or Whole Cell Immobilization for Efficient Biocatalysis: Focusing on Novel Supporting Platforms and Immobilization Techniques. In Frontiers in Bioengineering and Biotechnology (Vol. 9). Frontiers Media S.A. https://doi.org/10.3389/fbioe.2021.620292

Louati, I., Elloumi-Mseddi, J., Cheikhrouhou, W., Hadrich, B., Nasri, M., Aifa, S., Woodward, S., & Mechichi, T. (2020). Simultaneous cleanup of Reactive Black 5 and cadmium by a desert soil bacterium. Ecotoxicology and Environmental Safety, 190. https://doi.org/10.1016/j.ecoenv.2019.110103

Ma, J., Chen, D., Zhang, W., An, Z., Zeng, K., Yuan, M., & Shen, J. (2021). Enhanced performance and degradation of wastewater in microbial fuel cells using titanium dioxide nanowire photocathodes. RSC Advances, 11(4), 2242–2252. https://doi.org/10.1039/d0ra08747e

Mashkoor, F., & Nasar, A. (2020). Magsorbents: Potential candidates in wastewater treatment technology – A review on the removal of methylene blue dye. In Journal of Magnetism and Magnetic Materials (Vol. 500). Elsevier B.V. https://doi.org/10.1016/j.jmmm.2020.166408

Menon, S., Agarwal, H., & Shanmugam, V. K. (2021). Catalytical degradation of industrial dyes using biosynthesized selenium nanoparticles and evaluating its antimicrobial activities. Sustainable Environment Research, 31(1). https://doi.org/10.1186/s42834-020-00072-6

Meerbergen, K., Crauwels, S., Willems, K. A., Dewil, R., Van Impe, J., Appels, L., & Lievens, B. (2017). Decolorization of reactive azo dyes using a sequential chemical and activated sludge treatment. Journal of Bioscience and Bioengineering, 124(6), 668–673. https://doi.org/10.1016/j.jbiosc.2017.07.005

Mian, A. H., Qayyum, S., Zeb, S., Fatima, T., Jameel, K., & Rehman, B. (2024). Exploring Indigenous fungal isolates for efficient dye degradation: A comprehensive study on sustainable bioremediation in the total environment. Environmental Technology and Innovation, 34. https://doi.org/10.1016/j.eti.2024.103615

Mir-Tutusaus, J. A., Baccar, R., Caminal, G., & Sarrà, M. (2018). Can white-rot fungi be a real wastewater treatment alternative for organic micropollutants removal? A review. In Water Research (Vol. 138, pp. 137–151). Elsevier Ltd. https://doi.org/10.1016/j.watres.2018.02.056

Mishra, S., Lin, Z., Pang, S., Zhang, Y., Bhatt, P., & Chen, S. (2021). Biosurfactant is a powerful tool for the bioremediation of heavy metals from contaminated soils. Journal of Hazardous Materials, 418. https://doi.org/10.1016/j.jhazmat.2021.126253

Molinari, R., Lavorato, C., & Argurio, P. (2020). Application of hybrid membrane processes coupling separation and biological or chemical reaction in advanced wastewater treatment. In Membranes (Vol. 10, Issue 10, pp. 1–30). MDPI AG. https://doi.org/10.3390/membranes10100281

Montañez-Barragán, B., Sanz-Martín, J. L., Gutiérrez-Macías, P., Morato-Cerro, A., Rodríguez-Vázquez, R., & Barragán-Huerta, B. E. (2020). Azo dyes decolorization under high alkalinity and salinity conditions by Halomonas sp. in batch and packed bed reactor. Extremophiles, 24(2), 239–247. https://doi.org/10.1007/s00792-019-01149-w

Mu, Y., Rabaey, K., Rozendal, R. A., Yuan, Z., & Keller, J. (2009). Decolorization of azo dyes in bioelectrochemical systems. Environmental Science and Technology, 43(13), 5137–5143. https://doi.org/10.1021/es900057f

Nath, D., Chakraborty, I., & Ghangrekar, M. M. (2021). Integrating microbial electrochemical technologies for methane-to-bioelectricity and water-splitting to impart self-sustainability to wastewater treatment plants. In Bioresource Technology Reports (Vol. 13). Elsevier Ltd. https://doi.org/10.1016/j.biteb.2021.100644

Nawaz, M. S., & Ahsan, M. (2014). Comparison of physicochemical, advanced oxidation, and biological techniques for the textile wastewater treatment. Alexandria Engineering Journal, 53(3), 717–722. https://doi.org/10.1016/j.aej.2014.06.007

Nedjimi, B. (2021). Phytoremediation: a sustainable environmental technology for heavy metals decontamination. In SN Applied Sciences (Vol. 3, Issue 3). Springer Nature. https://doi.org/10.1007/s42452-021-04301-4

Neoh, C. H., Lam, C. Y., Lim, C. K., Yahya, A., Bay, H. H., Ibrahim, Z., & Noor, Z. Z. (2015). Biodecolorization of recalcitrant dye as the sole source of nutrition using Curvularia clavata NZ2 and decolorization ability of its crude enzymes. Environmental Science and Pollution Research, 22(15), 11669–11678. https://doi.org/10.1007/s11356-015-4436-4

Nguyen, H. A., Su, Y., & Lavie, A. (2016). Design and characterization of Erwinia chrysanthemi L-asparaginase variants with diminished L-glutaminase activity. Journal of Biological Chemistry, 291(34)

Pant, G., Garlapati, D., Agrawal, U., Prasuna, R. G., Mathimani, T., & Pugazhendhi, A. (2021). Biological approaches practiced using genetically engineered microbes for a sustainable environment: A review. In Journal of Hazardous Materials (Vol. 405). Elsevier B.V. https://doi.org/10.1016/j.jhazmat.2020.124631

Patel, H., Yadav, V. K., Yadav, K. K., Choudhary, N., Kalasariya, H., Alam, M. M., Gacem, A., Amanullah, M., Ibrahium, H. A., Park, J. W., Park, S., & Jeon, B. H. (2022). A Recent and Systemic Approach Towards Microbial Biodegradation of Dyes from Textile Industries. In Water (Switzerland) (Vol. 14, Issue 19). MDPI. https://doi.org/10.3390/w14193163

Patel, S., Majumder, S. K., Mondal, S., Das, P., & Ghosh, P. (2020). Treatment of a pharmaceutical industrial effluent by a hybrid process of advanced oxidation and adsorption. ACS Omega, 5(50), 32305–32317. https://doi.org/10.1021/acsomega.0c04139

Pietro-Souza, W., de Campos Pereira, F., Mello, I. S., Stachack, F. F. F., Terezo, A. J., Cunha, C. N. da, White, J. F., Li, H., & Soares, M. A. (2020). Mercury resistance and bioremediation mediated by endophytic fungi. Chemosphere, 240. https://doi.org/10.1016/j.chemosphere.2019.124874

Pimviriyakul, P., Wongnate, T., Tinikul, R., & Chaiyen, P. (2020). Microbial degradation of halogenated aromatics: molecular mechanisms and enzymatic reactions. In Microbial Biotechnology (Vol. 13, Issue 1, pp. 67–86). John Wiley and Sons Ltd. https://doi.org/10.1111/1751-7915.13488

Pineda Zapana, J. S., Arán, D. S., Bocardo, E. F., & Harguinteguy, C. A. (2022). Correction to: Treatment of tannery wastewater in a pilot scale hybrid constructed wetland system in Arequipa, Peru (International Journal of Environmental Science and Technology, (2020), 17, 11, (4419-4430), 10.1007/s13762-020-02797-8). In International Journal of Environmental Science and Technology (Vol. 19, Issue 5, p. 4613). Springer Science and Business Media Deutschland GmbH. https://doi.org/10.1007/s13762-020-03049-5

Pinheiro, L. R. S., Gradíssimo, D. G., Xavier, L. P., & Santos, A. V. (2022). Degradation of Azo Dyes: Bacterial Potential for Bioremediation. In Sustainability (Switzerland) (Vol. 14, Issue 3). MDPI. https://doi.org/10.3390/su14031510

Pirsaheb, M., Mohamadisorkali, H., Hossaini, H., Hossini, H., & Makhdoumi, P. (2020). The hybrid system successfully to consisting of activated sludge and biofilter process from hospital wastewater: Ecotoxicological study. Journal of Environmental Management, 276. https://doi.org/10.1016/j.jenvman.2020.111098

Prabhakar, Y., Gupta, A., & Kaushik, A. (2022). Using Indigenous bacterial isolate Nesterenkonia lacusekhoensis for removal of azo dyes: A low-cost eco-friendly approach for bioremediation of textile wastewaters. Environment, Development and Sustainability, 24(4), 5344–5367. https://doi.org/10.1007/s10668-021-01661-0

Prasanna, K., Sudarsan, J. S., & Nithiyanantham, S. (2017). Wastewater treatment using combined biological and constructed wetlands technique in paper mills. Sustainable Water Resources Management, 3(4), 431–439. https://doi.org/10.1007/s40899-017-0108-5

Punzi, M., Anbalagan, A., Aragão Börner, R., Svensson, B. M., Jonstrup, M., & Mattiasson, B. (2015). Degradation of a textile azo dye using biological treatment followed by photo-Fenton oxidation: Evaluation of toxicity and microbial community structure. Chemical Engineering Journal, 270, 290–299. https://doi.org/10.1016/j.cej.2015.02.042

Qiu, D., & Liu, S. (2021). Degradation of azo dye wastewater by the combination process of 3D BER and CW-MFC. IOP Conference Series: Earth and Environmental Science, 675(1). https://doi.org/10.1088/1755-1315/675/1/012017

Rafique, M. A., Jamal, A., Ali, Z., Kiran, S., Iqbal, S., Nosheen, S., Ansar, Z., & Hossain, M. B. (2022). Biologically Synthesized Copper Nanoparticles Show Considerable Degradation of Reactive Red 81 Dye: An Eco-Friendly Sustainable Approach. In BioMed Research International (Vol. 2022). Hindawi Limited. https://doi.org/10.1155/2022/7537955

Rafique, M. A., Kiran, S., Ashraf, A., Mukhtar, N., Rizwan, S., Ashraf, M., & Arshad, M. Y. (2022). Effective removal of Direct Orange 26 dye using copper nanoparticles synthesized from Tilapia fish scales. Global Nest Journal, 24(2), 311–317. https://doi.org/10.30955/gnj.004234

Rajesh, G., Kumar, P. S., Alanazi, A. K., Rangasamy, G., & Abo-Dief, H. M. (2023). Development of lattice defects and oxygen vacancies in Zn-doped CdAl2O4 nanoparticles for improving the photocatalytic efficiencies of brilliant green and brilliant blue dyes under visible illumination. Catalysis Communications, 183. https://doi.org/10.1016/j.catcom.2023.106762

Rajhans, G., Barik, A., Sen, S. K., & Raut, S. (2021). Degradation of dyes by fungi: an insight into mycoremediation. In Biotechnologia (Vol. 102, Issue 4, pp. 445–455). Termedia Publishing House Ltd. https://doi.org/10.5114/BTA.2021.111109

Sanjana, M., Prajna, R., Urvi, S. K., & Kavitha, R. V. (2024). Bioremediation - the recent drift towards a sustainable environment. In Environmental Science: Advances. Royal Society of Chemistry. https://doi.org/10.1039/d3va00358b

Saratale, R. G., Saratale, G. D., Chang, J. S., & Govindwar, S. P. (2011). Bacterial decolorization and degradation of azo dyes: A review. In Journal of the Taiwan Institute of Chemical Engineers (Vol. 42, Issue 1, pp. 138–157). https://doi.org/10.1016/j.jtice.2010.06.006

Sathishkumar, K., AlSalhi, M. S., Sanganyado, E., Devanesan, S., Arulprakash, A., & Rajasekar, A. (2019). Sequential electrochemical oxidation and bio-treatment of the azo dye congo red and textile effluent. Journal of Photochemistry and Photobiology B: Biology, 200. https://doi.org/10.1016/j.jphotobiol.2019.111655

Shah, M. P. (2014). Bioremedial Application of Bacillus Megaterium PMS82 in Microbial Degradation of Acid Orange Dye. International Journal of Environmental Bioremediation & Biodegradation, 2(3), 93–99. https://doi.org/10.12691/ijebb-2-3-1

Shah, V., Garg, N., & Madamwar, D. (2001). An integrated process of textile dye removal and hydrogen evolution using cyanobacterium, Phormidium valderianum. World Journal of Microbiology & Biotechnology, 17, 499–504.

Shanmugam, L., Ahire, M., & Nikam, T. (2020). Bacopa monnieri (L.) Pennell, a potential plant species for degradation of textile azo dyes. Environmental Science and Pollution Research, 27(9), 9349–9363. https://doi.org/10.1007/s11356-019-07430-x

Sheldon, R. A., Basso, A., & Brady, D. (2021). New frontiers in enzyme immobilization: Robust biocatalysts for a circular bio-based economy. In Chemical Society Reviews (Vol. 50, Issue 10, pp. 5850–5862). Royal Society of Chemistry. https://doi.org/10.1039/d1cs00015b

Shetty, K., & Krishnakumar, G. (2020). Algal and cyanobacterial biomass as potential dye biodecolorizing material: a review. In Biotechnology Letters (Vol. 42, Issue 12, pp. 2467–2488). Springer Science and Business Media B.V. https://doi.org/10.1007/s10529-020-03005-w

Singh, R. K., Tripathi, R., Ranjan, A., & Srivastava, A. K. (2019). Fungi as potential candidates for bioremediation. In Abatement of Environmental Pollutants: Trends and Strategies (pp. 177–191). Elsevier. https://doi.org/10.1016/B978-0-12-818095-2.00009-6

Srinivasan, S., & Sadasivam, S. K. (2021). Biodegradation of textile azo dyes by textile effluent non-adapted and adapted Aeromonas hydrophila. Environmental Research, 194. https://doi.org/10.1016/j.envres.2020.110643

Srivastava, A., Rani, R., & Kumar, S. (2022). Optimization, kinetics, and thermodynamics aspects in the biodegradation of reactive black 5 (RB5) dye from textile wastewater using isolated bacterial strain, Bacillus albus DD1. Water Science and Technology, 86(3), 610–624. https://doi.org/10.2166/wst.2022.212

Sudarshan, S., Bharti, V. S., Harikrishnan, S., Shukla, S. P., & RathiBhuvaneswari, G. (2022). Eco-toxicological effect of a commercial dye Rhodamine B on freshwater microalgae Chlorella vulgaris. Archives of Microbiology, 204(10). https://doi.org/10.1007/s00203-022-03254-5

Sudarshan, S., Harikrishnan, S., RathiBhuvaneswari, G., Alamelu, V., Aanand, S., Rajasekar, A., & Govarthanan, M. (2023). Impact of textile dyes on human health and bioremediation of textile industry effluent using microorganisms: current status and future prospects. In

Journal of Applied Microbiology (Vol. 134, Issue 2). Oxford University Press. https://doi.org/10.1093/jambio/lxac064

Sonwani, R. K., Swain, G., Jaiswal, R. P., Singh, R. S., & Rai, B. N. (2021). Moving bed biofilm reactor with immobilized low-density polyethylene–polypropylene for Congo red dye removal. Environmental Technology and Innovation, 23. https://doi.org/10.1016/j.eti.2021.101558

Tay, M. F., Liu, C., Cornelissen, E. R., Wu, B., & Chong, T. H. (2018). The feasibility of nanofiltration membrane bioreactor (NF-MBR)+reverse osmosis (RO) process for water reclamation: Comparison with ultrafiltration membrane bioreactor (UF-MBR)+RO process. Water Research, 129, 180–189. https://doi.org/10.1016/j.watres.2017.11.013

Thanavel, M., Kadam, S. K., Biradar, S. P., Govindwar, S. P., Jeon, B. H., & Sadasivam, S. K. (2019). Combined biological and advanced oxidation process for decolorization of textile dyes. SN Applied Sciences, 1(1). https://doi.org/10.1007/s42452-018-0111-y

Tiodar, E. D., Văcar, C. L., & Podar, D. (2021). Phytoremediation and microorganisms-assisted phytoremediation of mercury-contaminated soils: Challenges and perspectives. In International Journal of Environmental Research and Public Health (Vol. 18, Issue 5, pp. 1–38). MDPI AG. https://doi.org/10.3390/ijerph18052435

Van der Bruggen, B. (2018). Microfiltration, ultrafiltration, nanofiltration, reverse osmosis, and forward osmosis. In Fundamental Modeling of Membrane Systems: Membrane and Process Performance (pp. 25–70). Elsevier. https://doi.org/10.1016/B978-0-12-813483-2.00002-2

Varjani, S., Rakholiya, P., Ng, H. Y., You, S., & Teixeira, J. A. (2020). Microbial degradation of dyes: An overview. In Bioresource Technology (Vol. 314). Elsevier Ltd. https://doi.org/10.1016/j.biortech.2020.123728

Venkatesh, S., Venkatesh, K., & Quaff, A. R. (2017). Dye decomposition by combined ozonation and anaerobic treatment: Cost effective technology. Journal of Applied Research and Technology, 15(4), 340–345. https://doi.org/10.1016/j.jart.2017.02.006

Vikrant, K., Giri, B. S., Raza, N., Roy, K., Kim, K. H., Rai, B. N., & Singh, R. S. (2018). Recent advancements in bioremediation of dye: Current status and challenges. In Bioresource Technology (Vol. 253, pp. 355–367). Elsevier Ltd. https://doi.org/10.1016/j.biortech.2018.01.029

Vikrant, K., Kumar, V., Kim, K. H., & Kukkar, D. (2017). Metal-organic frameworks (MOFs): Potential and challenges for capture and abatement of ammonia. In Journal of Materials Chemistry A (Vol. 5, Issue 44, pp. 22877–22896). Royal Society of Chemistry. https://doi.org/10.1039/c7ta07847a

Wang, H., Zhang, Q., Qiu, M., & Hu, B. (2021). Synthesis and application of perovskite-based photocatalysts in environmental remediation: A review. In Journal of Molecular Liquids (Vol. 334). Elsevier B.V. https://doi.org/10.1016/j.molliq.2021.116029

Wang, H., Zheng, X. W., Su, J. Q., Tian, Y., Xiong, X. J., & Zheng, T. L. (2009). Biological decolorization of the reactive dyes Reactive Black 5 by a novel isolated bacterial strain Enterobacter sp. EC3. Journal of Hazardous Materials, 171(1–3), 654–659. https://doi.org/10.1016/j.jhazmat.2009.06.050

Wang, J., & Chen, H. (2020). Catalytic ozonation for water and wastewater treatment: Recent advances and perspective. In Science of the Total Environment (Vol. 704). Elsevier B.V. https://doi.org/10.1016/j.scitotenv.2019.135249

Wang, R., Li, H., Liu, Y., Chen, J., Peng, F., Jiang, Z., Liu, J., & Song, H. (2022). Efficient removal of azo dyes by Enterococcus faecalis R1107 and its application in simulated textile effluent treatment. Ecotoxicology and Environmental Safety, 238. https://doi.org/10.1016/j.ecoenv.2022.113577

Wang, Z. W., Liang, J. S., & Liang, Y. (2013). Decolorization of Reactive Black 5 by a newly isolated bacterium Bacillus sp. YZU1. International Biodeterioration and Biodegradation, 76, 41–48. https://doi.org/10.1016/j.ibiod.2012.06.023

Wuhrmann, K., Mechsner, K., & Kappeler, T. (1980). Microbiology and Investigation on Rate-Determining Factors in the Microbial Reduction of Azo Dyes. European J. Appl. Microbiol. Biotechnol, 9, 4.

Yang, C., Xu, W., Nan, Y., Wang, Y., Hu, Y., Gao, C., & Chen, X. (2020). Fabrication and characterization of a high-performance polyimide ultrafiltration membrane for dye removal. Journal of Colloid and Interface Science, 562, 589–597. https://doi.org/10.1016/j.jcis.2019.11.075

Zahirinejad, S., Hemmati, R., Homaei, A., Dinari, A., Hosseinkhani, S., Mohammadi, S., & Vianello, F. (2021). Nano-organic supports for enzyme immobilization: Scopes and perspectives. In Colloids and Surfaces B: Biointerfaces (Vol. 204). Elsevier B.V. https://doi.org/10.1016/j.colsurfb.2021.111774

Zhao, Y. (2020). A New Era of Metal-Organic Framework Nanomaterials and Applications. In ACS Applied Nano Materials (Vol. 3, Issue 6, pp. 4917–4919). American Chemical Society. https://doi.org/10.1021/acsanm.0c01241

Zhu, N., Zhang, B., & Yu, Q. (2020). Genetic Engineering-Facilitated Coassembly of Synthetic Bacterial Cells and Magnetic Nanoparticles for Efficient Heavy Metal Removal. ACS Applied Materials and Interfaces, 12(20), 22948–22957. https://doi.org/10.1021/acsami.0c04512

10

Eco-Clean: Harnessing Nature for Textile Dye Effluent Bioremediation

Tarali Kalita* and Priyanku Sarma

Cotton University, Department of Zoology, Panbazar, Dighalipukhuri Road, Guwahati - 781001, Assam, India

**Corresponding Author Email: tarali.kalita@cottonuniversity.ac.in*

Abstract

Textile dyes are the scaling environmental contaminants worldwide, and the relentless dumping of industrial effluents comprising synthetic dyes is soon becoming a global concern. Textile industries consume huge amounts of water and release substantial quantities of effluents enriched with toxic, harmful, and potentially carcinogenic organic and inorganic compounds including diverse kinds of dyes, salts, heavy metals, additives, odorants, surfactants, and other hazardous materials. It is estimated that around 10-15 % of synthetic dyes are lost during various textile industry operations. Bioremediation, a process that utilizes microorganisms and plants to degrade or detoxify pollutants, has emerged as a promising and sustainable approach to address this issue. Given its numerous benefits over traditional physico-chemical methods, bioremediation of dyes is an intriguing method for handling textile wastewater. The current chapter provides a comprehensive overview of the process of bioremediation of textile industries' waste, starting with the fundamentals of the process and its advantages over traditional treatment methods to exploring its latest advancements in the biodegradation of synthetic dyes in textile effluents. The applications of microbes, plants, yeast, algae, fungi, and enzymes and their role in the biodegradation of dyes are discussed. The main factors influencing the execution of biological treatment are briefly discussed. The chapter delves into the progress of emerging dye bioremediation treatment technologies such as bioreactors, microbial fuel cells, electro-bioremediation, and integrated/hybrid technologies. Lastly, the prospects of using bioremediation as a green sustainable technology, future research directions, and its challenges in textile wastewater treatment and purification are addressed in the current chapter.

Keywords: *effluent, synthetic dyes, biodegradation, bioreactors, textile, sustainable.*

1. Introduction

Dyes are compounds after addition to a substrate, for a moment give it a color through a process that modifies the colored substance's crystal structure (Bafana *et al.*, 2011). Such varieties of compounds with significant coloring capacity are widely employed in the textile, pharmaceutical, food, photographic, and paper industries (Zollinger *et al.*, 1987). Dyes can adhere to suitable surfaces through methods such as dissolving, forming covalent bonds, complexing with salts or metals, physically adsorbing, or being mechanically retained (Kirk and Othmer, 2004). The chemical structure and application of dyes determine their classification. The dye color is produced by a collection of atoms called chromophores. Chromophore centres, which are responsible for color, consist of various functional groups like azo, anthraquinone, methine, nitro, aryl methane, carbonyl, and others. Additionally, substituents that either donate or withdraw electrons to enhance or create the color of these chromophores are called auxochromes. Common examples of auxochromes include amine, carboxyl, sulfonate, and hydroxyl groups. There are three methods for dying textile materials: batch, continuous, and semi-continuous. The kind of material including yarn, fabric, fiber, clothing, and fabric structure as well as the general fiber size of the dye lot, type, and quality standards of the dyed fabric, all play a role in determining the dyeing method used. Among these techniques, the process called the batch procedure is the most frequently employed for dyeing textile materials. (Ogugbue *et al.*, 2011).

1.1. *Environmental Impact of Textile Dyes Effluent*

A high quantity of dyes is wasted on effluents each year as a result of the inefficient dyeing process in the textile industry's dyeing and finishing operations (Ogugbue *et al.*, 2011). Unfortunately, these dyes leak from conventional wastewater treatment processes and persist in the environment because of their high stability to various parameters like water, light, temperature, detergents, and many more (O'Neill *et al.*, 1999). Anti-microbial agents resistant to biological squalor are frequently used in the manufacture of textiles, specifically cotton (Couto, 2009). Due to their intricate aromatic composition and synthetic nature, these substances are less prone to biodegradation (Forgacs *et al.*, 2004). Considering the volume generated as well as the effluent composition, the effluent from textile plants is classified as the most polluting of all the industrial sectors, (Sen *et al.*, 2003, Ben *et al.*, 2012). Furthermore, because of the growing demand for textile products, the corresponding increase in their manufacturing, and the use of synthetic colors, dye wastewater has become one of the main causes of significant pollution problems in the modern day (Dos Santos *et al.*, 2007).

Chemical oxygen demand (COD), biochemical oxygen demand (BOD), pH, color, and salinity are some of the important parameters exhibited by textile effluents. The makeup of this wastewater is influenced by the diverse organic compounds, chemicals, and dyes employed during both dry and wet processing stages (Dos Santos *et al.*, 2007). Textile effluents typically contain a variety of pollutants, such as persistent organic compounds, dyes, toxic substances, surfactants, chlorinated chemicals, and salts(Ben *et al.*, 2012).Further, the existence of minute amounts of dyes (less than 1 milligram per liter for certain dyes) in water, which are still very noticeable, and the effects of other pollutants in textile wastewater harm the aesthetic quality and transparency of water bodies like lakes, rivers, and others, which damages the aquatic environment (Ibrahim *et al.*, 1996).

Table 1. Pollutants range in the textile raw effluent

Sl. No.	Parameter	Raw effluent
1	Surfactants (mg/l)	1.1
2	Color (A559)	1.202
3	Chemical oxygen demand (mg /l)	1.511
4	Biochemical oxygen demand (mg/l)	90.63
5	Hardness (mg/l as CaCO3)	86.5
6	Conductivity (mV)	109
7	pH	10.5
8	Cadmium (mg/l)	<0.2
9	Sodium	70%
10	Lead (mg/l)	<0.02
11	Boron (mg/l)	0.64
12	N-nitrate (mg/l)	2.0
13	Sulphate (mg/l)	345.3
14	Copper (mg/l)	0.2
15	Barium (mg/l)	<2.0
16	Aluminium (mg/l)	<0.01
17	Chromium (mg/l)	<0.0006
18	Fluoride (mg/l)	<0.01
19	Total iron (mg/l)	0.76
20	Cyanide (mg/l)	0.2
21	Arsenic (mg/l)	<0.005
22	Cobalt (mg/l)	<0.007
23	Phosphate (mg/l)	12
24	Phenolic compounds (mg/l)	0.077

Source: Ramachandran *et al.*, 2013

According to estimates, colorant losses to the environment during the dyeing process might range from 10% to 50% (Przystaś *et al.*, 2012, Vaidya and Datye, 1982). Notably, certain dyes are extremely lethal and they also decrease photosynthetic activity and light penetration, which results in O_2 shortages and restricts other beneficial uses like drinking water, irrigation, and recreation (Hubbe *et al.*, 2012).

Azo dyes make up a significant amount of all the organic natured dyes produced worldwide, making them the greatest class of colorants in terms of quantity and manufacturing volumes (Cariell *et al.*, 1998). The success of azo dyes can be attributed to their vast structural diversity, high molar extinction coefficient, medium-to-high fastness properties regarding light and moisture, and ease and cost-effectiveness of synthesis when compared to natural dyes (Seesuritachan *et al.*, 2007). Applications are present in a wide spectrum from textiles to pharmaceuticals and are also prevalent in food, paper, leather, and paints (Ben *et al.*, 2007). Some azo dyes are known to exhibit toxic properties, particularly associated with carcinogenic and mutagenic effects (Chung *et al.*, 1992). The harmful impacts of these azo dyes may arise from either the direct activity of the dyes themselves or from the aryl amine byproducts that are formed during the reductive biotransformation of the azo link (Rajaguru *et al.*, 1999). The azoreductases of intestinal microbes can convert azo dyes that enter the body by food into aromatic amines. The nitroreductase that the same microbes generate can metabolize the dyes if they are nitro (Umbuzeiro *et al*, 2005). Mammalian liver enzymes and other entities may also catalyze the reduction of the azo bond and the nitro group. In these processes, if N-hydroxyl amines are produced, they have the potential to damage DNA. The elimination of these compounds' color is one of the most challenging challenges met by textile wastewater treatment plants, primarily because these colored compounds are made to withstand biodegradation and hence last a long time in the environment (Hao *et al.*, 2000).

2. Principles of Bioremediation

Bioremediation involves utilizing living organisms, mainly microorganisms, to transform environmental pollutants into less harmful substances (Kensa, 2011). This process employs naturally occurring bacteria, fungi, or plants to break down or detoxify materials that pose risks to human health and the environment. Microorganisms used can either be native to the contaminated location or can be sourced from other areas and then introduced to the polluted sites. Living things undergo reactions as part of their metabolic processes that change contaminating substances. Multiple organisms are involved in the breakdown of a compound. Microorganisms are introduced to a polluted site to augment degradation, the process is known as bioaugmentation (Mueller, 1996). Microorganisms have to enzymatically process the pollutants and convert them into less harmful by-products. Bioremediation is effective only when surrounding conditions promote the growth and activity of microorganisms. Consequently, implementing this method often necessitates modifications to environmental factors to boost microbial proliferation and speed up the degradation process.

2.1. Advantages of Bioremediation over Traditional Techniques

Bioremediation methods are generally more cost-effective than conventional approaches like incineration. Additionally, they can often be applied directly at the contaminated site, minimizing risks to cleanup workers and the potential for broader exposure due to transportation accidents. Because bioremediation relies on natural processes, it is often viewed more favorably by the public compared to other technologies. The bioremediation process has its limitations, much like any other technology. Certain pollutants, including high chlorinated or aromatic hydrocarbon compounds, are impervious to microbial attack. There are currently no established guidelines to assess whether a contaminant can be degraded. As a result, predicting the cleanup rates for bioremediation efforts is challenging, since degradation may occur very slowly or not at all (Colberg, 1995).

2.2. Mechanism of Bioremediation in Textile Effluent

Table 2 presents biological treatment methods involving bacterial and fungal decolorization, as well as dye adsorption through bacterial and fungal biomass. Using a variety of microorganisms that can accumulate and break down various contaminants, including bacteria, yeasts, algae, and fungi, bioremediation systems are frequently used to remediate industrial effluents (Ramachandran *et al.*, 2013). The mechanism behind microbial degradation and decolorization of problematic textile dyes such as azo dyes which is also the most dominant compound in textile effluent, involves azo bond cleavage in azo compounds, causing color removal of azo compounds. Azo dyes are broken down into aromatic amines, which are mostly broken down in the presence of oxygen, after first undergoing reductive cleavage. The majority of dyes are mineralized into basic inorganic chemicals that pose no hazard to living things (Singh, 2017). The microbial electron transport chain includes reduced flavin nucleotides, which are oxidized by azo dye, resulting in the reduction of the dye itself during the oxidation of the flavin nucleotides. Following this, the reduced flavin nucleotides undergo a reoxidation process. Typically, this entire sequence requires an additional carbon source to enable decoupling and degradation to occur at a practical rate(Khandare and Govindwa, 2016).

Azo dyes are the only source of carbon and nitrogen for certain aerobic microorganisms, while oxygen-tolerant azo reductases are the only enzymes capable of catalyzing the reduction of azo groups in other strains. The carbon transforms into methane and carbon, after which electrons progress through the electron transport chain, eventually arriving at the azo dye that acts as the electron acceptor. This interaction between the electrons and the azo dye leads to the reduction of azo bonds, consequently resulting in the decolorization of the dye (Dilek *et al.*, 1999).

Table 2. Biological procedures of textile industry effluent

Biological Techniques	**Advantages**	**Disadvantages**
Biodegradation	Cost-effective biogas generation relies solely on anaerobic treatment, which depends on metabolic processes.	A consortium of mixed cultures utilizing a combined anaerobic & aerobic treatment system accomplished decolorization. Cells that were immobilized demonstrated higher efficiency in comparison to free cells, although the process was still slow, necessitating particular environmental conditions, regular maintenance, & a continuous supply of nutrients.
Biosorption	Applicable to both live & dead biomass without relying on metabolism, economically feasible, highly selective, low-cost to operate, & safe for microorganisms with no toxic effects.	Large-scale treatment remains largely unexplored, and the disposal of dye-saturated biomass poses challenges
Enzymatic degradation	Immobilized enzymes are highly effective for targeting specific compounds selectively.	Scale-up studies and analysis of reaction by-products require information on enzyme costs, stability, and potential enzyme inhibition factors.
Phytoremediation	Dyes that facilitate metal removal, interact with plants and native microflora, enhance soil fertility, and promote plant growth are environmentally friendly and suitable for large-scale applications.	Secondary pollution requires further investigation, as it becomes challenging to control during the monsoon season.
Orthodox biological action with aeration	Effective in reducing BOD, COD, and TSS levels while eliminating color.	Low total dissolved solids (TDS) levels and reduction of harmful heavy metals. Not effective in reducing non-biodegradable dyes.

***Source*:** Ramachandran *et al.*, 2013

3. Role of Microbes in Dye Metabolism and Degradation

As was mentioned in the previous section microbes play a great role in industrial dye metabolism and degradation. The whole foundation of bioremediation is based on the usage of a variety of microbes from bacteria to fungi. Likewise, microorganisms are crucial for decolorizing and cleaning up textile wastewater.

3.1. Bacterial Decolorisation and Degradation of Textile Dyes

The use of bacteria for the biodegradation of textile dyes is an interesting and low-cost method. Numerous studies have explored various bacterial species for their potential in bioremediating contaminants such as dyes, metals, soils, and water polluted by textile effluents. Several bacteria having the ability of decolorizing dyes in pure cultures have already been identified and documented (Table 3)(Pearce *et al.*, 2003). When it comes to the breakdown of colorants used in the textile industry, mixed microbial cultures are more advantageous than pure cultures. Different bacterial species may attack dye molecules directly, or a few strains may even be able to extrareduce the breakdown products produced by another strain (Junnarkar *et al.*, 2006).

3.1.1. Aerobic Bacteria

Aerobic bacteria particularly *Steptomycetes* which belong to the group Actinomycetes have shown promising results in the biodegradation of dyes in textile effluent. Ball *et al* (1989). screened several strains of actinomycetes, representing a wide spectrum of genera, for their capacity to decolorize the polymeric dyes. Only three strains were found to exhibit significant dye decolorization: *Streptomyces sp.* EC22, *Thermomonospora fusca* MT800, and *Streptomyces badius 252*.Given the chemicals' resistance to degradation by other bacteria under such circumstances, the discovery that actinomycetes can aerobically decolorize and degrade azo dyes was noteworthy. Actinomycetes produce extracellular peroxidases that are stated to help in the breakdown of dye compounds in effluent. It was suggested that peroxidases transformed the azo dye component into a cation radical that could be attacked by either H_2O or H_2O_2. This process led to the simultaneous cleavage of the azo bond in both symmetrical and asymmetrical manners, generating arbitrates that would undertake various redox reactions before yielding better stable compounds.

It has been challenging to separate bacteria that can aerobically mineralize colors and decolorize, particularly sulfonated azo dyes, with the notable exception of actinomycetes. Apart from azo dyes, there has been some interest in but limited success with bacteria's capacity to aerobically metabolize other dye classes.

Table 3. Strains of Bacteria degrading Synthetic dyes

Bacterial Strain	**Synthetic Dye**	**Reference**
Escherichia coli	Methyl Red, Ethyl Red	(Nakanishi *et al.*, 200)1
Sphingomonas xenophaga	Acid Orange	(Chen *et al.*, 2003)
Bacillus subtilis	Reactive azo dyes	(Horitsu *et al.*, 1997)
Aeromonas hydrophila	Reactive Blue 160, Acid Orange 7	(Coughlin *et al.*, 1999)
Corynebacterium glutamicum	Reactive Red	(Coughlin *et al.*, 1999)
Pseudomonas luteola	Reactive Azo dyes	(Hu *et al.*, 1992)
Rhizobium radiobacter	Reactive Yellow	(Won *et al.*, 2008)
Alcaligenes faecalis	Disperse dyes, phthalocyanine dyes, and Reactive azo dyes	(Nigam *et al.*, 1996)
Staphylococcus arlettae	Reactive Yellow 107, and Direct Blue 71, and Reactive Red 198,	(Singh and Singh., 2015)
Pseudomonas aeruginos	Direct Orange 39	(Singh and Singh., 2015)
Bacillus magaterium	Acid Red	(Ayman *et al.*, 2019)
Enterobacter agglomerans	Methyl Red	(Keharia *et al.*, 2003)

Source: Sharif *et al.*, 2020

3.1.2. Anaerobic Bacteria

Many bacteria under anaerobiotic conditions have been reported to readily decolorize azo dyes. Bacteria first induce the azo linkage to undergo reductive cleavage, which decolorizes the dye and produces colorless aromatic amines. Nevertheless, these substances are mutagenic, poisonous, and carcinogenic. Its impact on the environment is problematic in addition to its damaging impacts on anthropoid health (Chung *et al.* 1992).

The first stage in the bacterial metabolism of azo dyes under anaerobiotic conditions is the reductive breakdown of the azo bond. This reaction is facilitated by a range of soluble cytoplasmic enzymes with broad substratum specificity, commonly referred to as "azoreductases." In the absence of oxygen, these enzymes transfer electrons through soluble flavins to the azo dye, leading to its reduction. Several groups have utilized a two-stage action procedure to address the issue of the relative recalcitrance of azo dye breakdown products under anoxic circumstances (O'Neill *et al.* 2000; Oxspring *et al.* 1996). In the initial anaerobic phase, the azo dye is quickly transformed into its corresponding colorless aromatic amines, which are then metabolized with relative ease under subsequent aerobic conditions.

Table 4. Percentage efficacy of Bacterial species to decolorize Azo dyes

Bacteria Species	% Efficacy of Decolorization	Optimum pH
Chryseomonas luteola	42%	NA
Bacillus cereus	65%	7-9
Aeromonas hydrophila	90%	7
Staphylococcus aureus	76%	N/A
Klebsiella ozaenae	95.89%	7
Pseudomonas putida	90%	7
Pseudomonas aeruginosa	42.5%	N/A
Bacillus subtilis	54%	7
Bacillus mycoides	79%	N/A
Micrococcus sp.	77%	N/A

***Source*:** Sharif *et al.*, 2020

3.2. Fungal Decolorisation and Degradation of Textile Dyes (Mycoremediation)

The identification of fungal biosorption was based on the attachment of the solute to the biomass through mechanism that do not require transport or metabolic energy, but these processes may happen concurrently when living biomasses are utilized. Consequently, it can happen in biomass that is either living or dead (Fu, 2001). Deceased cells eliminate dyes through the machinery of biosorption which involves physicochemical exchanges such as deposition, adsorption, & ion exchange (Fu, 2002). The following categories can be used to group fungi according to how well they degrade: mycorrhizal fungal degradation, fungal biosorption, & ligninolytic fungal degradation (Hammal, 1996).

3.2.1. White Rot Fungi (WRF)

White-rot fungi are prominent microorganisms studied for dye decolorization. The ability of white-rot fungi to mineralize a wide variety of persistent organic contaminants sets them apart from biodegradative bacteria, which are often more substrate-specific. (Reddy 1995). The fundamental reason behind the degradative efficiency of white rot fungi to degrade substances such as aromatic compounds and pollutants is their lignin-degrading enzymes that are not substrate-specific (Hatakka, 1994). Lignin peroxidase produced by *Phanerochaete chrysosporium* has been shown to effectively degrade azo dyes into simpler substances (Toh *et al.*, 2003).

Unfortunately, despite the use of advanced analytical techniques, the degradative mechanisms of white-rot fungi other than *Phanerochaete* species have not been thoroughly explored due to the inherent complexity of both the dye molecules and the enzyme systems involved (Smyth *et al.* 1999).

Table 5. Decolorizing White Rot Fungi species (WRF)

WRF Species	**Dyes Decoloured**	**Reference**
Coriolus versicolor KR-65W	Methylene Blue, Poly R	(Kim *et al.*, 2007)
Lentinus tigrinus	Orange II, Reactive Blue 38	(Moreira *et al.*, 1996)
Phlebiabrevispora	Brilliant Green, Crystal Violet	(Arora *et al.*, 2002)
Phanerochaete chrysosporium	Methylene Blue, Congo Red	(Kim *et al.*, 2007)
Stereum rugosum	Remazol Brilliant Blue, Poly R	(Novotny *et al.*, 2004)
Funaliatrogii	Remazol Brillant Blue	(Ozsoy *et al.*, 2005)
Phanerochaete chrysosporium	Amido Black	(Senthikumar *et al.*, 2014)
Pleurotus ostreatus	Bromophenol Red, phenol Red	(Shrivastava *et al.*, 2005)
Pleurotus pulmonaris	Amido Black	(Tychanowicz *et al.*, 2004)
Coriolus versicolor KR-11W	Acid Green, Acid Red, Chrysophenine, Brilliant Yellow	(Knapp *et al.*, 1997)
Piptoporus betulinus	Methylene blue, Congo Red	(Kim *et al.*, 2007)
Bjerkandera adusta	Reactive Violet 5, Reactive Orange 96	(Heinfling *et al.*, 1997)

Fungi depend on their ability to adapt their metabolism in response to varying carbon and nitrogen levels. To facilitate this metabolic adjustment, they produce both intracellular and extracellular enzymes, which help in the breakdown of various resistant pollutants, including aromatic hydrocarbons, organic waste, and dye compounds. The capability of fungal enzyme systems to degrade a wide range of substances, regardless of their specific substrate, is a key factor in their success at breaking down colored and metallic effluents (Saratale, 2011). Similar enzymes have also been successfully used to breakdown dyes due to their low substrate specificity (lignin). Few studies have demonstrated that brown-rot fungi can eliminate different colors (Zhang *et al.*, 2006).

Table 6. Enzymes from different Fungi species that degrade effluents

Effluent	**Fungal sp.**	**Enzyme**	**Reference**
Bromophenol Dyes	*Plerotus ostreatus*	Manganese Peroxidase	(Ezeronye and Okerentugba, 1999)
Bleach plant effluents	*Pycnoporus sanguineis*	Laccase	(Limura *et al.*, 1996)
Non-Phenolic aromatic compounds	*Penicillium chrysogenum*	Lignin Peroxidase (Lip)	(Glenn *et al.*, 1986)
Phenolic compounds	*Penicillium chrysogenum*	Manganese Peroxidase	(Joshni *et al.*, 2011)
2,6-dimethoxyphenol (DMP)	*Bjerkandera adusta*	MnP	(Joshni *et al.*, 2011)

Azo dyes	*Pycnoporus sanguineis*	Laccase	(Pointing and Vrijmoed, 2000)
PCBs	*P.chrysosporium, Trametes versicolor*	Lip, MnP	(Novotony, 1997)
Solar Golden Yellow	*Scyzphyllum commune*	MnP, Laccase	(Ezeronye and Okerentugba, 1999)

3.3. Phytoremediation (Algal) of Textile Effluent

Phytoremediation refers to the process of using algae to remove or break down pollutants, such as textile dyes and xenobiotics, from residual water while promoting the growth of the algae biomass.

The algae are photoautotrophs and they have requirement of carbon source as such. Usage of algae for wastewater treatment is an efficient and cheap cost process. Even the effluents from the textile industry exhibit the growth of photosynthetic algae, making them a potential option for wastewater treatment (Dubey, 2011). Efficient color degradation can be attained by algal azo reductase enzymes that have the ability to break down azo dyes. Azo dyes can be degraded by various species of *Chlorella* and *Oscillatoria* into aromatic amines, which can subsequently be broken down into simpler organic compounds (Arora, 2002). Studies have shown that Oscillatoria sp. and Chlorella sp. can effectively decolorize and degrade more than thirty azo dyes into simpler aromatic amines (Zhang, 2006).

Table 7. Some of the enzymes from different Algal degrading effluents

Algae	Dye	% Efficiency	Reference
Algae	Malachite Green	85%	(Gajare and Menghani, 2012)
Green Algae	Azo Dyes	67%	(Omar, 2008)
Cosmarium sp	Malachite Green	92.4 %	(Daneshvar *et al.*, 2005)
Nostoc linckia	Methyl Red	88.3%	(Elisangel *et al.*, 2009)
Lyngbya sp	Textile Dye	97%	(Henciya *et al.*, 2013)
Lyngbya sp.	Blue dye	88.3%	(Brhmnbatt and Jasrai, 2016)
Spirogyra sp.	Blue dye	77.19%	(Brahmnbatt and Jasrai, 2016)
Chlorella sp.	Indigo	76.28%	(Brahmnbatt and Jasrai, 2016)
Phormidium sp.	Indigo dye	90%	(Dellamatricea *et al.*, 2017)
Synechococcus sp.	Remazol Dye	11.63%	(Dellamatricea *et al.*, 2017)
Lyngbya lagerlerimi.	Methyl Red	74%	(Meehan *et al.*, 2000)
Oscillatoria sp.	Blue Dye	35.61%	(El-Sheekh *et al.*, 2009)
Green algae	Indigo	87.97%	(Elisangel *et al.*, 2009)

3.4. Yeast Biodegradation of Textile Effluent

Yeast is an effective microorganism for the bioremediation of azo dyes due to its high capacity to absorb dyes, heavy metals, and colorants (Table 8). The primary methods by which yeast breaks down azo dyes are enzymatic degradation, adsorption, or a combination. Dye can be decolored by yeasts such as *Kluyveromyces marxianus* (Banat, 1999). Yeast belonging to *Kluyveromyces sp* actively growing under aerophilic conditions can mostly decolorize Remazol dye within 24 hours across a range of acidic pH levels. Additionally, two newly identified dye-decolorizing yeast strains, demonstrate strong efficiency in removing color from reactive azo dyes. One strain of *Saccharomyces cerevisiae* also effectively degrades methyl red through biodegradation, showing varying metabolic rates across different pH levels, with azo reductase playing a key role in the process (Jadhav, 2007).

Table 8. Textile dyes decolorize by Yeast species (Ascomycetes)

Yeast species	**Dyes Decolored**	**Reference**
Torulopsis candida	Brilliant red dye	(Yu and Wen, 2005)
Saccharomyces cerevisiae	Malachite Green	(Jadhav and Govindwar, 2007)
Candida kursei	Azo dye	(Charumathi and Das, 2010)
Pseudozyma rugulosa	Azo dye	(Charumathi and Das, 2010)
Candida zeylanoides	Azo benzenesulfonates	(Ramlho *et al.*, 2004)
Kluyveromyces marxianus	Remazol Dye	(Charumathi and Das, 2010)
Candida tropicalis	Reactive Blue 19	(Donmez and Asku, 2002)
Candida krusei	Azo dye	(Jadhav and Govindwar, 2007)
Trichosporon beigelli	Malachite green, Methyl violet	(Yu and Wen, 2005)

3.5. Dye degradation by Enzymes

Enzymes, which act as biological catalysts, are typically found in liquid form and are favored for applications where persistent pollutants are primarily eliminated through precipitation (Nguyen *et al.*, 2016). In recent times biological catalysts have become very prevalent for several industrial usage. Compared to traditional methods, enzymatic squalor of textile effluents offers several benefits, such as targeted substrate specificity, eco-friendly processes, accessibility, cost-effectiveness, and high efficiency (Kushch *et al.*, 2019). Added benefits of adopting enzymatic squalor processes include its milieu-friendly nature and potential for reusability. One of the main challenges in enzymatic degradation is the reduction in biocatalyst activity due to enzyme denaturation (Jun *et al.*, 2019). Table 9 sums up the enzymatic decolorization of dyes and dilapidation mechanisms. The effectiveness and breakdown of enzymatic dye decolorization are affected by several key factors, including pH, temperature, dye structure, the presence of redox mediators, as well as the concentrations of both the enzyme and dye (Teerapatsakul *et al.*, 2017).

Earlier studies have reported a wide spectrum of degradation mechanisms considering the nature of the enzymes used. Few common mechanisms includereduction, oxidation, aerobic and degradation hydroxylation. Both azo and non-azo dyes can be effectively degraded by enzymes (Ihsanullah *et al.*, 2020).

Table 9. Some of the Enzymes utilized for dye degradation

Enzyme	**Dye**	**Removal (%)**	**Mechanism**	**Reference**
Horseradish Peroxidase	Direct Blue dyes	83/79	Oxidation, reduction, hydroxylation	(Bayramoglu *et al.*, 2012)
Laccase from *Paraconiothyrium variabile*	Yellow 15, Disperse Blue 56	97/68.3/60.2/58	N/A	(Ashrafi *et al.*, 2013)
MnP	MG dye	96	Reduction of azo bond	(Yang *et al.*, 2016)
Calcium entrapped bitter gourd *(Momordica charantia)* peroxidase.	Disperse Brown, Disperse Red 17	98/71	Reduction and oxidation	(Veerasingam *et al.*, 2011)
Turnip peroxidase	Direct Red 23 (DR23)	88	Reduction, oxidation, hydrolysis	(Matto and Husain., 2009)
Tomato (*Lycopersicon esculentum*) peroxidase	DR23, Direct Blue 80	93/76	Reduction, oxidation	(Matto and Hussain, 2009)
Citrus limon peroxidase	Direct Yellow	89.47	Radical initiated ring opening	(Nouran *et al.*, 2017)
Trichosporon akiyoshidainum	Reactive Blue 221	72	Aerobic degradation	(Pajot *et al.*, 2011)

Source: Ihsanullah *et al.*, 2020

4. Bioremediation Techniques

Bioremediation techniques offer a workable and green approach to addressing the environmental challenges posed by textile dye effluents. Based on the elimination and transference of waste for treatment, bioremediation has two different types: *in-situ* and *ex-situ* (*Fig.* 1). For textile effluents, both in-situ and ex-situ methods are applied according to the convenience and area of remediation.

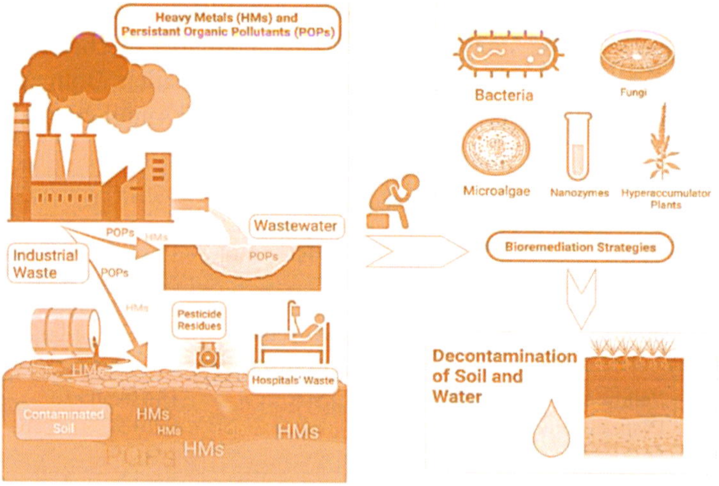

Fig. 1: Bioremediation Techniques (Adopted by Tufail *et al.*, 2022)

In situ remediation involves treating contaminants directly at the polluted site using biological agents, where microbes interact with the contaminants to achieve biotransformation. In contrast, *ex-situ* bioremediation, or off-site bioremediation, involves removing and transporting pollutants from one location to another for treatment. Both approaches rely on microorganisms for remediation. The choice of technique depends on factors like contaminant type, pollution site characteristics, pollution severity, and treatment cost.

4.1. Bioaugmentation and Bio-stimulation

Adding a cultured microbial population that can break down particular soil and groundwater pollutants is known as bioaugmentation. Bioaugmentation is now regarded as an effective technology for sites with low populations of microorganisms capable of biodegradation. Under optimal conditions, by adding an inoculum the biodegradation rate of contaminants could be amplified, which favors the prolonged activity and survival of the exogenic microbial population (Tyagi *et al.*, 2011). It was seen that this microbial population often struggled to compete effectively against the native microbes (Zouboulis *et al.*, 2019). Therefore, bioaugmentations process should be proceeded with knowledge of the specificity of the contaminants to be countered (Werheni-Ammeri *et al.*, 2022).

Bio-stimulation is an approach that involves altering the environment by adding essential nutrients and electron acceptors—such as phosphorus, nitrogen, oxygen, or carbon—to encourage the growth of native microbial populations responsible for bioremediation. By supplying nutrients directly to pollutants, bio-stimulation accelerates the action of these microbes, elevating their cell concentration and enhancing the rate at which they degrade contaminants (Popoola *et al.*, 2022). As a stimulant, a N_2 source is usually given (Werheni-Ammeri *et al.*, 2022). The technique evaluates the in-hand capacity of microbes to bio-biodegrade pesticides in crude oil spills and groundwater (Aldas-Vargas *et al.*, 2021).

A combination of these two techniques is done in the treatment of textile effluent. Tufail *et al.* (2022) demonstrated that indigo-contaminated textile effluents can be decolorized and efficiently treated by using a combination of bioaugmentation, bio-stimulation, and bio-ventilation. These methods provide a sustainable green alternative to the conventional treatment technologies generally applied to textile wastewater.

4.2. Bioreactors for Textile Effluents

Among various ex-situ bioremediation methods for treating textile waste, bioreactors stand out as the sole approach, with numerous reactor conFig. urations available for this purpose. It is classified into aerobic, anaerobic, and membrane bioreactors (MBRs).

1. ***Aerobic Bioreactors:*** These systems use oxygen-dependent microorganisms to break down the dyes. While effective for certain dyes, their efficiency can be limited by oxygen transfer rates and the presence of recalcitrant compounds.
2. ***Anaerobic Bioreactors:*** These bioreactors function in the absence of oxygen and are particularly effective for azo dyes, which can be reduced into colorless aromatic amines. Anaerobic processes often require subsequent aerobic treatment to fully mineralize intermediate compounds.
3. ***Membrane Bioreactors (MBRs):*** Combining biological treatment with membrane filtration, MBRs are highly effective in separating biomass and treated water, ensuring high-quality effluent. The hybrid system enhances dye degradation while reducing sludge production.

Biological sources commonly utilize bioreactors to decolorize and break down textile effluent. Various studies have documented combining, integrated anaerobic-aerobicor sequential, bioreactor systems for treating effluents containing azo dyes(Van Der Zees, 2005). Apart from the fundamental bioreactors mentioned, other specialized bioreactors are also used according to the contaminate present in the effluent or based on the microbial inoculum used in the bioreactor. Different reactor systems have been reported for producing peroxidase enzymes (lignin or manganese) by white rot fungi and the biological dye decolorization process. These reactor systems include stirred packed bed bioreactors, rotating disk reactors, silicone membrane reactors, tank reactors, fixed film reactors, airlift reactors or bubble columns, rotating biological contactors, immersion, and tray reactors.

4.3. Phytoremediation of Textile Wastewater

Phytoremediation involves using plants to treat contaminated soil, sediments, and water directly in the environment. Plants can withstand elevated levels of certain chemicals without experiencing negative effects. They also can absorb pollutants and, through the release of enzymes and root exudates, convert them into less harmful metabolites. Usage of engineered metal-accumulating plants for ecological clear-out is an emerging biotechnology (Rajendran and Gunasekaran, 2006). Various traditional methods are currently employed to remove contaminants from the textile industry and clean up the environment. These methods encompass lime-based coagulation, solvent extraction, ultrafiltration, chemical precipitation, aeration, chemical oxidation, reverse osmosis, electrolysis, and chlorination. However, most of them are expensive and bot of optimum conditioning (Tangahu *et al.*, 2011). Chemical methods raise expenses and produce significant volumes of sludge (Rakhshaee *et al.*, 1975). Both thermal and chemical approaches are costly and technically challenging. Each of these techniques has the potential to deteriorate the valuable element (Tangahu *et al.*, 2011). Amid all these approaches, phytoremediation through Aquatic Macrophyte Treatment Systems (AMATS) is a widely recognized environmental protection procedure effective in removing heavy metals and other pollutants. Freshwater vascular plants (FVP), combined with macroscopic algae, are collectively referred to as macrophytes. As noted by Hutchison (1975), macrophytes can concentrate elements from aquatic environments.

Table 10. Aquatic plant used for Phytoremediation

Common name	Scientific name	Trace element	Reference
Duckweed	*Lemna gibba L.*	Zn	(Fritioff and Greger *et al.*, 2008)
Mosquito fern	*Azolla caroliniana*	As	(Rai *et al.*, 2008)
Water hyacinth	*Eichhornia crassipes*	As, Fe, Cu, Zn, Pb	(Odjegba *et al.*, 2007; Alvarado *et al.*, 2008)
Watercress	*Lepidium sativum L.*	As	(Robinson *et al.*, 2005)
Waterweed/ Pondweed	*Elodea canadensis*	Cr, Zn, Cu, Cd	(Vardanyan and Ingole, 2006)
Water lily	*Mymphaea violacea*	Cd, Cu, Pb, Zn	(Frtioff & Greger, 2008)
Water lettuce	*Pistia stratiotes*	As, Cr, Pb, Ag, Cd,	(Miretzky *et al.*, 2004)
Star duckweed	*Lemna trisulca L*	Zn	(Huebert *et al.*, 1992)
Water fern	*Azolla filiculoides* *Azolla pinnata*	As, Hg, Cd	(Cardwell *et al.*, 2002)
Water spinach	*Ipomoea aquatic*	Pb,Cu, Zn	(Lee *et al.*, 1991; Gothberg *et al.*, 2002)
Indian Lotus	*Ludwigia perennis L*	Ba, Ti, Co, Pb	(Robinson *et al.*, 2005)
Water weed	*Hydrilla verticillata*	As, Pb, Zn, Cr	(Zhang *et al.*, 2008)
Butterfly fern	*Salvinia minima*	As,Cr, Pb, Hg (II), Cd,	(Banerjee and Sarker, 1997; Dhir, 2009)

4.3.1. Phytoremediation Mechanism of Wetland Plant

Two primary methods suggest using wetland plants to clean up wastewater from textile effluent and polluted water (Robinson *et al.*, 2005). The aquatic phytoremediation system has two main types: one involves entirely aquatic plants like water hyacinths, and the other uses submerged rhizospheres of terrestrial plants for metal pollutant removal, a process known as rhizofiltration (Robinson *et al.*, 2005).

A. *Eichhornia crassipes:* Differenttheoriesare made to rationalize the mechanism involved in *Eichhornia crassipes*-based treatment systems(Reddy, 1981;Reddy, 1983;Mandi 1994).Thepresenceofaquatic macrophytes in the water body alters the physio-chemical environment of the water body (Reddy, 1983). When other photosynthetic autotrophs are present in aquatic environments, they can lower dissolved CO_2 levels in the water during times of high photosynthetic activity. This process raises dissolved oxygen in the wastewater, subsequently leading to an increase in water pH(Reddy, 1981). Aquatic macrophytes, such as water hyacinths, absorb contaminants, and store them within their biomass. Such plants act as bio-accumulators because they gather contaminants in their tissues(Boyd, 1993). Itexhibits strong resistance to pollutants such as heavy metals and can accumulate substantial amounts. This process of removing heavy metals from contaminated water is known as phytoextraction. The uptake of contaminants is by three methods–

a) Root absorption: Roots take up water along with its residual contaminants. The carboxyl groups in the root system facilitate a substantial cation exchange across the cell membrane, which could be how heavy metals move through the roots where active absorption occurs. In sewage systems, the root structures of water hyacinths and similar aquatic plants create a favorable environment for aerobic bacteria. These bacteria consume nutrients, generating inorganic compounds that feed the plants. As the plants grow rapidly, they can be harvested to produce nutrient-rich compost. Water hyacinth is also used to remove or reduce nutrients, heavy metals, organic compounds, and pathogens from water.

b) Foliar absorption: Plants may absorb trace levels of certain pollutants through foliar absorption in addition to root absorption. They enter the body passively through stoma cells and cuticle fissures.

c) Adsorption: The root hair traps suspended solids and bacteria andalso provide accessory sites microbial growth. Contaminants adhere to the root surface due to the presence of bacteria, as well as an ionic imbalance across the cell membrane.

B. *Pistia stratiotes*: The removal of heavy metals from a solution by *Pistia stratiotes* occurs in two main stages. In the first stage, mechanisms such aschelation, adsorption, & ion exchange enable the plant to absorb heavy metals, with uptake levels varying between different *Pistia* species. The second stage involves root-induced precipitation of heavy metals (e.g., chromium), where both the species of *Pistia* and the specific metal affect the uptake process (Tripathi *et al.*, 2010). These plant helps in hyperaccumulation of many contaminants naturally (Thilakar *et al.*, 2012).

C. *Lemna minor*: Duckweed can treat wastewater through interactions with mutually aerobic and anaerobic bacteria. When duckweed forms a complete mat over the water surface, it creates three distinct precincts: an aerobic zone at the top, an anoxic zone in the middle, and another aerobic zone beneath(Skillicorn, 1993). In the aerobic zone, organic materials are oxidized by duckweed roots (Tchobanoglous and Burton, 1991). In anoxic conditions, bacteria carry out nitrification and also denitrification processes, converting compounds into ammonium and ortho-phosphate. These intermediates serve as nutrients for duckweed growth(Smith and Moelyowati, 2001).

Plants tolerant to polymeric dyes may be valuable for phytoremediation, as they can create a rhizosphere conducive to colonizing microbes with strong degradation abilities. Compounds from these plants may stimulate the production of fungal redox enzymes(Curreli, *et al*., 2001). Because of its high phenolic content and capacity to adsorb aromatic pollutants, *M. pulegium* clonal lines MPH-4 were initially screened and found to be a promising candidate for phytoremediation (Haudenschild *et al*., 2000).

5. Factors Affecting Bioremediation Efficiency

The process of bioremediation is affected by a range of factors, which determine its efficacy. In the case of bioremediation of textile effluents, factor spanning from pH to nutrients plays a vital role in how optimally the whole remediation process will function. Specific microorganism has its own specific degradation efficacy, restrictions, and individual working conditions.

5.1. Consequence of pH

The pH of the dye solution is crucial in biosorption, influencing how dyes interact with the biomass. Different types of dyes have varying pH requirements: basic dyes thrive under alkaline conditions, while reactive dyes need strongly acidic conditions for optimal biosorption.

The pH level greatly influences the effectiveness of dye decolorization by bacteria, with the optimal range for color removal being between 6 and 10. Adjusting the pH within 7 to 9.5, however, has minimal impact on the dye reduction process(Elisangel *et al*., 2009).The chemistry of dye molecules and fungal biomass is affected by the effluent's pH. Fungal biomass's ability to absorb acidic dyes rises as pH falls, while its capacity to biosorb basic dyes rises as pH rises (Mckay *et al*., 1999). The cell wall of fungi consists of lipo-polysaccharides, proteins-melanin complexes along with various other groups that can bind to dye molecules(Ozer *et al*., 2006). It was anticipated that the amino groups in cell wall of fungi, components would also become protonated at acidic pH levels. The increased dye absorption observed at lower pH levels may result from electrostatic attraction between the negatively charged dye anions and the positively charged cell surfaces(Waranusantigul, 2003). It was supported by Fu *et al*. (2002) that biosorption of some specific dyes by *Aspergillus niger* biomass could be reduced at optimum pH of 6.

5.2. Effect of Temperature

Temperature has only a minor impact on microbial biosorption within the 20-35°C range(Veglio and Beochini, 1997). Higher temperature usually enhances absorption due to the greater surface area activity and kinetic energy of the solute (Sag and Kutsal, 2000) and also the production of enzymes for decolorization enhances at optimal temperature. However, the exothermic characteristics of certain adsorption processes, along with rising temperatures, have been observed to decrease the biomass's biosorption capacity(Mameri *et al.*, 1994). Temperature plays a significant role on the absorption of dye effluents by microorganisms. Temperature on dye adsorption has been examined in thermophilic Cyanobacteria *Phormidium sp.* within a range of 25-45°C, showing a decrease in sorption capacity as the temperature exceeds 45°C (Aksu *et al.*, 2009).

Versatile textile dye effluent is generated at elevated temperatures, making temperature a crucial factor for the practical application of fungal biosorption. Fungal biomass can effectively remove dye at higher temperatures because it has more active surface sites and the dye molecules possess greater kinetic energy. Temperature pointedly affects the rate of dye adsorption based on the diverse dyes present in the dye medium. *Aspergillus foetidus* biomass demonstrated high efficiency in a decolorizing solution comprising a combination of colors, composite, and industrial effluent. The absorption capacity rose dramatically as the temperature rose from 30 to 50 °C, and it was highly temperature sensitive(Patel and Suresh, 2008).

5.3. Consequence of Nutrients

Nutrients play avital role in the dye decolorization process, higher quantity of nutrients pointedly effects the development activity of microorganisms and enhances the squalor of dyes in an aqueous medium. *Pseudomonas sp.* BSP-4, isolated from soil contaminated with azo dyes, can decolorize azo dye black E by using it as a nitrogen source, effectively removing it at concentrations up to 300 ppm within 36 hours(Sudhakar *et al.*, 2002). *Enterobacter agglomerans* decolorize methyl red in the presence of one, two, and four percent of glucose whereas in the absence of glucose, the decolorization rate of methyl red by *Enterobacter agglomerans* was less only (Bagramoglu, 2006). In the media, sodium nitrate can serve as a supply of nitrogen and accelerate the decolorization of methyl orange and malachite green. The nutrient composition of the medium significantly influences color removal, and the addition of natural supplements positively affects dye decolorization by fungi.

5.4. Impact of Dye Concentration

The concentration of dye significantly affects both the degree of dye elimination and the toxicity of the dye molecules. The proportion of dye removed rises over time, regardless of the initial concentration of the dye. However, the rate of decolorization decreases due to the dye's toxicity to bacteria, which primarily hampers metabolic processes or results in insufficient biomass to absorb higher dye concentrations. *Bacillus subtilis* potential to decolorize acid blue 113 was investigated at various dye concentrations ranging from 50 to 300 mg/l. As the dye concentration rose to 200 mg/l, the rate of decolorization accelerated (Gurulakshmi and Sudarmani, 2008). *Brevibacillus laterosporus* showed 81% dye removal within 48 hours under static conditions at the dye concentration of 50mg/l for some azo dyes (Jirasripongpun *et al.*, 2007).

Fungal growth is hampered by the persistence of a high concentration of dye. *C. versicolor* tolerated maximum dye (turquoise blue G, phthalocyanine dye) concentration of 1200 mg/l (Kapdan and Kargi, 2002).

6. Future Prospects and Challenges

Bioremediation, as it is seen is a better means of tackling the ongoing problem of textile wastewater and effluents. In recent times the various processes of bioremediation have seen a lot of advancements in refining, tuning, and increasing the efficiency of the techniques associated with bioremediation. To improve the outcome of remediation methods, wastewater is being treated with genetically engineered strains of microbes that are more effective.

6.1. Advances in Bioremediation Technologies

The traditional biological process offers a cost-effective and environmentally friendly approach to plummeting sludge production. It utilizes fewer chemical agents, conserves energy, and has a lower ecological impact. However, these conventional biological methods tend to be slow, with complete degradation taking longer than with alternative techniques. Additionally, biological treatment often requires more land and its effectiveness can be limited by sensitivity to chemical toxicity, operational type, and concentration levels. As a result, severalinvestigators are now exploring advanced techniques or modifications to existing biological treatments to develop solutions that are technically sound, environmentally sustainable, and economically viable(Vikrant *et al.*, 2018). Some of therecent advancements in wastewater bioremediation techniques are described below –

6.1.1. Genetic Modification for Dye Degradation

Numerous molecular biologists and microbiologists have recognized the probability of genetic engineering for use in the treatment of biodegradation (Mishra *et al.*, 2020). Genetically modified microorganisms (GMMs) or genetically engineered microorganisms (GEMs), altered through genetic modification techniques modelled after natural gene exchange among microorganisms, have proven to be a valuable tool for biodegradation(Kumar *et al.*, 2019). For bioremediation purposes, four primary methods are typically employed in the production of GEMs: (1) modifying enzyme specificities and affinities, (2) constructing and regulating metabolic pathways, (3) applying bio-affinity for chemical sensing bioreporters, toxicity reduction, and endpoint analysis and (4) optimizing growth, monitoring, and control of bioprocesses(Tahri *et al.*, 2012).GMOs can be developed by transferring genes between species or by modifying existing genes. In creating GMOs, functional genes from various bacterial strains—including *Bacillus idriensis*, *Sphingomonas desiccabilis*, *Escherichia coli*, *Pseudomonas putida*, *Mycobacterium marinum*, and *Ralstonia eutropha*—were utilized and introduced into other species (Saxena, 2019). A number compounds can be broken down by this method such as various chlorobenzene acids (Kumar *et al.*, 2013), toluene (Wang *et al.*, 2021), halogenated pesticides (Pimviriyakul, *et al.*, 2020), and toxic wastes(Pant *et al.*, 2021), etc. However, since dye degradation research employing this technology is still in its infancy, it is considered an emerging technique in this day and age.

6.1.2. Microbial Electrochemical Technologies (MET)

Microbial Electrochemical Technologies (MET) utilize microorganisms to convert chemical energy into electrical energy. This process occurs through electron transfer interactions between living microbial cells and electron-conducting materials, such as solid-state electrodes or naturally occurring minerals (Molognoni *et al.*, 2018). This property holds promise as an alternative to fossil fuels for energy production and as a safer substitute for harmful chemicals in wastewater treatment, particularly for removing pollutants from textile wastewater. The "Microbial Fuel Cell" (MFC) and "Bio-photovoltaic" (BPV) systems are among the best-known and most widely studied examples of microbial electrochemical technologies (MET) (Logan and Rabey, 2012). MFC is a method that produces electricity by using organic matter in wastewater as the substrate.

6.1.3. Nanomaterials in bioremediation

Nanotechnology has been used extensively in the food processing, bioremediation, pharmaceutical, and medical sectors for a number of decades (Liu and Dong, 2020). The swift advancement of nanobiotechnology has ushered in a new era in creating nano-structured materials that serve as platforms for immobilized enzymes (Zahirinejad *et al.*, 2021; Sheldon *et al.*,2021). Currently, research is increasingly focused on advanced functional materials, such as metal-organic frameworks (MOFs)for the elimination of synthetic dyes. MOFs interact with dye molecules (both anionic and cationic) through electrostatic attraction (Au, 2020). There are several studies reported for the removal of dyes using MOFs with excellent results (Ayati *et al.*, 2016; Karmakar *et al.*, 2109). Bio-nanomaterials are promising substances synthesized from a variety of biological components, including bacteria, peptides, nucleic acids, plants, and fungi(Zhao., 2020). Irrespective of their size, bio- nanomaterial possessing remarkable capabilities and biologically produced, biocompatible properties, these materials have drawn significant attention across diverse fields, including materials science, bioremediation, pharmacology, and biosensing (Menon *et al.*, 2021). According to recent studies (Khatoon and Sardar, 2017; Ahmad *et al.*, 2020), silver nanoparticles synthesized from certain strain of microbes were employed to degrade textile dyes under varying parameters.

6.2. Challenges and Future Research Directions

Microbial dye degradation offers promising potential as an effective water purification method; however, several challenges restrict its large-scale commercial use. Utilizing single cultures is often slow, whereas mixed microbial cultures accelerate degradation efficiency and improve pollutant removal efficiency. Prospective studies should focus on developing hybrid culture bioremediation methods, leveraging proteomics, and genetic engineering to enhance specific microbial traits. Additionally, research should assess the toxicity and potential applications of microbial degradation byproducts. Most studies thus far have focused on batch-scale applications, so investigating bioremediation's commercial viability as a cost-effective, efficient technique is essential. Microbial fuel cells (MFCs) show potential in this area, though limitations such as low power output and high costs need to be addressed. Evaluating degradation mechanisms, operational conditions, and the optimal growth environment for microorganisms is crucial. While research largely examines synthetic wastewater, further studies on real wastewater applications are necessary to gauge practical effectiveness. A multidisciplinary approach will be vital to overcome existing challenges and expand bioremediation's role in commercial water purification.

7. Conclusion

Various physicochemical treatment methods have been applied to mitigate dye pollution in water environments. However, conventional methods face limitations such as high operational and energy costs, harmful byproducts and extensive requirements for chemicals and energy. Bioremediation has emerged as a sustainable, cost effective and efficient alternative for treating textile effluent. Although there are several pressing challenges, future research may address many of these obstacles. Currently bioremediation methods remain mostly at the laboratory scale and scaling these systems for widespread commercial use possess a significant technological challenge. Multidisciplinary collaboration involving biotechnologists, engineers, material scientists, and biochemists isessential for advancing commercialization and process enhancements. Despite these challenges, recent advancements indicate a promising future for bioremediation in environmental clean-up and water treatment.

References

Ahmad M, Akanji MA, Usman ARA, Al-Farraj ASF, Tsang YF, Al-Wabel MI (2020). Turning date palm waste into carbon nanodots and nano zerovalent iron composites for excellent removal of methylthioninium chloride from water. Scientific Reports, 10(1), 1-15.

Aldas-Vargas, A., van der Vooren, T., Rijnaarts, H.H.M., Sutton, N.B., (2021). Biostimulation is a valuable tool to assess the pesticide biodegradation capacity of groundwater microorganisms. Chemosphere 280, 130793. https://doi.org/10.1016/ j. chemosphere.2021.130793

Alvarado, S., Guédez, M., Lué-Merú, M.P., Nelson, G., Alvaro, A., Jesús, A.C.,Gyula, Z, (2008). Arsenic removal from waters by bioremediation with the aquatic plants water hyacinth (Eichhornia crassipes) and lesser duckweed (Lemna minor). Bioresour. Technol., 99, 8436–8440.

Arlt VM, Glatt H, Muckel E, Pabel U, Sorg BL, Schmeiser HH, Phillips DH (2002). Metabolic activation of the environmental contaminant 3-nitrobenzanthrone by human acetyltransferases and sulfotransferase. Carcinogenesis, 23 (11), 1937-1945.

Arora DS, Chander M & Gill PK (2002). Involvement of lignin peroxidase, manganeseperoxidas and laccase in degradation and selective lihninolysis of wheat straw. Int BiodeterBiodegrad50, 115-120.

Arun Prasad AS, Bhaskara Rao KV, (2010). Physico-chemical characterization of textile effluent and screening for dye decolorizing bactéria. Global Journal of Biotechnology and Biochemistry 2010; 5(2), 80-86.

Ashrafi SD, Rezaei S, Forootanfar H, Mahvi AH, Faramarzi MA (2013). The enzymatic decolorization and detoxification of synthetic dyes by the laccase from a soil-isolated ascomycete, Paraconiothyriumvariabile.Int. Biodeterior. Biodegrad. 85, 173–181, https://doi.org/10.1016/j.ibiod.2013.07.006.

Au VK-M, (2020). Recent Advances in the Use of Metal-Organic Frameworks for Dye Adsorption.Frontiers in Chemistry, 8.

Ayman YI, Ewida A, El-Sesy ME & Zeid AA (2019). Complete degradation of azo dye acid red 337 by Bacillus megaterium KY848339.1 isolated from textile wastewater. Water Sci.,33,154–161.

Ayati A, Shahrak MN, Tanhaei B, Sillanpää M, (2016). Emerging adsorptive removal of azo dyeby metal-organic frameworks. Chemosphere, Vol. 160, p. 30–44.

Bafana A, Devi SS, Chakrabarti T, (2011). Azo dyes: past, present and the future. Environmental Reviews, 19 350–370.

Ben Mansour H, Corroler D, Barillier D, Ghedira K, Chekir L, Mosrati R, (2007). Evaluation of genotoxicity and pro-oxidant effect of the azo dyes: Acids yellow 17, violet 7 and orange 52, and their degradation products by Pseudomonas putida mt-2. Food and Chemical Toxicology, 45 (9) 1670-1677.

Banat IM, Mcmullan G, Mechan C, Kirby N, Nigam P, Smyth WF, Marchant R (2007). In Proceedings of the Industrial Waste Technical Conference Indian policy, 1- 6

Ben Mansour H, Houas I, Montassar F, Ghedira K, Barillier D, Mosrati R, Chekir-Ghedira L. Alteration of in vitro and acute in vivo toxicity of textile dyeing wastewater after chemical and biological remediation. Environmental Science and Pollution Research International 2012; DOI 10.1007/s11356-012-0802-7.

Bagramoglu G, Celik G, (2006). Africa.J Hazard Mater B, 137, 1689-97.

Ball AS, Betts WB, McCarthy AJ (1989) Degradation of lignin-related compounds by actinomycetes. Appl Environ Microbiol 55:1642–1644

Bieby VT, Siti Rozaimah SA,Hassan B, Mushrifah I, Nurina A,Muhammad M. A Review on HeavyMetals (As, Pb, and Hg) Uptake by Plants through Phytoremediation, International Journal of Chemical Engineering.,2011, 31.

Boyd, CE (1993). Vascular aquatic plants for mineral nutrient removal from polluted waters. Economic Botany, 24, 95–103.

Brahmbatt NH, Jasrai RT (2016). Role of Algae in bioremediation of textile effluents. Int J Eng Res Gen Sci 4: 443-453.

Carliell CM, Barclay SJ, Shaw C, Wheatley AD, Buckley C A, (1998). The effect of salts used in textile dyeing on microbial decolorization of a reactive azo dye. Environmental Technology, 19 (11) 1133-1137.

Cardwell, A.J., Hawker, D.W., Greenway, M., (2002). Metal accumulation in aquatic macrophytes from southeast Queensland, Australia. Chemosphere, 48, 653–663.

Charumathi D, Das N (2010). Bioaccumulation of synthetic dyes by Candida tropicalis growing in sugarcane bagasse extract medium. Adv Biol Res 4: 233-240.

Chen KC, Wu JY, Huang CC, Liang YM & Hwang SCJ (2003). Decolorization of azo dye using PVA-immobilized microorganisms. J biotechnol101: 241-252.

Cheriaa j, Bettaieb F, Denden I &Bakhrouf A (2009). Characterization of new algae isolated from the textile wastewater plant. J Food Agric Environ 7: 700-704.

Chung KT, Cerniglia CE (1992). Mutagenicity of azo dyes: Structure-activity relationships. Mutation Research, 277 (3) 201-220.

Chung KT, Stevens SE Jr, Cerniglia CR (1992). The reduction of azo dyes by the intestinal microflora. Crit Rev Microbiol, 18:175–190

Colberg PJS, Young, L.Y. (1995). Anaerobic degradation of nonhalogenated homocyclic aromatic compounds coupled with Nitrate, Iron, or Sulfate reduction. In: Microbial Transformation and Degradation ofToxic Organic Chemicals.Wiley-Liss, New York, pp. 307-330.

Couto SR. Dye removal by immobilized fungi. Biotechnology Advances 2009; 27(3) 227-235.

Coughlin MF, Kinkle BK & Bishop PL (1999). Degradation of azo dyes containing aminonaphthol by Sphingomonassp. strain 1CX. J Ind MicrobiolBiotechnol23: 341-346.

Daneshvar N, Ayazloo M, Khataee AR &Pourhassan M (2005). Biodegradation of the Textile Dye Malachite Green by Microalgae Cosmarium sp. Water and waste Treatment Research Laboratory, Department of Applied Chemistry, university of Tabriz, Tabriz, Iran.

Dellamatricea PM, Silva-Stenicob ME, Beraldo de Moraesc LA, Fioreb MF, Monteiro RTR (2017). Degradation of textile dyes by cyanobacteria. Brazilian J Microbiol48: 25–31.

Dhir, B.,(2017). Salvinia: an aquatic fern with potential use in Phytoremediation. Environ. Int. J. Sci. and Techno.,4, 23–27.

Dilek FB, Taplamacioglu HM, Tarlan E (1999), Color and AOX removal from pulping effluents by algae. Appl MicrobiolBiotechnol52: 585–591.

Donmez G &Asku Z (2002). Removal of chromium (VI) from saline wastewater by Dunaliellaspecies. Process Biochem38: 751-762.

Dos Santos AB. Reductive Decolourisation of Dyes by Thermophilic Anaerobic Granular Sludge. PhD-Thesis. Wageningen University, The Netherlands, 2005.

Dos Santos AB, Cervantes FJ, van Lier JB. Review paper on current technologies for decolorization of textile wastewaters: Perspectives for anaerobic biotechnology. Bioresource Technology 2007; 98 (12) 2369-2385.

Elisangel AF, Andrea Z, Fabio DG, Cristiano R, Regina D, Artur C, Inter BiodetBiodeg, 2009, 63, 280-288.

Elisangel AFZ, Andrea DG, Fabio R, Cristiano D & Regina C (2009). Biodegradation of textile azo dyes by a facultative Staphylococcus arlettaestrain VN-11 using a sequential microaerophilic/ aerobic process. Int BiodeterBiodeg 63: 280-288.

El-Sheekh MM, Gharieb MM & Abou-El-Souod GW (2009). Biodegradation of dyes by some green algae and cyanobacteria. Int BiodeterBiodegr 63: 699–704.

Ezeronye OU &Okerentugba PO (1999). Performance and efficiency of a yeast biofilter for the treatment of a Nigerian fertilizer plant effluent. World J Microbiol Biotechnol15: 515-516.

Forgacs E, Cserháti T, Oros G. Removal of synthetic dyes from wastewaters: a review. Environment International 2004; 30 (7) 953- 971

Fritioff, A., Greger, M., Aquatic and terrestrial plant species with potential to remove heavy metals from stormwater. Int. J. Phytorem., 2008, 5, 211–224.

Fu, Y., &Viraraghavan, T. (2001). Fungal decolorization of dye wastewaters: a review. Bioresource Technology, 79(3), 251–262. https://doi.org/10.1016/s0960-8524(01)00028-1

Fu, Y., &Viraraghavan, T. (2002). Dye biosorption sites in Aspergillus niger. Bioresource Technology, 82(2), 139–145. https://doi.org/10.1016/s0960-8524(01)00172-9

Gajare SM & Menghani S (2012). Biosorption of malachite green by naturally grown algal from Girna river, Jalgaon district, Maharashtra. J Algal Biomass Util13: 60–65.

Glenn JK, Akileswaran L & Gold MH (1986). Mn (II) oxidation is the principal function of the extracellular Mn-peroxidase from Phanerochaetechrysosporium. Arch BiochemBiophys251: 688–696.

Gothberg, A., Greger, M., Bengtsson, B.E., (2002). Accumulation of heavy metals in wate spinach (Ipomoea aquatica) cultivated in the Bangkok region.Thailand.Environ. Toxicol. Chem., 21, 1934–1939.

Hammel KE (1996). Extracellular free radical biochemistry of ligninolytic fungi. New J Chem 20: 195-198

Hao OJ, Kim H, Chiang PC, (2000). Decolorization of wastewater. Critical Reviews in Environmental Science and Technology, 30 (4) 449-505.

Hatakka A (1994). Lignin modifying enzymes from selected white-rot fungi: production and role in lignin degradation. FEMS Microbiol Rev 13: 125-135.

Heinfling A, Bergbauer M &Szewzyk U (1997). Biodegradation of azo and phthalocyanin dyes by Trametes versicolor and Bjerkanderaadusta. Appl MicrobiolBiotechnol48: 261–266.

Henciya S, Shankar MA & Malliga P (2013). Decolorization of Textile dye effluent by Marine cyanobacterium Lyngbya sp. BDU 9001 with coir pith. Int J environ sci 3: 644-651.

Homaeigohar S, (2020). The nanosized dye adsorbents for water treatment. Nanomaterials.10(2),295.

Horitsu H, Takada M, Idaka E, Tomoyeda M & Ogawa T (1997). Degradation of aminoazo benzene by Bacillus subtilis. European J Appl Microbiol4: 217 -224.

Hu TL (1992). Algal degradation and decolourization of textile effluents. Water Sci Technol26:357-366.

Hubbe MA, Beck KR, O'Neal WG, Sharma YC, (2012). Cellulosic substrates for removal of pollutants from aqueous systems: a review. 2. Dyes. Dye biosorption: Review. BioResources, 7 (2), 2592-2687.

Huebert, D.B., Shay, J.M., (1992). Zinc toxicity and its interaction with cadmium in the submerged aquatic macrophyte Lemna trisulca L. Environ. Toxicol. Chem.,11, 715–720.

Hutchinson,G.E,A Treatise on Limnology, 1975, 3,264-348.

Ibrahim MB, Poonam N, Datel S, Roger M, (1996). Microbial decolorization of textile dye-containing effluents: a review, Bioresource Technology, 58(3) 217-227.

Ihsanullah, I., Jamal, A., Ilyas, M., Zubair, M., Khan, G., & Atieh, M. A. (2020). Bioremediation of dyes: Current status and prospects. Journal of Water Process Engineering, 38, 101680. https://doi.org/10.1016/j.jwpe.2020.101680

Jadhav, J. P.,Govindwar, S. P. (2006). Biotransformation of malachite green by Saccharomyces cerevisiae MTCC 463. Yeast, 23(4), 315–323. https://doi.org/10.1002/yea.1356

Jadhav JP, Govindwar SP, (2007). Yeast decolorization of azo dyes. BioresourTechnol23: 316-323.

Jafari N, Soudi MR & Kasra- Kermanshahi R (2014). Biodegradation perspectives of azo dyes by yeasts. Microbiol83:484-497.

Joshni T, Chacko & Subramaniam K (2011). Enzymatic Degradation of Azo Dyes – A Review. Int J Environ Sci 1: 1250-1259.

Jun L.Y., Yon LS, Mubarak NM, Bing CH, Pan S, Danquah MK, Abdullah EC, Khalid M, (2019). An overview of immobilized enzyme technologies for dye and phenolic removal from wastewater, J. Environ. Chem. Eng., 7, https://doi.org/10.1016/j.jece.2019.102961.

Junnarkar N, Murty DS, Bhatt NS, Madamwar D, (2006). Decolorization of diazo dye remazol brilliant orange 81 by a novel bacterial consortium. World J MicrobiolBiotechnol, 22: 163-168.

Jirasripongpun K, Nasanit R, Niruntasook J, Chotikasatian B. Thinnest Int J Sc Tech, 2007, 12, 6-11.

Kapdan IK, Kargi F, Enzyme MicrobTechnol, 2002, 30, 195-199.

Karmakar S, Roy D, Janiak C, De S, (2019). Insights into multi-component adsorption of reactive dyes on MIL-101-Cr metal organic framework: Experimental and modeling approach. Separation and Purification Technology, 215,259-275.

Keharia H &Madamwar D (2003). Bioremediation concepts for the treatment of by containing water: A review. Indian J Exp Biol41: 1068-1074.

Kensa VM.,(2011). Bioremediation - An Overview. Jr. of Industrial Pollution Control 27(2), 161-168.

Khandare RV, Govindwa SP, (2016). Microbial Degradation Mechanism of Textile Dye, and Its Metabolic Pathway for Environmental Safety. Taylor & Francis Group, LL., 412.

Khatoon N, Sardar M., (2017). Efficient Removal of Toxic Textile Dyes using Silver. Nanocomposites,2,1-5.

Kim SU, Cheong YH, Seo DC, Hur JS, Heo JS & Cho JS (2007). Characterization of heavy metal tolerance and biosorption capacity of bacterium strain CPB4 (Bacillus spp.). Water Sci Technol55: 105–111.

Kirk-Othmer. Encyclopedia of Chemical Technology, v. 7, 5th Edition. Wiley-Interscience; 2004.

Knapp JS, Zhang F, Tapley KN (1997). Decolorization of orange II by a wood-rotting fungus. J chem TechnolBiotechnol69: 289-296.

Kumar A, Kumar A, Singh R, Singh R, Pandey S, Rai A,(2019). Genetically engineered bacteria for the degradation of dye and other organic compounds. In: Abatement ofEnvironmental Pollutants: Trends and Strategies, pp 331-350.

Kumar S, Dagar VK, Khasa YP, Kuhad RC, (2013). Genetically modified microorganisms (GMOs) for bioremediation.Biotechnology for Environmental Management andResource Recovery, 191-218.

Kushch OV, Zosenko IO, Shendrik AN, (2019). Comparison of N-Hydroxy compounds as mediators in laccase-catalysed decolorization of indigo carmine, ChemistrySelect, 4, 3905–3913.

Lee, C.K., Low, K.S., Hew, N.S., (1991). Accumulation of arsenic by aquatic plants. Sci.Total Environ.,103, 215–227.

Limura Y, Hartikainen P & Tatsumi K (1996). Microbial degradation of textile effluents. Appl MicrobiolBiotechnol45, 434–439.

Liu DM, Dong C, (2020). Recent advances in nano-carrier immobilized enzymes and their applications. Process Biochemistry,92,464-475.

Logan BE, Rabaey K, (2012). Conversion of wastes into bioelectricity and chemicals by using microbial electrochemical technologies. Science, 337(6095),686-690.

Logan BE, Rossi R, Ragab A, Saikaly PE, (2019). Electroactive microorganisms in bio-electrochemical systems. Nature Reviews Microbiology, 17(5),307-319.

Matto M, Q. Husain,(2009). Decolorization of direct dyes by immobilized turnip peroxidase in batch and continuous processes, Ecotoxicol. Environ. Saf., 72, 965–971,https://doi.org/10.1016/j.ecoenv.2008.02.019.

Matto M, Husain Q,(2008). Redox-mediated decolorization of Direct Red 23 and Direct Blue 80 catalyzed by bioaffinity-based immobilized tomato (Lycopersicon esculentum) peroxidase, Biotechnol. J., 3, 1224–1231, https://doi.org/ 10.1002/biot.200800049.

M.Gurulakshmi, D.N.P. Sudarmani, R. venba, Advanced Biotech, 2008, 12-18.

Mandi, L (1994). Marrakesh wastewater purification experiment using vascular aquatic plants Eichhornia crassipes and Lemna gibba. Water, Science & Technology, 29, 283- 287.

Meehan C, Banat IM, McMullan G, Nigam P, Symth F &Maarchant R (2000). Decolorization of Removal Black B using a thermotolerant Yeast, Kluyveromyces marxianus IMBS. Environ Int 26: 75 - 79.

Menon S, Agarwal H, Shanmugam VK, (2021). Catalytical degradation of industrial dyes using biosynthesized selenium nanoparticles and evaluating its antimicrobial activities. Sustainable Environmental Research,31(1),1-12.

Miretzky, P., Saralegui, A., Cirelli, A.F., (2004). Aquatic macrophytes potential for the simultaneous removal of heavy metals (Buenos Aires, Argentina). Chemosphere, 57, 997–1005.

Mishra B, Varjani S, Kumar G, Awasthi MK, Awasthi SK, Sindhu R,(2020). Microbial approaches for remediation of pollutants: Innovations, future outlook, and challenges. Energy and Environment.

Molognoni D, Chiarolla S, Cecconet D, Callegari A, Capodaglio AG, (2018). Industrial wastewater treatment with a bioelectrochemical process: Assessment of depuration efficiency and energy production. Water Science and Technology,77(1), 134-144.

Moreira MT, Mielgo I, Fejio G & Lema JM (1996). Evaluation of different fugal strains in the decolourization of synthetic dyes. Biotechnol Lett 18: 117-120.

McKay, G., Ho, Y. S., & Ng, J. C. Y. (1999). Biosorption of Copper from Waste Waters: A Review. Separation and Purification Methods, 28(1), 87–125. https://doi.org/10.1080/03602549909351645

N. Mameri, N. Boundries, L. Addou, D. Belhocine, H. Lounic, H. Gri, A, (1994). pauss, A Water Res, 33, 1347-54.

Nakanishi M, Yatome C, Ishida N &Kitade Y (2001). Putative ACP phosphodiesterase gene (acpD) encodes an azoreductase. J Biol Chem 276: 46394-46399.

Nguyen TA, Fu C, Juang RS, (2016). Effective removal of sulfur dyes from water by biosorption and subsequent immobilized laccase degradation on crosslinked chitosan beads, Chem. Eng. J., 304, 313–324.

Nigam P, Banat IM, Singh D & Marchant R (1996). Microbial process for the decolorization of textile effluent containing azo, diazo, and reactive dyes. Process Biochem31: 435–42.

Nouren S, Bhatti HN, Iqbal M, Bibi I, Kamal S, Sadaf S, Sultan M, Kausar A, Safa Y, (2017) . By-product identification and phytotoxicity of biodegraded Direct Yellow 4 dye, Chemosphere. 169, 474–484, https://doi.org/10.1016/j. chemosphere.2016.11.080.

Novotny C, Svobodova K, Kasinath A & Erbanova P (2004). Biodegradation of synthetic dyes by Irpexlacteusunder various growth conditions. Int BiodeterBiodegrad54: 215–223.

Novotny C, Vyas BRM, Erbanova P, Kubatova A & Sasek V (1997). Removal of PCBs by various white rot fungi in liquid cultures. Folia Microbiol42: 136-140.

O'Neill C, Hawkes FR, Hawkes DL, Lourenço ND, Pinheiro HM, Delée W, (1999). Colour in textile effluents – sources, measurement, discharge consents and simulation: a review. Journal of Chemical Technology and Biotechnology, 74 (11) 1009-1018.

O'Neill C, Lopez A, Esteves S, Hawkes FR, Hawkes DL, Wilcox S, (2000). Azo-dye degradation in an anaerobic-aerobic treatment system operating on simulated textile effluent. Appl MicrobiolBiotechnol 53:249–254.

Odjegba, V.J., Fasidi, I.O., (2007). Phytoremediation of heavy metals by Eichhornia crassipes. Environmentalist,27, 349–355.

Ogugbue CJ, Sawidis T, (2011). Bioremediation and Detoxification of Synthetic Wastewater Containing Triarylmethane Dyes by Aeromonas hydrophila Isolated from Industrial Effluent. Biotechnology Research International. DOI 10.4061/2011/967925.

Omar HH (2008). Algal Decolorization and Degradation of Monoazo and Diazo Dyes. Pak J Biol Sci 1: 1310-1316.

Oxspring DA, McMullan G, Smyth WF, Marchant R (1996) Decolourisation and metabolism of the reactive textile dye Remazol Black B, by an immobilized microbial consortium. Biotechnol Lett, 18:527–528.

Ozsoy HD, Unyayar A &Mazmanci MA (2005). Decolourization of reactive textile dyes Drimarene blue X3LR and ramzol brilliant blue R by FunaliatrogiiATCC 200800. Biodegradation 16: 195-204.

Pant G, Garlapati D, Agrawal U, Prasuna RG, Mathimani T, Pugazhendhi A, (2021). Biological approaches practised using genetically engineered microbes for a sustainable environment: A review. Journal of Hazardous Materials,124631.

Pajot HF, Fari˜na JI, De Fig.ueroa LIC, (2011). Evidence on manganese peroxidase and tyrosinase expression during decolorization of textile industry dyes by Trichosporon akiyoshidainum. Int. Biodeterior. Biodegradation 65, 1199–1207.

Patel, R., & Suresh, S. (2007). Kinetic and equilibrium studies on the biosorption of reactive black 5 dye by Aspergillus foetidus. Bioresource Technology, 99(1), 51–58. https://doi.org/10.1016/j.biortech.2006.12.003

Pearce CI, Lloyd JR, Guthrie JT, (2003). The Removal of Colour from Textile Wastewater Using Whole Bacterial Cells: A Review. Dyes Pigments, 58: 179–186.

Pimviriyakul P, Wongnate T, Tinikul R, Chaiyen P, (2020). Microbial degradation of halogenated aromatics: molecular mechanisms and enzymatic reactions. Microbial Biotechnology, 13(1),67-86.

Pointing &Vrijmoed (2000). Algal decolourization of azo dyes. World J MicrobiolBiotechnol16: 317–318.

Popoola, L.T., Yusuff, A.S., Adeyi, A.A., Omotara, O.O., (2022). Bioaugmentation and biostimulation of crude oil contaminated soil: process parameters influence. S. Afr. J. Chem. Eng. 39, 12–18. https://doi.org/10.1016/j.sajce.2021.10.003.

Przystaś W, Zabłocka-Godlewska E, Grabińska-Sota E, (2012). Biological Removal of Azo and Triphenylmethane Dyes and Toxicity of Process By-Products. Water Air Soil Pollut, 223 (4) 1581-1592.

R. Rakhshaee, M. Giahi,A. Pourahmad, (2012). Studying effect of cell wall's carboxyl carboxylate ratio changes of Lemna minor to remove heavy metals from aqueous solution, Journal of Hazardous Materials, 163, 165–173.

Rai, P.K., (2008). Phytoremediation of Hg and Cd from industrial effluents using an aquatic free floating macrophyte Azolla pinnata. Int. J. Phytorem., 10, 430–439.

Rajaguru P, Fairbairn LJ, Ashby J, Willington MA, Turner S, Woolford LA, Chinnasamy N, Rafferty JA, (1999). Genotoxicity studies on the azo dye Direct Red 2 using the in vivo mouse bone marrow micronucleus test. Mutation Research, 444(1) 175-180.

Ramachandran PV, Sundharam R, Palaniyappan J, Munusamy PA, (2013). Potential process implicated in bioremediation of textile effluents: A review. Advances in Applied Science Research, 4(1):131-145.

Ramalho PA, Cardoso MH, Cavaco-Paulo A & Ramalho MT (2004) Characterization of azo reduction activity in a novel ascomycete yeast strain. Appl Environ Microbiol70: 2279-2288.

Ranjana Julias Thilakar, Jeya Rathi. J and Prabha M. Pillai, (2012). Phytoaccumulation of Chromium and Copper by Pistia stratiotes L. and Salvinia natans (L.) J. Nat. Prod. Plant Resour., 2 (6),725-730.

Reddy CA (1995) The potential for white-rot fungi in the treatment of pollutants. Curr OpinBiotechnol 6:320–328

Reddy, K.R. Doil variations in physio-chemical parameters of water in selected aquatic systems. Hydrobiologia, 1981, 85, 201-207.

Reddy, K.R (1983). Fate of nitrogen and phosphorus in wastewater retention reservoir containing aquatic macrophytes. J. Environ. Qual., 12, 137- 141.

Rendón-Castrillón, L., Ramírez-Carmona, M., Ocampo-López, C., González-López, F., Cuartas-Uribe, B., & Mendoza-Roca, J. A. (2024). Efficient bioremediation of indigo-dye contaminated textile wastewater using native microorganisms and combined bioaugmentation-biostimulation techniques. Chemosphere, 141538. https://doi.org/10.1016/j.chemosphere.2024.141538

Robinson, B., Marchetti, M., Moni, C., Schroeter, L., van den Dijssel, C., Milne, G., Bolan, N., Mahimairaja, S., Arsenic accumulation by aquatic and terrestrial plants. In: Naidu, R., Smith, E., Owens, G., Bhattacharya, P., Nadebaum, P. (Eds.), Managing Arsenic in the Environment: From Soil to Human Health. CSIRO, Collingwood, Victoria, 2005, 235 247.

Sag Y, Kutsal T, (2000).Potential process implicated in bioremediation of textile effluents: A reviewBioChem Eng J, 2000, 6, 145-51.

Saratale RG, Saratale GD, Chang JS &Govindwar SP (2011). Bacterial decolorization and degradation of azo dyes: A review. J Tailand Inst Chem Eng 42: 138–157.

Saxena A, (2019). Striking balance between protein requirement and dietary phosphorus restrictions. Journal of Renal Nutrition and Metabolism, 5(1),1-1.

Seesuriyachan P, Takenaka S, Kuntiya A, Klayraung S, Murakami S, Aoki K, (2007). Metabolism of azo dyes by Lactobacillus casei TISTR 1500 and effects of various factors on decolorization. Water Research, 41(5) 985-992.

Sen S, Demirer GN, (2003). Anaerobic treatment of real textile wastewater with a fluidized bed reactor. Water Research, 37 (8) 1868-1878.

Senthilkumar S, Perumalsamy M & Prabhu HJ (2014). Decolourization potential of white-rot fungus

Sheldon RA, Basso A, Brady D, (2021). New frontiers in enzyme immobilisation: Robust biocatalysts for a circular bio-based economy. Chemical Society Reviews. https://doi.org/10.1039/d1cs00015b

Shrivastava R, Christian V & Vyas BRM (2005). Enzymatic decolorization of sulfonphthalein dyes. Enzyme MicrobTechnol36: 333–337.

Singh L & Singh VP (2015). Textile Dyes Degradation: A Microbial Approach for Biodegradation of Pollutants. Environ Sci Eng 2015:187-204.

Singh L, (2017). Biodegradation of synthetic dyes: a mycoremediation approach for degradation/ decolourization of textile dyes and effluents. J Appl BiotechnolBioeng,3: 430-435.

Skillicorn, P., W. Spira and W. Journey, Duckweed aquaculture: A new aquatic forming system for developing countries. The World Bank: Washington DC, USA.,1993.

Smith, M.D., I. Moelyowati, (2001). Duckweed based wastewater treatment (DWWT): design guidelines for hot climates. Water Sci. Technol., 43, 290-299.

Smyth WF, McClean S, O'Kane E, Banat I, McMullan G, (1999). Application of electrospray mass spectometry in the detectionand determination of Remazol textile dyes. J Chromatogr A 854:259–274.

Sudhakar P, Palaniappan R, Gowrie Shankar R, (2002). Asian J MicrobiolBiotechnol Environ, 2, 203-208.

Tahri N, Bahafid W, Sayel H, El Ghachtouli N, (2013). Biodegradation: Involved Microorganisms and Genetically Engineered Microorganisms. Biodegradation- Life of Science. https:// doi.org/10.5772/56194.In:

Tchobanoglous, G. and F.L. Burton, Wastewater engineering treatment, disposal, and reuse. 3rd Edn. McGraw Hill, New York,1991

Teerapatsakul C, Parra R, Keshavarz T, Chitradon L,(2017). Repeated batch for dye degradation in an airlift bioreactor by laccase entrapped in copper alginate, Int. Biodeterior. Biodegradation 120, 52–57.

Toh YC, Yen JJL, Obbard JP, Ting YP (2003). Decolourisation of azo dyes by White-rot fungi (WRF) isolated in Singapore. Enzyme Microb Tech 33: 569-575

Tripathi P, Kumar R, Sharma AK, Mishra A, Gupta R, (2010). Pistia stratiotes (jalkumbhi) pharmacogn rev.,4,153-160.

Tufail, M. A., Iltaf, J., Zaheer, T., Tariq, L., Amir, M. B., Fatima, R., Asbat, A., Kabeer, T., Fahad, M., Naeem, H., Shoukat, U., Noor, H., Awais, M., Umar, W., & Ayyub, M. (2022). Recent advances in bioremediation of heavy metals and persistent organic pollutants: A review. The Science of the Total Environment, 850, 157961. https://doi. org/10.1016/j.scitotenv.2022.157961

Tyagi, M., da Fonseca, M.M.R., de Carvalho, C.C.C.R., (2011). Bioaugmentation and biostimulation strategies to improve the effectiveness of bioremediation processes. Biodegradation 22, 231–241. https://doi.org/10.1007/S10532-010-9394-4/ TABLES/3.

Tychanowicz GK, Zilly A, Giatti C, de Souza M & Peralta R (2004). Decolorization of industrial dyes by solid-state cultures of Pleurotus pulmonarius. Process Biochem39: 855–859

Umbuzeiro GA, Freeman H, Warren SH, Kummrow F, Claxton LD, (2005). Mutagenicity evaluation of the commercial product CI Disperse Blue 291 using different protocols of the Salmonella assay. Food and Chemical Toxicology, 43 (1) 49-56.

Vaidya AA, Datye KV, (1982). Environmental pollution during chemical processing of synthetic fibers. Colourage, 14 3-10.

Vardanyan, L.G., Ingole, B.S., (2006) Studies on heavy metal accumulation in aquatic macrophytes from Sevan (Armenia) and Carambolim (India) lake systems. Environ. Int., 2007, 32, 208–218.

Van Der Zee FP, Villaverde S, Water Res, 2005, 39, 1425-1440.

Veerasingam S, Venkatachalapathy R, Sudhakar S, Raja P, Rajeswari V, (2011). Catalyzed degradation of disperse dyes by calcium alginate-pectin entrapped bitter gourd

(Momordica charantia) peroxidase, J. Environ. Sci. 23, 1129–1134, https://doi.org/10.1016/S1001-0742(10)60524-4.

Veglio F, Beochini F, (1997) Hydrometallurgy,44, 301-16.

Vikrant K, Giri BS, Raza N, Roy K, Kim KH, Rai BN, (2018). Recent advancements in bioremediation of dye: Current status and challenges. Bioresource Technology, 253(4):355–67.

Wang GH, Tsai TH, Kui CC, Cheng CY, Huang TL, Chung YC, (2021). Analysis of bioavailable toluene by using recombinant luminescent bacterial biosensors with different promoters. Journal of Biological Engineering, 15(1),1-9.

Waranusantigul P, Pokethitiyook P, Kruatrachhue M, Upatham ES, (2003). Environ Pollut, 125, 385-393.

Werheni AR., Di Rauso SG., Hidri Y., Abassi M.S., Mehri I., Costa S., Hassen, A., Rao, M.A., (2022). Combined bioaugmentation and biostimulation techniques in bioremediation of pentachlorophenol contaminated forest soil. Chemosphere 290, 133359. https://doi.org/10.1016/J. Chemosphere.2021.133359.

Won SW, Yun YS (2008). Biosorptive removal of reactive yellow 2 using waste biomass from lysine fermentation process. Dyes Pigment 76: 502- 507.

Yang X, Zheng J, Lu Y, Jia R, (2016). Degradation, and detoxification of the triphenylmethane dye malachite green catalyzed by crude manganese peroxidase from Irpexlacteus F17. Environ. Sci. Pollut. Res. 23, 9585–9597, https:// doi.org/10.1007/s11356-016-6164-9.

Yu Z & Wen X (2005). Screening and identification of yeasts for decoloring synthetic dyes in industrial wastewater. Int BiodeterBiodegr56: 109-114.

Z. Aksu, Ertugrul S, Domez, J Hazard Mater, 2009, 168, 310-318.

Zahirinejad S, Hemmati R, Homaei A, Dinari A, Hosseinkhani S, Mohammadi S,(2021). Nano-organic supports for enzyme immobilization: Scopes and perspectives. Colloids and Surfaces B: Biointerfaces,111774.

Zhang FM, Knapp JS, Tapley KN, (1999).Enzyme MicrobTechnol, 24, 48-53.

Zhang ZB, Liu CY, Wu ZZ, Xing L & Li PF (2006). Detection of nitric oxide in culture media and studies on nitric oxide formation by marine microalgae. Med Sci Monit 12: 75-85.

Zhang, X., Lin, A.-J., Zhao, F.-J., Xu, G.-Z., Duan, G.-L., Zhu, Y.-G., (2008). Arsenic accumulation by the aquatic fern Azolla: Comparison of arsenate uptake, speciation, and efflux by Azolla caroliniana and Azolla filiculoides. Environ. Pollut.,156, 1149–1155.

Zhao Y, (2020). A New Era of Metal–Organic Framework Nanomaterials and Applications. ACS Applied Nano Materials, 3(6): 4917-9.

Zimmermann T, Kulla HG, Leisinger T, (1982). Properties of purified orange II azoreductase, the enzyme initiating azo dye degradation by Pseudomonas KF46. Eur J Biochem, 129:197 203

Zollinger, H. Synthesis, (1987). Properties of Organic Dyes and Pigments. In: Color Chemistry. New York, USA: VCH Publishers, 92-102.

Zouboulis, Anastasios I., Moussas, Panagiotis A., Psaltou, S.G., Zouboulis, A.I., Moussas, P.A., (2019). Encyclopedia of Environmental Health, second ed. Groundwater and Soil Pollution: Bioremediation 2019–2022.

11

Role of Nano-Biotechnology in Management of Oil Spills: A Bioremediation Approach

***Prerna Salvi*[1], *Khushboo Jain*[1*] *and Avinash Marwal*[2]**

[1]*Dept. of Biotechnology, Faculty of Science, MVSC Campus, Janardan Rai Nagar Rajasthan Vidyapeeth (Deemed to be University), Udaipur-313001, Rajasthan, India*

[2]*Nanoparticle Synthesis & Bioinformatics Laboratory, Department of Biotechnology Vigyan Bhawan, Block-B, New Campus, Mohanlal Sukhadia University Udaipur-313001, Rajasthan, India*

**Corresponding Author Email: khushboojain.1791@gmail.com*

Abstract

Large-scale oil spills have been happening for decades, receiving a lot of media attention and posing a serious threat to the environment. As a result, governments and organizations are under pressure to act quickly to address the spill. However, over that period, cleanup methods have not improved much at all, and sometimes, quicker response times may not be the most efficient or environmentally beneficial option. Techniques for responding to oil spills include mechanical, thermal, chemical, and natural methods. Each of these approaches is reliant on environmental variables such the location, weather, and viscosity, content, and amount of the oil. One of the most effective ways to remove oil is through bioremediation, and nanoscience can help to enhance the natural process. This chapter discusses the function of nanoscience, which may provide a long-term way to improve post-oil spill response technologies because nanosized materials can be functionalized for specialized uses like the creation of magnetically responsive sorbents or Pickering emulsions. By using this novel method, oil spill sites can anticipate less long-term negative effects on the environment and economy, as well as enhanced biodegradation of the spilled oil and improved oil recovery. Moreover, integrating this emerging topic with the thoroughly researched discipline of bioremediation could potentially close the gap between bench scale and commercial scale applications.

Keywords: *Bioremediation, Nano-Biotechnology, Nanoparticles, Oil Spills, Environment, Marine Pollution, Case Studies*

1. Introduction

Oil spills are defined as the unintentional release of liquid petroleum hydrocarbons into the environment, predominantly impacting marine ecosystems. This form of pollution, often caused by human activities, can also occur on terrestrial landscapes. Spills typically result from the discharge of crude oil from sources such as tankers, offshore platforms, drilling rigs, and wells, as well as from refined petroleum products like gasoline and diesel (Smith *et al.*, 2019). While such incidents are primarily accidental, they may arise from various circumstances, including inadequate storage or poor maintenance, leading to container leaks. Historically, one of the earliest documented oil spills occurred in 1907 in Batum, a region then part of the Russian Empire (modern-day Batumi, Georgia). This incident, involving a collision between an oil-laden ship and another vessel in the Black Sea, marked the beginning of recorded oil spill events, though historical records from that time remain limited (Johnson *et al.*, 2021). The consequences of oil spills are profound, affecting both the environment and the economy. On the ocean's surface, oil disrupts aquatic ecosystems by blocking sunlight penetration and depleting dissolved oxygen levels, which are vital for marine life. Crude oil compromises the insulating and waterproofing properties of fur and feathers, leading to hypothermia and death in oil-coated birds and marine mammals. Additionally, ingesting oil can cause severe toxicity, while damage to habitats and reproductive systems often delays the long-term recovery of affected species. Coastal vegetation, particularly in ecosystems such as saltwater marshes and mangroves, is also highly vulnerable to oil contamination. Furthermore, many animals inadvertently ingest oil while grooming or consuming contaminated prey, leading to poisoning and often death. The economic repercussions of oil spills extend to tourism and commerce, particularly in cases where beaches and populated shorelines are fouled. Industries dependent on seawater, such as power plants and desalination facilities, are also adversely affected. The fishing industry is one of the hardest hits; major oil spills typically result in the suspension of commercial fishing activities to protect equipment, vessels, and public health by preventing the sale of contaminated seafood. While the immediate environmental impacts of oil spills are evident, their long-term effects on ecological systems remain challenging to evaluate. Nonetheless, the substantial financial costs of compensating affected individuals and communities have prompted efforts to minimize the occurrence of such disasters (Kumar and Banerjee, 2018; Patel *et al.*, 2020).

The methods used for cleaning up oil spills depend on several factors, including the size and location of the spill, the rate of oil release, the type of oil involved, and the water's temperature and chemical properties. The primary goal of oil spill response strategies is to contain the oil, remove sufficient quantities to mitigate environmental and economic damage, and allow the natural recovery processes of the ecosystem to take effect. One of the most common containment methods involves deploying floating booms around the spill source or at strategic locations such as harbour entrances and channels. These barriers limit the spread of the oil slick across the water's surface. Skimming is another widely used technique that physically separates the oil from the water, collecting it into tanks for removal. Skimming, like the use of booms, is most effective in calm waters. Sorbents, including materials like straw, volcanic ash, and polyester-derived plastics, are also employed to absorb oil from the water. In some cases, chemical surfactants and solvents are applied to the oil slick to promote its dispersion into the sea, accelerating its natural breakdown. When oil reaches the shoreline, cleanup efforts become more labour-intensive. Workers often use hand tools and heavy machinery to remove oil-soaked sand, rocks, and other debris from beaches and rocky shores. Despite these efforts, oil spills are not permanent environmental disasters (Roy *et al.*, 2018). Over time, natural processes such as wind, waves, and evaporation help disperse and reduce the oil. Additionally, naturally occurring microorganisms play a critical role in biodegrading the oil, breaking it down into less harmful substances. However, these natural processes are slow and often insufficient on their own, necessitating careful monitoring and supplementary interventions to ensure effective cleanup and ecological restoration. A natural remediation process called bioremediation, where marine microorganisms break down spilled oil as a primary energy source, has shown promise. Recently, the introduction of nanoparticles to oil-contaminated areas has gained attention as a means to enhance the efficiency of this process. Through the strategic use of nanoparticles, it is possible to increase the bioavailability of oil, optimize microbial activity, and overcome barriers to natural degradation, presenting a more sustainable approach to oil spill management. The objective of this chapter is to explore the role of nano-biotechnology in enhancing bioremediation processes for managing oil spills. By integrating nanoparticles with bioremediation strategies, we aim to examine their potential to improve microbial degradation of oil, overcome the challenges posed by traditional remediation techniques, and ultimately contribute to more effective and sustainable oil spill management.

2. Case studies and Insights on Oil Spill Incidents

Oil extraction and transportation can sometimes lead to environmental disasters due to factors such as aging or faulty equipment, human mistakes, and unforeseen circumstances. These incidents occasionally result in oil being released into the environment, with the most severe spills spilling tens of millions of gallons. Such spills have caused significant damage, including contaminated coastlines, polluted fisheries, harmed wildlife, and lost tourism revenue. Below are the top nine most catastrophic oil spills; (Rafferty *et al.*, 2023).

2.1. The Amoco Cadiz Oil Spill (1978)

On March 16, 1978, the Amoco Cadiz, a Very Large Crude Carrier (VLCC), ran aground on the shallow rocks off the coast of Brittany, France, while navigating the rough waters of the English Channel. The vessel, carrying approximately 69 million gallons (about 260,000 cubic meters) of light crude oil, suffered damage to its rudder and hydraulic system due to a large wave. Rescue attempts to secure towlines failed, and the ship was pushed toward the Brittany coast by strong winds. Upon impact with underwater rocks, the ship's hull was ruptured, releasing the oil. The oil spill affected about 200 miles (321 km) of coastline, leading to the destruction of marine life, including the death of approximately 20,000 birds and millions of invertebrates like mollusks and crustaceans. Additionally, oyster beds were heavily contaminated. In 1990, Amoco Corporation agreed to compensate French claimants \$120 million, as well as \$35 million to Royal Dutch Shell for the lost crude oil.

2.2. The Atlantic Empress Oil Spill (1979)

On July 19, 1979, the Atlantic Empress collided with the Aegean Captain in the Atlantic Ocean, releasing an estimated 90 million gallons (340,000 cubic meters) of oil. The collision occurred off the coast of Trinidad and Tobago during a tropical storm, causing both vessels to catch fire. While the Atlantic Empress burned for two weeks before sinking, the fire on the Aegean Captain was extinguished, and the ship was towed to Trinidad. Despite the massive volume of oil released, the spill caused minimal environmental damage to the nearby beaches due to strong winds pushing the oil out to sea, where it dispersed. However, the disaster led to the deaths of 27 sailors and left a lasting impact on maritime safety regulations.

2.3. The Castillo de Bellver Oil Spill (1983)

In August 1983, the Castillo de Bellver, an oil tanker carrying an estimated 53.5 million gallons (202,000 cubic meters) of crude oil, capsized following a fire. The tanker was located 70 miles off the coast of Cape Town, South Africa, in the South Atlantic Ocean when the fire broke out. The tanker split into two sections, and the stern, still containing about 110,000 tons of oil, drifted towards the coast and eventually sank in deep waters. Most of the oil slick was carried away by the Benguela Current before it dispersed into the open ocean, resulting in minimal environmental damage. The fire did burn a portion of the oil, but only a small fraction of the overall spill was directly involved in the blaze. Estimates vary, with some sources citing 79 million gallons of oil on board at the time of the fire.

2.4. The Persian Gulf War Oil Spill (1991)

The Persian Gulf War Oil Spill is the largest known intentional oil spill in history. In 1990, as part of the Iraqi occupation of Kuwait, Iraqi forces deliberately released between 380 million and 520 million gallons (1.4 to 2 billion litres) of oil into the northern Persian Gulf. This act was part of a larger strategy to prevent a U.S.-led coalition from advancing into southern Iraq. In addition to the massive release of oil from Kuwait's Sea Island terminal, Iraqi forces ignited hundreds of oil wells in Kuwait, which burned for months, contributing to an extensive environmental disaster. The spill caused widespread damage to marine ecosystems, including coral reefs and fish populations, and is considered one of the most devastating environmental acts of the 20th century.

2.5. The Mingbulak (Fergana Valley) Oil Spill (1992)

On March 2, 1992, the Mingbulak Oil Spill, caused by a blowout at an oil well in Uzbekistan's Fergana Valley, became the largest land-based oil spill in history. More than 88 million gallons (3,33,000 cubic meters) of oil were released into the valley, and the oil ignited, burning for two months before the well pressure subsided. Despite efforts to contain the spill with berms and dikes, large portions of the oil were lost to fire, and the surrounding land, including agricultural fields, was significantly affected by the spill. This disaster remains the largest oil spill in terms of land-based contamination.

2.6. The Kolva River Spill (1994)

The Kolva River Spill occurred in the Russian Arctic when a corroded pipeline ruptured, spilling approximately 84 million gallons (318,000 cubic meters) of oil into the Kolva River. This spill, which lasted for several months, contaminated about 186 squares kilometres (72 square miles) of tundra and wetlands. The region had been plagued by continuous oil pipeline leaks, contributing to persistent environmental damage in the area. Efforts to contain the oil involved the construction of dikes around the spill site, but these containment measures were eventually overwhelmed, leading to a significant release of oil into the surrounding environment.

2.7. BP's Deepwater Horizon Oil Spill (2010)

The Deepwater Horizon disaster, which began on April 20, 2010, is considered the largest accidental oil spill in history. The explosion of the BP-operated rig resulted in the release of approximately 134 million gallons (5,05,000 cubic meters) of oil into the Gulf of Mexico. Some estimates suggest the actual amount may have been as high as 206 million gallons. The oil slick covered more than 2,100 kilometres of coastline, affecting ecosystems across the Gulf Coast. The spill caused severe damage to marine life, fisheries, and coastal habitats, leading to the deaths of thousands of animals and long-term environmental degradation. BP faced numerous lawsuits and paid over $65 billion in compensation for the economic, environmental, and health impacts caused by the spill.

3. Methods to Remove Oil Spill

Oil spills represent a critical hazard to marine and coastal environments, necessitating the implementation of rapid and efficient remediation strategies. A diverse range of techniques has been designed to address the ecological damage caused by such incidents, encompassing physical, chemical, biological, and advanced technological methods. The selection of an appropriate approach is influenced by various factors, including the scale of the spill, its geographical context, and the specific characteristics of the oil involved (Fig. 1).

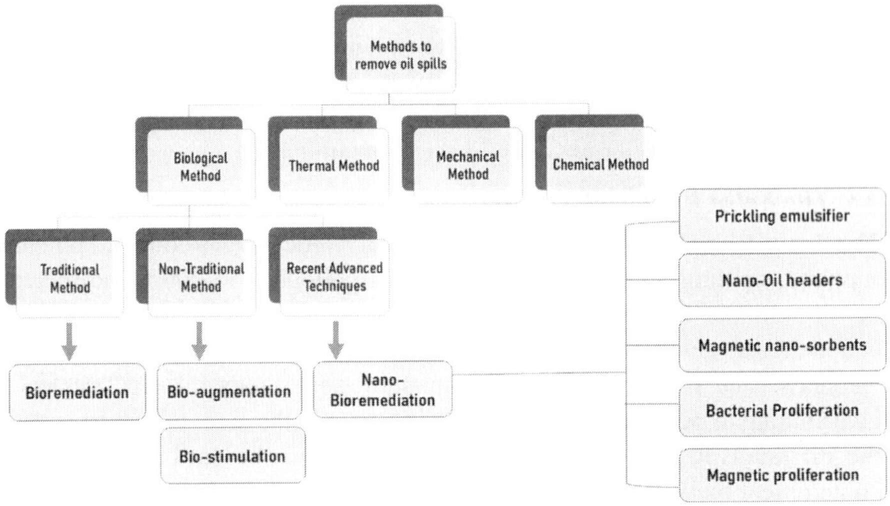

Fig. 1: Different methodology used to remove oil spills: from traditional to emerging technologies

3.1. Mechanical Method

Mechanical techniques are commonly employed as containment strategies to prevent the spread of oil after a spill. Floating booms, constructed from various materials, are widely used to contain oils of different viscosities and thicknesses, facilitating the prompt recovery or removal of the oil. However, booms are costly and labour-intensive, requiring careful handling to ensure the proper positioning of the net below the water surface. While numerous types of booms exist, they all serve the same purpose of physically containing the oil, and they necessitate supplementary methods for oil removal. The containment provided by booms enhances the effectiveness of other mechanical techniques, such as skimming and sorbents. Fire-resistant booms can also be used to contain oil for in situ burning. Mechanical methods have been a crucial part of oil spill response since the 1969 Santa Barbara spill, demonstrating effectiveness in preventing the spread of oil. However, they do have limitations; localized containment can increase the impact on the surrounding marine ecosystem, although rapid removal can mitigate this issue. These methods are also highly dependent on favourable weather conditions. Calm waters are essential to avoid clogging of skimmers and to ensure that sorbents remain on the surface. Additionally, mechanical methods require further treatment of the collected oil, such as separating it from debris and water, or employing physical or chemical treatments. Alternatively, thermal and chemical methods may offer quicker, less labour-intensive options for removal without requiring additional treatment (Hossain *et al.*, 2021; Lee *et al.*, 2017).

3.2. Thermal Methods

In situ burning is a thermal method that involves minimal equipment: fire-resistant booms to contain the oil and an igniter. Oil trapped within the booms is set alight using various igniting tools, such as surface-deployed lighters, torches, weed burners, or fuel-soaked sorbents. In situ burning is a highly efficient and rapid method for oil removal, though its effectiveness is limited by the natural emulsification of the oil over time. For successful ignition, the oil slick must be approximately 3 mm thick, and the water content must be below 25%. When employed in calm waters before emulsification, in situ burning can achieve removal rates of up to 95% (Li *et al.*, 2020). However, the technique has several drawbacks, including significant air pollution, the sinking of burned residue, and the potential for secondary fires or explosions.

3.3. Chemical Methods

Chemical methods, which involve the use of chemical agents, are applied to either prevent the spreading of oil or minimize its environmental impact.

Emulsifiers and dispersants break down the oil slick into smaller droplets, while oil herders aggregate the oil into a thicker, water-insoluble slick that can be ignited for in situ burning. These chemical methods are particularly useful in rough or icy waters, where traditional mechanical techniques may fail. Surfactants, such as emulsifiers or dispersants, function by reducing the oil-water interfacial tension and stabilizing the resulting emulsion, thus enhancing the bioavailability of the oil for microbial degradation. Corexit, a dispersant used extensively during the Deepwater Horizon spill, has been studied for its effectiveness in oil slick dispersion. However, the presence of Corexit has been shown to inhibit bioremediation and poses health risks to marine organisms, with evidence of reduced cell viability and DNA damage in exposed human bronchial cells. Some studies dispute these findings, citing variations in laboratory conditions and threshold levels, and questioning the reliability of self-reported data. Despite these concerns, dispersing oil into smaller droplets is an effective oil spill remediation method, promoting biodegradation, and has been considered a primary response strategy in large spills. Oil herders, typically silicone-based and non-biodegradable, aggregate oil into a thicker slick, facilitating continued in situ burning (Miller and Thompson, 2016). However, the environmental and biological impacts of silicone-based herders remain uncertain. The use of chemical methods has been controversial, particularly after the Deepwater Horizon spill, where the extensive use of dispersants raised concerns about their long-term ecological effects.

3.4. Biological Methods

3.4.1. Bioaugmentation

Bioaugmentation, first employed in the 1970s, is a bioremediation strategy aimed at enhancing the degradation of hydrocarbons by introducing hydrocarbonoclastic (oil-degrading) microorganisms into affected environments. This technique is particularly useful in areas where native microbial communities lack the capacity to metabolize all hydrocarbons present or where indigenous species exhibit extended lag phases due to oil exposure. The introduced microorganisms may be isolated and cultured from previous oil spill sites or genetically modified to improve their biodegradation capabilities. Despite its advantages, bioaugmentation presents certain challenges. Introduced species often compete with native microbial communities for essential nutrients, potentially disrupting the local ecosystem. For effective implementation, detailed knowledge of the oil's composition, the indigenous microbial community, and their hydrocarbon degradation abilities is essential. Additionally, understanding which introduced species can establish synergistic relationships with native microbes is critical for success (Chen *et al.*, 2022).

Bioaugmentation has proven effective in accelerating bioremediation, as hydrocarbonoclastic species are naturally ubiquitous but can represent as little as 0.003% to nearly 100% of the microbial community. In environments where these populations are at the lower end of this spectrum, supplementation can significantly enhance biodegradation. Advances in bioaugmentation techniques, such as the genetic modification or immobilization of cells in laboratory and mesocosm studies, have demonstrated promising results in improving the efficiency and reliability of this approach.

3.4.2. Bio-stimulation of Oil Spills

In situ bioremediation techniques for oil spills are heavily influenced by environmental factors such as pH, temperature, and nutrient availability, particularly nitrogen and phosphorus. Biostimulation is a method that involves the addition of limiting nutrients to enhance the growth of indigenous microbial communities capable of degrading hydrocarbons. Typically, oleophilic fertilizers are used to supply essential inorganic nutrients like nitrogen, phosphorus, and potassium, promoting the proliferation of oil-degrading microorganisms.

Bio-stimulation has proven effective in enhancing the biodegradation of oil. Following the Exxon Valdez oil spill, shorelines were treated with oleophilic fertilizers such as Inipol EAP 22, which contained 7.4% nitrogen and 0.7% phosphorus, and Customblen, a slow-release fertilizer with 28% nitrogen and 3.5% phosphorus. The success of bio-stimulation can lead to changes in the microbial consortia of an affected area, with its effectiveness being highly site-specific. Factors such as oil composition, the indigenous microbial community, and local environmental conditions play crucial roles in determining the overall efficacy of bio-stimulation as a remediation strategy (Bai *et al.*, 2019).

3.4.3. Bioremediation of Oil Spills

The use of microorganisms to degrade hydrocarbons has a long history, particularly in wastewater treatment and environmental remediation. The targeted application of microbes for oil spill cleanup is credited in part to George Robinson, who utilized marine microorganisms to break down oil following the 1969 Santa Barbara oil spill in California. Since then, bioremediation has emerged as a viable and effective technique for addressing oil spills. Its potential scalability has led many experts to suggest that bioremediation may be the only feasible method for mitigating large-scale oil spills.

Crude oil is a complex mixture of hydrocarbons, including alkanes, aromatic compounds, and resins. Each microbial species involved in oil degradation can metabolize only specific types of hydrocarbons, necessitating a diverse microbial community for efficient remediation. The relative abundance of these species evolves during the bioremediation process. For example, during the Deepwater Horizon spill, alkane-degrading microorganisms were initially dominant. As the concentration of alkanes declined in subsequent weeks, polycyclic aromatic hydrocarbon (PAH)-degrading species became more prevalent. The effectiveness of bioremediation is influenced by the metabolic capabilities of native microbial communities and their ability to adapt to hydrocarbons as the primary carbon and energy source. Factors such as oil composition, concentration, weather conditions, salinity, temperature, and nutrient availability further shape microbial abundance and activity.

Traditional oil spill remediation methods, including mechanical containment, in situ burning, and chemical treatments, often present challenges such as environmental damage, secondary fires, residual contaminants, and negative effects on marine life and human health. These conventional methods have not evolved in ways that significantly reduce their ecological impact or improve the safety of those implementing them. Consequently, there is a growing demand for cost-effective, environmentally friendly, and more efficient approaches to manage future oil spills. Bioremediation, although effective, often requires months or even years for the complete degradation of oil spills. While advancements have been made in bioremediation techniques, integrating nanoscience with this process has the potential to accelerate its efficiency and effectiveness.

4. Nanotechnology Based Bioremediation

Nanobiotechnology is an emerging field that combines the principles of nanotechnology and biotechnology to create innovative devices, materials, and systems at the nanoscale with wide range of applications. This interdisciplinary approach focuses on utilizing nanoscale materials, which exhibit unique properties such as increased surface area, high reactivity, and extremely small size, to advance bioremediation for oil spills (Fig. 2). Nanoparticles, initially used as magnetic sorbents, have demonstrated the ability to act as emulsifiers, facilitating the breakdown of oil. By increasing the bioavailability of oil, these nanoparticles provide a surface for microbes to attach to and promote their growth. The emulsification process expands the oil-water interfacial area, often leading to increased microbial degradation of oil. Nano-enhanced bioremediation has proven effective across a range of oil types and is largely unaffected by weather conditions. As such, it presents a sustainable and environmentally friendly alternative to conventional oil spill remediation methods, such as chemical dispersants, sorbents, and *In situ* burning, which can be toxic and less effective. Recent innovations in this field include the use of natural Pickering emulsifiers to promote bacterial proliferation on oil droplets, the incorporation of magnetic particles for easy bacterial removal, and the development of nutrient-loaded particles for targeted release at the oil-water interface (Khan *et al.*, 2022) (Table 1).

Table 1. Treatment of oil spills through Nanomaterials

S. No	Type of Nanomaterial	Type of Oil Treated	Quantity removed	References
1	Surface-Modified Magnetite	Crude Oil	90-95%	Abdullah *et al.*, 2023
2	Nanosponges	Crude Oil and Tar	85-90%	Li and Ma *et al.*, 2000
3	Magnetic Nanoparticles	Oil-Contaminated Water	94%	Bi *et al.*, 2025
4	Electrospun Nanocomposite Fiber Mats	Oil-in-Water Emulsions	85-88%	Barroso-Solares *et al.*, 2018
5	Carbon Nanomaterials	Diesel Oil	86-89%	Pandey and Saxena *et al*, 2024
6	Magnetic Nanomaterials	Heavy Crude Oil	90%	Sizyakov, 2020
7	Nanotechnology-Based Methods	Marine Oil Spills	92%	Prajapat *et al.*, 2024
8	Nano-Enhanced Bioremediation Agents	Oil Spills	85-90%	Pete *et al.*, 2021
9	Magnetic Micelle Arrays	Hydrophobic Organic Compounds	80%	Shah *et al.*, 2022
10	Magnetic Nanoparticle Adsorbents	Emerging Organic Contaminants	89-92%	Rasheed *et al.*, 2022
11	Carbon-Based Magnetic Nanocomposites	Various Oils	80%	Mehmood *et al.*, 2021
12	Graphene Aerogels	Liquid Pollutants	90-95%	Gorgolis *et al.*, 2024
13	Carbon Nanotubes	Oil-Contaminated Water	92%	Mnaty, *et al.*, 2023
14	EDTA-Functionalized Magnetic Nanoparticles	Heavy Metal Contaminants	85%	Huang and Keller, 2015
15	Advanced Oil Skimmer with Optimized Geometry	Seawater Oil Spills	93%	Broje and Keller 2006

4.1. Pickering Emulsifiers

A Pickering emulsion is a type of emulsion that is stabilized by solid particles, such as nanoparticles or microorganisms, which adsorb onto the oil-water interface and prevent the coalescence of the droplets. These solid particles form a monolayer at the interface, creating a physical barrier that stabilizes the emulsion and enhances its stability. The key to the formation of a Pickering emulsion lies in the wettability of the particles; they should be partially hydrophobic with a contact angle greater than 90° to ensure that they do not remain in the aqueous phase. This wettability condition allows the particles to irreversibly adsorb onto the oil-water interface, reducing interfacial energy and preventing droplet fusion. The stability of a Pickering emulsion is influenced by several factors, including the size, shape, and surface area of the particles. Additionally, the concentration of particles and the degree of interfacial coverage are crucial for the emulsion's stability. As the surface coverage of the oil-water interface increases, resistance to coalescence also increases, because the particles create a steric barrier that prevents the droplets from merging. The size and shape of the particles affect the surface area that is covered at the interface, further influencing the overall stability of the emulsion. The mechanism behind the stabilization of Pickering emulsions involves both hydrophobic interactions and electrostatic forces (Zhang and Liu, 2021). The oil-water interface is typically negatively charged, and positively charged particles experience a stronger electrostatic attraction to the interface, enhancing their adsorption. This results in the formation of stable emulsions that are resistant to coalescence, which is crucial in a variety of applications, such as enhanced oil recovery, drug delivery systems, and bioremediation, where controlling the stability of emulsions is essential for improving efficiency and effectiveness.

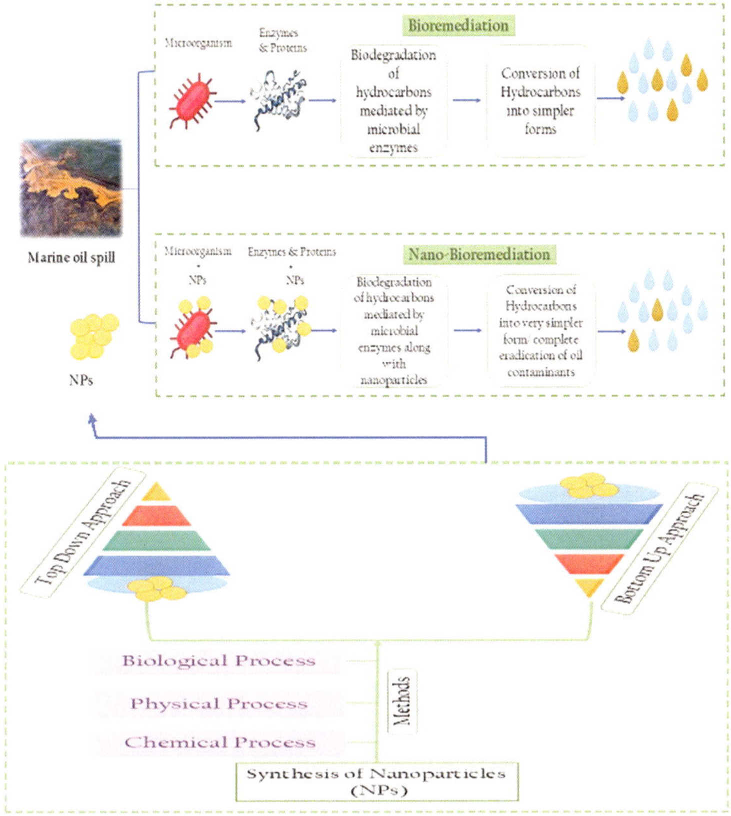

Fig. 2: Mechanism of Nano- Bioremediation for Oil Spill

4.2. Nanoparticles Based on Oil Headers

Nanoparticles based on oil heads refer to engineered nanomaterials that are designed to interact specifically with oil and water interfaces, often used in applications such as enhanced oil recovery, oil spill remediation, and environmental clean-up. These nanoparticles typically consist of amphiphilic structures, which possess both hydrophilic (water-attracting) and hydrophobic (oil-attracting) properties. These dual properties allow the nanoparticles to adsorb effectively at the oil-water interface, stabilizing emulsions and improving the efficiency of oil recovery or remediation processes. The behaviour of nanoparticles at the oil-water interface is governed by their surface chemistry and morphology. Nanoparticles are often functionalized with specific surface groups, such as surfactants or polymers, that enhance their affinity for either oil or water. This selective affinity allows for the manipulation of interfacial tension and facilitates the dispersion or aggregation of oil droplets, thus enhancing the separation or degradation of oils in contaminated environments.

In oil spill remediation, nanoparticles with oil-attracting "oil heads" are particularly effective in capturing and concentrating oil from water surfaces, facilitating its removal through various mechanical, chemical, or biological processes. They also play a critical role in emulsifying the oil, increasing its surface area and bioavailability to microorganisms involved in bioremediation. Additionally, some nanoparticles possess magnetic properties, which allow them to be easily separated from the environment after they have completed their task, making them reusable & minimizing secondary environmental impacts.

4.3. Magnetic Nanosorbents

Magnetic nanosorbents are advanced materials designed to effectively remove oil from various surfaces, including water. Sorbents, in general, are materials that are insoluble in both water and oil, and they facilitate oil recovery through absorption, adsorption, or a combination of both processes. Absorbents soak up and retain oil, often expanding to more than one and a half times their original size. Adsorbents, on the other hand, typically feature porous or rough surfaces, which enhance their adsorption capacity. In practice, materials that function by either absorption, adsorption, or a combination of both are commonly referred to as sorbents.

The effectiveness of sorbents is influenced by several factors, including the rate of sorption, oil retention, and ease of application and removal. Traditional sorbents, however, are often limited by low oil retention and a lack of selectivity, as they tend to absorb both oil and water. Nanotechnology has enabled the development of enhanced sorbents with specific properties to overcome these limitations. Nanoparticles, due to their large surface-area-to-volume ratio, exhibit a significantly higher sorption capacity and improved oil retention compared to conventional sorbents. Additionally, these nanoparticles can be engineered to possess hydrophobic characteristics that enhance their selective uptake of oil while minimizing the absorption of water. This selectivity ensures that the sorbent remains buoyant and effective in oil spill remediation applications, thus preventing it from sinking in aquatic environments. Magnetic nanoparticles, when incorporated into sorbents, further improve their efficiency by facilitating easy recovery from the contaminated area using external magnetic fields, reducing the need for complex collection processes. This combination of magnetic properties and engineered surface characteristics makes nanoparticle-based sorbents a promising solution for more efficient and targeted oil spill cleanup (Patra *et al.*, 2020).

4.4. Bacterial Proliferation

Surfactants are commonly used in oil spill remediation to break down oil into smaller droplets, thereby enhancing degradation by marine microorganisms.

However, many chemical dispersants are toxic to marine life and can exacerbate the toxicity of the oil, hindering the bioremediation process. Pickering emulsifiers, which are nanoparticles synthesized from natural precursors, present an environmentally friendly alternative. Research has demonstrated that suspended particles can significantly enhance oil biodegradation by marine microorganisms. For instance, particles that stabilize oil droplets have been shown to increase biodegradation rates by five times compared to non-dispersed oil, with a 3% increase in polycyclic aromatic hydrocarbon (PAH) degradation over 32 days.

The emulsification of oil into droplets increases the surface area available for microbial degradation compared to a surface oil slick, which in turn improves the bioavailability of oil to microorganisms in the water column. Additionally, hydrocarbonoclastic bacteria, which form biofilms on hydrophobic surfaces, show a preference for adhering to particle-stabilized droplets. This preference occurs because the surface energy of the bacteria is often lower than that of the seawater-air interface, leading to an attraction between the microbes and the hydrophobic droplets. The attachment of bacteria to these particle-stabilized droplets promotes bacterial growth in the presence of hydrocarbons, enhancing hydrocarbon degradation. This effect is more pronounced when compared to traditional surfactant-stabilized oil-in-water emulsions, suggesting that nanoparticle-enhanced emulsification can significantly improve the efficacy of bioremediation in oil spill scenarios (Li *et al.*, 2020; Zhang *et al.*, 2021).

4.5. Magneto-Responsive Bacteria

Marine microorganisms can contribute to oil removal not only through biodegradation but also by utilizing modified cell surfaces that enhance their interaction with oil. One promising strategy involves modifying the surfaces of bacteria to increase their hydrophobicity, which improves their adhesion to the oil-water interface. Additionally, these bacteria can be engineered to become magnetically responsive, further facilitating oil removal. By modifying the cell surfaces to be magnetically responsive, the bacteria can be attracted to and interact with magnetic nanoparticles. This enables their separation from the environment using an external magnetic field. An example of this approach involves the magnetic modification of *Alcanivorax borkumensis*, a hydrocarbon-degrading bacterium, using cationic polyelectrolyte-modified magnetite nanoparticles. These nanoparticles, with a 70–100 nm thick shell, are loosely bound to the cell surface and induce magneto-responsiveness. This modification led to nearly complete oil removal, but it also significantly hindered the bacterial growth rate, indicating that while this approach enhances oil recovery, it may limit the bacteria's ability to proliferate (Patra *et al.*, 2020; Chen *et al.*, 2022).

4.6. Targeted Bio-stimulation of Oil Spills

Nanotechnology offers promising solutions for the targeted bio-stimulation of oil spills, particularly through the use of nanoparticles loaded with growth-limiting nutrients. These nutrients can be released specifically at the oil-water interface, where they are needed most, thereby reducing the risk of nutrient dilution and competition from non-oil-degrading microorganisms. Mesoporous silica nanoparticles, designed to be hydrophobically responsive, are ideal for this application. These particles can be loaded with essential nutrients, such as nitrogen, phosphorus, and potassium, and feature a gated system that prevents nutrient release unless the particles come into contact with the oil-water interface. The gate is activated when the nanoparticles encounter lipophilic alkanes, such as those found in oil, which solvate the surface modification, releasing the encapsulated nutrients. This targeted release ensures that nutrients are delivered precisely where they are needed, thereby enhancing the growth of oil-degrading microorganisms and preventing the growth of non-oil-degrading species. This system has the potential for scalable and adaptable applications, making it a promising tool for bioremediation efforts (Roy *et al.*, 2018).

The combination of nanomaterials to enhance bioremediation represents an emerging field in oil spill response. Early research suggests that these technologies can improve the efficiency of biodegradation, facilitate the growth of oil-degrading microorganisms, and enable more effective oil removal. Additionally, the targeted release of growth-limiting nutrients enhances the overall bioremediation process. However, while evidence supports the potential of nanomaterials to enhance bioremediation, more research is needed. Future studies should focus on large-scale trials, exploring the biological and environmental impacts of introducing nanoparticles into marine ecosystems. Additionally, scalability, cost analysis, and long-term environmental safety should be considered to determine whether nano-enhanced bioremediation can provide a truly eco-friendly and cost-effective alternative to conventional oil spill remediation methods.

5. Conclusion

An oil spill is a type of pollution that occurs when human activity releases petroleum hydrocarbons into the environment. Because oil spills have the potential to release dangerous quantities of oil into crucial marine and coastal areas, they can have a major negative impact on the environment. Oil-based product accumulation changes biodiversity and pollutes the environment;

therefore, nano-biotechnology is a green and efficient way to remove the accumulation using biological systems, a process called bioremediation. This chapter's primary objective is to outline the significance, workings, uses, and constraints of nano-biotechnology in oil spill remediation. Traditional methods don't provide economical and environmentally beneficial alternatives. The use of microbe-based inhibitors in oil spill cleaning, commonly referred to as bioremediation, is a result of biotechnology advancements. Bioremediation may now be carried out successfully thanks to nanobiotechnology. Nanoparticles have enormous commercial potential. However, more research is necessary to ascertain the dangerous effects of nanoparticles in marine ecosystems due to environmental concerns.

References

Abdullah, M. M. S., Ali, M. S., & Al-Lohedan, H. A. (2023). Oil spill cleanup employing surface-modified magnetite nanoparticles using two new polyamines. Journal of Chemistry, 2023, Article 7515345, 1–11. https://doi.org/10.1155/2023/7515345

Bai, L., Lv, S., & Xiang, W. (2019). Enhancing oil biodegradation through biostimulation: A case study. Marine Pollution Bulletin, 145, 56-63. https://doi.org/10.1016/j.marpolbul.2019.05.018

Barroso-Solares, S., Pinto, J., Nanni, G., Fragouli, D., & Athanassiou, A. (2018). Enhanced oil removal from water in oil-stable emulsions using electrospun nanocomposite fiber mats. RSC Advances, 8(14), 7641–7650. https://doi.org/10.1039/C8RA00252E

Bi, H., Mulligan, C. N., Lee, K., Zhang, B., Chen, Z., & An, C. (2025). Nanotechnology for oil spill response and cleanup in coastal regions. Environmental Science: Nano.

Broje, V., & Keller, A. A. (2006). Improved mechanical oil spill recovery using an optimized geometry for the skimmer surface. Environmental Science & Technology, 40(24), 7914–7918. https://doi.org/10.1021/es060672u

Chen, Z., Wang, H., & Zhao, Y. (2022). Nanotechnology-enhanced bioremediation strategies for oil spill cleanup. Environmental Nanotechnology, 13, 102345. https://doi.org/10.1016/j.envnano.2022.102345

Gorgolis, G., Paterakis, G., & Galiotis, C. (2024). Synthesis of graphene aerogels toward the absorption of liquid and gas pollutants. arXiv. https://doi.org/10.48550/arXiv.2407.16392

Hossain, M., Ahmed, N., & Chowdhury, R. (2021). Mechanical techniques for oil spill response. Ocean Engineering, 234, 109123. https://doi.org/10.1016/j.oceaneng.2021.109123

Huang, Y., & Keller, A. A. (2015). EDTA-functionalized magnetic nanoparticle sorbents for cadmium- and lead-contaminated water treatment. Water Research, 80, 159–168. https://doi.org/10.1016/j.watres.2015.04.035

Khan, N., Roy, P., & Das, S. (2022). Nanotechnology and environmental sustainability: Focus on oil spills. Sustainable Chemistry, 35, 101025. https://doi.org/10.1016/j.suschem.2022.101025

Kumar, R., & Banerjee, S. (2018). Environmental impacts of oil spills: An integrated review. Ecotoxicology, 27(3), 312-321. https://doi.org/10.1007/s10646-018-1921-9

Lee, K., Blaker, J. J., & Murakami, R. (2017). Advances in fire-resistant booms for oil containment. Journal of Hazardous Materials, 341, 142-150. https://doi.org/10.1016/j.jhazmat.2017.08.035

Li, D., & Ma, M. (2000). Nanosponges for water purification. Clean Products and Processes, 2(2), 112–116. https://doi.org/10.1007/s100980000061

Li, Y., Zhang, T., & Chen, G. (2020). In-situ burning: Efficiency and environmental considerations. Environmental Science & Technology, 54(15), 9212-9221. https://doi.org/10.1021/acs.est.0c01975

Mehmood, A., Khan, F. S. A., Mubarak, N. M., Mazari, S. A., Jatoi, A. S., Khalid, M., Tan, Y. H., Karri, R. R., Walvekar, R., Abdullah, E. C., & Nizamuddin, S. (2021). Carbon- and polymer-based magnetic nanocomposites for oil-spill remediation: A comprehensive review. Environmental Science and Pollution Research, 28(39), 54477–54496. https://doi.org/10.1007/s11356-021-16045-0

Miller, S., & Thompson, J. (2016). Evaluating Corexit's environmental impact: Lessons from the Deepwater Horizon spill. Marine Pollution Bulletin, 113(1), 145-153. https://doi.org/10.1016/j.marpolbul.2016.06.024

Mnaty, K. H., & Saleem, A. M. (2023, July). Using the nanotechnology method to remove oil spill pollution of water. In AIP Conference Proceedings (Vol. 2809, No. 1). AIP Publishing.

Pandey, S., & Saxena, M. (2024). Carbon nanomaterial for oil spill clean-up. In A. Bachheti, R. K. Bachheti, & A. Husen (Eds.), Carbon-Based Nanomaterials (pp. 1–20). Springer. https://doi.org/10.1007/978-981-97-0240-4_15

Patel, D., Singh, R., & Mehta, N. (2020). Addressing economic repercussions of oil spills. Energy Policy, 145, 111744. https://doi.org/10.1016/j.enpol.2020.111744

Patra, S., Das, B., & Ray, S. (2020). Development of magnetic nanosorbents for oil spill cleanup. Materials Today, 34, 96-104. https://doi.org/10.1016/j.mattod.2020.09.010

Pete, A. J., Bharti, B., & Benton, M. G. (2021). Nano-enhanced bioremediation for oil spills: A review. ACS ES&T Engineering, 1(6), 928–946. https://doi.org/10.1021/acsestengg.1c00230

Prajapat, R., Yadav, H., Shaik, A. H., *et al.* (2024). A review of the prospects, efficacy, and sustainability of nanotechnology-based approaches for oil spill remediation. Waste Management & Research. https://doi.org/10.1177/0734242X241257095

Rafferty, J. P. (2023, September 1). 9 of the Biggest Oil Spills in History. Encyclopedia Britannica. https://www.britannica.com/story/9-of-the-biggest-oil-spills-in-history

Rasheed, T. (2022). Magnetic nanomaterials: Greener and sustainable alternatives for the adsorption of hazardous environmental contaminants. Journal of Cleaner Production, 362, Article 132338. https://doi.org/10.1016/j.jclepro.2021.132338

Roy, A., Singh, P., & Kumar, V. (2018). Nanotechnology applications in oil spill remediation. Journal of Environmental Management, 224, 96-104. https://doi.org/10.1016/j.jenvman.2018.05.034

Shah, M. U. H., Reddy, A. V. B., & Moniruzzaman, M. (2022). Ionic liquid–based surfactants for oil spill remediation. In Ionic Liquid-Based Technologies for Environmental Sustainability (pp. 257–268). Elsevier. https://doi.org/10.1016/B978-0-12-823851-7.00012-8

Sizyakov, M. I. (2020). New nanomaterials and dispersants for oil spill response. Problems of Gathering, Treatment and Transportation of Oil and Oil Products, 100, 100–109. https://doi.org/10.17122/ntj-oil-2020-5-100-109

Smith, J., Wilson, T., & O'Brien, K. (2019). Historical analysis of oil spills and their environmental impacts. Environmental Science Reviews, 12, 213-228. https://doi.org/10.1016/j.esrev.2019.01.010

Zhang, X., & Liu, J. (2021). Pickering emulsions in oil spill remediation: A comprehensive review. Advances in Colloid and Interface Science, 293, 102446. https://doi.org/10.1016/j.cis.2021.102446

12

Phycoremediation- An Emerging Technology to Mitigate Water Pollutants

***Premlata Vikal*[1], *Pragya Dadhich*[2*] *and Preeti*[3]**

[1]*Department of Botany, S.M.B. Govt. P.G. College, Nathdwara, Rajasthan, India*
[2]*Department of Botany, University College of Science, Udaipur, Rajasthan, India*
[3]*Department of Botany, University College of Science, Udaipur, Rajasthan, India*
**Corresponding Author Email: pragyadadhich7@gmail.com*

Abstract

In a world where water pollution poses a growing threat to ecosystems and human health, a new beacon of hope has emerged in the form of algae. This chapter delves into the transformative technology of phycoremediation, a ground-breaking method that harnesses the potential of algae to remove, degrade, or neutralize contaminants from wastewater and has emerged as a promising and sustainable technology for mitigating water pollution. This approach harnesses the natural biological processes of various algal species to effectively treat a diverse array of pollutants, including heavy metals, excess nutrients, organic compounds, and pathogens. Studies have shown that rapid growth rates, high pollutant uptake capacities, and the ability of algae to produce valuable biomass position phycoremediation as a cost-effective and environmental friendly alternative to conventional water treatment methods. Additionally, the process contributes to carbon sequestration and can be integrated with biofuel production, enhancing its economic viability and ecological benefits. This multifunctional approach not only addresses the critical issue of water pollution but also contributes to the circular economy by converting waste into resources. As global demand for sustainable water management solutions continues to rise, phycoremediation is poised to play a crucial role in addressing both water pollution challenges and resource recovery opportunities. Future research and development efforts such as scaling up the technology, improving the economic feasibility and expanding the applications to include diverse wastewater streams, such as industrial effluents and agricultural runoff likely will make phycoremediation a cornerstone of sustainable water treatment and environmental protection.

Keywords: *Phycoremediation, algae, water pollution, sustainability, environment*

1. Introduction

Water is the most essential resource for life, supporting the survival of all living beings, including humans, and playing a crucial role in food production and economic progress. Currently, many cities around the world are facing significant water shortages. About 40 % of global food production depends on irrigation, and countless industrial operations rely heavily on water. The availability and quality of water, whether surface or groundwater, have a major impact on environmental health, economic development, and societal growth. Human activities like urban expansion, population increase, industrial output, and climate shifts are causing a decline in water quality. As nations develop, industrial activities increasingly lead to the pollution of water sources, resulting in negative consequences such as widespread disease and diminished beauty of water bodies. This pollution is a significant threat to both the planet and its inhabitants. Higher temperatures are interfering with the metabolic processes of aquatic life, while changes in water pH create unstable conditions. Additionally, increased cloudiness in water reduces oxygen levels, affecting deep-water species and contributing to health issues for both animals and humans. Polluted water also hinders growth and development in all living organisms Haldar *et al.*, 2015 and Arif *et al.*, 2020. A study conducted by Halder *et al.* (2015) highlighted severe health challenges faced by local communities due to pollution in the Turag River, Dhaka, caused by the discharge of untreated wastewater into the river. It was revealed that elevated pH levels in the water, resulting from the use of caustic chemicals such as soda ash and caustic soda in industrial processes, were associated with skin irritation and sores. Dietary exposure to polluted crops and fish was found to contribute to stomach ailments, including gastric ulcers. Concerns were raised about potential groundwater contamination through the infiltration of industrial waste. Microbial contamination was identified as a likely cause of diarrhoea and dysentery. The use of river water for bathing and laundry was observed to facilitate the spread of waterborne diseases. Outbreaks of yellow fever, cholera, dengue, malaria, and other epidemics were noted as prevalent in the area. Respiratory problems and foul odours were attributed to air pollution caused by industrial discharge. Maternal and child health in riverbank slums found to be vulnerable, emphasizing the urgent need for interventions to address these public health concerns.Wastewater produced by industries, agriculture, and domestic activities contains high levels of nutrients, heavy metals, and chemicals, which harm the environment by polluting land, air, and water, ultimately posing risks to human and animal health. Traditional remediation methods are often ineffective and environmentally harmful, whereas alternative approaches are more affordable, efficient, and eco-friendly. Bioremediation, the use of biological organisms to restore natural resources, has emerged as a sustainable solution. Phycoremediation, a specialized bioremediation method, utilizes algae to eliminate harmful pollutants such as hydrocarbons, pesticides, radioactive substances, and heavy metals (Koul *et al.*, 2022).

Imagine harnessing some tiny, green superheroes to overcome the challenge of remediating waste water as they are capable of absorbing, neutralizing, and breaking down harmful pollutants from water bodies and calling the super power as phycoremediation. In search for an eco-, as well as, budget friendly remedy, phycoremediation has emerged as a worthy solution. Phycoremediation is an innovative technology that taps into the remarkable power of algae to combat water pollution. By using microalgae or macroalgae, this process tackles contaminants like heavy metals, excess nutrients from agricultural runoff, and toxic organic compounds, all while generating valuable biomass that can be transformed into biofuels, fertilizers, or animal feed. As the world struggles with increasing water contamination from industrial discharge, urban waste, and agricultural practices, phycoremediation shines as a beacon of sustainable innovation. Unlike conventional chemical treatments that risk creating secondary pollution, this natural method utilizes sunlight and photosynthesis to purify water, aligning perfectly with the principles of green technology. It's a promising step forward in our quest for cleaner water and a healthier planet. Algae also act as a carbon sink and, due to their high surface-to-volume ratio, are effective in absorbing, concentrating, and transforming pollutants through their metal-binding properties (Koul*et al.*, 2022). Many surface water bodies around the world are suffering from issues such as eutrophication, contamination, and depletion. Accomplishing sustainable treatment methods is crucial for enhancing the effectiveness of nutrient and pollutant removal in algal systems. Compared to traditional treatment methods, phycoremediation offers potential for commercially viable solutions by reducing costs and providing renewable bioenergy options. This process not only provides advanced bio-treatment but also generates valuable biomass, which can be used in various applications. Moreover, phycoremediation can eliminate heavy metals and harmful organic compounds without causing secondary contamination. Additionally, the algal biomass produced during phycoremediation has high protein and lipid content, making it suitable for use in biofuels, food, and animal feed industries (Priyadharshini *et al.*, 2021).

2. Phycoremediation

Phycoremediation involves utilizing microalgae or macroalgae to eliminate or transform pollutants, such as nutrients and hazardous chemicals, from wastewater. Several applications of phycoremediation (Table-1) hold promise for small-scale applications and could gain commercial viability in the future (Kumar and Sweety, 2024):

Table 1. Applications of algae-based technologies in wastewater treatment and environmental management (Kumar and Sweety, 2024).

Application	Description
Nutrient Extraction	Removes nutrients from municipal wastewater and organic-rich effluents.
Pollutant Absorption	Utilizes algae-based materials to capture nutrients and synthetic chemicals.
Acidic and Metal-Contaminated Water	Treats wastewater infected with high metal or acid content.
Carbon Capture	Captures carbon di oxide
Pollutant Degradation	Breaks down and transforms synthetic contaminants.
Toxin Detection	Employs algae-powered biosensors to identify toxic substances

Algae, among the earliest photosynthetic organisms, play a key role in utilizing sunlight to convert carbon dioxide, nutrients, and water into energy while enriching the atmosphere with oxygen. Their growth depends on the availability of nitrogen, carbon, phosphorus, and trace nutrients in the environment. Phycoremediation leverages algae to treat wastewater sustainably, offering a cost-effective and eco-friendly alternative to conventional methods that often involve high costs and risks of secondary pollution (Ogundele *et al.*,2023). Among the various strategies of phytoremediation, phycoremediation is unique in its specific focus on using micro- and macroalgae for environmental clean-up. Microalgae possess an exceptional ability to absorb and accumulate pollutants due to their large surface area relative to volume and efficient uptake systems. By sequestering contaminants within their biomass, they reduce the pollutant load in their surroundings. Additionally, certain microalgae species are equipped with specialized enzymes that enable them to break down and metabolize organic pollutants like hydrocarbons, pesticides, and pharmaceuticals. These features make microalgae valuable tools for detoxifying and degrading pollutants in both aquatic and terrestrial ecosystems. Phycoremediation represents an eco-friendly and sustainable remediation method that relies on biological processes, reducing the dependence on complex infrastructure or chemical treatments. Moreover, the biomass produced during this process offers diverse opportunities for reuse. For instance, microalgal biomass can be converted into biofuels, used as livestock and aquaculture feed, or serve as a resource for producing bioactive compounds (Barra and Greco, 2023). Biosorption, bioaccumulation and EPS(Extracellular polymeric substance) are some of the mechanisms involved in phycoremediation of heavy metals. Bioaccumulation is metabolism dependent process under which metal ions are taken up inside the cells. It occurs only in living state of algal cells (Machado and Valiaparampil, 2024). Biosorption on the other hand is a passive process, works at faster rate. It includes several processes such as adsorption, chelation, ion exchange and surface precipitation (Bilal *et al.*, 2018). Literature review has showed that production of EPS by cyanobacteria make them suitable for metal capturing and hence for phycoremediation.

3. Types of Pollutants Targeted by Phycoremediation

Phycoremediation effectively targets diverse pollutants using algae-based systems. These include excessive nutrients such as nitrogen and phosphorus commonly found in municipal and agricultural wastewater, preventing eutrophication in water bodies. Heavy metals like zinc, cadmium, lead, and mercury, often present in industrial effluents, are also removed, mitigating their toxic effects. Synthetic chemicals, including pesticides and pharmaceuticals, are broken down or absorbed, reducing their environmental persistence. Additionally, phycoremediation addresses pollutants in acidic and metal-contaminated waters, neutralizing acidity and binding toxic metals. Carbon dioxide is captured through algal growth, contributing to carbon sequestration, while organic contaminants are degraded or transformed into less harmful compounds. The chapter now moves on to some researches that has been done to remove pollutants from waste water. Study by Zhou *et al.* (2014) demonstrated the efficacy of four freshwater green microalgae *Chlamydomonas reinhardtii, Scenedesmus obliquus, Chlorella pyrenoidosa,* and *Chlorella vulgaris* in reducing a range of contaminants in wastewater. Result of their study showed that these algae exhibited the ability to simultaneously remove total nitrogen, total phosphorus, metals, and various organic compounds. Their study suggested that microalgae could serve as an alternative or complementary approach for treating wastewater containing a mixture of inorganic and organic pollutants. Abreu *et al.* (2012) performed mixotrophic cultivation of microalgae *Chlorella vulgaris* using hydrolysed cheese whey as a supplement. Result showed enhanced growth and higher algal biomass productivity, including lipids, starch, and proteins. Cheese whey boosted carbohydrate consumption due to its growth-promoting nutrients, outperforming pure glucose and galactose mixtures. Their strategy not only reduces production costs but also mitigates environmental issues related to dairy waste disposal. Research on phycoremediation of tannery wastewater (TWW) using *Scenedesmus* sp. By Ajayan *et al.* (2015) demonstrated the effectiveness of the algae in reducing pollutants under laboratory conditions. Over the 12 days of their study a significant removal of heavy metals (Cr, Cu, Pb, Zn) and nutrients (NO_3, PO_4) was achieved. TWW from wet-blue tanning processes was also found to support dense algal growth, making the algae a sustainable medium for phycoremediation and biomass production. Ugya *et al.* (2021) studied the use of microalgal biofilms for the remediation of petroleum-contaminated water demonstrate their potential in removing various pollutants. The biofilms, grown in nutrient-rich environments, showed significant reductions in parameters such as turbidity, total petroleum hydrocarbons, and several physicochemical factors, including sulfate, chloride, and metals. Characterization of the biofilm, in their study revealed its effectiveness in pollutant adsorption, attributed to its large surface area and extracellular polymeric substances (EPS). Phytochemical analyses from their study suggested that the biofilm's chemical composition plays a crucial role in both pollutant degradation and antioxidative responses to the contaminants. A study on the application of *Botryococcus* sp. for phycoremediation of bathroom greywater from village houses by Mohamed *et al.* (2017) demonstrated its effectiveness in reducing nutrients and organic pollutants. The result of their study achieved high reductions in biological oxygen demand (85.3–98%) and chemical oxygen demand (71.22–85.47%), alongside significant removal of nitrates, ammonia, orthophosphates, and potassium. The system they studied proved to be a cost-effective, eco-friendly, and energy-efficient method for treating greywater with no harmful by-products, highlighting its suitability for low-resource settings. Zhang *et al.* (2011) performed a study on *Chlamydomonas reinhardtii*, a model unicellular alga, and explored its response to the herbicide fluroxypyr, commonly used for weed and insect control. The research found that the alga accumulated significant amounts of fluroxypyr, which was rapidly degraded within the cells, with over 57% degradation observed by day 5 of the study. The exposure also induced reactive oxygen species production, leading to DNA degradation and the activation of antioxidant enzymes, suggesting an adaptive mechanism to mitigate fluroxypyr toxicity. The study provided insights into the ecological risks and biological responses of algae to herbicide contamination in aquatic ecosystems. Kausalya and Rao (2015) evaluated marine algae *Sargassum polycystum* and *Sargassum tenerrimum* for their effectiveness against a range of pathogens, including both Gram-positive and Gram-negative bacteria, as well as fungal species. The study underlined the potential of marine algae as a rich source of bioactive compounds, opening avenues for innovative drug development aimed at addressing microbial infections and benefiting public health. Inthorn *et al.* (1996) observed that among 17 cyanobacterial strains the filamentous *Tolypothrix tenius* exhibits a high level of Cd tolerance. Awasthi *et al.* (2004) compared the adsorption ability of alginate-immobilized cyanobacteria (*Anacystis nidulans*) and two green algae (*Chlorella vulgaris* and *Scenedesmus quadricauda*) for zinc, nickel and cadmium and concluded that *Anacystis nidulans* more efficiently adsorbed cadmium and nickel while zinc was better adsorbed by *Chlorella vulgaris*. Immobilized algal cells provide an alternative to conventional techniques of metal recovery. Kaur and Goyal (2018) investigated the accumulation of heavy metals (Cu, Zn, Cr and Pb) from coal fly ash by cyanobacterial strains such as *Nostoc muscorum, Anabaena variabilis, Tolypothrix tenuis,* and *Aulosira fertilissimia. Cyanobacteria* were grown in suitable growth medium (BG-11) and distilled water to observe their growth and metal accumulation. *Nostoc muscorum* showed maximum uptake of Cr, Pb. *Anabaena variabilis* showed maximum uptake of Cu and Cr and Zn. Table-2 provides identity of some more algal species involved in heavy metal removal form waste water.

Table 2. Algae utilized for heavy metal remediation

Algae	Heavy metal	References
Anacystis nidulans, Chlorella vulgaris and *Scenede mus quadricauda*	Ni, Zn and Cd	Awasthi and Rai, 2004
Spirulina sp.	Pb^{2+} and Zn^{2+}	Anejaet *al.*, 2010
Nostoc muscorum	Zn^{2+} and Cu^{2+}	Goswamiet *al.*, 2015
Oedogonium sp. And *Nostoc* sp.	Pb^{2+}	Gupta and Rastogi, 2008
Aphanothece halophytica	Hg, As and Cd	Laloknamet *al.*, 2009
Microcystis sp.	Zn	Pradhanet *al.*, 1998
Lyngbya taylorii	Pb	Klimmeket *al.*, 2001
Microcystis aeruginosa	Ni	Pradhanet *al.*, 1998
Oscillatoria angustissima	Cu	Mohapatra and Gupta, 2005
Spirulina sp.	Hg	Chojnacka, Chojnacki, and Gorecka, 2004

4. Advantages of Phycoremediation

Phycoremediation has several advantages that makes it an eco-friendly and cost-effective approach for environmental management. Here's a breakdown:

- Eco-Friendly and Sustainable
    - Natural Process: Uses naturally occurring organisms, reducing reliance on chemical treatments.
    - Carbon Sequestration: Algae absorb CO_2 during photosynthesis, mitigating greenhouse gas emissions.

- Cost-Effective
  - Low Input Requirement: Algae grow with minimal nutrients, utilizing sunlight and carbon dioxide as energy sources.
  - Self-Sustaining: Once established, algal systems can often self-propagate, reducing maintenance costs.
- Effective Pollutant Removal
  - Heavy Metals: Algae can remove heavy metals like zinc, lead, cadmium, and mercury from wastewater.
  - Nutrients: They effectively remove excess nutrients such as nitrogen and phosphorus, preventing eutrophication in water bodies.
  - Organic Pollutants: Some algal species degrade hydrocarbons, dyes, and other organic contaminants.
- Versatility
  - Wide Application: Works for industrial effluents, municipal wastewater, agricultural runoff, and even air purification.
  - Adaptability: Algal strains can be genetically modified to target specific pollutants.
- Byproduct Utilization
  - Biomass for Bioenergy: Algal biomass harvested during remediation can be converted into biofuels.
  - High-Value Products: Extracts from algae can be used for pharmaceuticals, cosmetics, and animal feed.
- Improves Biodiversity
  - Algal systems create habitats for other organisms, enhancing the biodiversity of aquatic ecosystems.
- Scalability
  - Small to Large Systems: Can be implemented in small ponds or scaled to treat large industrial wastewater facilities.
- Low Energy Requirement
  - Relies on natural processes, reducing the need for energy-intensive equipment compared to conventional treatment methods.

By leveraging these advantages, phycoremediation presents a promising alternative for achieving sustainable environmental remediation while fostering resource recovery and conservation.

5. Environmental and Economic Benefits of Phycoremediation

Algae play a vital role in environmental sustainability and economic growth by capturing carbon dioxide and producing biofuels. Through carbon sequestration, algae absorb significant amounts of CO_2 from the atmosphere and industrial emissions, helping to mitigate climate change. Economically, algae serve as a renewable resource for biofuel production, offering a cleaner alternative to fossil fuels. This dual functionality makes algae a key player in advancing green technologies and fostering a sustainable economy. Algae have a crucial role in the global carbon cycle and offer significant potential for CO_2 biofixation. Study by Tsai *et al.* (2015) explored the carbon utilization capability of microalgae in a simulated freshwater ecosystem. Mass cultivation of microalgae demonstrated efficient CO_2 biofixation, directly utilizing over 100% of the atmospheric carbon source. Environmental variables were monitored in their study, and key factors influencing growth in an unsupplemented freshwater medium were mathematically analysed by them to understand the ecological complexity of algae. The results of their research highlighted that atmospheric CO_2 served as the primary carbon source, with statistical analyses confirming its major role in algal growth systems.

Biofuels offer a sustainable alternative to limited and polluting fossil fuels, with algae providing a highly efficient raw material for their production. Algal farming avoids competing with agricultural land for food crops and requires less complex processing than lignocellulosic biomass. Many microalgae species are versatile, capable of simultaneously generating energy and treating wastewater. Cultivation, a critical stage in biofuel production, can be conducted in controlled photobioreactors or simpler, cost-effective open systems. Conversion methods include thermal, chemical, and biological processes, with biodiesel derived through lipid modification, bioalcohol from fermentation, and biohydrogen via various photosynthetic and fermentative pathways (Voloshin*et al.*, 2016).

6. Challenges and Future Directions in Phycoremediation

To effectively address environmental contamination, it is crucial to delve deeper into the underlying molecular and biochemical processes that drive phycoremediation. Such insights could lead to the creation of more advanced and affordable techniques for addressing environmental contamination (Pandey *et al.*, 2019).

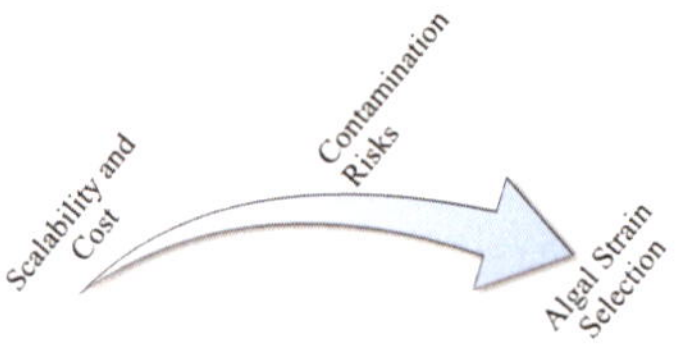

Fig. 1: Challenges in phycoremediation

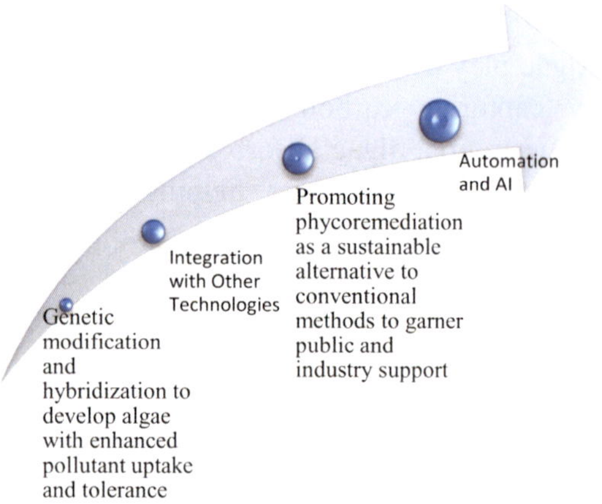

Fig. 2: Future directions in phycoremediation

To effectively address environmental contamination, it is crucial to delve deeper into the underlying molecular and biochemical processes that drive phycoremediation. Such insights could lead to the creation of more advanced and affordable techniques for addressing environmental contamination (Pandey *et al.*, 2019). Even though heavy metal phycoremediation is still a lab-scale process, understanding the mechanisms at action can help increase efficiency, selectivity, and lower processing costs. The performance of bioremediation has recently been shown to be improved by physicochemical and molecular changes made to the microbial cell surface. Indeed, a number of microorganisms with novel cell surface characteristics have undergone bioengineering. On the other hand, there are still few instances of microalgae being genetically modified to remove heavy metals. Large-scale usage of phycoremediation procedures will be made possible by more research in the areas of genetic engineering, pre-treatment, immobilisation techniques, and a combination of other physicochemical strategies (Danouche *et al.*, 2021). It has recently been demonstrated that physicochemical and molecular modifications to the algal cell surface are viable strategies for enhancing bioremediation efficacy. Therefore, using new chemical agents to bioengineer the cell surface in order to produce advantageous cell surface properties may aid in the biosorption process (Pradhan *et al.*, 2022).

7. Conclusion

Phycoremediation emerges as a sustainable, eco-friendly solution for mitigating water pollution through the use of algae. It effectively targets diverse pollutants, including heavy metals, nutrients, and organic contaminants, while generating valuable byproducts like biofuels and fertilizers. The approach offers numerous advantages such as cost-effectiveness, scalability, and carbon sequestration. Despite its potential, challenges like scalability, economic feasibility, and limited research on genetic engineering need to be addressed. Advancing this technology through innovations in genetic modification, bioengineering, and process optimization can unlock its full potential, establishing phycoremediation as a cornerstone of future environmental management and resource recovery.

References

Abreu, A. P., Fernandes, B., Vicente, A. A., Teixeira, J., andDragone, G. (2012). Mixotrophic cultivation of Chlorella vulgaris using industrial dairy waste as organic carbon source. Bioresource technology, 118, 61-66.

Ajayan, K. V., Selvaraju, M., Unnikannan, P., and Sruthi, P. (2015).Phycoremediation of tannery wastewater using microalgae Scenedesmus species. International journal of phytoremediation, 17(10), 907-916.

Aneja, R. K., Chaudhary, G., Ahluwalia, S. S., & Goyal, D. (2010).Biosorption of Pb 2+ and Zn 2+ by non-living biomass of Spirulina sp. Indian journal of microbiology, 50, 438-442.

Arif, A., Malik, M. F., Liaqat, S., Aslam, A., Mumtaz, K., Afzal, A., andJaved, R. (2020). Water pollution and industries. Pure and applied Biology, 9(4), 2214-2224.

Awasthi, M. and Rai, L. 2004. Adsorption of nickel, zinc and cadmium by immobilized green algae and cyanobacteria: a comparative study. Annals of Microbiology. 54: 257-268.

Barra, L., and Greco, S. (2023). The potential of microalgae in phycoremediation.Microalgae - Current and Potential Applications.

Bilal, M., Rasheed, T., Sosa-Hernández, J. E., Raza, A., Nabeel, F., andIqbal, H. M. (2018). Biosorption: an interplay between marine algae and potentially toxic elements—a review. Marine drugs, 16(2), 65.

Chojnacka, K., Chojnacki, A., andGórecka, H. (2004). Trace element removal by Spirulina sp. from copper smelter and refinery effluents. Hydrometallurgy, 73(1-2), 147-153.

Danouche, M., El Ghachtouli, N., and El Arroussi, H. (2021). Phycoremediation mechanisms of heavy metals using living green microalgae: physicochemical and molecular approaches for enhancing selectivity and removal capacity. Heliyon, 7(7).

Goswami, S., Diengdoh, O. L., Syiem, M. B., Pakshirajan, K., andKiran, M. G. (2015). Zn (II) & Cu (II) removal by Nostoc muscorum: A cyanobacterium isolated from a coal mining pit in Chiehruphi, Meghalaya, India. Canadian Journal of Microbiology, 61(3), 209-215.

Gupta, V. K., andRastogi, A. (2008).Biosorption of lead (II) from aqueous solutions by non-living algal biomass Oedogonium sp. and Nostoc sp.—a comparative study. Colloids and Surfaces B: Biointerfaces, 64(2), 170-178.

Halder, J. N., and Islam, M. N. (2015).Water pollution and its impact on the human health. Journal of environment and human, 2(1), 36-46.

Inthorn, D., Nagase, H., Isaji, Y., Hirata, K., and Miyamoto, K. 1996. Removal of cadmium from aqueous solution by the filamentous cyanobacterium Tolypothrix tenuis. Journal of Fermentation And Bioengineering. 82(6): 580-584.

Kaur, R. and Goyal, D. 2018.Heavy metal accumulation from coal fly ash by cyanobacterial biofertilizers. Particulate Science and Technology. 36(4): 513-516.

Kausalya, M., andRao, G. N. (2015).Antimicrobial activity of marine algae. J. Algal Biomass Utln, 6(1), 78-87.

Klimmek, S., Stan, H. J., Wilke, A., Bunke, G., and Buchholz, R. (2001). Comparative analysis of the biosorption of cadmium, lead, nickel, and zinc by algae. Environmental science & technology, 35(21), 4283-4288.

Koul, B., Sharma, K., and Shah, M. P. (2022). Phycoremediation: A sustainable alternative in wastewater treatment (WWT) regime. Environmental Technology & Innovation, 25, 102040.

Kumar, A., andSweety. (2024). Phycoremediation of Wastewater - An Approach Used Microalgae for Bioremediation. International Journal of Current Microbiology and Applied Sciences, 13(7), 28–34.

Laloknam, S., Sirisopana, S., Phornphisutthimas, S., Takabe, T., and Incharoensakdi, A. (2009). Removal of mercury, arsenic and cadmium in synthetic wastewater by cyanobacterium Aphanothece halophytica. NU Science Journal, 6, 96-104.

Machado, A. A.,and Valiaparampil, J. G. (2024).Unlocking the Potential of Algae for Heavy Metal Remediation. Water, Air, & Soil Pollution, 235(10), 629.

Mohamed, R. M., Al-Gheethi, A. A., Aznin, S. S., Hasila, A. H., Wurochekke, A. A., andKassim, A. H. (2017). Removal of nutrients and organic pollutants from household greywater by phycoremediation for safe disposal. International Journal of Energy and Environmental Engineering, 8, 259-272.

Mohapatra, H., and Gupta, R. (2005). Concurrent sorption of Zn (II), Cu (II) and Co (II) by Oscillatoria angustissima as a function of pH in binary and ternary metal solutions. Bioresource Technology, 96(12), 1387-1398.

Ogundele, O. D., Adewumi, A. J., andOyegoke, D. A. (2023). Phycoremediation: Algae as an Effective Agent for Sustainable Remediation and Waste Water Treatment. Environmental & Earth Sciences Research Journal, 10(1).

Pandey, A., Singh, M. P., Kumar, S., andSrivastava, S. (2019). Phycoremediation of persistent organic pollutants from wastewater: retrospect and prospects. Application of Microalgae in Wastewater Treatment: Volume 1: Domestic and Industrial Wastewater Treatment, 207-235.

Pradhan, B., Bhuyan, P. P., Nayak, R., Patra, S., Behera, C., Ki, J. S., and Jena, M. (2022). Microalgal phycoremediation: A glimpse into a sustainable environment. Toxics, 10(9), 525.

Pradhan, S., Singh, S., Rai, L. C., and Parker, D. L. (1998). Evaluation of metal biosorption efficiency of laboratory-grown Microcystis under various environmental conditions. Journal of Microbiology and Biotechnology, 8(1), 53-60.

Priyadharshini, S. D., Babu, P. S., Manikandan, S., Subbaiya, R., Govarthanan, M., andKarmegam, N. (2021). Phycoremediation of wastewater for pollutant removal: a green approach to environmental protection and long-term remediation. Environmental Pollution, 290, 117989.

Tsai, D. D. W., Chen, P. H., Chou, C. M. J., Hsu, C. F., andRamaraj, R. (2015). Carbon sequestration by alga ecosystems. Ecological Engineering, 84, 386-389.

Ugya, Y. A., Hasan, D. U. B., Tahir, S. M., Imam, T. S., Ari, H. A., andHua, X. (2021). Microalgae biofilm cultured in nutrient-rich water as a tool for the phycoremediation of petroleum-contaminated water. International Journal of Phytoremediation, 23(11), 1175-1183.

Voloshin, R. A., Rodionova, M. V., Zharmukhamedov, S. K., Veziroglu, T. N., and Allakhverdiev, S. I. (2016). Biofuel production from plant and algal biomass. International Journal of Hydrogen Energy, 41(39), 17257-17273.

Zhang, S., Qiu, C. B., Zhou, Y., Jin, Z. P., and Yang, H. (2011). Bioaccumulation and degradation of pesticide fluroxypyr are associated with toxic tolerance in green alga Chlamydomonas reinhardtii. Ecotoxicology (London, England), 20(2), 337–347.

Zhou, G. J., Ying, G. G., Liu, S., Zhou, L. J., Chen, Z. F., andPeng, F. Q. (2014).Simultaneous removal of inorganic and organic compounds in wastewater by freshwater green microalgae. Environmental Science: Processes & Impacts, 16(8), 2018-2027.

13

Microbial Bioremediation of Pharmaceutical Pollutants: A Bioengineering Approach to Sustainable Water Treatment

Ankur Das Talukdar[2], Juhi Goyal[1], Priti Yadav[1], Priyank Upadhyay[1] and Nitish Rai[2]*

[1]*Department of Biotechnology, Mohanlal Sukhadia University, Udaipur, Rajasthan India*

[2]*Department of Zoology, University of Lucknow, Lucknow, Uttar Pradesh, India*

**Corresponding Author Email: rai_nitish@lkouniv.ac.in*

Abstract

The increasing presence of pharmaceutical pollutants in water system poses a significant challenge to environmental health and safety. Conventional waste water treatment methods often fall short in effectively removing these contaminants, necessitating the exploration of innovative bioremediation strategies. Harnessing the capabilities of microbes, particularly through bioengineering, presents a promising avenue for the degradation and detoxification of pharmaceutical waste in aquatic environment. This chapter highlights the role of microorganisms, including bacteria, fungi, and microalgae, in bioremediation processes, emphasizing how bioengineered strains can enhance pollutant removal efficiently. Through a comprehensive examination of microbial interactions with pharmaceutical compounds, we outline the mechanism of bioremediation and explore the transformative potential of these methods for sustainable water treatment. Furthermore, the chapter explores the future prospects and challenges in optimizing microbial bioremediation technologies, offering insights into strategies for successful implementation in diverse ecological contexts.

Keywords: *Pharmaceutical Pollutants, Microbial Bioremediation, Bioengineered Microbes, Wastewater Treatment, Sustainable Water Management*

1. Introduction

Pharmaceutical pollution has emerged as a significant global concern, raising alarms about its potential impacts on ecosystems and human health.

The journey of pharmaceutical compounds from their point of use to contamination of water bodies reveals a complex and interconnected web of anthropogenic activities. From manufacturing plants to individual households, these pollutants persistently infiltrate aquatic systems, creating a growing environmental crisis (Jamur, 2024; Nguyen *et al.*, 2024). The origins of pharmaceutical pollution are multifaceted, reflecting the diversity of modern pharmaceutical use. Wastewater treatment plants, though pivotal in managing domestic and industrial effluents, are often ineffective in completely removing pharmaceutical residues (Massima Mouele *et al.*, 2021). Improper disposal methods, such as flushing unused medications down toilets, further exacerbate the issue. Healthcare facilities and pharmaceutical manufacturing units contribute significantly, with untreated or partially treated effluents releasing a cocktail of active pharmaceutical ingredients (Insani *et al.*, 2020). Additionally, agricultural runoff, laden with veterinary antibiotics and growth hormones, introduces these contaminants into surface and groundwater systems. Studies have identified alarming concentrations of pharmaceuticals and personal care products (PPCPs) in Indian wastewater and water bodies. Antibiotics like ciprofloxacin were found at levels up to 40 times higher than in other countries, with concentrations reaching 14 mg/L in treatment plant effluents (Balakrishna *et al.*, 2017). Other commonly detected compounds include antidepressants, antihypertensives, and analgesics like ibuprofen (Subedi *et al.*, 2017). The contamination extends to surface water, groundwater, and even drinking water sources, raising concerns about antibiotic resistance development (Fick *et al.*, 2009). In some cases, pharmaceutical metabolites were found at concentrations up to 7 times higher than their parent compounds (Subedi *et al.*, 2017). These findings highlight the urgent need for improved wastewater management in India's pharmaceutical production and urban areas.

Pharmaceutical pollutants in aquatic ecosystems pose significant risks to non-target organisms, even at low concentrations (ng/L to μg/L) (Fabbri, 2015). These contaminants enter water bodies through various routes, including wastewater treatment plants, agricultural runoff, and aquaculture. The continuous presence of pharmaceuticals can lead to unexpected and complex ecotoxicological effects on aquatic fauna (Mezzelani *et al.*, 2018). Two major concerns are the promotion of antimicrobial resistance and endocrine disruption, such as the feminization of male fish (Nguyen *et al.*, 2024). Additionally, pharmaceuticals can disrupt natural chemical information flows in aquatic environments, interfering with essential processes like mate location, feeding, and predator avoidance (Van Donk *et al.*, 2016). The effects of these pollutants can cascade through communities, indirectly affecting multiple species. Beyond direct toxicity, pharmaceutical pollutants contribute to broader ecological disturbances. Increased biological oxygen demand (BOD) and eutrophication have been observed in contaminated water bodies, exacerbating the degradation of aquatic habitats (Chakrabarti, 2018). Bioaccumulation further magnifies the risks, as pharmaceutical residues traverse the food chain, potentially impacting higher trophic levels, including humans (Du *et al.*, 2014). To combat pharmaceutical pollution, several remediation techniques have been explored. Advanced oxidation processes, adsorption, and microbial treatments show promise in degrading complex pharmaceutical compounds (Mansouri *et al.*, 2021). However, the removal efficiencies of conventional wastewater treatment plants remain inconsistent, with some pharmaceutical metabolites detected at concentrations exceeding their parent compounds (Subedi *et al.*, 2017).

Bioremediation, a sustainable and cost-effective approach, is gaining attention for its potential to address pharmaceutical pollutants. Utilizing microorganisms to metabolize and break down pollutants into less harmful substances offers a path toward mitigating contamination. However, significant challenges remain, including the development of more effective wastewater treatment technologies and the implementation of robust monitoring programs (Corpus *et al.*, 2024). Pharmaceutical pollution represents a pressing environmental challenge that requires concerted global efforts. Enhanced regulatory frameworks, public awareness campaigns, and innovative research are essential to mitigate the risks posed by these contaminants (Li *et al.*, 2024). The development of decentralized treatment systems and sustainable practices within the pharmaceutical industry will be pivotal in addressing this multifaceted issue. By investing in advanced water treatment technologies and fostering international collaboration, the global community can work towards safeguarding aquatic ecosystems and public health. The management of pharmaceutical pollutants in the environment has become an urgent challenge, requiring innovative and sustainable approaches. Among these, bioremediation has emerged as a promising solution, leveraging biological processes to degrade, immobilize, or transform pharmaceutical contaminants in soil and water (Alex *et al.*, 2025). This eco-friendly technology offers an alternative to conventional methods, providing both environmental and economic advantages.

Bioremediation employs a range of biological agents, including bacteria, fungi, algae, and plants, to treat pharmaceutical waste. These organisms utilize enzymatic and metabolic pathways to break down complex pharmaceutical compounds, transforming them into less harmful substances (Silva *et al.*, 2019). For instance, bacteria like *Pseudomonas putida* and *Bacillus subtilis* rely on oxidative and reductive pathways mediated by enzymes such as cytochrome P450 to degrade pollutants. Similarly, fungi such as *Trametes versicolor* and

Phanerochaete chrysosporium employ ligninolytic enzymes like laccases and peroxidases to mineralize recalcitrant pharmaceuticals, including diclofenac and carbamazepine (Narayanan *et al.*, 2023). Algae, including *Chlorella vulgaris* and *Scenedesmus obliquus*, contribute to pharmaceutical pollutant removal through biosorption and photosynthetic processes, offering an additional layer of remediation (Larsen *et al.*, 2019). The choice of bioremediation strategy depends on the pollutant type, environmental conditions, and the specific capabilities of the microorganisms involved.

Bioremediation offers significant advantages over traditional methods, such as incineration, chemical treatments, and advanced oxidation processes. These conventional approaches are often energy-intensive, costly, and capable of producing secondary pollutants, such as toxic byproducts or greenhouse gases (Ayilara & Babalola, 2023). In contrast, bioremediation utilizes natural or bioengineered organisms to achieve complete mineralization of pollutants into harmless end products, such as water, carbon dioxide, and biomass. This process minimizes the risk of toxic residues and reduces chemical oxygen demand (COD) in pharmaceutical effluents (Vidali, 2007). Moreover, bioremediation is cost-effective, versatile, and adaptable to a wide range of environmental conditions, enabling the treatment of both biodegradable and recalcitrant pharmaceutical compounds (Kuppan *et al.*, 2024).

Advances in bioengineering have significantly enhanced the capabilities of bioremediation. Genetically modified microbes are now tailored to target specific pharmaceutical pollutants with greater efficiency and specificity. These bioengineered strains optimize metabolic pathways, enhancing the production of key enzymes such as laccases and cytochrome P450, and ensuring faster and more complete degradation of contaminants (Rafeeq *et al.*, 2023). Moreover, engineered microbes can tolerate harsh environmental conditions, such as high salinity or extreme pH levels, and incorporate novel biosynthetic pathways to degrade otherwise non-biodegradable substances. Their ability to minimize harmful byproducts and utilize biosensors for real-time monitoring further underscores their potential in pharmaceutical waste management (Ayilara & Babalola, 2023).

This chapter explores the potential of bioremediation as an efficient and sustainable solution for treating pharmaceutical wastewater, emphasizing its role in addressing the persistent issue of pharmaceutical contamination. It examines the advantages and limitations of conventional treatment techniques, providing a foundation for developing integrated and advanced solutions tailored to modern pharmaceutical effluent management challenges. The chapter also highlights various eco-friendly bioremediation strategies, including biotransformation, enzyme-mediated degradation, biofilm formation, biostimulation, and biodegradation, showcasing their effectiveness in removing pharmaceutical pollutants from wastewater. Additionally, it underscores the critical need for future research and innovation, particularly in the application of bioengineered microbes, which offer enhanced efficiency, specificity, and adaptability. These advanced approaches pave the way for a more comprehensive and effective strategy to mitigate pharmaceutical contamination in the environment.

2. Source of Pharmaceutical Derived Pollutants

Water pollution occurs due to overaccumulation of the harmful substances such as organic and inorganic chemicals as well as microorganisms in the water. The chemical contaminants from industries, agriculture, uncontrolled discharge of sewerage, oil and chemical spills, and often natural disaster contribute towards the water pollutants. Over the past few decades, the use of chemicals has increased in industries, agriculture, mining, refining, petroleum, cosmetics, personal care, and healthcare, which has led to their increased runoff in water bodies (Naidu *et al.*, 2021). Among the various chemicals, pharmaceuticals, in particular, have contributed most to enhancing the quality of life. The pharmaceutical industry is responsible for the production of drugs for human and animal healthcare management, disease diagnosis, prevention, and treatment. The pharmaceutical industry has gained a significant increase in terms of market value, available drugs, production, availability to every individual, and consumption (González Peña *et al.*, 2021). Current pharmaceutical research is mainly market-driven, with a greater emphasis on drug efficacy in patients and their production. Consequently, their environmental impacts and degradability in the environment are often minimally explored (González Peña *et al.*, 2021). The increasing demand for pharmaceuticals and personal care products, and their unprescribed use has led to their misuse and environmental release (González Peña *et al.*, 2021; Khan & Barros, 2023). The pharmaceutical compounds are one of the major part of contaminants of emerging concern (CEC). United States Environment Protection Agency (EPA) defines CECs as pollutants, mainly unregulated, which are often present in environment whose presence and significance are only now being evaluated (US EPA, 2015). CECs are currently not under the routine monitoring and may become a candidate for future regulation.

A large number of pharmaceuticals released in the environment are sourced directly from manufacturers. The regulation on the emissions of such industries varies between countries and sites, therefore many industries fail to meet the environment standard. The release of the active pharmaceutical ingredients are rarely regulated, thus it is difficult to monitor their presence in the industrial discharge. The direct and indirect release of the industrial discharge pose a great threat to the environment. The problem is more serious in developing countries which lack proper treatment facilities (Patel *et al.*, 2019; Rehman *et al.*, 2015). High pharmaceuticals in the effluent near the major industrial sites have been reported across the world in developing countries such as India (Balakrishna *et al.*, 2017; Fick *et al.*, 2009; Larsson *et al.*, 2007; Sharma *et al.*, 2019), Pakistan (Ashfaq *et al.*, 2017; Hussain *et al.*, 2016; Scheurell *et al.*, 2009), China (Li *et al.*, 2018; Liu *et al.*, 2020; Zhu *et al.*, 2024; Zhu *et al.*, 2024), as well as developed countries such as Canada (Kleywegt *et al.*, 2019), USA The pharmaceutical production involves multiple steps between acquiring the raw materials and production and supply of finished product, and each step poses a risk for the environmental release of chemical waste unless appropriate measures are taken. Despite the risks, the lack of appropriate regulations and protocols for waste disposal makes it difficult for industries to control the pollutants released. Therefore new eco-innovation in production and drugs along with industrial and government regulations became important factors for the pharmaceutical production (Fatta-Kassinos *et al.*, 2011; Li & Hamblin, 2016).

According to a report 'The Global Use of Medicines 2024: Outlook to 2028', by IQVIA Institute (*The Global Use of Medicines 2024*) the global use of pharmaceuticals increased by 14% between the year 2018 and 2024 and is expected to increase by 12% by the year 2028. Although pharmaceuticals have revolutionized healthcare, their misuse, and improper disposal pose significant risks to human health and the environment (Nepal *et al.*, 2020). The improper disposal of medications in household or municipal waste can contaminate the environment. Most households dispose of unused medicines in the garbage or sewage. Despite recommendation for disposal by returning to the pharmacist, it has been reported that due to lack of proper knowledge, instruction, and awareness, most of pharmaceuticals end up in garbage or sewage (Nepal *et al.*, 2020; Paut Kusturica *et al.*, 2017). It has also been observed that higher concentrations of pharmaceuticals are found in hospital sewage compared to municipal however the total load of pharmaceuticals in hospital sewage disposal is lower than the municipal sewage disposal (Mackuľak *et al.*, 2021; Schuster *et al.*, 2008; Thomas & Langford, 2010). Pharmaceuticals in municipal sewage primarily originate from human excretion and the improper disposal of unused medications. A substantial portion of a pharmaceutical remains unmetabolized after administration and excreted unchanged. Pharmaceuticals can leach into soil and groundwater through various routes Fig. 1, polluting water sources and potentially harming aquatic life. This environmental contamination can have long-term consequences, impacting both current and future generations. Therefore, there is need of cleaner methods for manufacturing companies, along with proper regulation on industrial effluent release in environment. The development of proper disposal methods and increase awareness among the people is crucial for the controlling the pharmaceutical release in environment.

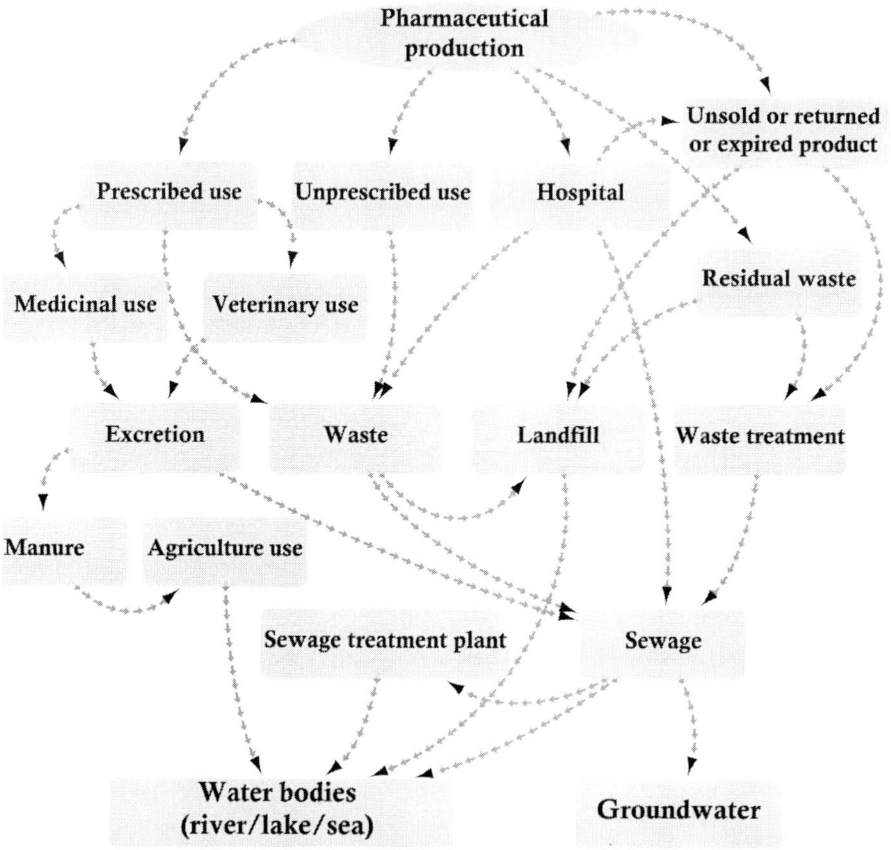

Fig. 1. Source of pharmaceutical pollutants and their routes leading to release in water

3. Effect of Pharmaceutical Pollutants on Aquatic Ecosystem

Although pharmaceutical pollutants are present in the aquatic ecosystem at minimal concentrations (ng/l), they tend to bio-accumulate due to their slow degradation, widespread usage, and continuous discharge (Hussain *et al.*, 2023). This build-up causes significant changes in the aquatic ecosystem which has been discussed in the following sections:

3.1. Behavioural changes

Several studies have reported that pharmaceutical pollutants in aquatic ecosystems can significantly affect fish behavior and physiology. Prolonged exposure to pharmaceutical pollutants, especially fluoxetine, an anti-depressant drug, can profoundly affect fish behavior across various life stages. Thoré *et al.* (2023) demonstrated that fluoxetine exposure alters geotaxic behavior in adult killifish while Wiles *et al.* (2020) reported increased coercive mating behavior in male guppies. In addition to fluoxetine, another anti-depressant venlafaxine was also found to disrupt locomotor responses during light/dark transitions in zebrafish (Hong *et al.*, 2021). Moreover, in brown trout, venlafaxine reduced larval growth and survival and influenced behavior in both larvae and juveniles (Ziegler *et al.*, 2021). Zebrafish exposed to venlafaxine displayed inhibited courtship behavior, altered brain monoamine levels, and changes in gene expression related to serotonin and dopamine systems (Tang *et al.*, 2022). These behavioral alterations are likely due to the neurotoxic effects of antidepressants, which can disrupt neurotransmitter systems that are evolutionarily conserved across species (Salahinejad *et al.*, 2022). Psychiatric drugs such as benzodiazepines and oxazepam can modify fish behavior by increasing activity levels and boldness, potentially disturbing natural predator-prey relationships (Saaristo *et al.*, 2019). Such pharmaceuticals in aquatic environments potentially threaten fish populations and overall ecosystem health (Aich *et al.*, 2024).

3.2. Organ Damage

Pharmaceutical pollutants in aquatic ecosystems present considerable risks to fish health, particularly by affecting liver function and overall physiology. These contaminants, which include antidepressants, anti-hypertensives, and anti-inflammatory drugs, can accumulate in fish tissues and trigger a range of biochemical responses (Duarte *et al.*, 2020). Exposure to these pharmaceuticals can induce DNA damage, oxidative stress, and changes in antioxidant enzymes and energy metabolism (Gallego-Ríos *et al.*, 2021). The liver, as the primary organ for detoxification, is particularly susceptible to these pollutants, exhibiting both structural and functional alterations (Topić Popović *et al.*, 2023). Ciprofloxacin, both alone and in combination with paracetamol was found to induce liver damage, oxidative stress, and, genotoxicity in adult zebrafish at environmentally relevant concentrations (Elizalde-Velázquez *et al.*, 2022). Recently, Gao *et al.* (2024) reported that amoxicillin treatment during the embryonic stage increases the risk of non-alcoholic fatty liver disease (NAFLD) in zebrafish larvae. Studies have shown that antibiotics can impair immune function, trigger oxidative stress, and cause cellular damage in fish (Limbu *et al.*, 2021). In a recent investigation, exposure to antibiotics resulted in cyto-genotoxic effects, erythrocyte deformities, and histological changes in vital organs such as the liver and kidneys in stinging catfish (Islam *et al.*, 2022). Several antibiotics, including quinolones, sulfonamides, tetracyclines, and macrolides, have been shown to bioaccumulate in fish tissues and aquatic environments. These antibiotics cause developmental, metabolic, and cardiovascular damage, while also altering immune and antioxidant responses in fish (Yang *et al.*, 2020). Even short-term exposure to sublethal concentrations of antibiotics like florfenicol can induce oxidative stress in fish hepatocytes (Shiroma *et al.*, 2020). The toxicity of antibiotics in fish raises concerns about bioaccumulation and potential risks to human health through consumption (Yang *et al.*, 2020).

3.3. Metabolic dysregulations and Microbial community imbalance

Exposure to pharmaceutical pollutants such as antibiotics (tetracycline, cefotaxime sodium, sulfamonomethoxine, and enrofloxacin) at ecological concentrations has been found to interrupt early development and metabolic systems in zebrafish larvae including steroid biosynthesis and other metabolic pathways (Qiu *et al.*, 2020). The sublethal effect of Non-steroidal anti-inflammatory drugs (NSAIDs) at environmental amounts has been observed on freshwater invertebrates (Parolini, 2020). López-Doval *et al.* (2020) demonstrated that NSAIDs can disrupt prostaglandin production in aquatic insects. Antibiotics in aquatic ecosystems alter microbial communities and disturb metabolic pathways, including amino acid and carbohydrate metabolism (Kang *et al.*, 2022). A reduction of bacterial diversity in water and sediment was observed upon exposure to high concentrations of enrofloxacin (Dai *et al.*, 2023). Similarly, ofloxacin was found to alter prokaryotic community structure and impair nitrogen fixation and photosynthetic capacity (Deng *et al.*, 2022).

3.4. Endocrine Disruption and Reproductive Issues

Studies have indicated that antibiotics such as triclosan can interfere with the hypothalamic-pituitary-gonadal axis, alter sex hormone levels, and increase vitellogenin production in male fish (Wang, 2024). These imbalances may result in reproductive impairments, oxidative stress, and histopathological changes (Adhikari *et al.*, 2023). Even in minimal amounts, antibiotics can be toxic to fish, causing a range of physiological and immunological disorders (Bojarski *et al.*, 2020). *In-silico* studies have revealed potential interactions between antibiotics and nuclear hormone receptors, with sulfonamides exhibiting the highest endocrine-disrupting potential (Song *et al.*, 2021). Certain chemicals have been reported to interfere with reproductive physiology, lowering reproductive rates and developing female characteristics in male fish (Delbes *et al.*, 2022).

3.5. Antibiotic Resistance

The prevalence of antibiotics in the environment and chronic exposure to sub-therapeutic doses present significant ecotoxicological risks and drive the proliferation of antibiotic-resistant genes in bacterial populations (Okoye *et al.*, 2022). Long-term exposure to antibiotics in aquaculture sediments revealed varied responses based on application history, with microbial communities in sediments exposed for 30 years showing greater resistance to oxytetracycline and sulfadiazine (Deng *et al.*, 2022). This issue is a global concern, with pharmaceutical manufacturing facilities in both developed and developing nations contributing to ecosystem contamination (Anjanapriya *et al.*, 2021).

3.6. Oxygen depletion and Sunlight blockage

The presence of pharmaceutical pollutants and their by-products raises Biological Oxygen Demand, depleting oxygen levels and leading to mortality among aquatic organisms (A. Pandey *et al.*, 2023). Pharmaceuticals impact species across trophic levels, from diatoms to fish, with the potential to interrupt entire ecosystems. The effect of pharmaceutical pollutants on primary producers like diatoms poses a major risk as they play key roles in oxygen production and as bio-indicators (Kock *et al.*, 2023). Aggregates of pharmaceutical residues can reduce sunlight penetration, limiting its availability to algae and cyanobacteria essential for food production within the ecosystem. This disruption adversely impacts the health of the ecosystem's natural flora and fauna (Kayode-Afolayan *et al.*, 2022).

4. Microbial Bioremediation of Pharmaceutical Waste

Microbial bioremediation is a transformative strategy in environmental biotechnology, leveraging the natural metabolic processes of microorganisms to degrade pharmaceutical pollutants in wastewater (Maglione *et al.*, 2024). Among the most effective microbial agents in this domain are bacteria, fungi, & microalgae, each offering distinct and complementary mechanisms for pollutant removal (Pereira *et al.*, 2023; Thirumalaivasan *et al.*, 2024) (Fig. 2).

Bacteria play a pivotal role in bioremediation due to their metabolic versatility and adaptability to diverse environments. They employ enzymatic pathways to break down complex pharmaceutical compounds into simpler, less harmful substances. Bacteria such as *Pseudomonas*, *Acinetobacter*, and *Bacillus* have shown high efficiency in degrading antibiotics, hormones, and analgesics (Narayanan *et al.*, 2023). These microorganisms can metabolize pollutants directly or use co-metabolism, where degradation occurs incidentally while the bacteria metabolize other compounds. Furthermore, bacterial biofilms, which are aggregates of bacteria adhering to surfaces, provide enhanced pollutant removal through their stable and resilient communities (Muhammad *et al.*, 2020). This makes bacteria an indispensable tool in wastewater treatment plants and engineered bioreactors designed for pharmaceutical waste remediation.

Fungi, particularly ligninolytic fungi like white-rot fungi, contribute uniquely to the degradation of persistent and complex pharmaceutical pollutants. Their enzymatic systems, including lignin peroxidase, manganese peroxidase, and laccase, enable them to break down structurally complex and recalcitrant molecules such as non-steroidal anti-inflammatory drugs (NSAIDs), antibiotics, and endocrine-disrupting compounds (Kumar & Chandra, 2020). The mycelial structure of fungi allows for extensive contact with pollutants, enhancing their adsorption and degradation capabilities. Fungi are also used in bioreactor systems, where they offer advantages such as high tolerance to toxicity and the ability to operate in varying environmental conditions (Dinakarkumar *et al.*, 2024). The potential of fungal systems to efficiently treat wastewater underscores their growing application in pharmaceutical waste management.

Microalgae provide a dual benefit in bioremediation through pollutant removal and biomass recovery. Algal species like *Chlorella* and *Scenedesmus* not only metabolize pharmaceutical pollutants but also offer oxygenation to the system through photosynthesis, fostering aerobic degradation by other microorganisms (Liberti *et al.*, 2024). Microalgae can remove a wide range of pharmaceuticals, including beta-blockers, antidepressants, and antibiotics, by assimilating them as nutrients or transforming them enzymatically (Miazek & Brozek-Pluska, 2019). In symbiotic setups, such as microalgal-bacterial consortia, the combined action of these organisms significantly enhances pollutant removal efficiency (Vo *et al.*, 2024). Additionally, microalgae offer the advantage of resource recovery, as their biomass can be harvested and repurposed for biofuel production or other value-added applications (Bora *et al.*, 2024).

The application of bacteria, fungi, and microalgae in bioremediation highlights their role as game changers in addressing pharmaceutical pollution. These microorganisms not only degrade pollutants but also contribute to the sustainability of water treatment processes by reducing chemical usage and supporting carbon-neutral operations (Kaur *et al.*, 2024). Their integration into engineered systems, such as photobioreactors and biofilters, exemplifies the potential of microbial bioremediation to revolutionize wastewater treatment and protect aquatic ecosystems.

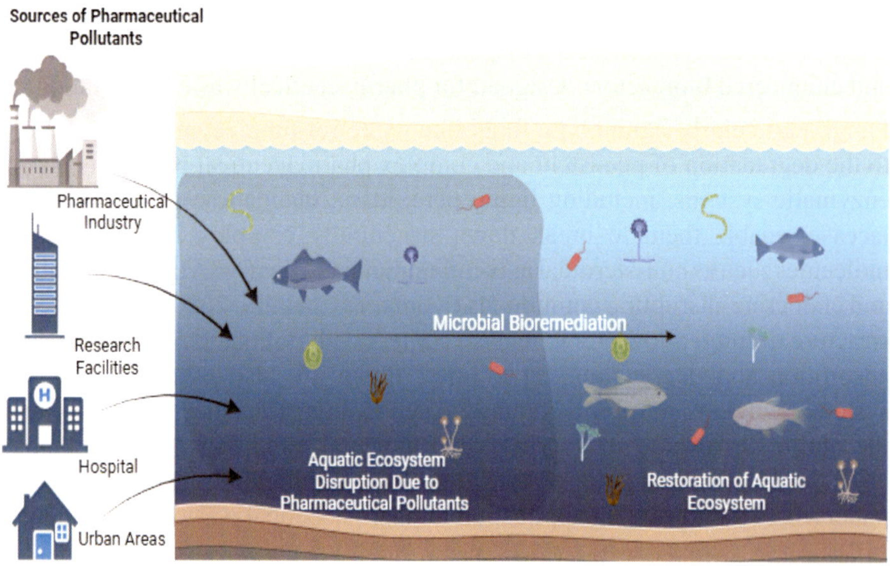

Fig. 2. Microbial bioremediation of pharmaceutical pollutants in aquatic system. Created with Biorender.com.

5. Bacterial Bioremediation of Pharmaceutical Pollutants in Wastewater

Conventional wastewater treatment plants often struggle to effectively eliminate pharmaceutical contaminants, driving the need for alternative remediation approaches (González-González, Parra-Arroyo, *et al.*, 2022). Bioremediation, which leverages microorganisms to break down organic pollutants, has emerged as a sustainable, cost-efficient, and environmentally friendly method for addressing pharmaceutical waste (Bala *et al.*, 2022). A wide range of bacterial species has shown promise in removing pharmaceutical pollutants. Researchers are investigating various bioremediation systems, such as anaerobic and aerobic reactors, to enhance degradation efficiency. The use of naturally occurring degrading bacteria is particularly well-suited for applications in natural ecosystems (Lakhani *et al.*, 2022).

Many studies have explored the bacterial-aided removal of non-steroidal anti-inflammatory drugs (NSAIDs) in wastewater. Show *et al.* (2023) highlighted the potential of *Microbacterium paraoxydans*, isolated from pharmaceutical industrial wastewater, in the degradation of ibuprofen. The data showed that upon optimization of five parameters such as yeast extract, pH, inoculum dose, temperature, agitation speed, and incubation time while keeping concentration and time constant, *M. paraoxydans* is a promising candidate for biodegradation of pharmaceutical pollutants and can mitigate harmful environmental effects.

In another study, Chopra & Kumar (2022) examined the microbial species in wastewater samples to identify the principal degrader of ibuprofen through ibuprofen-enriched culture. The 16S rRNA sequencing revealed *Bacillus siamensis* strain DSI-1 as the crucial biodegrader of ibuprofen and holds potential for *in-situ* bioremediation. In another study Rutere *et al.* (2020) investigated the ibuprofen removal by the bacterial community of hyporheic zone river sediments. The 16S rRNA gene analyses identified numerous previously unknown bacterial taxa that responded to ibuprofen such as Actinobacteria, Gemmatimonadetes, Acidobacteria, Bacteroidetes, Proteobacteria, and Latescibacteria. Furthermore, from ibuprofen-rich sediments, two strains *Novosphingobium* sp. strain CN1 and *Pseudomonas thivervalensis* strain MAH1 were isolated which displayed significant ibuprofen consumption (400 and 300 μM).

Rhodococcus sp. D32 has demonstrated significant potential in dexamethasone biodegradation, uncovering a novel steroid nucleus-opening pathway and mitigating ecotoxicity in aquatic ecosystems (Zhang *et al.*, 2024). Zhu *et al.* (2023) examined the microbial population of sediment samples collected from the coast of Xiao Shi Island, China by utilizing an artificial microcosm model developed in the laboratory via enrichment culture (Cu (II) and lincomycin stress) for 255 days. The 16S rRNA analysis revealed genus *Rossellomorea*. Strain ZC255 exhibited significant bioremoval of Cu and lincomycin and suggested its potential application in pollutant elimination. Furthermore, Table 1 provides an overview of the bacteria, the pollutants they remediate, and the associated mechanisms involved.

Table 1. Microbial bioremediation: Pollutants remediated and the underlying mechanism.

S. No.	Microorganism	Pharmaceutical pollutant remediated	Probable mechanism underlying remediation	References
BACTERIA				
1.	*Rhodococcus sp.* D32	Dexamethasone	Biotransformation	(Zhang *et al.*, 2024)
2.	*Acetobacterium, Petrimonas, Desulfovibrio, Geobacter, Nocardiaceae, and Pandoraea*	Ibuprofen and naproxen		(Wang *et al.*, 2022)
3.	*Novosphingobium* sp. strain CN1 and *Pseudomonas thivervalensis* strain MAH1	Ibuprofen		(Rutere *et al.*, 2020)
4.	*Labrys portucalensis* F11	Ofloxacin, ciprofloxacin, and norfloxacin		(Amorim *et al.*, 2014)
5.	*Bacillus ligniniphilus L1*	Vancomycin, tobramycin, and linezolid	Enzyme-mediated degradation	(Nawaz *et al.*, 2024)
6.	Strain ZC255 (Genus *Rossellomorea*)	Lincomycin	Biofilm formation	(Zhu *et al.*, 2023)
7.	*Vulgati bacter incomptus* and *Chthoniobacter flavus*	Pentadecane and hexadecane	Biostimulation	(Sharma *et al.*, 2023)
8.	*Planococcus sp.* S5 (immobilised strain)	Naproxen	Biodegradation	(Dzionek *et al.*, 2024)
9.	*Bacillus licheniformis* strain PPY-2	Paracetamol		(Chopra & Kumar, 2023)
10.	*Pseudomonas moorei* KB4 *and Bacillus thuringiensis* B1	Naproxen, ibuprofen, diclofenac, and paracetamol		(Marchlewicz *et al.*, 2023)
11.	*Rhizobium daejeonense* IBU_18 & *Stenotrophomonas humi* DIC_5	Diclofenac and ibuprofen		(Pápai *et al.*, 2023)
12.	*Microbacterium paraoxydans*	Ibuprofen		(Show *et al.*, 2023)
13.	*Streptomyces rochei*	Diclofenac & clarithromycin		(Ercoli *et al.*, 2023)
14.	*Streptomyces chrestomyceticus* & *flavofuscus*	Acetaminophen		(Embarez *et al.*, 2023)

15.	*Serratia marcescens* strain WW1	Tetracycline	Biodegradation	(Bhatt *et al.*, 2022)
16.	*Bacillus siamensis* strain DSI-1	Ibuprofen		(Chopra & Kumar, 2022)
17.	*Pseudomonas silesiensis* F6a	Sulfamethoxazole		(Liu *et al.*, 2022)
18.	*Rhodoccoccus* sp. ED55	Bisphenol-A , 17β-estradiol, 17α-ethynylestradiol, and bisphenol-S (BPS)		(Moreira *et al.*, 2021)
19.	Genus *Pseudomonas* and *Acinetobacter*	Bezafibrate or paroxetine		(Fernandes *et al.*, 2020)
20.	*Klebsiella* sp. WAH1	Diclofenac sodium		(Sharma *et al.*, 2021)
MICROALGA–BACTERIA CONSORTIA				
1.	*Pseudorhizobium, Sphingopyxis,* and the Cyanobacterium *Cyanobium*	Androstenedione and β-estradiol	Biotransformation	(Guevara *et al.*, 2024)
FUNGUS				
1.	*Fomes fomentarius*	Bleomycin, Vincristine	Bioabsorption and Biotransformation	(Vaksmaa *et al.*, 2023)
2.	*Tinea versicolor*	Diclofenac, Naproxen, Indomethacin, Ibuprofen, Fenoprofen		(Vaksmaa *et al.*, 2023)
3.	*Trametes versicolor*	Ibuprofen, Diclofenac, Carbamazepine	Biotransformation	(Marco-Urrea *et al.*, 2009)
4.	*Mucor hiemalis*	Acetaminophen		(Esterhuizen-Londt *et al.*, 2016)
5.	*Trichoderma harzianum*	Clarithromycin, Carbamazepine		(Buchicchio *et al.*, 2016)
6.	*Penicillium oxalicum*	Sulfamethoxazole, Diclofenac		(Olicón-Hernández *et al.*, 2019)
7.	*Rhodotorula mucilaginosa*	Phenanthrene		(Martínez-Ávila *et al.*, 2021)

8.	*Gymnopilus luteofolius*	Lopromide, Carbamazepine, Diclofenac	Biodegradation	(Amobonye *et al.*, 2023)
9.	*Trichoderma longibrachiatum*	Carbamazepine, Diclofenac, Ibuprofen		(Kasonga *et al.*, 2023)
10.	*Phanerochaete chrysosporium*	Tetracycline, Sulfamethoxazole, Paracetamol		(Rodarte-Morales *et al.*, 2011)
11.	*Aspergillus niger*	Acetaminophen, Ciprofloxacin		(Bodin *et al.*, 2016)
12.	*Pleurotus ostreatus*	Polycyclic aromatic hydrocarbons (PAHs)		(Buchicchio *et al.*, 2016)
13.	*Rhizopus microsporus*	Carbamazepine		(Kasonga *et al.*, 2021)
14.	*Mucor circinelloides*	Ibuprofen		(Kasonga *et al.*, 2021)
MICROALGAE				
1.	*Chlamydomonas reinhardtii, Scenedesmus obliquus, Chlorella pyrenoidosa,* and *Chlorella vulgaris*	4-Nonylphenol, 4-OHA, Bentazone, lincomycin, trimethoprim, 4-hydroxy-androst-4-ene-17-dione (4-OHA), Paracetamol, Salinomycin, Progesterone, Testosterone, Salinomycin, and other	Biodegradation, and Biosorption	(Zhou *et al.*, 2014)
2.	*Selenastrum capricornutum* and *Chlamydomonas*	β-estradiol (E2) and 17α-ethinylestradiol (EE2)	Biodegradation and Biosorption	(Hom-Diaz *et al.*, 2015)
3.	*Scenedesmus obliquus*	Tramadol	Biosorption	(Ali *et al.*, 2018)
4.	*Cladophora* spp.	Triclosan, Methyl-triclosan and triclocarban	Bio-uptake	(Coogan *et al.*, 2007)
5.	*Chlorella pyrenoidosa*	Triclosan	Bio-uptake and Biodegradation	(Wang *et al.*, 2013)

6.	*Chlorella sorokiniana, C. vulgaris, C. saccharophila, Coelastrella sp., Coelastrum astroideum, Desmodesmus sp.,* and *Scenedesmus obliquus*	Hydroxyzine, Mirtazapine, Diphenhydramine, Memantine, Orphenadrine, Bupropion, Biperiden, Flecainide, Trihexyphenidyl, Clomipramine, Amitriptyline, Biperiden, Trihexyphenidyl, Clomipramine & Amitriptyline	Bio-uptake	(Gojkovic *et al.*, 2019)
7.	*Chlorella* spp., *Chlamydomonas* spp., and *Mychonastes* spp.	7-amino cephalosporanic acid	Biosorption and Biodegradation	(Guo *et al.*, 2016)
8.	*Chlorella pyrenoidosa*	Amoxicillin and cefradine	Biodegradation	(Li *et al.*, 2015)
9.	*Chlorella sorokiniana*	Paracetamol & Salicylic acid	Biodegradation	(Escapa *et al.*, 2017b)
10.	*Chaetoceros muelleri* and biochar	Carbamazepine, sulfame-thazine, and tramadol	Biosorption and biodegradation	(Mojiri *et al.*, 2020)
11.	*Nannochloris* spp.	17β-estradiol, estrone, 17α-ethinylestradiol, testosterone, bisphenol A, triclosan, and salicylic acid	Bio-uptake and biosorption	(Bai & Acharya, 2019)
12.	*Chlamydomonas mexicana* and *Chlorella vulgaris*	Bisphenol A	Bio-uptake and biodegradation	(Ji *et al.*, 2014)
13.	*Chlorella sp. and Scenedesmus sp.*	Caffeine, ibuprofen, galaxolide, tributyl phosphate, 4-octylphenol, tris(2-chloroethyl) phosphate & carbamazepine	Bio-uptake and biodegradation	(Matamoros *et al.*, 2016)
14.	*Microcystis aeruginosa*	Amoxicillin	Biodegradation	(Liu *et al.*, 2015)
15.	*Chlamydomonas mexicana*	Ciprofloxacin and its co-metabolites	Biodegradation	(Xiong *et al.*, 2017)

16.	*Mixed population of wild freshwater green algal*	Atenolol, bispropol, metoprolol, clarithromycine, bupropion, atracurium, diltiazem, and terbutaline	Biosorption and biodegradation	(Gentili & Fick, 2017)
17.	*Chlorella pyrenoidosa*	Cefradine	Biodegradation	(Chen *et al.*, 2015)
18.	*Cymbella sp. and Scenedesmus quadricauda*	Naproxen	Biodegradation	(Ding *et al.*, 2017)
19.	*Tetradesmus obliquus*	Fluoranthene	Biodegradation	(Torbati *et al.*, 2024)
20.	Dead biomass of *Scenedesmus obliquus*	Salicylic acid and ibuprofen	Biosorption	(Silva *et al.*, 2020)
21.	*Chlorella vulgaris*	Norfloxacin, sulfamethazine, and roxithromycin	Biosorption, bio-uptake and biodegradation	(Long *et al.*, 2024)
22.	*Mixed population of wild freshwater green algal*	Tinidazole	Photodegradation	(Ahmad *et al.*, 2024)
23.	*Aerobic activated sludge system*	Metformin, phenformin, nateglinide, repaglinide, dapagliflozin, sitagliptin, and gliclazide	Biodegradation	(Wu *et al.*, 2024)
24.	*Chlorella vulgaris and Tetradesmus obliquus*	Salicylic acid, paracetamol	Biodegradation and biosorption	(Escapa *et al.*, 2017a)
25.	*Chlorella vulgaris and Chlorella sorokiniana*	Diclofenac	Biodegradation	(Tahraoui *et al.*, 2024)
26.	Non-Living Microalgae Biomass of *Synechocystis* sp. and *Scenedesmus* sp	Diclofenac	Bio-uptake	(Coimbra *et al.*, 2018)
27.	Dead biomass of *Phaeodactylum tricornutum*	Oxytetracycline	Biosorption	(Santaeufemia *et al.*, 2016)
28.	Dead biomass of the *Phaeodactylum tricornutum*	Ibuprofen	Biosorption	(Santaeufemia *et al.*, 2018)
29.	*Chlorella sorokiniana*	Diclofenac, ibuprofen, paracetamol, metoprolol, carbamazepine & trimethoprim	Biosorption and Biodegradtion	(de Wilt *et al.*, 2016)

5.1. Bioengineered Bacteria to the Rescue

Engineered bacteria present a promising solution, leveraging metabolic engineering and recombinant DNA technology to improve bioremediation efficiency (Sanghvi *et al.*, 2020). Genetic engineering has transformed bioremediation by introducing innovative methods to enhance bacterial performance in treating wastewater. Through genetic modifications, researchers can develop bioengineered bacterial strains that offer several advantages such as increased effectiveness in pharmaceutical pollutant breakdown, tolerating extreme environmental conditions, and addressing a wider spectrum of contaminants. These advancements promise to increase the efficiency and sustainability of wastewater treatment systems. Genetically engineered bacteria are eco-friendly solutions for the restoration of aquatic ecosystem balance. By enhancing the ability of bacteria to degrade pharmaceutical pollutants, engineered bacteria pose an effective alternative to traditional cleanup methods (Rafeeq *et al.*, 2023).

5.1.1. Advantages of Bioengineered Bacteria in the Bioremediation Process

The bioengineered bacteria offer several advantages (Figure 3) as follows.

5.1.1.1. Advancing Degradation Mechanisms

A key application of genetic engineering in bioremediation is the optimization of metabolic pathways for pollutant breakdown. Bacteria can be modified to effectively decompose persistent and complex pharmaceutical pollutants by integrating or refining genes that code for specialized enzymes. Studies have reported that incorporating genes for enzymes like laccases, dehalogenases, or peroxidases into bacterial genomes enhances their ability to degrade resistant pollutants. Furthermore, pathway reconstruction could be achieved through a synthetic biology approach which enables the design and assembly of entire metabolic pathways, allowing bacteria to convert complex pollutants into simpler, non-toxic compounds efficiently (Rebello *et al.*, 2021).

5.1.1.2. Broadening Substrate Utilization

Genetic engineering equips bacteria with the ability to metabolize pollutants beyond their natural range by introducing genes from other organisms with specialized catabolic functions. This approach enhances their versatility in degrading diverse contaminants. Horizontal gene transfer mediates the incorporation of genes from extremophiles or other pollutant-degrading microbes into common wastewater bacteria, such as *Pseudomonas spp.* or *Escherichia coli*, enabling these strains to address previously untreatable contaminants. Furthermore, studies have demonstrated that the design of enzymes with expanded substrate specificities empowers bacteria to degrade a wider array of chemical compounds commonly found in wastewater (Yang *et al.*, 2024).

5.1.1.3. Enhancing Environmental Resilience

Wastewater environments often pose harsh conditions, such as extreme pH, salinity, or toxic substances. Bioengineered bacteria can strengthen bacterial resilience by adding stress resistance genes such as those that confer resistance to oxidative stress or high salinity enhance bacterial adaptability. Furthermore, bacteria are engineered to produce biofilms that improve their survival and operational stability in wastewater treatment systems (Maqsood *et al.*, 2023).

5.1.1.4. Minimizing Ecotoxicity

Some pollutants are converted into equally or more toxic byproducts during natural bacterial degradation. Genetic engineering addresses this concern by incorporating genes to convert toxic intermediates into harmless end products reducing ecological harm. The bioengineering approach helps achieve co-metabolism optimization i.e. degradation of multiple pollutants simultaneously to minimize the accumulation of toxic byproducts (Maqsood *et al.*, 2023).

5.1.1.5. Biosensors for Monitoring

Genetically engineered bacteria can also function as biosensors to detect pollutants in wastewater. By integrating reporter genes linked to pollutant-responsive pathways, these organisms provide real-time monitoring of contamination levels, ensuring timely and efficient treatment (Liu *et al.*, 2022).

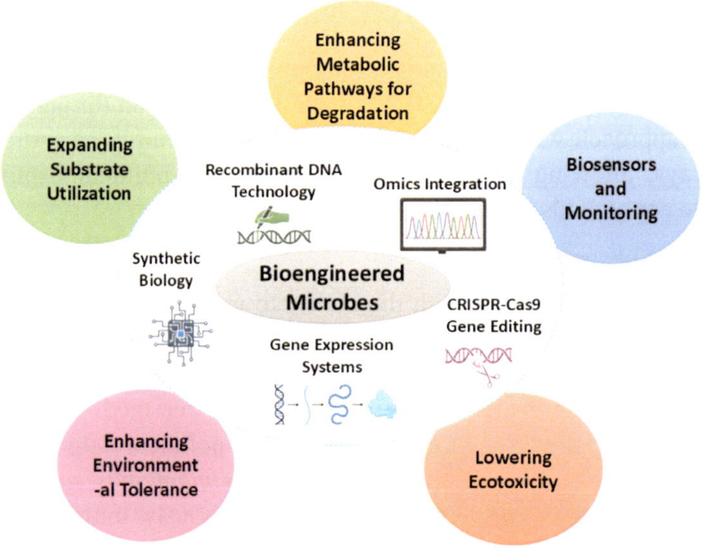

Fig. 3: Techniques employed for the development of bioengineered microbes & their advantages

5.1.2. Techniques for the development of bioengineered bacteria

Genetic engineering techniques enable the development of bioengineered bacteria with enhanced proficiencies for wastewater bioremediation. By leveraging molecular biology and synthetic biology tools, researchers can create bacterial strains that degrade complex pollutants, tolerate extreme environmental conditions, and ensure safe application (González-González, Flores-Contreras, *et al.*, 2022). Following are the key techniques involved in the creation of genetically engineered bacteria (Fig. 3).

5.1.2.1. Recombinant DNA Technology

Involves the insertion of target genes, such as those encoding pollutant-degrading enzymes, into bacterial genomes using plasmids and optimized expression systems (Sharma *et al.*, 2024).

5.1.2.2. Horizontal Gene Transfer

Transfers genes from pollutant-degrading microorganisms to bacteria, expanding their substrate range and metabolic capabilities (Michaelis & Grohmann, 2023).

5.1.2.3. CRISPR-Cas9 Gene Editing

Allows precise genome modifications to introduce or remove genes, facilitating the creation of bacteria capable of efficiently breaking down contaminants (Yang *et al.*, 2024).

5.1.2.4. Synthetic Biology

Enables the design and assembly of new metabolic pathways and genetic circuits, equipping bacteria with advanced degradation mechanisms (Jiménez-Díaz *et al.*, 2022; Webster *et al.*, 2024).

5.1.2.5. Omics Integration

Incorporates data from transcriptomics, proteomics, and metabolomics to identify and engineer genes involved in pollutant degradation (Sharma *et al.*, 2022).

5.1.2.6. Metabolic Engineering

Optimizes metabolic pathways to enhance degradation efficiency and ensure compatibility with bacterial cellular processes (Luo *et al.*, 2023).

5.1.2.7. Gene Expression Systems

Employs inducible or constitutive promoters and feedback loops to dynamically regulate the expression of degradation-related genes (Ravikumar *et al.*, 2017).

These advancements allow the construction of specialized bacterial strains capable of addressing complex wastewater challenges, contributing to sustainable and efficient water treatment solutions (Singh *et al.*, 2022).

Wang *et al.* (2022) demonstrated enhanced efficiency of electrostimulated biofilm in the removal of pharmaceutical pollutants, naproxen, and ibuprofen. The microbial assessment of biofilm revealed that the increased efficiency of pollutant removal is due to the synergistic effects of microbial consortia containing fermenters (*Acetobacterium, Petrimonas*), electro-active bacteria (*Desulfovibrio, Geobacter*), and degraders (*Nocardiaceae, Pandoraea*). The study integrated microbial processes (biological systems) with electrical stimulation (an engineering method) to improve the transformation of, ibuprofen and naproxen and showcased a bioengineering approach.

6. Fungal Bioremediation of Pharmaceutical Pollutants in Wastewater

Fungal bioremediation has emerged as a promising solution for the removal of pharmaceutical pollutants in wastewater, leveraging the unique metabolic and enzymatic capabilities of fungi. Pharmaceuticals, including antibiotics, analgesics, anti-inflammatory drugs, and endocrine-disrupting compounds, are often detected in aquatic environments due to their incomplete removal by conventional wastewater treatment systems (Ortúzar *et al.*, 2022). Fungi, particularly filamentous and ligninolytic species, are highly effective in degrading and transforming these pollutants due to their ability to produce extracellular enzymes and withstand extreme environmental conditions (Rosa *et al.*, 2024; Vaksmaa *et al.*, 2023).

Fungi degrade pharmaceutical pollutants primarily through enzymatic processes involving extracellular oxidoreductases such as laccase, lignin peroxidase (LiP), and manganese peroxidase (MnP) (Okal *et al.*, 2023). These enzymes are non-specific, enabling fungi to break down a wide range of structurally diverse and recalcitrant pharmaceutical compounds. Laccase, for example, catalyzes the oxidation of phenolic and non-phenolic substrates, making it effective against antibiotics like tetracycline and anti-inflammatory drugs such as diclofenac (Chmelová *et al.*, 2024). Similarly, LiP and MnP can degrade complex aromatic structures, including synthetic hormones and certain chemotherapeutic agents. In addition to enzymatic degradation, fungi also employ adsorption mechanisms. The fungal cell wall, composed of chitin, glucans, and proteins, acts as a biosorbent, binding pharmaceutical pollutants and removing them from aqueous systems (Legorreta-Castañeda *et al.*, 2020). This dual mechanism of degradation and adsorption makes fungi particularly versatile in wastewater treatment.

Several fungal species have demonstrated efficacy in the bioremediation of pharmaceutical pollutants. White-rot fungi, such as *Phanerochaete chrysosporium*, *Pleurotus ostreatus*, and *Trametes versicolor*, are among the most studied due to their potent ligninolytic enzyme systems (Dinakarkumar *et al.*, 2024). For instance, *Trametes versicolor* has shown significant potential in the removal of antibiotics, including amoxicillin and ciprofloxacin, and anti-inflammatory drugs like ibuprofen (Chmelová *et al.*, 2024). Other fungi, such as *Aspergillus* and *Penicillium* species, have been reported to degrade non-steroidal anti-inflammatory drugs and selective serotonin reuptake inhibitors (SSRIs) (Olicón-Hernández *et al.*, 2020). Fungi can also be employed in co-culture systems, where they work synergistically with bacteria to enhance pollutant degradation. Such systems combine the complementary metabolic pathways of both microorganisms, improving the efficiency and range of pharmaceutical removal (Espinosa-Ortiz *et al.*, 2022).

Fungal bioremediation can be implemented in various wastewater treatment conFig.urations, including bioreactors, constructed wetlands, and immobilized fungal systems. For example, fungal-packed bed reactors have been used to treat pharmaceutical-laden wastewater, demonstrating high removal efficiencies for compounds like diclofenac and carbamazepine (Dalecka *et al.*, 2020). Similarly, fungal immobilization on solid supports, such as alginate beads or biochar, enhances pollutant degradation by increasing fungal stability and activity under operational conditions (Gao *et al.*, 2022). Constructed wetlands incorporating fungi have also gained attention for their ability to treat pharmaceutical pollutants in natural or semi-natural settings. The integration of fungi into these systems not only reduces pharmaceutical concentrations but also enhances the overall microbial diversity, creating a more robust and effective treatment system (Malik *et al.*, 2023). Fungi offer several advantages over other microbial systems for pharmaceutical bioremediation. Their extracellular enzyme production enables the degradation of pollutants that are resistant to bacterial metabolism (Dinakarkumar *et al.*, 2024). Moreover, fungi can tolerate and thrive in environments with high pollutant concentrations and varying pH levels, making them suitable for industrial and municipal wastewater treatment (Sar *et al.*, 2024). Their ability to function in non-sterile conditions further enhances their practicality for large-scale applications. Advances in fungal bioremediation research are focused on optimizing enzyme production, improving fungal immobilization techniques, and developing hybrid systems that integrate fungal and bacterial processes (Dinakarkumar *et al.*, 2024; Thirumalaivasan *et al.*, 2024). Genetic engineering and omics-based approaches are also being explored to enhance the biodegradation capabilities of fungi and uncover novel pathways for pharmaceutical degradation (Miglani *et al.*, 2022). Fungal bioremediation represents a sustainable and efficient approach to addressing the challenges posed by pharmaceutical pollutants in wastewater. By harnessing the unique metabolic potential of fungi, it is possible to develop innovative treatment solutions that safeguard aquatic ecosystems and public health.

6.1. Bioengineered Fungus to the Rescue

Fungal bioremediation, while highly promising, has certain limitations when it comes to the efficient degradation of pharmaceutical pollutants. These challenges include the specificity of fungal enzymes to certain pollutants, and the slow degradation rate of recalcitrant compounds (Mousavi *et al.*, 2021). Bioengineered fungi offer a powerful solution to these limitations by enhancing the natural capabilities of fungi through genetic and metabolic modifications (Garg *et al.*, 2024). Natural fungi are recognized for their ability to produce a diverse range of extracellular enzymes capable of degrading pharmaceutical pollutants. However, their application in bioremediation is limited by several inherent constraints (Garg *et al.*, 2024). One major limitation is their low degradation efficiency, as many natural fungal strains can only partially break down pollutants. This partial degradation often leads to the accumulation of intermediate metabolites, which may still pose significant environmental risks (Mao & Guan, 2016). Additionally, natural fungi exhibit sensitivity to extreme environmental conditions commonly found in industrial wastewater. Factors such as high salinity, extreme pH levels, and elevated pollutant concentrations can significantly hinder their growth and secondary metabolites production (Thangavelu *et al.*, 2006). Furthermore, the extracellular enzymes secreted by fungi often lack stability under operational conditions, leading to rapid degradation (Ayala *et al.*, 2008). This instability limits their effectiveness in large-scale bioremediation systems, thereby necessitating strategies to enhance their robustness and performance for practical applications.

Bioengineering has revolutionized the capabilities of fungi in bioremediation by addressing their inherent limitations and enhancing their efficiency in degrading pharmaceutical pollutants. Through genetic modifications, bioengineered fungal strains exhibit improved enzyme production, expanded substrate specificity, and increased environmental resilience (Akerman-Sanchez & Rojas-Jimenez, 2021). For example, genetic engineering techniques have enabled the overexpression of key enzymes such as laccase, manganese peroxidase, and lignin peroxidase (Janusz *et al.*, 2013). Promoter regions that boost transcription rates of laccase genes have significantly enhanced the degradation efficiency of pharmaceuticals like diclofenac and ibuprofen. Moreover, introducing genes from other microorganisms has broadened the substrate range of fungi, allowing them to metabolize complex pollutants like carbamazepine and certain chemotherapeutics (K.b. *et al.*, 2023). Additionally, bioengineering has fortified fungi against environmental stresses by incorporating genes that confer resistance to extreme pH, temperature, and toxic by-products. Also, immobilized bioengineered fungi have shown sustained enzyme activity in continuous flow bioreactor systems, ensuring consistent pharmaceutical pollutant removal in large-scale applications (Yamaguchi & Miyazaki, 2024). These advancements position bioengineered fungi as a powerful tool for addressing pharmaceutical pollution.

The bioengineering of fungi involves a systematic process comprising several critical steps to enhance their capabilities for bioremediation. The first step is gene selection, where specific genes responsible for improved enzyme production, pollutant degradation pathways, or stress resistance are identified (Wang *et al.*, 2020). These genes can originate from the fungi themselves or from other organisms, such as bacteria or plants, broadening the range of potential modifications. Next, advanced gene editing and transformation techniques, such as CRISPR-Cas9, Agrobacterium-mediated transformation, or electroporation, are employed to insert, delete, or modify the target genes within the fungal genome (Verma *et al.*, 2023). Once the genetic modifications are made, the engineered fungal strains undergo validation and optimization. In controlled laboratory conditions, these strains are tested for their degradation efficiency and environmental resilience. Promising strains are further refined for large-scale application through iterative bioengineering and selection processes (Rebello *et al.*, 2021). This stepwise approach ensures the development of robust fungal strains tailored for effective bioremediation in diverse settings.

7. Microalgal Bioremediation of Pharmaceutical Pollutants in Waste Water

There is a vital need for development of effective wastewater treatment methods for the is effective removal of pharmaceuticals. Microalgae have a huge potential for removing organic and inorganic pollutants from wastewater. Many large-scale treatment technologies based on microalgae are already in operation. They employ three primary mechanisms for bioremediating pharmaceuticals, biosorption, bio-uptake, and biodegradation (Sutherland & Ralph, 2019). The algae based full-scale enhanced pond systems (EPS) have been developed to treat the municipal wastewater through winter, spring and summer. Such EPS are capable of removing the nutrient removal such as nitrogen, phosphorus and carbon from diverse waterbodies (Abdelfattah *et al.*, 2023; Sutherland *et al.*, 2018). In recent years, microalgae have emerged as promising agents for detoxifying a diverse array of toxicants released from agricultural, industrial, and domestic sources. Various microalgae species have been successfully employed in the treatment of pharmaceuticals in wastewater, as shown in Table 1.

Biosorption, one of the prominent methods for pharmaceutical removal, is the passive process for removal depending on the interaction between microalgal cell walls and pharmaceutical. The ability of microalgae for pharmaceutical removal depends on their chemical nature. The surfaces of microalgal cells possess various functional groups, including carboxyl, hydroxyl, sulfate, and other charged groups, each exhibiting different affinities for different compounds. Biosorption occurs via various chemical processes such as adsorption reactions, ion exchange reactions functional groups on the microalgal surface, surface complexation reactions, chelation and microprecipitation. The rate of the adsorption process is influenced by the physicochemical properties of the surrounding environment, including factors like temperature, redox conditions, and pH (Al-Mashhadani *et al.*, 2023; Çetinkaya Dönmez *et al.*, 1999; Schmitt *et al.*, 2001; Silva *et al.*, 2019). Table 1 shows the biosorption of some prominent pharmaceuticals by microalgae. The rate of biosorption varies among different reports, for example, de Wilt *et al.* (2016) reported low adsorption <20% of six pharmaceuticals by *Chlorella sorokiniana* while in contrast Guo *et al.* (2016) reported of 100% biosorption of 7-amino cephalosporanie acid by *Chlorella* sp., *Chlamydomonas* sp., and *Mychonastes sp.*. The major challenge with biosorption is its non-selective nature. In the presence of multiple contaminants binding sites get saturated which interferes with efficiency of adsorption rate. Although various studies have highlighted the use of live microalgae for pharmaceuticals removal, only a few studies employee non-living microalgae biomass for the adsorptive removal of pharmaceuticals (Coimbra *et al.*, 2018; Santaeufemia *et al.*, 2016, 2018).

Another method by which microalgae detoxify the contaminates, is through bio-uptake. Bio-uptake is a process where unlike biosorption the contaminates are transported across the cell wall and get accumulated with in the cell. The contaminants are transported through the cell wall into the cell via passive diffusion, or facilitated diffusion, or active transport. Bio-uptake has been reported as main mechanism for removal some lipophilic pharmaceuticals, for example, Coogan *et al.* (2007) reported bio-uptake as main mechanism for removal of Triclosan, Methyl-triclosan and triclocarban by *Cladophora* spp. Similarly, Wang *et al.* (2013) reported bio-uptake of triclosan by *Chlorella pyrenoidosa*.

Biodegradation or biotransformation of pharmaceutical by microalgae, is an emerging field of study. Unlike other method which simply remove contaminants biodegradation involves metabolic breakdown of these contaminates into simple forms, generating the algal biomass. Unlike other methods, biodegradation solves the issues related to disposal of biomass after waste treatment. There are two principle mechanisms of microalgal biodegradation, first where the contaminates act as a carbon source electron donor/acceptor, for the microalga other where the contaminates are degraded by cellular enzymes (Tiwari *et al.*, 2017). Study conducted by Zhou *et al.* (2014) reported the use of four freshwater green microalgae species, *C. reinhardtii, S. obliquus, C. pyrenoidosa, and C. vulgaris* for effective removal of the both organic and inorganic contaminants from the wastewater by both biosorption and biodegradation. They effectively removed more than 50% of organic contaminants, including pharmaceuticals, herbicides, and pesticides, and reduced the estrogenic activity by up to 81% from the treated effluent.

8. Future Prospect and Challenges

Bioengineered fungi have emerged as an advanced solution to the limitations of natural fungal bioremediation, enhancing the ability of fungi to degrade pharmaceutical pollutants in wastewater (Akerman-Sanchez & Rojas-Jimenez, 2021). Natural fungi, while effective, are constrained by low enzyme production, limited substrate specificity, and sensitivity to extreme environmental conditions such as high pollutant concentrations and pH variability. These challenges often reduce their efficiency in breaking down recalcitrant pharmaceutical compounds like antibiotics, synthetic hormones, and non-steroidal anti-inflammatory drugs (Megharaj *et al.*, 2011). To overcome these limitations, bioengineering techniques enhance the enzymatic capacity, resilience, and pollutant degradation capabilities of fungi. Advanced tools such as CRISPR-Cas9 and Agrobacterium-mediated transformation enable precise modifications of fungal genomes, resulting in improved enzyme stability and increased tolerance to challenging environments, including high-salinity wastewater (Thangavelu *et al.*, 2006; Verma *et al.*, 2023). Field applications of bioengineered fungi have demonstrated their effectiveness in bioreactors and constructed wetland systems, where they maintain higher enzymatic activity over extended periods (Gebru & Werkneh, 2024). Immobilized fungal systems in packed bed reactors have shown superior removal rates for pharmaceuticals, while synergistic microbial consortia combining bioengineered fungi with bacteria further enhance the degradation of complex pollutant mixtures (Zheng *et al.*, 2021). Advancements in synthetic biology and multi-omics approaches, including genomics and proteomics, have optimized these fungal systems for large-scale applications, improving wastewater management and safeguarding aquatic ecosystems (Kadri *et al.*, 2023).

Despite these advancements, several challenges impede the widespread implementation of microbial bioremediation for pharmaceutical pollutants. One significant hurdle is maintaining the stability and activity of introduced microbial communities in diverse and often harsh environmental conditions typical of wastewater treatment systems (Ayilara & Babalola, 2023). Factors such as fluctuating pH, temperature variations, and the presence of co-contaminants can adversely affect microbial performance. Moreover, the potential risks associated with the release of genetically modified organisms (GMOs) into the environment necessitate stringent regulatory frameworks and public acceptance, which can delay or restrict practical applications (Hussain *et al.*, 2018). Another challenge lies in the incomplete degradation of complex pharmaceutical compounds, leading to the formation of potentially toxic metabolites that may pose environmental and health risks (Ortúzar *et al.*, 2022). Therefore, comprehensive assessments of degradation pathways and by-products are essential to ensure the safety and effectiveness of bioremediation efforts.

To address these challenges, future research should focus on developing robust microbial consortia capable of withstanding environmental stresses and effectively degrading a broad spectrum of pharmaceutical pollutants (Kumar Srivastava, 2024). Employing adaptive laboratory evolution techniques can enhance microbial resilience and degradation efficiency. Additionally, advancements in containment strategies for GMOs, such as the development of synthetic auxotrophy or kill-switch mechanisms, can mitigate ecological risks associated with their environmental release (Rottinghaus *et al.*, 2022). Furthermore, integrating bioremediation with other treatment technologies, such as advanced oxidation processes or membrane filtration, may offer synergistic effects, leading to more comprehensive removal of pharmaceutical contaminants (Babu Ponnusami *et al.*, 2023). Additionally, implementing real-time monitoring systems with biosensors can optimize bioremediation processes, ensuring consistent reliability and improved efficiency.

9. Conclusion

The increasing prevalence of pharmaceutical pollutants in aquatic environments poses a critical threat to ecological balance and human health, underscoring the urgent need for innovative remediation strategies. This chapter has explored the transformative potential of microbes in addressing these challenges, emphasizing the role of bioengineering in enhancing their bioremediation capabilities. Through advancements in microbial genomics and metabolic engineering, the degradation of complex pharmaceutical pollutants has become increasingly feasible and efficient.

Various microbial species, including bacteria, fungi, and microalgae, have shown significant promise in the removal of pharmaceutical pollutants, alleviating their toxic effects on ecosystems. Bioengineering approaches have further refined these microbial systems, enabling customized and effective solutions for diverse pollutant elimination. The integration of bioremediation into conventional wastewater treatment processes represents a pivotal step toward developing comprehensive and sustainable water management strategies. Nonetheless, challenges such as antimicrobial resistance, ecological risks of genetically modified organisms, and environmental variability necessitate continued research and robust regulatory frameworks. Future efforts should prioritize advancing biotechnological innovations, scaling up bioremediation systems, and fostering interdisciplinary collaboration to effectively address pharmaceutical pollution.

In summary, microbial bioremediation offers a promising pathway toward mitigating pharmaceutical contamination in water bodies. By harnessing the power of bioengineering and fostering sustainable solutions, aquatic ecosystems can be safeguarded and ultimately preserve public health for generations to come.

Acknowledgements

The author acknowledges Rashtriya Uchchatar Shiksha Abhiyan (RUSA) 2.0 [Grant No. F./RUSA/Gen/MLSU/2020/6397] and Department of Health Research [Grant No. R. 11014/08/2023-GIA/HR]

References

Abdelfattah, A., Ali, S. S., Ramadan, H., El-Aswar, E. I., Eltawab, R., Ho, S.-H., Elsamahy, T., Li, S., El-Sheekh, M. M., Schagerl, M., Kornaros, M., & Sun, J. (2023). Microalgae-based wastewater treatment: Mechanisms, challenges, recent advances, and future prospects. Environmental Science and Ecotechnology, 13, 100205. https://doi.org/10.1016/j.ese.2022.100205

Adhikari, A., Das, B. K., Ganguly, S., Nag, S. K., Sadhukhan, D., & Raut, S. S. (2023). Emerging contaminant triclosan incites endocrine disruption, reproductive impairments and oxidative stress in the commercially important carp, Catla (Labeo catla): An insight through molecular, histopathological and bioinformatic approach. Comparative Biochemistry and Physiology Part C: Toxicology & Pharmacology, 268, 109605. https://doi.org/10.1016/j.cbpc.2023.109605

Ahmad, W., Kaur, N., & Parashar, A. (2024). Algal organic matter induced photodegradation of tinidazole. Nucleosides, Nucleotides & Nucleic Acids, 43(7), 607–618. https://doi.org/10.1080/15257770.2023.2289466

Aich, U., Polverino, G., Yazdan Parast, F., Melo, G. C., Tan, H., Howells, J., Nosrati, R., & Wong, B. B. M. (2024). Long-term effects of widespread pharmaceutical pollution on trade-offs between behavioural, life-history and reproductive traits in fish. Journal of Animal Ecology. https://doi.org/10.1111/1365-2656.14152

Akerman-Sanchez, G., & Rojas-Jimenez, K. (2021). Fungi for the bioremediation of pharmaceutical-derived pollutants: A bioengineering approach to water treatment. Environmental Advances, 4, 100071. https://doi.org/10.1016/j.envadv.2021.100071

Alex, A. M., Vivekanadam, S., Periyasamy, V., & Vairaperumal, T. (2025). Chapter 12—Bioremediation of pharmaceuticals and antibiotics emerging contaminants from wastewater. In A. L. Srivastav, I. Zinicovscaia, & L. Cepoi (Eds.), Biotechnologies for Wastewater Treatment and Resource Recovery (pp. 159–169). Elsevier. https://doi.org/10.1016/B978-0-443-27376-6.00025-6

Ali, M. E. M., Abd El-Aty, A. M., Badawy, M. I., & Ali, R. K. (2018). Removal of pharmaceutical pollutants from synthetic wastewater using chemically modified biomass of green alga Scenedesmus obliquus. Ecotoxicology and Environmental Safety, 151, 144–152. https://doi.org/10.1016/j.ecoenv.2018.01.012

Al-Mashhadani, E. S. M., Al-Mashhadani, M. K. H., & Al-Maari, M. A. (2023). Biosorption of Ciprofloxacin (CIP) using the Waste of Extraction Process of Microalgae: The Equilibrium Isotherm and Kinetic Study. Iraqi Journal of Chemical and Petroleum Engineering, 24(4), 1–15. https://doi.org/10.31699/IJCPE.2023.4.1

Amobonye, A., Aruwa, C. E., Aransiola, S., Omame, J., Alabi, T. D., & Lalung, J. (2023). The potential of fungi in the bioremediation of pharmaceutically active compounds: A comprehensive review. Frontiers in Microbiology, 14, 1207792. https://doi.org/10.3389/fmicb.2023.1207792

Amorim, C. L., Moreira, I. S., Maia, A. S., Tiritan, M. E., & Castro, P. M. L. (2014). Biodegradation of ofloxacin, norfloxacin, and ciprofloxacin as single and mixed substrates by Labrys portucalensis F11. Applied Microbiology and Biotechnology, 98(7), 3181–3190. https://doi.org/10.1007/s00253-013-5333-8

Anjanapriya, S., SulaimanMumtaz, M., Mohideen, M. H. A. K., Radha, A., Sasirekha, N., Sawicka, B., & Tamizhazhagan, V. (2021). Pharmaceutical Pollution Crisis in the World: A Menace to Ecosystem. Entomology and Applied Science Letters, 8(1–2021), Article 1–2021. https://doi.org/10.51847/iUGgphofKK

Ashfaq, M., Nawaz Khan, K., Saif Ur Rehman, M., Mustafa, G., Faizan Nazar, M., Sun, Q., Iqbal, J., Mulla, S. I., & Yu, C.-P. (2017). Ecological risk assessment of pharmaceuticals in the receiving environment of pharmaceutical wastewater in Pakistan. Ecotoxicology and Environmental Safety, 136, 31–39. https://doi.org/10.1016/j.ecoenv.2016.10.029

Ayala, M., Pickard, M. A., & Vazquez-Duhalt, R. (2008). Fungal enzymes for environmental purposes, a molecular biology challenge. Journal of Molecular Microbiology and Biotechnology, 15(2–3), 172–180. https://doi.org/10.1159/000121328

Ayilara, M. S., & Babalola, O. O. (2023). Bioremediation of environmental wastes: The role of microorganisms. Frontiers in Agronomy, 5. https://doi.org/10.3389/fagro.2023.1183691

Babu Ponnusami, A., Sinha, S., Ashokan, H., V Paul, M., Hariharan, S. P., Arun, J., Gopinath, K. P., Hoang Le, Q., & Pugazhendhi, A. (2023). Advanced oxidation process (AOP) combined biological process for wastewater treatment: A review on advancements, feasibility and practicability of combined techniques. Environmental Research, 237, 116944. https://doi.org/10.1016/j.envres.2023.116944

Bai, X., & Acharya, K. (2019). Removal of seven endocrine disrupting chemicals (EDCs) from municipal wastewater effluents by a freshwater green alga. Environmental Pollution (Barking, Essex: 1987), 247, 534–540. https://doi.org/10.1016/j.envpol.2019.01.075

Bala, S., Garg, D., Thirumalesh, B. V., Sharma, M., Sridhar, K., Inbaraj, B. S., & Tripathi, M. (2022). Recent Strategies for Bioremediation of Emerging Pollutants: A Review for a Green and Sustainable Environment. Toxics, 10(8), 484. https://doi.org/10.3390/toxics10080484

Balakrishna, K., Rath, A., Praveenkumarreddy, Y., Guruge, K. S., & Subedi, B. (2017). A review of the occurrence of pharmaceuticals and personal care products in Indian water bodies. Ecotoxicology and Environmental Safety, 137, 113–120. https://doi.org/10.1016/j.ecoenv.2016.11.014

Bhatt, P., Jeon, C.-H., & Kim, W. (2022). Tetracycline bioremediation using the novel Serratia marcescens strain WW1 isolated from a wastewater treatment plant. Chemosphere, 298, 134344. https://doi.org/10.1016/j.chemosphere.2022.134344

Bojarski, B., Kot, B., & Witeska, M. (2020). Antibacterials in Aquatic Environment and Their Toxicity to Fish. Pharmaceuticals, 13(8), 189. https://doi.org/10.3390/ph13080189

Bora, A., Thondi Rajan, A. S., Ponnuchamy, K., Muthusamy, G., & Alagarsamy, A. (2024). Microalgae to bioenergy production: Recent advances, influencing parameters, utilization of wastewater – A critical review. Science of The Total Environment, 946, 174230. https://doi.org/10.1016/j.scitotenv.2024.174230

Çetinkaya Dönmez, G., Aksu, Z., Öztürk, A., & Kutsal, T. (1999). A comparative study on heavy metal biosorption characteristics of some algae. Process Biochemistry, 34(9), 885–892. https://doi.org/10.1016/S0032-9592(99)00005-9

Chakrabarti, S. (2018). Eutrophication - A Global Aquatic Environmental Problem: A Review. https://www.semanticscholar.org/paper/Eutrophication-A-Global-Aquatic-Environmental-A-Chakrabarti/17cd1178d0ba18b78b29d18afd598ce1bf5fa1df

Chen, J., Zheng, F., & Guo, R. (2015). Algal Feedback and Removal Efficiency in a Sequencing Batch Reactor Algae Process (SBAR) to Treat the Antibiotic Cefradine. PLoS ONE, 10(7), e0133273. https://doi.org/10.1371/journal.pone.0133273

Chmelová, D., Ondrejovič, M., & Miertuš, S. (2024). Laccases as Effective Tools in the Removal of Pharmaceutical Products from Aquatic Systems. Life, 14(2), 230. https://doi.org/10.3390/life14020230

Chopra, S., & Kumar, D. (2022). Characterization and biodegradation of ibuprofen by Bacillus siamensis strain DSI-1 isolated from wastewater. Rendiconti Lincei. Scienze Fisiche e Naturali, 33(3), 643–652. https://doi.org/10.1007/s12210-022-01085-6

Chopra, S., & Kumar, D. (2023). Characterization and biodegradation of paracetamol by biomass of Bacillus licheniformis strain PPY-2 isolated from wastewater. Rendiconti Lincei. Scienze Fisiche e Naturali, 34(2), 491–501. https://doi.org/10.1007/s12210-023-01140-w

Coimbra, R. N., Escapa, C., Vázquez, N. C., Noriega-Hevia, G., & Otero, M. (2018). Utilization of Non-Living Microalgae Biomass from Two Different Strains for the Adsorptive Removal of Diclofenac from Water. Water, 10(10), Article 10. https://doi.org/10.3390/w10101401

Coogan, M. A., Edziyie, R. E., La Point, T. W., & Venables, B. J. (2007). Algal bioaccumulation of triclocarban, triclosan, and methyl-triclosan in a North Texas wastewater treatment plant receiving stream. Chemosphere, 67(10), 1911–1918. https://doi.org/10.1016/j.chemosphere.2006.12.027

Corpus, R. M. B., Bayani, M. S., Aguilar, J. L., Aguilar, J. C. B., & Aguilar, H. B. (2024). Emerging pollutants in waste water: Challenges and advancements in treatment technology. IOP Conference Series: Earth and Environmental Science, 1372(1), 012037. https://doi.org/10.1088/1755-1315/1372/1/012037

Dai, Y., Peng, J.-J., Zhang, T.-Y., Xie, X.-P., Luo, S.-S., Liu, W.-C., & Ma, Y. (2023). Effects of antibiotics (enrofloxacin) on microbial community of water and sediment in an aquatic ecological model. Frontiers in Veterinary Science, 10, 1151988. https://doi.org/10.3389/fvets.2023.1151988

Dalecka, B., Oskarsson, C., Juhna, T., & Kuttava Rajarao, G. (2020). Isolation of Fungal Strains from Municipal Wastewater for the Removal of Pharmaceutical Substances. Water, 12(2), 524. https://doi.org/10.3390/w12020524

De Wilt, A., Butkovskyi, A., Tuantet, K., Leal, L. H., Fernandes, T. V., Langenhoff, A., & Zeeman, G. (2016). Micropollutant removal in an algal treatment system fed with source separated wastewater streams. Journal of Hazardous Materials, 304, 84–92. https://doi.org/10.1016/j.jhazmat.2015.10.033

Delbes, G., Blázquez, M., Fernandino, J. I., Grigorova, P., Hales, B. F., Metcalfe, C., Navarro-Martín, L., Parent, L., Robaire, B., Rwigemera, A., Van Der Kraak, G., Wade, M., & Marlatt, V. (2022). Effects of endocrine disrupting chemicals on gonad development: Mechanistic insights from fish and mammals. Environmental Research, 204, 112040. https://doi.org/10.1016/j.envres.2021.112040

Deng, Y., Debognies, A., Zhang, Q., Zhang, Z., Zhou, Z., Zhang, J., Sun, L., Lu, T., & Qian, H. (2022). Effects of ofloxacin on the structure and function of freshwater microbial communities. Aquatic Toxicology, 244, 106084. https://doi.org/10.1016/j.aquatox.2022.106084

Dinakarkumar, Y., Ramakrishnan, G., Gujjula, K. R., Vasu, V., Balamurugan, P., & Murali, G. (2024). Fungal bioremediation: An overview of the mechanisms, applications and future perspectives. Environmental Chemistry and Ecotoxicology, 6, 293–302. https://doi.org/10.1016/j.enceco.2024.07.002

Ding, T., Lin, K., Yang, B., Yang, M., Li, J., Li, W., & Gan, J. (2017). Biodegradation of naproxen by freshwater algae Cymbella sp. And Scenedesmus quadricauda and the comparative toxicity. Bioresource Technology, 238, 164–173. https://doi.org/10.1016/j.biortech.2017.04.018

Du, B., Haddad, S. P., Luek, A., Scott, W. C., Saari, G. N., Kristofco, L. A., Connors, K. A., Rash, C., Rasmussen, J. B., Chambliss, C. K., & Brooks, B. W. (2014). Bioaccumulation and trophic dilution of human pharmaceuticals across trophic positions of an effluent-dependent wadeable stream. Philosophical Transactions of the Royal Society B: Biological Sciences, 369(1656), 20140058. https://doi.org/10.1098/rstb.2014.0058

Duarte, I. A., Reis-Santos, P., Novais, S. C., Rato, L. D., Lemos, M. F. L., Freitas, A., Pouca, A. S. V., Barbosa, J., Cabral, H. N., & Fonseca, V. F. (2020). Depressed, hypertense and sore: Long-term effects of fluoxetine, propranolol and diclofenac exposure in a top predator fish. Science of The Total Environment, 712, 136564. https://doi.org/10.1016/j.scitotenv.2020.136564

Dzionek, A., Wojcieszyńska, D., Marchlewicz, A., Smułek, W., Potocka, I., Jałowiecki, Ł., Borgulat, J., Płaza, G., & Guzik, U. (2024). Naproxen as environmental pollution, its effect on bacteria metabolism and degradation mechanism in immobilized Planococcus sp. S5. Chemical Engineering Journal, 481, 148174. https://doi.org/10.1016/j.cej.2023.148174

Elizalde-Velázquez, G. A., Rosas-Ramírez, J. R., Raldua, D., García-Medina, S., Orozco-Hernández, J. M., Rosales- Pérez, K., Islas-Flores, H., Galar-Martínez, M., Guzmán-García, X., & Gómez-Oliván, L. M. (2022). Low concentrations of ciprofloxacin alone and in combination with paracetamol induce oxidative stress, upregulation of apoptotic-related genes, histological alterations in the liver, and genotoxicity in Danio rerio. Chemosphere, 294, 133667. https://doi.org/10.1016/j.chemosphere.2022.133667

Embarez, D. H., Razek, A. S. A., Basalious, E. B., Mahmoud, M., & Nadia M., H. (2023). Acetaminophen-traces bioremediation with novel phenotypically and genotypically characterized 2 Streptomyces strains using chemo-informatics, in vivo, and in vitro experiments for cytotoxicity and biological activity. Journal of Genetic Engineering and Biotechnology, 21(1), 171. https://doi.org/10.1186/s43141-023-00602-w

Ercoli, L., Rossetto, R., Di Giorgi, S., Raffaelli, A., Nuti, M., & Pellegrino, E. (2023). Effective bioremediation of clarithromycin and diclofenac in wastewater by microbes and Arundo donax L. Environmental Science and Pollution Research, 30(31), 77193–77209. https://doi.org/10.1007/s11356-023-27660-4

Escapa, C., Coimbra, R. N., Paniagua, S., García, A. I., & Otero, M. (2017a). Comparison of the culture and harvesting of Chlorella vulgaris and Tetradesmus obliquus for the removal of pharmaceuticals from water. Journal of Applied Phycology, 29(3), 1179–1193. https://doi.org/10.1007/s10811-016-1010-5

Escapa, C., Coimbra, R. N., Paniagua, S., García, A. I., & Otero, M. (2017b). Paracetamol and salicylic acid removal from contaminated water by microalgae. Journal of Environmental Management, 203(Pt 2), 799–806. https://doi.org/10.1016/j.jenvman.2016.06.051

Espinosa-Ortiz, E. J., Rene, E. R., & Gerlach, R. (2022). Potential use of fungal-bacterial co-cultures for the removal of organic pollutants. Critical Reviews in Biotechnology, 42(3), 361–383. https://doi.org/10.1080/07388551.2021.1940831

Fabbri, E. (2015). Pharmaceuticals in the environment: Expected and unexpected effects on aquatic fauna. Annals of the New York Academy of Sciences, 1340(1), 20–28. https://doi.org/10.1111/nyas.12605

Fatta-Kassinos, D., Meric, S., & Nikolaou, A. (2011). Pharmaceutical residues in environmental waters and wastewater: Current state of knowledge and future research. Analytical and Bioanalytical Chemistry, 399(1), 251–275. https://doi.org/10.1007/s00216-010-4300-9

Fernandes, J. P., Duarte, P., Almeida, C. M. R., Carvalho, M. F., & Mucha, A. P. (2020). Potential of bacterial consortia obtained from different environments for bioremediation of paroxetine and bezafibrate. Journal of Environmental Chemical Engineering, 8(4), 103881. https://doi.org/10.1016/j.jece.2020.103881

Fick, J., Söderström, H., Lindberg, R. H., Phan, C., Tysklind, M., & Larsson, D. G. J. (2009). Contamination of surface, ground, and drinking water from pharmaceutical production. Environmental Toxicology and Chemistry, 28(12), 2522–2527. https://doi.org/10.1897/09-073.1

Gallego-Ríos, S. E., Peñuela, G. A., & Martínez-López, E. (2021). Updating the use of biochemical biomarkers in fish for the evaluation of alterations produced by pharmaceutical products. Environmental Toxicology and Pharmacology, 88, 103756. https://doi.org/10.1016/j.etap.2021.103756

Gao, P., Chang, C., Liang, J., Du, F., & Zhang, R. (2024). Embryonic Amoxicillin Exposure Has Limited Impact on Liver Development but Increases Susceptibility to NAFLD in Zebrafish Larvae. International Journal of Molecular Sciences, 25(5), 2744. https://doi.org/10.3390/ijms25052744

Gao, Y., Shah, K., Kwok, I., Wang, M., Rome, L. H., & Mahendra, S. (2022). Immobilized fungal enzymes: Innovations and potential applications in biodegradation and biosynthesis. Biotechnology Advances, 57, 107936. https://doi.org/10.1016/j.biotechadv.2022.107936

Garg, S., Kim, M., & Romero-Suarez, D. (2024). Current advancements in fungal engineering technologies for Sustainable Development Goals. Trends in Microbiology. https://doi.org/10.1016/j.tim.2024.11.001

Gebru, S. B., & Werkneh, A. A. (2024). Applications of constructed wetlands in removing emerging micropollutants from wastewater: Occurrence, public health concerns, and removal performances – a review. South African Journal of Chemical Engineering, 48, 395–416. https://doi.org/10.1016/j.sajce.2024.03.004

Gentili, F. G., & Fick, J. (2017). Algal cultivation in urban wastewater: An efficient way to reduce pharmaceutical pollutants. Journal of Applied Phycology, 29(1), 255–262. https://doi.org/10.1007/s10811-016-0950-0

Gojkovic, Z., Lindberg, R. H., Tysklind, M., & Funk, C. (2019). Northern green algae have the capacity to remove active pharmaceutical ingredients. Ecotoxicology and Environmental Safety, 170, 644–656. https://doi.org/10.1016/j.ecoenv.2018.12.032

González Peña, O. I., López Zavala, M. Á., & Cabral Ruelas, H. (2021). Pharmaceuticals Market, Consumption Trends and Disease Incidence Are Not Driving the Pharmaceutical Research on Water and Wastewater. International Journal of Environmental Research and Public Health, 18(5), Article 5. https://doi.org/10.3390/ijerph18052532

González-González, R. B., Flores-Contreras, E. A., Parra-Saldívar, R., & Iqbal, H. M. N. (2022). Bio-removal of emerging pollutants by advanced bioremediation techniques. Environmental Research, 214, 113936. https://doi.org/10.1016/j.envres.2022.113936

González-González, R. B., Parra-Arroyo, L., Parra-Saldívar, R., Ramirez-Mendoza, R. A., & Iqbal, H. M. N. (2022). Nanomaterial-based catalysts for the degradation of endocrine-disrupting chemicals – A way forward to environmental remediation. Materials Letters, 308, 131217. https://doi.org/10.1016/j.matlet.2021.131217

Guevara, G., Espinoza Solorzano, J. S., Vargas Ramírez, M., Rusu, A., & Navarro Llorens, J. M. (2024). Characterizing A21: Natural Cyanobacteria-Based Consortium with Potential for Steroid Bioremediation in Wastewater Treatment. International Journal of Molecular Sciences, 25(23), Article 23. https://doi.org/10.3390/ijms252313018

Guo, W.-Q., Zheng, H.-S., Li, S., Du, J.-S., Feng, X.-C., Yin, R.-L., Wu, Q.-L., Ren, N.-Q., & Chang, J.-S. (2016). Removal of cephalosporin antibiotics 7-ACA from wastewater during the cultivation of lipid-accumulating microalgae. Bioresource Technology, 221, 284–290. https://doi.org/10.1016/j.biortech.2016.09.036

Hom-Diaz, A., Llorca, M., Rodríguez-Mozaz, S., Vicent, T., Barceló, D., & Blánquez, P. (2015). Microalgae cultivation on wastewater digestate: β-estradiol and 17α-ethynylestradiol degradation and transformation products identification. Journal of Environmental Management, 155, 106–113. https://doi.org/10.1016/j.jenvman.2015.03.003

Hong, X., Zhao, G., Zhou, Y., Chen, R., Li, J., & Zha, J. (2021). Risks to aquatic environments posed by 14 pharmaceuticals as illustrated by their effects on zebrafish behaviour. Science of The Total Environment, 771, 145450. https://doi.org/10.1016/j.scitotenv.2021.145450

Hussain, A., Ashique, S., Zaheen Hassan, Mohd., Afzal, O., Asiri, Y. I., Kumar, P., Dua, K., Webster, T. J., Altamimi, A. S. A., & Altamimi, M. A. (2023). Pharmaceutical contaminants in aquatic systems, conventional and green strategies, recent updates, challenges and policies, and potential outcomes. Journal of Molecular Liquids, 389, 122905. https://doi.org/10.1016/j.molliq.2023.122905

Hussain, I., Aleti, G., Naidu, R., Puschenreiter, M., Mahmood, Q., Rahman, M. M., Wang, F., Shaheen, S., Syed, J. H., & Reichenauer, T. G. (2018). Microbe and plant assisted-remediation of organic xenobiotics and its enhancement by genetically modified organisms and recombinant technology: A review. The Science of the Total Environment, 628–629, 1582–1599. https://doi.org/10.1016/j.scitotenv.2018.02.037

Hussain, S., Naeem, M., & Chaudhry, M. N. (2016). Estimation of Residual Antibiotics in Pharmaceutical Effl uents and their Fate in Affected Areas. Polish Journal of Environmental Studies, 25(2), 607–614. https://doi.org/10.15244/pjoes/61229

Insani, W. N., Qonita, N. A., Jannah, S. S., Nuraliyah, N. M., Supadmi, W., Gatera, V. A., Alfian, S. D., & Abdulah, R. (2020). Improper disposal practice of unused and expired pharmaceutical products in Indonesian households. Heliyon, 6(7), e04551. https://doi.org/10.1016/j.heliyon.2020.e04551

Islam, M. T., Mostakim, G. M., Azom, M. G., Rahman, U. O., Khan, M. M., Quader Khan, M. G., & Islam, M. S. (2022). Effect of an amalgamated antibiotic and its connection to cyto-genotoxicity and histo-architectural malformations in stinging catfish. Emerging Contaminants, 8, 381–390. https://doi.org/10.1016/j.emcon.2022.09.001

Jamur, Jasim. M. S. (2024). Analytical techniques in pharmaceutical pollution of the world's rivers; a review. ChemChemTech, 67(5), 6–16. https://doi.org/10.6060/ivkkt.20246705.7017

Janusz, G., Kucharzyk, K. H., Pawlik, A., Staszczak, M., & Paszczynski, A. J. (2013). Fungal laccase, manganese peroxidase and lignin peroxidase: Gene expression and regulation. Enzyme and Microbial Technology, 52(1), 1–12. https://doi.org/10.1016/j.enzmictec.2012.10.003

Ji, M. K., Kabra, A. N., Choi, J., Hwang, J. H., Kim, J. R., Abou-Shanab, R. A. I., Oh, Y. K., & Jeon, B. H. (2014). Biodegradation of bisphenol A by the freshwater microalgae Chlamydomonas mexicana and Chlorella vulgaris. Ecological Engineering, 73, 260–269. https://doi.org/10.1016/j.ecoleng.2014.09.070

Jiménez-Díaz, V., Pedroza-Rodríguez, A. M., Ramos-Monroy, O., & Castillo-Carvajal, L. C. (2022). Synthetic Biology: A New Era in Hydrocarbon Bioremediation. Processes, 10(4), Article 4. https://doi.org/10.3390/pr10040712

K.b., A., Madhavan, A., Tarafdar, A., Sirohi, R., A.n., A., Kuriakose, L. L., Awasthi, M. K., Binod, P., Varjani, S., & Sindhu, R. (2023). Filamentous fungi for pharmaceutical compounds degradation in the environment: A sustainable approach. Environmental Technology & Innovation, 31, 103182. https://doi.org/10.1016/j.eti.2023.103182

Kadri, M. S., Singhania, R. R., Haldar, D., Patel, A. K., Bhatia, S. K., Saratale, G., Parameswaran, B., & Chang, J.-S. (2023). Advances in Algomics technology: Application in wastewater treatment and biofuel production. Bioresource Technology, 387, 129636. https://doi.org/10.1016/j.biortech.2023.129636

Kang, J., Zhang, Z., Chen, Y., Zhou, Z., Zhang, J., Xu, N., Zhang, Q., Lu, T., Peijnenburg, W. J. G. M., & Qian, H. (2022). Machine learning predicts the impact of antibiotic properties on the composition and functioning of bacterial community in aquatic habitats. Science of The Total Environment, 828, 154412. https://doi.org/10.1016/j.scitotenv.2022.154412

Kasonga, T. K., Kamika, I., & Ngole-Jeme, V. M. (2023). Ligninolytic enzyme activity and removal efficiency of pharmaceuticals in a water matrix by fungus Rhizopus sp. Isolated from cassava. Environmental Technology. https://www.tandfonline.com/doi/abs/10.1080/09593330.2021.2024885

Kaur, B., Choudhary, R., Sharma, G., & Brar, L. K. (2024). Sustainable and effective microorganisms method for wastewater treatment. Desalination and Water Treatment, 319, 100419. https://doi.org/10.1016/j.dwt.2024.100419

Kayode-Afolayan, S. D., Ahuekwe, E. F., & Nwinyi, O. C. (2022). Impacts of pharmaceutical effluents on aquatic ecosystems. Scientific African, 17, e01288. https://doi.org/10.1016/j.sciaf.2022.e01288

Khan, A. H. A., & Barros, R. (2023). Pharmaceuticals in Water: Risks to Aquatic Life and Remediation Strategies. Hydrobiology, 2(2), Article 2. https://doi.org/10.3390/hydrobiology2020026

Kleywegt, S., Payne, M., Ng, F., & Fletcher, T. (2019). Environmental loadings of Active Pharmaceutical Ingredients from manufacturing facilities in Canada. Science of The Total Environment, 646, 257–264. https://doi.org/10.1016/j.scitotenv.2018.07.240

Kock, A., Glanville, H. C., Law, A. C., Stanton, T., Carter, L. J., & Taylor, J. C. (2023). Emerging challenges of the impacts of pharmaceuticals on aquatic ecosystems: A diatom perspective. Science of The Total Environment, 878, 162939. https://doi.org/10.1016/j.scitotenv.2023.162939

Kumar Srivastava, S. (2024). Recent advances in removal of pharmaceutical pollutants in wastewater using metal oxides and carbonaceous materials as photocatalysts: A review. RSC Applied Interfaces, 1(3), 340–429. https://doi.org/10.1039/D3LF00142C

Kumar, A., & Chandra, R. (2020). Ligninolytic enzymes and its mechanisms for degradation of lignocellulosic waste in environment. Heliyon, 6(2), e03170. https://doi.org/10.1016/j.heliyon.2020.e03170

Kuppan, N., Padman, M., Mahadeva, M., Srinivasan, S., & Devarajan, R. (2024). A comprehensive review of sustainable bioremediation techniques: Eco friendly solutions for waste and pollution management. Waste Management Bulletin, 2(3), 154–171. https://doi.org/10.1016/j.wmb.2024.07.005

Lakhani, S., Acharya, D., Sakariya, R., Sharma, D., Patel, P., Shah, M., & Prajapati, M. (2022). A comprehensive study of bioremediation for pharmaceutical wastewater treatment. Cleaner Chemical Engineering, 4, 100073. https://doi.org/10.1016/j.clce.2022.100073

Larsen, C., Yu, Z. H., Flick, R., & Passeport, E. (2019). Mechanisms of pharmaceutical and personal care product removal in algae-based wastewater treatment systems. Science of The Total Environment, 695, 133772. https://doi.org/10.1016/j.scitotenv.2019.133772

Larsson, D. G. J., de Pedro, C., & Paxeus, N. (2007). Effluent from drug manufactures contains extremely high levels of pharmaceuticals. Journal of Hazardous Materials, 148(3), 751–755. https://doi.org/10.1016/j.jhazmat.2007.07.008

Legorreta-Castañeda, A. J., Lucho-Constantino, C. A., Beltrán-Hernández, R. I., Coronel-Olivares, C., & Vázquez-Rodríguez, G. A. (2020). Biosorption of Water Pollutants by Fungal Pellets. Water, 12(4), Article 4. https://doi.org/10.3390/w12041155

Li, H., Pan, Y., Wang, Z., Chen, S., Guo, R., & Chen, J. (2015). An algal process treatment combined with the Fenton reaction for high concentrations of amoxicillin and cefradine. RSC Advances, 5(122), 100775–100782. https://doi.org/10.1039/C5RA21508K

Li, S., Shi, W., Li, H., Xu, N., Zhang, R., Chen, X., Sun, W., Wen, D., He, S., Pan, J., He, Z., & Fan, Y. (2018). Antibiotics in water and sediments of rivers and coastal area of Zhuhai City, Pearl River estuary, south China. Science of The Total Environment, 636, 1009–1019. https://doi.org/10.1016/j.scitotenv.2018.04.358

Li, X., & Hamblin, D. (2016). Factors impacting on cleaner production: Case studies of Chinese pharmaceutical manufacturers in Tianjin, China. Journal of Cleaner Production, 131, 121–132. https://doi.org/10.1016/j.jclepro.2016.05.066

Li, X., Shen, X., Jiang, W., Xi, Y., & Li, S. (2024). Comprehensive review of emerging contaminants: Detection technologies, environmental impact, and management strategies. Ecotoxicology and Environmental Safety, 278, 116420. https://doi.org/10.1016/j.ecoenv.2024.116420

Liberti, D., Pinheiro, F., Simões, B., Varela, J., & Barreira, L. (2024). Beyond Bioremediation: The Untapped Potential of Microalgae in Wastewater Treatment. Water, 16(19), Article 19. https://doi.org/10.3390/w16192710

Limbu, S. M., Chen, L., Zhang, M., & Du, Z. (2021). A global analysis on the systemic effects of antibiotics in cultured fish and their potential human health risk: A review. Reviews in Aquaculture, 13(2), 1015–1059. https://doi.org/10.1111/raq.12511

Liu, N., Jin, X., Feng, C., Wang, Z., Wu, F., Johnson, A. C., Xiao, H., Hollert, H., & Giesy, J. P. (2020). Ecological risk assessment of fifty pharmaceuticals and personal care products (PPCPs) in Chinese surface waters: A proposed multiple-level system. Environment International, 136, 105454. https://doi.org/10.1016/j.envint.2019.105454

Liu, X., Chen, J., Liu, Y., Wan, Z., Guo, X., Lu, S., & Qiu, D. (2022). Sulfamethoxazole degradation by Pseudomonas silesiensis F6a isolated from bioelectrochemical technology-integrated constructed wetlands. Ecotoxicology and Environmental Safety, 240, 113698. https://doi.org/10.1016/j.ecoenv.2022.113698

Liu, Y., Feng, J., Pan, H., Zhang, X., & Zhang, Y. (2022). Genetically engineered bacterium: Principles, practices, and prospects. Frontiers in Microbiology, 13, 997587. https://doi.org/10.3389/fmicb.2022.997587

Liu, Y., Wang, F., Chen, X., Zhang, J., & Gao, B. (2015). Cellular responses and biodegradation of amoxicillin in Microcystis aeruginosa at different nitrogen levels. Ecotoxicology and Environmental Safety, 111, 138–145. https://doi.org/10.1016/j.ecoenv.2014.10.011

Long, S., Hamilton, P. B., Wang, C., Li, C., Xue, X., Zhao, Z., Wu, P., Gu, E., Uddin, M. M., Li, B., & Xu, F. (2024). Bioadsorption, bioaccumulation and biodegradation of antibiotics by algae and their association with algal physiological state and antibiotic physicochemical properties. Journal of Hazardous Materials, 468, 133787. https://doi.org/10.1016/j.jhazmat.2024.133787

López-Doval, J. C., Serra-Compte, A., Rodríguez-Mozaz, S., Barceló, D., & Sabater, S. (2020). Diet quality and NSAIDs promote changes in formation of prostaglandins by an aquatic invertebrate. Chemosphere, 257, 126892. https://doi.org/10.1016/j.chemosphere.2020.126892

Luo, Q., Ding, N., Liu, Y., Zhang, H., Fang, Y., & Yin, L. (2023). Metabolic Engineering of Microorganisms to Produce Pyruvate and Derived Compounds. Molecules, 28(3), Article 3. https://doi.org/10.3390/molecules28031418

Mackuľak, T., Cverenkárová, K., Vojs Staňová, A., Fehér, M., Tamáš, M., Škulcová, A. B., Gál, M., Naumowicz, M., Špalková, V., & Bírošová, L. (2021). Hospital Wastewater-Source of Specific Micropollutants, Antibiotic-Resistant Microorganisms, Viruses, and Their Elimination. Antibiotics (Basel, Switzerland), 10(9), 1070. https://doi.org/10.3390/antibiotics10091070

Maglione, G., Zinno, P., Tropea, A., Mussagy, C. U., Dufossé, L., Giuffrida, D., & Mondello, A. (2024). Microbes' role in environmental pollution and remediation: A bioeconomy focus approach. AIMS Microbiology, 10(3), 723–755. https://doi.org/10.3934/microbiol.2024033

Malik, S., Bora, J., Nag, S., Sinha, S., Mondal, S., Rustagi, S., Hazra, R., Kumar, H., Rajput, V. D., Minkina, T., Sadier, N. S., & Almutary, A. G. (2023). Fungal-Based Remediation in the Treatment of Anthropogenic Activities and Pharmaceutical-Pollutant-Contaminated Wastewater. Water, 15(12), Article 12. https://doi.org/10.3390/w15122262

Mansouri, F., Chouchene, K., Roche, N., & Ksibi, M. (2021). Removal of Pharmaceuticals from Water by Adsorption and Advanced Oxidation Processes: State of the Art and Trends. Applied Sciences, 11(14), Article 14. https://doi.org/10.3390/app11146659

Mao, J., & Guan, W. (2016). Fungal degradation of polycyclic aromatic hydrocarbons (PAHs) by Scopulariopsis brevicaulis and its application in bioremediation of PAH-contaminated soil. Acta Agriculturae Scandinavica, Section B — Soil & Plant Science. https://www.tandfonline.com/doi/abs/10.1080/09064710.2015.1137629

Maqsood, Q., Sumrin, A., Waseem, R., Hussain, M., Imtiaz, M., & Hussain, N. (2023). Bioengineered microbial strains for detoxification of toxic environmental pollutants. Environmental Research, 227, 115665. https://doi.org/10.1016/j.envres.2023.115665

Marchlewicz, A., Guzik, U., Hupert-Kocurek, K., & Wojcieszyńska, D. (2023). Evaluation of the Defined Bacterial Consortium Efficacy in the Biodegradation of NSAIDs. Molecules, 28(5), Article 5. https://doi.org/10.3390/molecules28052185

Massima Mouele, E. S., Tijani, J. O., Badmus, K. O., Pereao, O., Babajide, O., Zhang, C., Shao, T., Sosnin, E., Tarasenko, V., Fatoba, O. O., Laatikainen, K., & Petrik, L. F. (2021). Removal of Pharmaceutical Residues from Water and Wastewater Using Dielectric Barrier Discharge Methods—A Review. International Journal of Environmental Research and Public Health, 18(4), 1683. https://doi.org/10.3390/ijerph18041683

Matamoros, V., Uggetti, E., García, J., & Bayona, J. M. (2016). Assessment of the mechanisms involved in the removal of emerging contaminants by microalgae from wastewater: A laboratory scale study. Journal of Hazardous Materials, 301, 197–205. https://doi.org/10.1016/j.jhazmat.2015.08.050

Megharaj, M., Ramakrishnan, B., Venkateswarlu, K., Sethunathan, N., & Naidu, R. (2011). Bioremediation approaches for organic pollutants: A critical perspective. Environment International, 37(8), 1362–1375. https://doi.org/10.1016/j.envint.2011.06.003

Mezzelani, M., Gorbi, S., & Regoli, F. (2018). Pharmaceuticals in the aquatic environments: Evidence of emerged threat and future challenges for marine organisms. Marine Environmental Research, 140, 41–60. https://doi.org/10.1016/j.marenvres.2018.05.001

Miazek, K., & Brozek-Pluska, B. (2019). Effect of PHRs and PCPs on Microalgal Growth, Metabolism and Microalgae-Based Bioremediation Processes: A Review. International Journal of Molecular Sciences, 20(10), 2492. https://doi.org/10.3390/ijms20102492

Michaelis, C., & Grohmann, E. (2023). Horizontal Gene Transfer of Antibiotic Resistance Genes in Biofilms. Antibiotics, 12(2), Article 2. https://doi.org/10.3390/antibiotics12020328

Miglani, R., Parveen, N., Kumar, A., Ansari, Mohd. A., Khanna, S., Rawat, G., Panda, A. K., Bisht, S. S., Upadhyay, J., & Ansari, M. N. (2022). Degradation of Xenobiotic Pollutants: An Environmentally Sustainable Approach. Metabolites, 12(9), 818. https://doi.org/10.3390/metabo12090818

Mojiri, A., Baharlooeian, M., Kazeroon, R. A., Farraji, H., & Lou, Z. (2020). Removal of Pharmaceutical Micropollutants with Integrated Biochar and Marine Microalgae. Microorganisms, 9(1), 4. https://doi.org/10.3390/microorganisms9010004

Moreira, I. S., Lebel, A., Peng, X., Castro, P. M. L., & Gonçalves, D. (2021). Sediments in the mangrove areas contribute to the removal of endocrine disrupting chemicals in coastal sediments of Macau SAR, China, and harbour microbial communities capable of degrading E2, EE2, BPA and BPS. Biodegradation, 32(5), 511–529. https://doi.org/10.1007/s10532-021-09948-9

Mousavi, S. M., Hashemi, S. A., Iman Moezzi, S. M., Ravan, N., Gholami, A., Lai, C. W., Chiang, W.-H., Omidifar, N., Yousefi, K., & Behbudi, G. (2021). Recent Advances in Enzymes for the Bioremediation of Pollutants. Biochemistry Research International, 2021, 5599204. https://doi.org/10.1155/2021/5599204

Muhammad, M. H., Idris, A. L., Fan, X., Guo, Y., Yu, Y., Jin, X., Qiu, J., Guan, X., & Huang, T. (2020). Beyond Risk: Bacterial Biofilms and Their Regulating Approaches. Frontiers in Microbiology, 11, 928. https://doi.org/10.3389/fmicb.2020.00928

Naidu, R., Biswas, B., Willett, I. R., Cribb, J., Kumar Singh, B., Paul Nathanail, C., Coulon, F., Semple, K. T., Jones, K. C., Barclay, A., & Aitken, R. J. (2021). Chemical pollution: A growing peril and potential catastrophic risk to humanity. Environment International, 156, 106616. https://doi.org/10.1016/j.envint.2021.106616

Narayanan, M., Kandasamy, S., Lee, J., & Barathi, S. (2023). Microbial degradation and transformation of PPCPs in aquatic environment: A review. Heliyon, 9(8), e18426. https://doi.org/10.1016/j.heliyon.2023.e18426

Nawaz, M. Z., Khalid, H. R., Mirza, M. U., Xu, L., Haider, S. Z., Al-Ghanim, K. A., Barceló, D., & Zhu, D. (2024). Elucidating the bioremediation potential of laccase and peroxidase enzymes from Bacillus ligniniphilus L1 in antibiotic degradation: A computationally guided study. Bioresource Technology, 413, 131520. https://doi.org/10.1016/j.biortech.2024.131520

Nepal, S., Giri, A., Bhandari, R., Chand, S., Nepal, S., Aryal, S., Khanal, P., Moktan, J. B., & Shastry, C. S. (2020). Poor and Unsatisfactory Disposal of Expired and Unused Pharmaceuticals: A Global Issue. Current Drug Safety, 15(3), 167–172. https://doi.org/10.2174/1574886315666200626164001

Nguyen, M.-K., Lin, C., Bui, X.-T., Rakib, Md. R. J., Nguyen, H.-L., Truong, Q.-M., Hoang, H.-G., Tran, H.-T., Malafaia, G., & Idris, A. M. (2024). Occurrence and fate of pharmaceutical pollutants in wastewater: Insights on ecotoxicity, health risk, and state–of–the-art removal. Chemosphere, 354, 141678. https://doi.org/10.1016/j.chemosphere.2024.141678

Okal, E. J., Heng, G., Magige, E. A., Khan, S., Wu, S., Ge, Z., Zhang, T., Mortimer, P. E., & Xu, J. (2023). Insights into the mechanisms involved in the fungal degradation of plastics. Ecotoxicology and Environmental Safety, 262, 115202. https://doi.org/10.1016/j.ecoenv.2023.115202

Okoye, C. O., Nyaruaba, R., Ita, R. E., Okon, S. U., Addey, C. I., Ebido, C. C., Opabunmi, A. O., Okeke, E. S., & Chukwudozie, K. I. (2022). Antibiotic resistance in the aquatic environment: Analytical techniques and interactive impact of emerging contaminants. Environmental Toxicology and Pharmacology, 96, 103995. https://doi.org/10.1016/j.etap.2022.103995

Olicón-Hernández, D. R., Ortúzar, M., Pozo, C., González-López, J., & Aranda, E. (2020). Metabolic Capability of Penicillium oxalicum to Transform High Concentrations of Anti-Inflammatory and Analgesic Drugs. Applied Sciences, 10(7), 2479. https://doi.org/10.3390/app10072479

Ortúzar, M., Esterhuizen, M., Olicón-Hernández, D. R., González-López, J., & Aranda, E. (2022). Pharmaceutical Pollution in Aquatic Environments: A Concise Review of Environmental Impacts and Bioremediation Systems. Frontiers in Microbiology, 13, 869332. https://doi.org/10.3389/fmicb.2022.869332

Pandey, A., Rai, N., Tripathi, S. M., & Mishra, S. (2023). Pharmaceutical Waste: Risks & Challenges Faced by Aquatic Ecosystem. 147–157. International Conference on Frontiers in Desalination, Energy, Environment and Material Sciences for Sustainable Development & Annual Congress of InDA. https://doi.org/10.21467/proceedings.161.17

Pápai, M., Benedek, T., Táncsics, A., Bornemann, T. L. V., Plewka, J., Probst, A. J., Hussein, D., Maróti, G., Menashe, O., & Kriszt, B. (2023). Selective enrichment, identification, and isolation of diclofenac, ibuprofen, and carbamazepine degrading bacteria from a groundwater biofilm. Environmental Science and Pollution Research, 30(15), 44518–44535. https://doi.org/10.1007/s11356-022-24975-6

Parolini, M. (2020). Adverse Effects Induced by Nonsteroidal Anti-inflammatory Drugs on Freshwater Invertebrates. In L. M. Gómez-Oliván (Ed.), Non-Steroidal Anti-Inflammatory Drugs in Water: Emerging Contaminants and Ecological Impact (pp. 147–160). Springer International Publishing. https://doi.org/10.1007/698_2020_547

Patel, M., Kumar, R., Kishor, K., Mlsna, T., Pittman, C. U. Jr., & Mohan, D. (2019). Pharmaceuticals of Emerging Concern in Aquatic Systems: Chemistry, Occurrence, Effects, and Removal Methods. Chemical Reviews, 119(6), 3510–3673. https://doi.org/10.1021/acs.chemrev.8b00299

Paut Kusturica, M., Tomas, A., & Sabo, A. (2017). Disposal of Unused Drugs: Knowledge and Behavior Among People Around the World. In P. de Voogt (Ed.), Reviews of Environmental Contamination and Toxicology Volume 240 (pp. 71–104). Springer International Publishing. https://doi.org/10.1007/398_2016_3

Pereira, A., de Morais, E. G., Silva, L., Pena, A., Freitas, A., Teixeira, M. R., Varela, J., & Barreira, L. (2023). Pharmaceuticals Removal from Wastewater with Microalgae: A Pilot Study. Applied Sciences, 13(11), Article 11. https://doi.org/10.3390/app13116414

Qiu, W., Liu, X., Yang, F., Li, R., Xiong, Y., Fu, C., Li, G., Liu, S., & Zheng, C. (2020). Single and joint toxic effects of four antibiotics on some metabolic pathways of zebrafish (Danio rerio) larvae. Science of The Total Environment, 716, 137062. https://doi.org/10.1016/j.scitotenv.2020.137062

Rafeeq, H., Afsheen, N., Rafique, S., Arshad, A., Intisar, M., Hussain, A., Bilal, M., & Iqbal, H. M. N. (2023). Genetically engineered microorganisms for environmental remediation. Chemosphere, 310, 136751. https://doi.org/10.1016/j.chemosphere.2022.136751

Rafeeq, H., Afsheen, N., Rafique, S., Arshad, A., Intisar, M., Hussain, A., Bilal, M., & Iqbal, H. M. N. (2023). Genetically engineered microorganisms for environmental remediation. Chemosphere, 310, 136751. https://doi.org/10.1016/j.chemosphere.2022.136751

Ravikumar, S., Baylon, M. G., Park, S. J., & Choi, J. (2017). Engineered microbial biosensors based on bacterial two-component systems as synthetic biotechnology platforms in bioremediation and biorefinery. Microbial Cell Factories, 16(1), 62. https://doi.org/10.1186/s12934-017-0675-z

Rebello, S., Nathan, V. K., Sindhu, R., Binod, P., Awasthi, M. K., & Pandey, A. (2021). Bioengineered microbes for soil health restoration: Present status and future. Bioengineered, 12(2), 12839–12853. https://doi.org/10.1080/21655979.2021.2004645

Rehman, M. S. U., Rashid, N., Ashfaq, M., Saif, A., Ahmad, N., & Han, J.-I. (2015). Global risk of pharmaceutical contamination from highly populated developing countries. Chemosphere, 138, 1045–1055. https://doi.org/10.1016/j.chemosphere.2013.02.036

Rosa, F. M., Mota, T. F. M., Busso, C., Arruda, P. V. de, Brito, P. E. M., Miranda, J. P. M., Trentin, A. B., Dekker, R. F. H., & Cunha, M. A. A. da. (2024). Filamentous Fungi as Bioremediation Agents of Industrial Effluents: A Systematic Review. Fermentation, 10(3), Article 3. https://doi.org/10.3390/fermentation10030143

Rottinghaus, A. G., Ferreiro, A., Fishbein, S. R. S., Dantas, G., & Moon, T. S. (2022). Genetically stable CRISPR-based kill switches for engineered microbes. Nature Communications, 13(1), 672. https://doi.org/10.1038/s41467-022-28163-5

Rutere, C., Knoop, K., Posselt, M., Ho, A., & Horn, M. A. (2020). Ibuprofen Degradation and Associated Bacterial Communities in Hyporheic Zone Sediments. Microorganisms, 8(8), Article 8. https://doi.org/10.3390/microorganisms8081245

Saaristo, M., Lagesson, A., Bertram, M. G., Fick, J., Klaminder, J., Johnstone, C. P., Wong, B. B. M., & Brodin, T. (2019). Behavioural effects of psychoactive pharmaceutical exposure on European perch (Perca fluviatilis) in a multi-stressor environment. Science of The Total Environment, 655, 1311–1320. https://doi.org/10.1016/j.scitotenv.2018.11.228

Salahinejad, A., Attaran, A., Meuthen, D., Chivers, D. P., & Niyogi, S. (2022). Proximate causes and ultimate effects of common antidepressants, fluoxetine and venlafaxine, on fish behavior. Science of The Total Environment, 807, 150846. https://doi.org/10.1016/j.scitotenv.2021.150846

Sanghvi, G., Thanki, A., Pandey, S., & Singh, N. K. (2020). 17—Engineered bacteria for bioremediation. In V. C. Pandey & V. Singh (Eds.), Bioremediation of Pollutants (pp. 359–374). Elsevier. https://doi.org/10.1016/B978-0-12-819025-8.00017-X

Santaeufemia, S., Torres, E., & Abalde, J. (2018). Biosorption of ibuprofen from aqueous solution using living and dead biomass of the microalga Phaeodactylum tricornutum. Journal of Applied Phycology, 30(1), 471–482. https://doi.org/10.1007/s10811-017-1273-5

Santaeufemia, S., Torres, E., Mera, R., & Abalde, J. (2016). Bioremediation of oxytetracycline in seawater by living and dead biomass of the microalga Phaeodactylum tricornutum. Journal of Hazardous Materials, 320, 315–325. https://doi.org/10.1016/j.jhazmat.2016.08.042

Sar, T., Marchlewicz, A., Harirchi, S., Mantzouridou, F. T., Hosoglu, M. I., Akbas, M. Y., Hellwig, C., & Taherzadeh, M. J. (2024). Resource recovery and treatment of wastewaters using filamentous fungi. Science of The Total Environment, 951, 175752. https://doi.org/10.1016/j.scitotenv.2024.175752

Scheurell, M., Franke, S., Shah, R. M., & Hühnerfuss, H. (2009). Occurrence of diclofenac and its metabolites in surface water and effluent samples from Karachi, Pakistan. Chemosphere, 77(6), 870–876. https://doi.org/10.1016/j.chemosphere.2009.07.066

Schmitt, D., Müller, A., Csögör, Z., Frimmel, F. H., & Posten, C. (2001). The adsorption kinetics of metal ions onto different microalgae and siliceous earth. Water Research, 35(3), 779–785. https://doi.org/10.1016/s0043-1354(00)00317-1

Schuster, A., Hädrich, C., & Kümmerer, K. (2008). Flows of Active Pharmaceutical Ingredients Originating from Health Care Practices on a Local, Regional, and Nationwide Level in Germany—Is Hospital Effluent Treatment an Effective Approach for Risk Reduction? Water, Air, & Soil Pollution: Focus, 8(5), 457–471. https://doi.org/10.1007/s11267-008-9183-9

Sharma, B. M., Bečanová, J., Scheringer, M., Sharma, A., Bharat, G. K., Whitehead, P. G., Klánová, J., & Nizzetto, L. (2019). Health and ecological risk assessment of emerging contaminants (pharmaceuticals, personal care products, and artificial sweeteners) in surface and groundwater (drinking water) in the Ganges River Basin, India. Science of The Total Environment, 646, 1459–1467. https://doi.org/10.1016/j.scitotenv.2018.07.235

Sharma, P., Nadda, A. K., & Kumar, S. (2023). Microbial community profiling in bio-stimulated municipal solid waste for effective removal of organic pollutants containing endocrine disrupting chemicals. Microbiological Research, 267, 127273. https://doi.org/10.1016/j.micres.2022.127273

Sharma, P., Singh, S. P., Iqbal, H. M. N., & Tong, Y. W. (2022). Omics approaches in bioremediation of environmental contaminants: An integrated approach for environmental safety and sustainability. Environmental Research, 211, 113102. https://doi.org/10.1016/j.envres.2022.113102

Sharma, S., Pathania, S., Bhagta, S., Kaushal, N., Bhardwaj, S., Bhatia, R. K., & Walia, A. (2024). Microbial remediation of polluted environment by using recombinant E. coli: A review. Biotechnology for the Environment, 1(1), 8. https://doi.org/10.1186/s44314-024-00008-z

Sharma, S., Setia, H., & Toor, A. P. (2021). Assessing the bioremediation potential of indigenously isolated Klebsiella sp. WAH1 for diclofenac sodium: Optimization, toxicity and metabolic pathway studies. World Journal of Microbiology and Biotechnology, 37(2), 33. https://doi.org/10.1007/s11274-021-02998-4

Shiroma, L. S., Soares, M. P., Cardoso, I. L., Ishikawa, M. M., Jonsson, C. M., & Queiroz, S. C. N. (2020). Evaluation of health and environmental risks for juvenile tilapia (Oreochromis niloticus) exposed to florfenicol. Heliyon, 6(12). https://doi.org/10.1016/j.heliyon.2020.e05716

Show, S., Sarkar, P., Barman, S., & Halder, G. (2023). Microbial remediation of ibuprofen contaminated water using novel isolate Microbacterium paraoxydans. Chemical Papers, 77(1), 517–531. https://doi.org/10.1007/s11696-022-02499-0

Silva, A., Coimbra, R. N., Escapa, C., Fig.ueiredo, S. A., Freitas, O. M., & Otero, M. (2020). Green Microalgae Scenedesmus Obliquus Utilization for the Adsorptive Removal of Nonsteroidal Anti-Inflammatory Drugs (NSAIDs) from Water Samples. International Journal of Environmental Research and Public Health, 17(10), 3707. https://doi.org/10.3390/ijerph17103707

Silva, A., Delerue-Matos, C., Fig.ueiredo, S. A., & Freitas, O. M. (2019). The Use of Algae and Fungi for Removal of Pharmaceuticals by Bioremediation and Biosorption Processes: A Review. Water, 11(8), Article 8. https://doi.org/10.3390/w11081555

Singh, D., Goswami, R. K., Agrawal, K., Chaturvedi, V., & Verma, P. (2022). Bio-inspired remediation of wastewater: A contemporary approach for environmental clean-up. Current Research in Green and Sustainable Chemistry, 5, 100261. https://doi.org/10.1016/j.crgsc.2022.100261

Song, C., Wu, Q., Sun, J., Zhang, R., Chen, J., Wang, X., Fang, L., Liu, Z., Shan, X., & Yin, Y. (2021). In silico evaluation of interactions between antibiotics in aquaculture and nuclear hormone receptors. Aquaculture Environment Interactions, 13, 389–397. https://doi.org/10.3354/aei00414

Subedi, B., Balakrishna, K., Joshua, D. I., & Kannan, K. (2017). Mass loading and removal of pharmaceuticals and personal care products including psychoactives, antihypertensives, and antibiotics in two sewage treatment plants in southern India. Chemosphere, 167, 429–437. https://doi.org/10.1016/j.chemosphere.2016.10.026

Sutherland, D. L., & Ralph, P. J. (2019). Microalgal bioremediation of emerging contaminants—Opportunities and challenges. Water Research, 164, 114921. https://doi.org/10.1016/j.watres.2019.114921

Sutherland, D. L., Heubeck, S., Park, J., Turnbull, M. H., & Craggs, R. J. (2018). Seasonal performance of a full-scale wastewater treatment enhanced pond system. Water Research, 136, 150–159. https://doi.org/10.1016/j.watres.2018.02.046

Tahraoui, H., Belhadj, A.-E., Amrane, A., Toumi, S., Jaouadi, B., & Zhang, J. (2024). Maximizing diclofenac bioremoval efficiency using Chlorella vulgaris strain H1 and Chlorella sorokiniana strain H2: Unveiling the impact of acetic acid on microalgae. Journal of the Taiwan Institute of Chemical Engineers, 165, 105783. https://doi.org/10.1016/j.jtice.2024.105783

Tang, Y., Fan, Z., Yang, M., Zhang, S., Li, M., Fang, Y., Li, J., & Feng, X. (2022). Low concentrations of the antidepressant venlafaxine affect courtship behaviour and alter serotonin and dopamine systems in zebrafish (Danio rerio). Aquatic Toxicology, 244, 106082. https://doi.org/10.1016/j.aquatox.2022.106082

Thangavelu, V., Tang, J., Ryan, D., & Valix, M. (2006). Effect of saline stress on fungi metabolism and biological leaching of weathered saprolite ores. Minerals Engineering, 19(12), 1266–1273. https://doi.org/10.1016/j.mineng.2006.02.007

The Global Use of Medicines 2024: Outlook to 2028. (n.d.). Retrieved January 5, 2025, from https://www.iqvia.com/insights/the-iqvia-institute/reports-and-publications/reports/the-global-use-of-medicines-2024-outlook-to-2028

Thirumalaivasan, N., Gnanasekaran, L., Kumar, S., Durvasulu, R., Sundaram, T., Rajendran, S., Nangan, S., & Kanagaraj, K. (2024). Utilization of fungal and bacterial bioremediation techniques for the treatment of toxic waste and biowaste. Frontiers in Materials, 11. https://doi.org/10.3389/fmats.2024.1416445

Thomas, K. V., & Langford, K. H. (2010). Point Sources of Human Pharmaceuticals into the Aquatic Environment. In K. Kümmerer & M. Hempel (Eds.), Green and Sustainable Pharmacy (pp. 211–223). Springer. https://doi.org/10.1007/978-3-642-05199-9_14

Thoré, E. S. J., Vanden Berghen, B., Brendonck, L., & Pinceel, T. (2023). Long-term exposure to a pharmaceutical pollutant affects geotaxic behaviour in the adult but not juvenile life stage of killifish. Science of The Total Environment, 876, 162746. https://doi.org/10.1016/j.scitotenv.2023.162746

Tiwari, B., Sellamuthu, B., Ouarda, Y., Drogui, P., Tyagi, R. D., & Buelna, G. (2017). Review on fate and mechanism of removal of pharmaceutical pollutants from wastewater using biological approach. Bioresource Technology, 224, 1–12. https://doi.org/10.1016/j.biortech.2016.11.042

Topić Popović, N., Lorencin, V., Strunjak-Perović, I., & Čož-Rakovac, R. (2023). Shell Waste Management and Utilization: Mitigating Organic Pollution and Enhancing Sustainability. Applied Sciences, 13(1), Article 1. https://doi.org/10.3390/app13010623

Torbati, S., Atashbar Kangarloei, B., & Asalpisheh, Z. (2024). Fluoranthene biotreatment using prominent freshwater microalgae: Physiological responses of microalgae and artificial neural network modeling of the bioremoval process. International Journal of Phytoremediation, 26(7), 1038–1048. https://doi.org/10.1080/15226514.2023.2288900

US EPA, O. (2015, August 18). Contaminants of Emerging Concern including Pharmaceuticals and Personal Care Products [Reports and Assessments]. https://www.epa.gov/wqc/contaminants-emerging-concern-including-pharmaceuticals-and-personal-care-products

Vaksmaa, A., Guerrero-Cruz, S., Ghosh, P., Zeghal, E., Hernando-Morales, V., & Niemann, H. (2023). Role of fungi in bioremediation of emerging pollutants. Frontiers in Marine Science, 10. https://doi.org/10.3389/fmars.2023.1070905

Van Donk, E., Peacor, S., Grosser, K., De Senerpont Domis, L. N., & Lürling, M. (2016). Pharmaceuticals May Disrupt Natural Chemical Information Flows and Species Interactions in Aquatic Systems: Ideas and Perspectives on a Hidden Global Change. In P. De Voogt (Ed.), Reviews of Environmental Contamination and Toxicology Volume 238 (Vol. 238, pp. 91–105). Springer International Publishing. https://doi.org/10.1007/398_2015_5002

Verma, V., Batta, A., Singh, H. B., Srivastava, A., Garg, S. K., Singh, V. P., & Arora, P. K. (2023). Bioengineering of fungal endophytes through the CRISPR/Cas9 system. Frontiers in Microbiology, 14, 1146650. https://doi.org/10.3389/fmicb.2023.1146650

Vidali, M. (2007). Bioremediation. An overview. https://doi.org/10.1351/pac200173071163

Vo, T. P., Danaee, S., Chaiwong, C., Pham, B. T., Poddar, N., Kim, M., Kuzhiumparambil, U., Songsomboon, C., Pernice, M., Ngo, H. H., Ralph, P. J., & Vo, P. H. N. (2024). Microalgae-bacteria consortia for organic pollutants remediation from wastewater: A critical review. Journal of Environmental Chemical Engineering, 12(6), 114213. https://doi.org/10.1016/j.jece.2024.114213

Wang, F. (2024). Reproductive endocrine disruption effect and mechanism in male zebrafish after life cycle exposure to environmental relevant triclosan. Aquatic Toxicology, 270, 106899. https://doi.org/10.1016/j.aquatox.2024.106899

Wang, Q., Zhong, C., & Xiao, H. (2020). Genetic Engineering of Filamentous Fungi for Efficient Protein Expression and Secretion. Frontiers in Bioengineering and Biotechnology, 8. https://doi.org/10.3389/fbioe.2020.00293

Wang, S., Wang, X., Poon, K., Wang, Y., Li, S., Liu, H., Lin, S., & Cai, Z. (2013). Removal and reductive dechlorination of triclosan by Chlorella pyrenoidosa. Chemosphere, 92(11), 1498–1505. https://doi.org/10.1016/j.chemosphere.2013.03.067

Wang, Z., Liang, B., Hou, Y., Li, S., Xie, L., Peng, L., Zhang, P., Wang, A., Yun, H., & Li, X. (2022). Weak electrostimulation enhanced the microbial transformation of ibuprofen and naproxen. Science of The Total Environment, 835, 155522. https://doi.org/10.1016/j.scitotenv.2022.155522

Webster, L. J., Villa-Gomez, D., Brown, R., Clarke, W., & Schenk, P. M. (2024). A synthetic biology approach for the treatment of pollutants with microalgae. Frontiers in Bioengineering and Biotechnology, 12. https://doi.org/10.3389/fbioe.2024.1379301

Wiles, S. C., Bertram, M. G., Martin, J. M., Tan, H., Lehtonen, T. K., & Wong, B. B. M. (2020). Long-Term Pharmaceutical Contamination and Temperature Stress Disrupt Fish Behavior. Environmental Science & Technology, 54(13), 8072–8082. https://doi.org/10.1021/acs.est.0c01625

Wu, L., Luo, H., Xu, J., Yu, L., Xiong, J., Liu, Y., Huang, X., & Zou, X. (2024). Vital role of CYP450 in the biodegradation of antidiabetic drugs in the aerobic activated sludge system and the mechanisms. Journal of Hazardous Materials, 469, 134056. https://doi.org/10.1016/j.jhazmat.2024.134056

Xiong, J.-Q., Kurade, M. B., Kim, J. R., Roh, H.-S., & Jeon, B.-H. (2017). Ciprofloxacin toxicity and its co-metabolic removal by a freshwater microalga Chlamydomonas mexicana. Journal of Hazardous Materials, 323(Pt A), 212–219. https://doi.org/10.1016/j.jhazmat.2016.04.073

Yamaguchi, H., & Miyazaki, M. (2024). Bioremediation of Hazardous Pollutants Using Enzyme-Immobilized Reactors. Molecules, 29(9), 2021. https://doi.org/10.3390/molecules29092021

Yang, C., Song, G., & Lim, W. (2020). A review of the toxicity in fish exposed to antibiotics. Comparative Biochemistry and Physiology Part C: Toxicology & Pharmacology, 237, 108840. https://doi.org/10.1016/j.cbpc.2020.108840

Yang, P., Condrich, A., Lu, L., Scranton, S., Hebner, C., Sheykhhasan, M., & Ali, M. A. (2024). Genetic Engineering in Bacteria, Fungi, and Oomycetes, Taking Advantage of CRISPR. DNA, 4(4), Article 4. https://doi.org/10.3390/dna4040030

Zhang, J.-N., Gong, X.-F., Ying, G., Wang, A.-G., Dong, M.-J., Liu, Y., Deng, T.-C., & Xu, M.-Y. (2024). Innovative bioremediation of dexamethasone in aquatic ecosystems using Rhodococcus sp. D32: Pathway discovery and reduction of ecotoxicity. Journal of Cleaner Production, 444, 141319. https://doi.org/10.1016/j.jclepro.2024.141319

Zheng, Z., Ali, A., Su, J., Huang, T., Wang, Y., & Zhang, S. (2021). Fungal pellets immobilized bacterial bioreactor for efficient nitrate removal at low C/N wastewater. Bioresource Technology, 332, 125113. https://doi.org/10.1016/j.biortech.2021.125113

Zhou, G.-J., Ying, G.-G., Liu, S., Zhou, L.-J., Chen, Z.-F., & Peng, F.-Q. (2014). Simultaneous removal of inorganic and organic compounds in wastewater by freshwater green microalgae. Environmental Science: Processes & Impacts, 16(8), 2018–2027. https://doi.org/10.1039/C4EM00094C

Zhu, H.-S., Liang, X., Liu, J.-C., Zhong, H.-Y., Yang, Y.-H., Guan, W.-P., Du, Z.-J., & Ye, M.-Q. (2023). Antibiotic and Heavy Metal Co-Resistant Strain Isolated from Enrichment Culture of Marine Sediments, with Potential for Environmental Bioremediation Applications. Antibiotics, 12(9), Article 9. https://doi.org/10.3390/antibiotics12091379

Zhu, L., Wu, Q., Zeng, J., Gao, S., Zhang, X., Linghu, K., & Wu, P. (2024). Co-tracing of groundwater emerging and conventional contaminants in karst area of Southwest China: Distribution, source, and source-specific health risks. Journal of Hydrology, 645, 132172. https://doi.org/10.1016/j.jhydrol.2024.132172

Zhu, S., Xiao, Y., Xia, L., Li, J., Lei, S., Liu, J., & Liu, L. (2024). Occurrence, spatiotemporal distribution, and health risk of antibiotics in the Wuhan section of the Yangtze River, China. Environmental Science and Pollution Research, 31(55), 64152–64170. https://doi.org/10.1007/s11356-024-35513-x

Ziegler, M., Banet, M., Bauer, R., Köhler, H.-R., Stepinski, S., Tisler, S., Huhn, C., Zwiener, C., & Triebskorn, R. (2021). Behavioral and Developmental Changes in Brown Trout After Exposure to the Antidepressant Venlafaxine. Frontiers in Environmental Science, 8. https://doi.org/10.3389/fenvs.2020.586584

14

Bioremediation of Microplastics in Wastewater: The Need of the Hour

Mridul Jain[1], Manishita Das Mukherji[1], Rana Mukherji[2], Manmohan Singh[1] and Ravneet Chug[1*]

[1]*Amity Institute of Biotechnology, Amity University Rajasthan, Jaipur, India*

[2]*ICFAI Tech, The ICFAI University, Jaipur, India*

**Corresponding Author Email: rchug@jpr.amity.edu*

Abstract

Plastic products are extensively used in our day-to-day life due to their convenience, affordability, lightweight nature, and durability. Their increased usage has led to a rapid increase in their production. However, microplastics are highly resistant to degradation, leading to persistence of plastic waste in the environment for decades to centuries. Improperly managed plastic waste is the primary source of microplastics resulting in their presence in almost all environmental spheres. Extensive research and studies have been conducted on the fate and occurrence of microplastics in the environment. Their persistent presence in food chains has raised various ecotoxicological and health concerns. Their omnipresence and harmful impacts underline the need for their sequestration from the environment. Numerous methods have been proposed in the past for the degradation of microplastics. Bioremediation is one of the efficient and promising approaches which involves the use of a plethora of biological agents including bacteria, fungi, and even higher eukaryotes. This review throws light on successful employment of bioremediation as a promising technique for degradation of microplastics in wastewater treatment process

Keywords: *Bioremediation, Microplastics, Degradation, Wastewater Treatment, Ecotoxicology*

1. Introduction

Since plastic was discovered, global plastic production for industrial scale usages has grown exponentially, causing a tremendous increase in plastic waste from a few tons in the 1950's to almost 400 million tons in 2018 (Syberg *et al.* 2021). Plastic is non-biodegradable and ubiquitous.

Environmental degradation of plastic operates through a mechanism of physical fragmentation, photodegradation and microbial degradation (Wright *et al.*, 2020), resulting in fragmentation of larger plastics into smaller plastic particles, also called microplastics and nanoplastics (MNPs). These debris are serious environmental threats (Hartmann *et al.*, 2019).

Microplastics (MPs), which are plastic fragments below 5 mm (Ragusa *et al.*, 2021), have received considerable attention related to their associated ecological and health risks (Ma *et al.*, 2020; Senathirajah *et al.*, 2021). Holes are Plastics in the environment are popularly classified by their size: MPs (less than 0.5 cm), mesoplastics (0.5–5 cm), macroplastics (5–50 cm), and megaplastics (>50 cm) (Lebreton *et al.*, 2018). MP's are split into primary and secondary types. Example of primary MPs are directly exuded into the environment via industrial products (e.g., pre-formed MP pellets for use in production process) and microbeads in personal care products (e.g. toothpaste, cosmetics, facial cleansers) (Gregory, 1978; Barnes *et al.*, 2009; Fendall & Sewell, 2009).

Further classification based on size can be made of MPs through the continued breakdown; these include large MPs (1–3 mm), small MPs (0.3–1 mm), and nanoplastics (<0.3 mm) (Claessens *et al.*, 2011; Andrady, 2011; Hidalgo-Ruz *et al.*, 2012; Horton *et al.*, 2017). MPs are classified according to their size into microbeads, nurdles, fibers, foam, and fragments (Hidalgo-Ruz *et al.*, 2012). The parent materials of microplastics include (but are not limited to) polyethylene (PE), polystyrene (PS), polypropylene (PP), polyester (PES), polyvinyl chloride (PVC), polyamide (PA), acrylic polymers (AC), polyether (PT), cellophane (CP), polyurethane (PU), and other undefined types (De Sá *et al.*, 2018)

Although MNPs are biochemically inert, changes in their physicochemical properties increase their bioavailability and toxicity (Ma *et al.*, 2020). They form aggregates with organic and inorganic materials, disrupting food chains and causing biomagnification (Da Costa, 2018). MNPs contaminate marine, freshwater, and terrestrial ecosystems through ingestion by organisms like bivalves, copepods, and polychaetes, leading to oxidative stress, cellular disruption, and pollutant transport (Hu & Palić, 2020; Ding *et al.*, 2021). Additionally, plastics leach toxic chemicals, such as DDT, which bioaccumulate and pose health risks to humans (Jin *et al.*, 2021).

Recent studies highlight the aerial transport of MNPs, resulting in their extensive dispersion and raising concerns about exposure risks, as noted by Evangeliou *et al.* (2020). Despite their widespread presence, the long-term chronic effects of MNPs remain poorly understood. This review aims to explore the environmental impacts of MNPs, along with mitigation strategies and remediation technologies, offering valuable insights for researchers, policymakers, and stakeholders to address this critical issue effectively.

2. Environmental Impacts of Microplastics: A Global Challenge for Ecosystem and Health

2.1. Aquatic Ecosystem

MNP pollution is prevalent in all aquatic, terrestrial, and atmospheric environments that are strongly interconnected through complex source-pathway-sink networks influencing their movement and retention (Bergmann *et al.*, 2019; Zhang *et al.*, 2020a). While it is ubiquitous, most of the studies have been on aquatic environments rather than land and atmospheric environments. Aquatic MNP debris is mainly composed microplastics, including polyethylene (PE), polypropylene (PP), polystyrene (PS), and polyvinyl chloride (PVC) that interacts with organic matter, inorganic elements, and microorganisms (Mammo *et al.*, 2020). These plastics adsorb toxic chemicals and transport them through habitats. Thus, there are significant ecological risks posed to the aquatic food web by their consumption by aquatic species (Wu *et al.*, 2019; Yuan *et al.*, 2020).

MNPs also affect aquatic life; they kill microorganisms, zooplankton, and even vertebrates such as fish and seabirds. They inhibit the growth of microbes, block digestive systems, and cause malnutrition and death in marine organisms like mussels and oysters (Nomura *et al.*, 2016; Lee *et al.*, 2013). These plastics also pose health risks to humans by bioaccumulating into the food web (Sana *et al.*, 2020).

2.2. Terrestrial Ecosystem

MNP pollution has been recognized as one of the most persistent to cause anthropogenic impacts in terrestrial ecosystems. Most plastic waste originates from land-based sources and is eventually transported into aquatic habitats. Terrestrial environments act as reservoirs for MNPs, in which they accumulate and pose extreme environmental and ecological risks. Some of these risks have included geochemical alteration in soils, environmental toxicity introduced, and exposure pathways provided for terrestrial systems' biota (Allouzi *et al.*, 2021).

The distribution and concentration of MNPs in soil are very different and not restricted to soil but also present in wastewater treatment plants. This indicates a disturbing increase in the rate of MNP contamination, especially in the regions with high plastic use and improper waste management. MNP accumulation in terrestrial habitats poses very serious environmental and ecological implications. MNPs last hundreds of years in the environment, interacting with organic matter and toxic materials in the soil. The resultant interactions alter the physicochemical properties of soils, contribute to deterioration in the quality of groundwaters, and generally hinder plant growth and productivity (Wahl *et al.*, 2021; Li *et al.*, 2019).

The effects extend to the soil organisms, such as earthworms and nematodes. These organisms play critical roles in nutrient cycling and energy flow. Exposure to MNPs leads to bioaccumulation, oxidative stress, genotoxicity, and reproductive toxicity in these organisms, reducing their survival, reproduction, and ecological functions (Wang *et al.*, 2021). Such disruptions may cascade through the ecosystem, impairing processes like litter decomposition and nutrient availability. MNPs may also act as carriers of pathogens, heavy metals, and organic pollutants imposing another ecological threat to the environment. Their hydrophobic nature and their big surface area-to-volume ratio allow them to connect with these harmful substances; thus, they can carry through the soil, into plants, and up the food chain. These, in turn, increase the threats since the microorganisms attached to MNPs may introduce invasive species and pathogens into new environments that exacerbate their effects on terrestrial ecosystems and human health (Atugoda *et al.*, 2021; Chai *et al.*, 2020).

2.3. Atmosphere

Recent studies have determined the atmosphere as an important reservoir and source of microplastic and nanoplastic (MNP) pollution. MNPs have been found in urban, suburban, and rural air with a potential to travel large distances from their origin sites and accumulate in terrestrial and aquatic environments, thus posing risks to the biosphere (Mbachu *et al.*, 2020). The least-studied environmental compartment relative to MNP occurrence and distribution is the atmosphere, although quite important (González-Pleiter *et al.*, 2021). The main threat of airborne MNPs lies in their direct inhalation, so an increase in the available knowledge about their concentrations, sources, and risks is badly needed (Chen *et al.*, 2020).

Atmospheric MNPs interact with other environmental factors through dynamic processes. It is known that MNPs become airborne through wind dispersion and their subsequent deposition in terrestrial as well as aquatic systems may lead to source-sink dynamics of plastic pollution (Chen *et al.*, 2020). Distance travelled by MNPs could be more than 100 km; from Arctic and Swiss Alps (Allen *et al.*, 2019; Bergmann *et al.*, 2019).

Transportation of MNPs is affected by the wind direction, rainfall, size of particles, and also by human activities (Zhang *et al.*, 2020b). For instance, the change in the wind direction in Hamburg resulted in increased MNP contamination (Klein & Fischer, 2019), and smaller MNPs (<25 μm) were transported more than larger MNPs. As such, given the widespread nature of the airborne MNPs and associated risks, integrating their analysis into routine assessments of air quality becomes necessary to counter environmental and health impacts.

3. *Bioremediation: A Sustainable Solution to Pollution*

Conventional methods for removing MPs involve processes such as filtration, disinfection, and coagulation/sedimentation. Research has highlighted that coagulation and filtration techniques are particularly effective in reducing MP levels. In water treatment systems, ultraviolet (UV)-based methods are commonly employed for oxidation and disinfection purposes (Pivokonsky *et al.*, 2018; Ma *et al.*, 2019; Zhou *et al.*, 2021). Filtration provides a rapid solution for MP removal; however, its efficiency largely depends on the type of filtration system used. While larger MPs can be effectively filtered out, smaller MPs and nanoplastics (NPs) are more challenging to eliminate (Li *et al.*, 2021). Eco-friendly materials like biochar have shown promise as adsorbents and flocculants for coagulation-based MP removal. Although higher doses of coagulants can improve MP removal efficiency, they may also pose toxicity risks (Wang *et al.*, 2020; Zhou *et al.*, 2021). Currently, chemical methods using coagulants, oxidation catalysts, and absorbents dominate MP removal strategies. However, these approaches are often associated with drawbacks such as high costs and inconsistent efficiency under certain conditions (Gao *et al.*, 2022).

An environmentally friendly alternative is bioremediation, which utilizes microorganisms, including fungi, bacteria, and enzymes, to degrade MPs in soil, water, and sediments (Gadd, 2004). Bacteria employed in bioremediation can vary from highly specific strains isolated from environments like soil or sludge to unspecific strains equipped with specialized enzymes for degradation (Semrany *et al.*, 2012). Genetically engineered bacteria, developed through genome modifications, are also utilized in this process (Todd, 2014). The two most widely used bioremediation strategies are bioaugmentation and bio stimulation. Bioaugmentation involves introducing specific microbial strains to accelerate the degradation of pollutants, while biostimulation focuses on enhancing the activity of native microbial communities by altering environmental conditions (Das & Kumar, 2014; Kumari *et al.*, 2016). Conevtional and sustainble methods for microplastic removal are depicetd in Fig. 1.

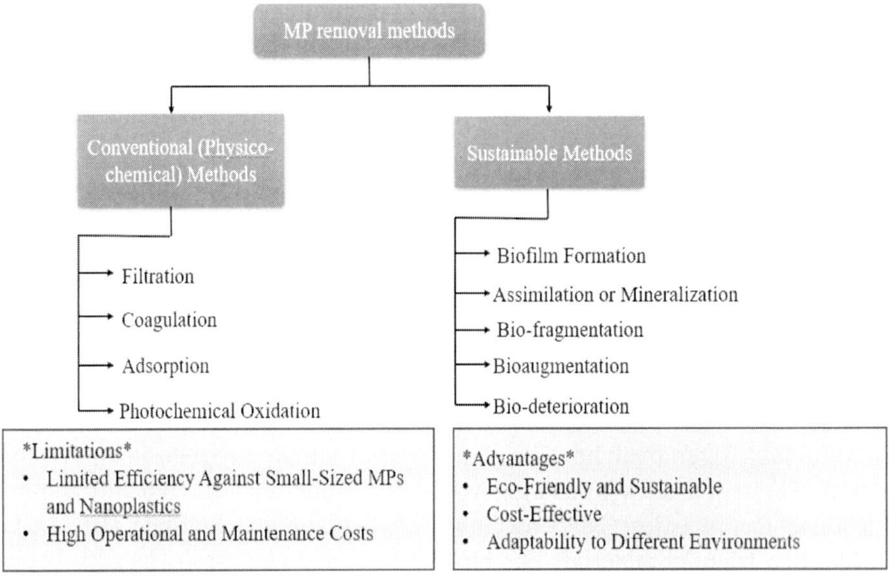

Fig. 1: Methods of Microplastic Removal, their limitations and advantages

Microbial degradation of MPs is not well understood but generally occurs in two phases: reduction of polymer molecular weight, allowing smaller molecules to penetrate cells and interact with enzymes (Albertsson *et al.*, 1995). The degradation rate depends on environmental factors and the physiological state of microorganisms (Speight, 2018). More than 400 microbial species have been reported to be capable of biodegrading plastics, but there is still much to be done to degrade MPs such as PP, PS, and PVC. The surface area of the polymer, hydrophobicity, and shape influence biodegradation (Artham and Doble, 2008; Sangeetha Devi *et al.*, 2015). Pretreatment can weaken the structure of MPs, thus hastening microbial degradation. Microbes can degrade MPs either directly by utilizing them as growth substrates or indirectly through metabolic products. Mixed cultures give better results than pure strains (Skariyachan *et al.*, 2018).

MPs are degraded mainly through biological mechanism which occurs through enzymatic activities of microorganisms, resulting in different changes in the polymer structure of the microplastic. Temperature, moisture, and pH are some of the factors that have a great impact on the biological degradation of organic pollutants (Sivan, 2011).

3.1. Bio-deterioration

In this degradation stage, enzymes attack the microplastic, breaking down the polymer into monomers. This breakdown occurs both within and on the surface of the plastic material (Lucas *et al.*, 2008). The enzymatic degradation of microplastics depends on the types of enzymes produced by microorganisms, such as urease, protease, and lipases (Sivan, 2011).

3.2. Bio-fragmentation

Biofragmentation is the enzymatic breakage or cleavage of the polymeric chains of microplastics (Kjeldsen *et al.*, 2018). Hydrolases and oxidoreductases break down plastic's polymeric chains and some others catalyse oxidation reactions that release free radicals. Oxidations create polarities within a polymer chain, causing a carbonyl and a hydroxyl group in a polymer chain to come into existence (Thakur *et al.*, 2023).

3.3. Assimilation or Mineralization

In this process, monomers are assimilated by microbes to produce biomass, CO_2, and CH_4. The monomers provided the source of carbon for microbial cells for energy production (Kjeldsen *et al.*, 2018). Even so, some plastic monomers were not easily absorbed due to the semi-permeable characteristic of microbial cell membranes. Microbial cells overcome this by using the non-assimilated monomers through biotransformation, whereby enzymes secreted by microorganisms transform these chemical compounds into end products that can later be assimilated by the same or different microorganisms (Ali *et al.*, 2021).

Microorganisms and their degradation potential for degrading different types of microplastics have been listed in table 1.

Table 1. Various microorganisms and their degradation potential degrading different types of microplastics

Type of Microplastic	Microorganism	Degradation potential (%)	Reference
Polyethylene	*Pseudomonas fluorescens*	18 %	(Thomas *et al.*, 2015)
	Aspergillus flavus	16.2 %	(Nandi and Joshi, 2013)
	Oscillatoria subbrevis	30 %	(Sarmah and Rout, 2018)
Low density polyethylene	*Aneurinibaccilus sps.*	58.2 %	(Skariyachan *et al.*, 2018)
	Aspergillus japonicus	11.11 %	(Raaman *et al.*, 2012
High density polyethylene	*Aneurinibacillus sps.*	45.7 %	(Skariyachan *et al.*, 2018)
	Penicillium oxalicum NS4	55.34 %	(Ojha *et al.*, 2017)
Polystyrene	*Penicillium sps.*	8.4 %	(Oviedo-Anchundia *et al.*, 2021)
	Bacillus subtilis	20 %	(Asmita *et al.*, 2015)
Polypropylene	*Aspergillus fumigatus*	18.08 %	(Oliya *et al.*, 2020)
	Brevibaccilus sps.	56 %	(Skariyachan *et al.*, 2018)

4. Conclusion and Future Prospects

The persistence of microplastics (MPs) in various ecosystems carry substantial impacts to environmental and human health. While conventional methods such as filtration, coagulation, and UV treatment have been used for MP removal, these methods have various limitations, such as, high costs, inefficiency in eliminating smaller plastic particles, and potential environmental side effects. As a solution to these problems, bioremediation has emerged as a promising and sustainable alternative, utilizing microorganisms like bacteria and fungi to break down microplastics into harmless compounds. This method offers an eco-friendlier approach to plastic waste management, especially when integrated in the existing wastewater treatment plants.

However, despite several advantages of bioremediation, various challenges uphold. The pathway through which microorganisms degrade microplastics are not yet fully studied, and the efficiency of biodegradation is influenced by different external factors, including, temperature, pH, and the chemical composition of the plastic. To overcome these challenges, upcoming research should focus on identifying and optimizing microbial strains and consortia that are highly efficient in breakdown of a broad spectrum of microplastic. Various approaches can also be used to increase the specificity and activity of plastic-degrading enzymes. Additionally, technologies such as genomics, proteomics, and metabolomics can provide detailed insights on the metabolic pathways and regulatory mechanisms involved in plastic degradation.

In conclusion, the development and optimization of integrated bioremediation microorganisms that can be scaled up for real-world applications is crucial. While bioremediation is still in its developmental phase, it holds immense promise as a sustainable and effective solution for mitigating microplastic pollution in wastewater and beyond.

References

Albertsson, A. C., Barenstedt, C., Karlsson, S., & Lindberg, T. (1995). Degradation product pattern and morphology changes as means to differentiate abiotically and biotically aged degradable polyethylene. Polymer, 36(16), 3075-3083.

Ali, S. S., Elsamahy, T., Koutra, E., Kornaros, M., El-Sheekh, M., Abdelkarim, E. A., ... & Sun, J. (2021). Degradation of conventional plastic wastes in the environment: A review on current status of knowledge and future perspectives of disposal. Science of the Total Environment, 771, 144719.

Allen, S., Allen, D., Phoenix, V. R., Le Roux, G., Durántez Jiménez, P., Simonneau, A., ... & Galop, D. (2019). Atmospheric transport and deposition of microplastics in a remote mountain catchment. Nature Geoscience, 12(5), 339-344.

Allouzi, M. M. A., Tang, D. Y. Y., Chew, K. W., Rinklebe, J., Bolan, N., Allouzi, S. M. A., & Show, P. L. (2021). Micro (nano) plastic pollution: The ecological influence on soil-plant system and human health. Science of the Total Environment, 788, 147815.

Andrady, A. L. (2011). Microplastics in the marine environment. Marine Pollution Bulletin, 62(8), 1596-1605.
Artham, T., & Doble, M. (2008). Biodegradation of aliphatic and aromatic polycarbonates. Macromolecular bioscience, 8(1), 14-24.
Asmita, K., Shubhamsingh, T., & Tejashree, S. (2015). Isolation of plastic degrading micro-organisms from soil samples collected at various locations in Mumbai, India. International Research Journal of Environment Sciences, 4(3), 77-85.
Atugoda, T., Vithanage, M., Wijesekara, H., Bolan, N., Sarmah, A. K., Bank, M. S., ... & Ok, Y. S. (2021). Interactions between microplastics, pharmaceuticals and personal care products: Implications for vector transport. Environment International, 149, 106367.
Barnes, D. K., Galgani, F., Thompson, R. C., & Barlaz, M. (2009). Accumulation and fragmentation of plastic debris in global environments. Philosophical transactions of the royal society B: biological sciences, 364(1526), 1985-1998.
Bergmann, M., Mützel, S., Primpke, S., Tekman, M. B., Trachsel, J., & Gerdts, G. (2019). White and wonderful? Microplastics prevail in snow from the Alps to the Arctic. Science advances, 5(8), eaax1157.
Chai, B., Wei, Q., She, Y., Lu, G., Dang, Z., & Yin, H. (2020). Soil microplastic pollution in an e-waste dismantling zone of China. Waste Management, 118, 291-301.
Chen, G., Feng, Q., & Wang, J. (2020). Mini-review of microplastics in the atmosphere and their risks to humans. Science of the Total Environment, 703, 135504.
Claessens, M., De Meester, S., Van Landuyt, L., De Clerck, K., & Janssen, C. R. (2011). Occurrence and distribution of microplastics in marine sediments along the Belgian coast. Marine pollution bulletin, 62(10), 2199-2204.
da Costa, J. P. (2018). Micro-and nanoplastics in the environment: research and policymaking. Current Opinion in Environmental Science & Health, 1, 12-16.
Das, M. P., & Kumar, S. (2014). Microbial deterioration of low density polyethylene by Aspergillus and Fusarium sp. Int J Chem Tech Res, 6(1), 299-305.
De Sá, L. C., Oliveira, M., Ribeiro, F., Rocha, T. L., & Futter, M. N. (2018). Studies of the effects of microplastics on aquatic organisms: what do we know and where should we focus our efforts in the future?. Science of the total environment, 645, 1029-1039.
Ding, Y., Zhang, R., Li, B., Du, Y., Li, J., Tong, X., ... & Zhang, Y. (2021). Tissue distribution of polystyrene nanoplastics in mice and their entry, transport, and cytotoxicity to GES-1 cells. Environmental Pollution, 280, 116974.
Evangeliou, N., Grythe, H., Klimont, Z., Heyes, C., Eckhardt, S., Lopez-Aparicio, S., & Stohl, A. (2020). Atmospheric transport is a major pathway of microplastics to remote regions. Nature communications, 11(1), 3381.
Fendall, L. S., & Sewell, M. A. (2009). Contributing to marine pollution by washing your face: microplastics in facial cleansers. Marine pollution bulletin, 58(8), 1225-1228.
Gadd, G. M. (2004). Microbial influence on metal mobility and application for bioremediation. Geoderma, 122(2-4), 109-119.
Gao, W., Zhang, Y., Mo, A., Jiang, J., Liang, Y., Cao, X., & He, D. (2022). Removal of microplastics in water: Technology progress and green strategies. Green Analytical Chemistry, 3, 100042.
González-Pleiter, M., Edo, C., Aguilera, Á., Viúdez-Moreiras, D., Pulido-Reyes, G., González-Toril, E., ... & Rosal, R. (2021). Occurrence and transport of microplastics sampled within and above the planetary boundary layer. Science of the Total Environment, 761, 143213.
Gregory, M. R. (1978). Accumulation and distribution of virgin plastic granules on New Zealand beaches. New Zealand Journal of Marine and Freshwater Research, 12(4), 399-414.

Hartmann, N. B., Huffer, T., Thompson, R. C., Hassellov, M., Verschoor, A., Daugaard, A. E., ... & Wagner, M. (2019). Are we speaking the same language? Recommendations for a definition and categorization framework for plastic debris.

Hidalgo-Ruz, V., Gutow, L., Thompson, R. C., & Thiel, M. (2012). Microplastics in the marine environment: a review of the methods used for identification and quantification. Environmental science & technology, 46(6), 3060-3075.

Horton, A. A., Walton, A., Spurgeon, D. J., Lahive, E., & Svendsen, C. (2017). Microplastics in freshwater and terrestrial environments: Evaluating the current understanding to identify the knowledge gaps and future research priorities. Science of the total environment, 586, 127-141.

Hu, M., & Palić, D. (2020). Micro-and nano-plastics activation of oxidative and inflammatory adverse outcome pathways. Redox biology, 37, 101620.

Jin, M., Wang, X., Ren, T., Wang, J., & Shan, J. (2021). Microplastics contamination in food and beverages: Direct exposure to humans. Journal of Food Science, 86(7), 2816-2837.

Kjeldsen, A., Price, M., Lilley, C., Guzniczak, E., & Archer, I. (2018). A review of standards for biodegradable plastics. Ind. Biotechnol. Innov. Cent, 33(1).

Klein, M., & Fischer, E. K. (2019). Microplastic abundance in atmospheric deposition within the Metropolitan area of Hamburg, Germany. Science of the Total Environment, 685, 96-103.

Kumari, S., Ghosh, P., Thakur, I. S., & Chaudhuri, S. R. (2016). Bioaugmentation for degradation of chlorophenols using industrial waste as inoculum source. Environmental Science and Pollution Research, 23(12), 11620-11631.

Lebreton, L., Slat, B., Ferrari, F., Sainte-Rose, B., Aitken, J., Marthouse, R., ... & Reisser, J. (2018). Evidence that the Great Pacific Garbage Patch is rapidly accumulating plastic. Scientific reports, 8(1), 1-15.

Lee, K. W., Shim, W. J., Kwon, O. Y., & Kang, J. H. (2013). Size-dependent effects of micro polystyrene particles in the marine copepod Tigriopus japonicus. Environmental science & technology, 47(19), 11278-11283.

Li, J., Wang, B., Chen, Z., Ma, B., & Chen, J. P. (2021). Ultrafiltration membrane fouling by microplastics with raw water: Behaviors and alleviation methods. Chemical Engineering Journal, 410, 128174.

Li, Q., Wu, J., Zhao, X., Gu, X., & Ji, R. (2019). Separation and identification of microplastics from soil and sewage sludge. Environmental Pollution, 254, 113076.

Lucas, N., Bienaime, C., Belloy, C., Queneudec, M., Silvestre, F., & Nava-Saucedo, J. E. (2008). Polymer biodegradation: Mechanisms and estimation techniques–A review. Chemosphere, 73(4), 429-442.

Ma, B., Xue, W., Hu, C., Liu, H., Qu, J., & Li, L. (2019). Characteristics of microplastic removal via coagulation and ultrafiltration during drinking water treatment. Chemical Engineering Journal, 359, 159-167.

Ma, H., Pu, S., Liu, S., Bai, Y., Mandal, S., & Xing, B. (2020). Microplastics in aquatic environments: toxicity to trigger ecological consequences. Environmental Pollution, 261, 114089.

Mbachu, O., Jenkins, G., Pratt, C., & Kaparaju, P. (2020). A new contaminant superhighway? A review of sources, measurement techniques and fate of atmospheric microplastics. Water, Air, & Soil Pollution, 231, 1-27.

Nandi, R. G., & Joshi, M. (2013). Biodegradation of pretreated polythene by different species of Aspergillus isolated from garbage soil. Biosci. Biotech. Res. Comm, 6, 199-201.

Nomura, T., Tani, S., Yamamoto, M., Nakagawa, T., Toyoda, S., Fujisawa, E., ... & Konishi, Y. (2016). Cytotoxicity and colloidal behavior of polystyrene latex nanoparticles toward filamentous fungi in isotonic solutions. Chemosphere, 149, 84-90.

Nupur Ojha, N. O., Neha Pradhan, N. P., Surjit Singh, S. S., Anil Barla, A. B., Anamika Shrivastava, A. S., Pradip Khatua, P. K., ... & Sutapa Bose, S. B. (2017). Evaluation of HDPE and LDPE degradation by fungus, implemented by statistical optimization.

Oliya, Pramila, Singh, Sanjeev, Goel, Navodit, Singh, U.P., & Srivastava, A.K. (2020). Polypropylene degradation potential of microbes isolated from solid waste dumping site. Pollut Res, 39(2), 268-277.

Oviedo-Anchundia, R., del Castillo, D. S., Naranjo-MorÃ, J., Francois, N., AlarcÃ³n, A., Villafuerte, J. S., & Barcos-Arias, M. (2021). Analysis of the degradation of polyethylene, polystyrene and polyurethane mediated by three filamentous fungi isolated from the Antarctica. African Journal of Biotechnology, 20(2), 66-76.

Pivokonsky, M., Cermakova, L., Novotna, K., Peer, P., Cajthaml, T., & Janda, V. (2018). Occurrence of microplastics in raw and treated drinking water. Science of the total environment, 643, 1644-1651.

Raaman, N., Rajitha, N., Jayshree, A., & Jegadeesh, R. (2012). Biodegradation of plastic by Aspergillus spp. isolated from polythene polluted sites around Chennai. J. Acad. Indus. Res, 1(6), 313-316.

Sana, S. S., Dogiparthi, L. K., Gangadhar, L., Chakravorty, A., & Abhishek, N. (2020). Effects of microplastics and nanoplastics on marine environment and human health. Environmental Science and Pollution Research, 27, 44743-44756.

Sangeetha Devi , Kannan, V. R., Natarajan, K., Nivas, D., Kannan, K., Chandru, S., & Antony, A. R. (2016). The role of microbes in plastic degradation. Environ waste manage, 341, 341-370.

Sarmah, P., & Rout, J. (2018). Efficient biodegradation of low-density polyethylene by cyanobacteria isolated from submerged polyethylene surface in domestic sewage water. Environmental Science and Pollution Research, 25, 33508-33520.

Semrany, S., Favier, L., Djelal, H., Taha, S., & Amrane, A. (2012). Bioaugmentation: possible solution in the treatment of bio-refractory organic compounds (Bio-ROCs). Biochemical Engineering Journal, 69, 75-86.

Senathirajah, K., Attwood, S., Bhagwat, G., Carbery, M., Wilson, S., & Palanisami, T. (2021). Estimation of the mass of microplastics ingested–A pivotal first step towards human health risk assessment. Journal of Hazardous Materials, 404, 124004.

Sivan, A. (2011). New perspectives in plastic biodegradation. Current opinion in biotechnology, 22(3), 422-426.

Skariyachan, S., Patil, A. A., Shankar, A., Manjunath, M., Bachappanavar, N., & Kiran, S. (2018). Enhanced polymer degradation of polyethylene and polypropylene by novel thermophilic consortia of Brevibacillus sps. and Aneurinibacillus sp. screened from waste management landfills and sewage treatment plants. Polymer Degradation and Stability, 149, 52-68.

Speight, J. G. (2018). Reaction mechanisms in environmental engineering: analysis and prediction. Butterworth-Heinemann.

Syberg, K., Nielsen, M. B., Clausen, L. P. W., Van Calster, G., Van Wezel, A., Rochman, C., ... & Hansen, S. F. (2021). Regulation of plastic from a circular economy perspective. Current opinion in green and sustainable chemistry, 29, 100462.

Thakur, B., Singh, J., Singh, J., Angmo, D., & Vig, A. P. (2023). Biodegradation of different types of microplastics: Molecular mechanism and degradation efficiency. Science of The Total Environment, 877, 162912.

Thomas, B. T., Olanrewaju-Kehinde, D. S. K., Popoola, O. D., & James, E. S. (2015). Degradation of plastic and polythene materials by some selected microorganisms isolated from soil. World Appl Sci J, 33(12), 1888-1891.

Todd, E. C. D. (2014). Safety of Food and beverages: safety of genetically modified foods.

Wahl, A., Le Juge, C., Davranche, M., El Hadri, H., Grassl, B., Reynaud, S., & Gigault, J. (2021). Nanoplastic occurrence in a soil amended with plastic debris. Chemosphere, 262, 127784.

Wang, Q., Adams, C. A., Wang, F., Sun, Y., & Zhang, S. (2022). Interactions between microplastics and soil fauna: a critical review. Critical Reviews in Environmental Science and Technology, 52(18), 3211-3243.

Wang, Z., Lin, T., & Chen, W. (2020). Occurrence and removal of microplastics in an advanced drinking water treatment plant (ADWTP). Science of the Total Environment, 700, 134520.

Wright, S. L., Ulke, J., Font, A., Chan, K. L. A., & Kelly, F. J. (2020). Atmospheric microplastic deposition in an urban environment and an evaluation of transport. Environment international, 136, 105411.

Wu, J., Jiang, R., Lin, W., & Ouyang, G. (2019). Effect of salinity and humic acid on the aggregation and toxicity of polystyrene nanoplastics with different functional groups and charges. Environmental Pollution, 245, 836-843.

Yuan, J., Ma, J., Sun, Y., Zhou, T., Zhao, Y., & Yu, F. (2020). Microbial degradation and other environmental aspects of microplastics/plastics. Science of the Total Environment, 715, 136968.

Zhang, F., Man, Y. B., Mo, W. Y., Man, K. Y., & Wong, M. H. (2020 a). Direct and indirect effects of microplastics on bivalves, with a focus on edible species: A mini-review. Critical reviews in environmental science and technology, 50(20), 2109-2143.

Zhang, Y., Kang, S., Allen, S., Allen, D., Gao, T., & Sillanpää, M. (2020b). Atmospheric microplastics: A review on current status and perspectives. Earth-Science Reviews, 203, 103118.

Zhou, G., Wang, Q., Li, J., Li, Q., Xu, H., Ye, Q., ... & Zhang, J. (2021). Removal of polystyrene and polyethylene microplastics using PAC and FeCl3 coagulation: Performance and mechanism. Science of the Total Environment, 752, 141837.

15

From Plastic Disaster to Sustainable Solutions: Innovative Microbial Techniques for Bioremediation and PHA Bioproduction

Kevanshi M. Dave[1], Shivani V. Raval[1,3] Shreyas A. Bhatt[2] and Vikram H. Raval[1*]

[1]*Department of Microbiology and Biotechnology, University School of Sciences Gujarat University, Ahmedabad*

[2]*Department of Life sciences, Hemchandracharya North Gujarat University, Patan*

[3]*Department of Biotechnology, M G Science Institute (Autonomous), Navrangpura - 380009, Ahmedabad*

**Corresponding Author Email: vhraval@gujaratuniversity.ac.in; vikramhraval@gmail.com*

Abstract

Plastic is a wonderful gift of science to society as having unparalleled multiple properties which make it the universal choice of material in almost every step of life. However, their non-degradability creates plastic disasters where plastics and microplastics have directly and deadly impacting humans and the environment Plastic recycling is even more dangerous as it utilizes more chemicals and toxins also liberating microplastics. Other management methods such as incineration, landfills, pyrolysis, and physicochemical ways of degradation are not successful also are not eco-friendly. These issues can be solved by a biological approach to plastic degradation, called "Bioremediation". Various types of microorganisms and their consortia may able to convert various types of plastics into gages, biomass, water, and inorganic matter thus considered as ecofriendly and effective tools to degrade plastics. This review gives an upgraded and comprehensive outlook of plastic bioremediation by different types of microbes, metabolic pathways, and enzymes involved, as well as advanced techniques such as meta omics and engineering to accelerate degradation rate and efficiency. Advantages, challenges, and future guidelines are also described. However, without plastic life cannot be expected as it is a basic necessity of society.

So along with research oriented on plastic degradation, the development of alternative and safer substitutes of plastic should be focused. As a green plastic, PHA(Polyhydroxyalkanoates) has biodegradable, biocompatible, and sustainable advantages along with similar properties to synthetic plastics making them a perfect replacement for it. Here, the insides of PHA bioproduction and biodegradation are illustrated. Overall, this review highlights plastic bioremediation and plastic bioproduction aspects.

Keywords: *plastic, microplastic, bioremediation, enzymes, microorganisms, bioplastic, PHA*

1. Introduction

The term “plastic” is derived from the Greek word “Plasticos" which means to be capable of molding into various sizes and shapes. Daily used synthetic plastics are categorized based on their physical and chemical properties, such as thermoplastic and thermosetting. (i) **Thermoplastics**: As their names suggest, thermoplastic polymers melt and turn into a plastic or rubbery state when exposed to heat. They have various numbers of polymeric chains and are the most widely used plastics in day-to-day life. Examples of thermoplastics are Polystyrene(PS), Polyvinylchloride (PVC), Polypropylene (PP), Polyethylene Terephthalate (PET), Polyethylene (PE) etc. (Massy, 2017), (Sharma *et al.*, 2017). (ii) **Thermosetting polymers**: As the name suggests, they set or harden permanently when heated (Massy, 2017). Due to their strong crosslinked structure, they are non-recyclable polymers. Examples include Polyurethanes(PUR), Phenol formaldehyde, Melamine formaldehyde, and Polyester Resin (PR) (Sharma *et al.*, 2017). Different polymers and their applications are illustrated in Fig. 1.

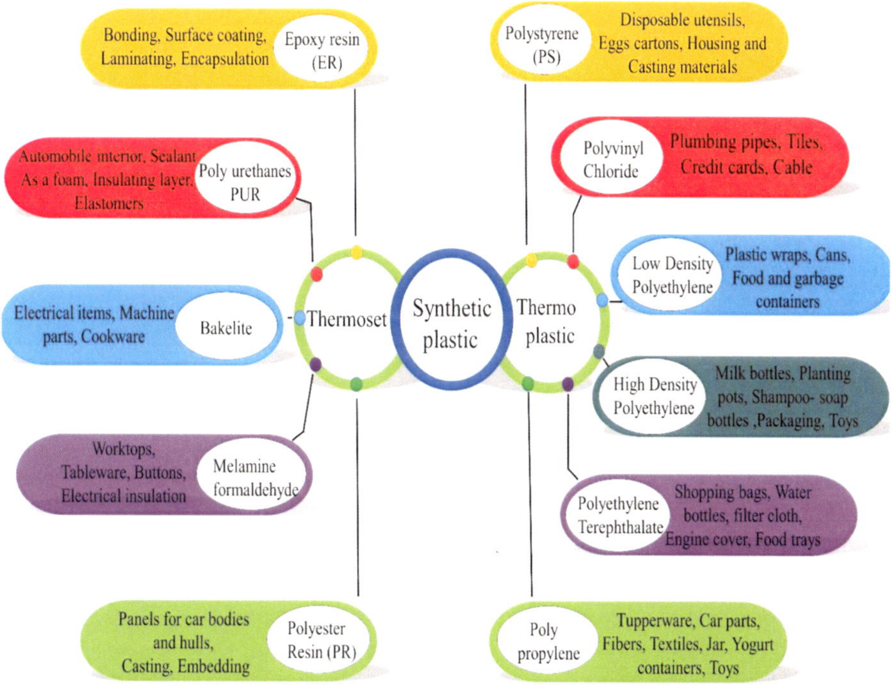

Fig. 1: Classification of Synthetic plastic and their common applications

The main reason for Plastic dominancy in everyday life is its properties such as durability, lightweight with high strength, rigidity or flexibility, stability, molding capacity, transparency or opacity, water resistance, resistance to heat and chemicals, and most importantly inexpensiveness (Israni and Shivakumar, 2019).

However, its non-degradability, unsustainable manufacture, use, and disposal have surpassed its advantages and have turned into a global plastic crisis. Worldwide plastic output was 368 million metric tons (Mt) in 2019, which is assumed to double in the next 20 years(Walker and Fequet, 2023). Around 9% of plastic garbage has ever been recycled worldwide, 12% has been burned, and 79% has accumulated in natural ecosystems(Geyer *et al.*, 2017). A growing volume of plastic waste enters into various streams and undergoes degradation through different mechanisms such as abrasion, exposure to ultraviolet light, hydrolysis, oxidation, and microbial breakdown. These processes eventually break the primary polymer chains, forming smaller plastic fragments, known as microplastics (MPs), which are particles measuring less than 5 mm in size(Lokesh *et al.*, 2023). MP particles are found in environments coming from a wide range of different sources and they vary in their shape, size, density, and chemical structure. Microplastics are classified as primary microplastics (PMPs) and secondary microplastics (SMPs). PMPs are made up of polypropylene, polystyrene, polyethylene, and polyurethane, and they are used in commercial applications such as beauty and personal care products as toothpaste, hand, and facial cleaners, industrial and biomedical products, insect repellent, etc. While SMP is produced from the fragmentation of plastic wastes through biological, mechanical, and physical forces (Malankowska *et al.*, 2021).

1.1. Fate and Effects of Plastic and Microplastics

1.1.1. Air pollution

Plastic in the air is present in the form of micro and nanoplastics(NPs) and responsible for respiratory and oral exposures. Airborne MPs that have been found so far worldwide range in size from 2 μm to 1 mm, and the most common polymers are rayon, PA, PE, PP, polystyrene (PS), PET, PVC, polyurethane (PUR), and polyacrylonitrile (PAN)(Enyoh *et al.*, 2019). Plastics used for packaging purposes such as e PS, PE, PET, and biothene degrade faster in the air compared to water and produce MPs(Biber *et al.*, 2019). When plastic garbage is weathered and exposed to environmental conditions, the risk factors linked to MPs rise. Oceanic waves (sea spray), landfills, waste burning, building materials, plastic pieces from furniture, traffic molecules (tire/brake wear), industrial emissions, mulching materials in soil, and atmospheric fallout from wastewater sludge through land applications or other matrices can all expose people to aerial MPs or NPs (Sridharan *et al.*, 2021). Studies revealed that characteristic carbonyl bonds (C=O) formed into aged airborne plastics which are highly active and elevate their adsorption behavior thus co-existing pollutants such as polycyclic aromatic hydrocarbons (PAHs), polychlorinated biphenyls (PCBs), antibiotics, and also leach out additives like bisphenol A (BPA), phthalates, and other endocrine disruptors spread in air (Mao *et al.*, 2020),(Akhbarizadeh *et al.*, 2021). Data shows that the airborne NPs transferred across the pulmonary cell barrier to secondary organs such as the placenta (Fournier *et al.*, 2020). Airborne PS MPs have toxicological effects including change in morphology and retardation of human alveolar cell proliferation (Goodman *et al.*, 2021). Atmospheric MPs cause serious health hazards such as immune disorders, inflammation, and neurodegeneration(Prata *et al.*, 2021). High microplastic particulates (<10 μm) in the air cause high death numbers due to air pollution (EPA AQI Guidelines, 2014).

1.1.2. Greenhouse gas and Climate change

An increased plastic demand leads to rapid growth of plastic industries which contributes to global green house gas emissions and climate change. According to data by calculating the weighted average carbon intensity of oil well energy production in 90 countries and 8966 on-stream oil fields worldwide, greenhouse gas emissions from well-to-refinery in 2015 were estimated to be 68 million tons of CO_2 equivalents (CO_2e), based on the distribution of roughly 4% crude oil as the raw material for plastics (Masnadi *et al.*, 2018).Various routs associated with greenhouse gas emission include the manufacture and transportation of raw plastic material, methods of plastic waste management, and the release of plastics into the environment. Research revealed that 79% of the carbon in one ton of plastic packaging trash is typically flammable, releasing 790 kg of carbon dioxide (or around 2.9 tons of CO_2) into the environment (Shen *et al.*, 2020). 850 million tons of greenhouse gasses are already released annually from the production and burning of plastics, and by 2050, that amount is expected to rise to 2.8 gigatons of carbon dioxide, which equals 615 coal-fired power plants. Methane and ethane are released by the aging of landfilled and ocean-dumped plastic through exposure to sunlight and heat(Haque and Fan, 2023). The study revealed that ocean carbon cycle disturbance due to plastic. In marine ecosystems presence of MPs adversely affected the microalgal growth and their photosynthetic activity resulted in their carbon fixation rate alteration (Shen *et al.*, 2020). Microplastics in the air can change the Earth's energy balance by absorbing and scattering radiation (Zhao *et al.*, 2023) Certain microplastics, such as the colorless fibers and pieces released by synthetic fabrics, absorb infrared light and scatter visible and ultraviolet light. Microplastics in the air have a slight cooling impact when they are close to the Earth's surface and a slight warming effect when they are farther up in the atmosphere (Revell *et al.*, 2021). The massive production of plastics in the plastics and petrochemical industries could destabilize the climate, making it challenging to keep the rise in global temperatures to 1.5 or even 2°C (Shen *et al.*, 2020). MPs influence the terrestrial ecosystem with alteration of soil carbon storage and soil carbon cycle (Rillig *et al.*, 2021). MPs on soil also alter the nutrient cycle such as the nitrogen and phosphorus cycle. Oxidative stress and nitrogen imbalance affect soil productivity, which results in poor food crop growth (Wang *et al.*, 2021).

1.1.3. Biodiversity and Food Chain

Plastic garbage affects terrestrial biodiversity through ingestion and entanglements of land animals that mistake it for food, nesting materials, or shelter(Blettler and Mitchell, 2021). Microplastics present in soil accumulate, ingested, and incorporated into tissues of soil organisms resulting in limiting their functions and biodiversity (Browne *et al.*, 2008). Through ontogenic transference, microplastics can enter the terrestrial biosphere. The life cycle of terrestrial insects can transfer microplastics from their larval stages, which develop in freshwater sources, to their terrestrial adult stages in Mayflies, mosquitoes, midges, and dragonflies(Al-Jaibachi *et al.*, 2018). Studies have shown that MPs present in soil decrease the population of nematodes and microarthropods(Lin *et al.*, 2020). There have been documented instances where terrestrial mammals, such as cattle, camels, zebras, hyenas, and elephants, have accidentally consumed plastic garbage, leading to their demise. Wildlife that becomes entangled in plastic debris may suffer injuries, become immobile, and become more vulnerable to predators and environmental dangers (Anunobi, 2022). Plastic in rivers, sediments, and inland water sources affects freshwater biodiversity. There are reports of one or more individuals of numerous fish species ingesting plastic in rivers and other inland habitats worldwide(Urbanski *et al.*, 2020). This is because fish may serve as vectors for the mobilization and transmission of plastic items across food webs, as they are a significant source of food for various aquatic and terrestrial organisms(Azevedo-Santos *et al.*, 2021).Freshwater algae and plants uptake MPs which causes risk of transfer towards secondary consumers and further food chain(Chae *et al.*, 2018). Crustaceans in freshwater are vulnerable to interactions with many kinds of plastic. Plastic materials can harm crustaceans in ways that are either fatal or sublethal. In the laboratory detection, the effects of polystyrene on cladocerans showed altered reproduction to decreased survival (Ziajahromi *et al.*, 2018). Plastic ingestion causes morphological changes such as altered head structure in *Chironomus riparius* by exposure to six different polymers (Stanković *et al.*, 2020). More than 150 freshwater Fish species reported that ingested plastics(Azevedo-Santos *et al.*, 2021).MPs and NPs are present in the base of food chain such as zooplanktons and transferred to higher levels of the food chain. Consumption of seafood such as Crustaceans, sea fish, oysters, and mussels which contain MPs and NPs in their tissues causes a higher risk of plastic transfer in humans (Wang *et al.*, 2020).

1.1.4. Human health risks

The extensive presence of microplastics in the environment and their documented harmful effects make them a possible health risk to humans. The effectiveness of MPs' absorption via the gastrointestinal, alveolar, and cutaneous epithelium is influenced by their size. Through endocytosis, MPs of 0.1–10 μm can pass through the placenta, blood-brain barriers, and gastrointestinal epithelium to enter the bloodstream. MPs about 5 μm can interact with blood circulation, lymph nodes, macrophages, and spleen. MPs sized more than 10 μm can accumulate in the liver, kidney, and brain through the gastrointestinal tract(Prietl *et al.*, 2014). Mostly plastic in the form of MPs and NPs espouses humans through skin and hair contact, oral intact, and respiration(Prata *et al.*, 2020). Data have shown the presence of MPs with sizes from 20.34 to

307.29 μm in the human placenta (Zhu *et al.*, 2023) and 2–12 μm sized MPs in human breastmilk (Ragusa *et al.*, 2022) which may affect the developing fetus. Obesity, diabetes, cardiovascular disease, and other chronic diseases are more likely to occur in individuals who have experienced negative effects during their early development(Barker, 1995). The cytotoxic activity of MPs causes oxidative stress and membrane damage resulting in cell death (Rubio *et al.*, 2020). Through one-time used food containers and other sources such as seafood or packed drinks MPs are ingested and cause damage to intestinal walls, which can lead to inflammation, oxidative stress, and alteration of gut microbiota (Osman *et al.*, 2023). Chronic-obstructive pulmonary disease (COPD) asthma and other respiratory illness caused by inhalation of MPs (Lu *et al.*, 2022). According to several studies, environmental microplastics can interfere with melatonin production and circadian rhythms, which may cause sleep disturbances and increase the risk of sleep disorders (Campanale *et al.*, 2020). Microplastics may interfere with hormone balance and metabolism, which could lead to weight gain obesity, and reproductive disorders (Landrigan *et al.*, 2023).

1.2. Plastic waste management

Traditional waste management methods such as recycling, incineration, and landfilling have been utilized for handling plastic-based waste.

1.2.1. Landfill

After being used in various dustbins, plastics are disposed of in landfills. People frequently construct landfills with mound pits or natural pits to fill them to easily and effectively dispose of vast amounts of waste plastic(Ware *et al.*, 2017). Several safety precautions should be taken throughout this manual removal process to prevent secondary side effects including soil degradation and groundwater pollutants that may arise from inadequate processing(Evode *et al.*, 2021). It requires a lot of communal work, such as excavating a lengthy trench or burying waste far below and allowing it to break down. This procedure is carried out extremely slowly it may take over a year(Liang *et al.*, 2021). The degradation of various plastic wastes can take a long time because of their unique biochemical characteristics and ambient or climatic circumstances, such as sunlight, wind, etc. (Zhou *et al.*, 2014). In landfill processing, plastic bags and other long polymer wastes can take ten to one hundred years to break down (Thiounn and Smith, 2020). The fact that it contributes to climate change and ignites methane as a flammable gas is among its drawbacks. Also, it damages wildlife and contaminates water ecosystems and soil (Zhou *et al.*, 2014). Thus landfills are not a viable option for managing plastic trash in the long run.

1.2.2. Incineration

Waste incineration is complete combustion, which is the chemical term for burning trash in oxygen. Combustion produces a vast amount of energy. Thus, incineration has been seen as an easy way to address the widespread contamination of land-based plastics by effectively managing plastic pollution, it can also produce usable heat and energy(Shen *et al.*, 2020). However, burning plastic garbage can release greenhouse gasses, most often CO_2. Other harmful chemicals such as HCL, HCN, and heavy metals are generated from incineration (Pan *et al.*, 2020). As a result, burning waste plastics is not the best option because it necessitates a comprehensive system for treating and purifying pollutants.

1.2.3. Recycling

Recycling is a management technique that gathers waste and transforms it into raw materials that may be used again to create other useful products. Thermosetting waste polymers can be recycled directly as extra materials following physical recycling known as mechanical pulverization because of their superior mechanical, electrical, and chemical stability (Chen *et al.*, 2019). Mechanical crushing and hot extrusion combined with an adhesive can yield the value of reuse for the majority of thermoplastic waste plastics, such as recycled rubber and plastic bottle refurbishing (Briassoulis *et al.*, 2013). The physical management approach is inexpensive, easy to use, and produces high production efficiency with minimal secondary pollution but, the remanufactured items have low economic benefits & poor performance (Pan *et al.*, 2020).

1.2.4. Pyrolysis

The process of converting gasses and fatty oils to recover crude petrochemicals and produce hydrocarbons is known as pyrolysis. It is even utilized to turn plastic trash into renewable energy and recover crude petrochemicals(Anuar Sharuddin *et al.*, 2016). Based on the quantity of heat energy required to break down plastic linkages, the pyrolysis process is divided into three primary categories: high, medium, and low. The temperature range utilized to break down the plastic structure determines the use of medium and high temperatures(Evode *et al.*, 2021). This scientific approach to trash management is effective, despite the potential hefty initial investment.

To overcome plastic pollution and manage plastic waste, proper biological techniques can become a boon. Such as Bioremediation techniques to degrade plastics as a plastic waste management solution and the Production of Bioplastic for a sustainable and green future. Compared to traditional plastic waste management methods, the plastic biodegradation technique using microorganisms is known as the effective, environmentally friendly, low-cost, and easy-to-operate and control method for waste. This chapter gives a brief view and understanding of Plastic impact, bioremediation of plastics, biodegradation mechanisms of different synthetic plastics, Bioproduction, and biodegradation of Bioplastic (PHA).

2. Bioremediation

Bioremediation is the process by which waste is decomposed by microorganisms. Another way to describe bioremediation is as the area of biotechnology that uses microorganisms to biodegrade all-natural substances that can be handled under the biodegradation of bacteria, fungi, algae, and plants. Its primary goals are detoxification and decontamination(Asgher *et al.*, 2020). When plastic polymers are exposed to molecules, such as nitrogen or oxygen, and a carbon = carbon double bond is present, they can separate and biodegrade. This makes it easier for the trash to start biodegrading; extracellular enzymes are also used to break down plastic polymers. Enzymes serve as chemical catalysts in the bioremediation process by lowering the activation energy and changing the substrate into the product (Jaiswal *et al.*, 2020). Microbial oxidoreductase, microbial oxygenase, laccases, peroxidases, microbial lignin peroxidases, hydrolases, microbial manganese peroxidases, & microbial lipase are among the enzymes that are most likely to be employed in the process (Qamar *et al.*, 2021).

2.1. Influencing Factors for Plastic Remediation

The main elements influencing the microbial biodegradation process are environmental variables and polymer properties. Morphology, functional group, flexibility, additives, molecular weight, cross-linking, co-polymers, crystalline structure, and blend are examples of polymer characteristics (Varjani and Upasani, 2017).

The first-order structures (chemical structure, molecular weight, and molecular weight distribution), the high-order structures (glass transition temperature, melting temperature, modulus of elasticity, crystallinity, and crystal structure), and the surface conditions (surface area, hydrophilic, and hydrophobic properties) of polymers all have significant effects on the biodegradation processes(Tokiwa *et al.*, 2009). Polyesters having side chains tend to be less assimilated than those without. Since the molecular weight affects a number of the polymer's physical characteristics, it is also crucial for biodegradability. The degradability of the polymer was reduced as its molecular weight increased. *Rhizopus delmar* lipase (endo-cleavage type) destroyed PCL with greater molecular weight ($Mn > 4{,}000$) more slowly than that with a lower Mn(Tokiwa and Suzuki, 1978). Furthermore, polymer morphology significant impact on biodegradation rates. The degree of crystallinity has a significant impact on biodegradability since enzymes primarily target a polymer's amorphous domains. The amorphous region is more prone to deterioration because of the loose packing of its molecules. The polymers' crystalline portion is more resilient than their amorphous portion. PLA's rate of breakdown lowers as the polymer's crystallinity increases(Tsuji and Miyauchi, 2001). The enzymatic breakdown of polymers is significantly impacted by the polyesters' melting temperature (Tm). The polymer's biodegradation decreases as the Tm increases(Tokiwa *et al.*, 2009). pH, temperature (temperature degradation), moisture content (hydrolytic breakdown), enzymes, and the kind of microbial strain are examples of environmental influences. These elements cause variations in polymers' bulk, crystallinity, & color (Varjani and Upasani, 2017)

3. Mechanism of Bioremediation of Plastic/MPs

Biodegradation refers to the process by which complex polymeric materials are decomposed by microorganisms, including archaea, bacteria, and fungi, into non-toxic products that can be reintegrated into biogeochemical cycles. The four primary factors contributing to plastic/ microplastic degradation are biodegradation, hydrolysis, thermo oxidative degradation, photodegradation, and (Ghosh *et al.*, 2013). Caused of the availability of water in aquatic habitats, it is hypothesized that hydrolysis will play a more prominent role in affecting plastic biodegradation. Conversely, in terrestrial settings, the ability of organisms to degrade plastic polymers is believed to be more influenced by heat (thermo-oxidative) and light (photodegradation)(Amobonye *et al.*, 2021a). The most common mechanism proposed for the bioremediation of plastic/microplastic includes biodeterioration, bio-fragmentation, assimilation, and mineralization, all of which are carried out by different bond cleavages and enzyme activity(Amobonye *et al.*, 2021b). Biodegradation of polymer goes through three steps:

i. Microbial adherence to the polymer surface
ii. Growth of microbial community, utilizing polymer as a nutrient by extracellular activity
iii. Deterioration of polymer by various microbial actions (Singh and Rawat, 2019).

The biodeterioration process of micro/nano plastics (MNPs), similar to other plastic materials, begins with microorganisms adhering to the polymer's surface, which aids in surface attachment and colonization. As this attachment occurs, the chemical and physical interactions of the microorganisms, boosted by the influence of additional biological agents and external environmental factors (such as chemicals, light, and temperature), lead to changes in the properties of the MNPs(Habib *et al.*, 2020). The bacterium utilizes the carbon resources present in the plastic to support and perform additional metabolic activities after it has adhered to the surface of the polymer(Lokesh *et al.*, 2023). Subsequently, organisms liberate extracellular enzymes and depolymerize the plastic by generating free radicles. Enzymatic breakdown of the main polymer chain yields the component oligomers or monomers, including ethylene glycol and terephthalic acid, that can subsequently be assimilated by the cell(Pathak and Navneet, 2017). Active and passive transport processes facilitate the incorporation of simpler intermediates into various microorganisms, as has been reported(Amobonye *et al.*, 2021a). Numerous biomolecules have been linked to the assimilative transport of plastic degradative intermediates, including monooxygenases, porins, and the ATP-binding cassette family of proteins(Amobonye *et al.*, 2021b). In the final step assimilated intracellular molecules completely break down into CH_4, CO_2, H_2O, and N_2 by action of intracellular enzymes. Well, the final product depends upon the respiration condition of microorganisms (Priyanka and Archana, 2011). Fig. 2 illustrates the general bioremediation concept.

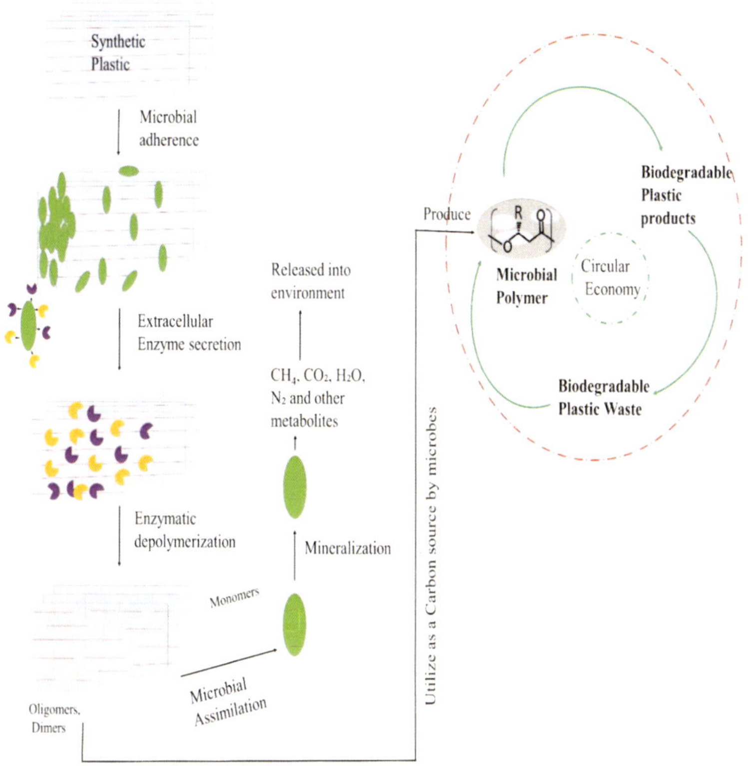

Fig. 2: Microbial Plastic degradation with Bioplastic production as a circular economy

1. Aerobic Biodegradation: Also known as aerobic respiration in which polymer compounds break down with the aid of oxygen. Under aerobic conditions, oxygen acts as the electron acceptor in this respect, reducing O_2 and forming CO_2 and H_2O(Brennecke *et al.*, 2016).
2. Anaerobic Biodegradation: Microorganisms decompose plastic polymers when oxygen is not available. Compounds such as Sulfate ($SO_4$2-), nitrate (NO_3), metallic elements like iron (Fe3+) and manganese (Mn4+), or even carbon dioxide (CO_2) may serve as the terminal electron acceptors. Additionally, depending on the electron acceptor, the side-products of anaerobic respiration may include reduced versions of alloys, methane (CH_4), hydrogen sulfide (H_2S), nitrogen gas (N_2)(Lokesh *et al.*, 2023).

4. Role of Microbial Enzymes in Plastic Remediation

Microbial enzymes participate in bio-transformation and bio-degradation activities, play a crucial role in environmental remediation and ecological health through biodegradation, and combine the benefits of efficiency and functionality to make bio-catalytic processes more competitive, safe, clean, and environmentally friendly(Carmen, 2021). The enzymes that microorganisms release aid in the degradation of plastics. Enzymes start the breakdown of polymeric chains, while microorganisms metabolize the hydrolyzed compounds. It is well known that polymers are resistant to biodegradation; the powerful C–C bonds, high molar weight, hydrophobic outer surface, and degree of crystallinity of the polymer are the components that cause this behavior. Nevertheless, only a small number of extracellular enzymes have been identified and documented to be able to break down synthetic polymers(Show *et al.*, 2021). The plastic degradation process involves two major groups of Enzymes. One is responsible for surface modification and the second is responsible for enzymatic hydrolysis (Othman *et al.*, 2021).Enzymes alter the polymer surface by increasing hydrophilicity, this modification is responsible for increased susceptibility to degradation process (Show *et al.*, 2021). Hydrolase enzymes such as lipase, cutinase, protease, and carboxyl esterase are generally responsible for these modifications. Enzymes that can depolymerize polymers into monomers are involved in polymer hydrolysis. Examples of them include esterases, hydrolases, peroxidases, amidases, laccases, and oxidases(Gaur *et al.*, 2022). PETase the enzymes that belong to the esterase family is a highly studied enzyme responsible for the degradation of PET. PETase breaks down PET into mono(2-hydroxyethyl) terephthalic acid (MHET) and small amounts of bis(2-hydroxyethyl) terephthalic acid (BHET) and terephthalic acid were obtained as a secondary product (Yoshida *et al.*, 2016). Further, the MHETase enzyme converts MHET into terephthalic acid

(TPA) and ethylene glycol(EG). TPA is utilized in the form of protocatechuate (PCA) in the metabolic pathway and EG is converted into acetate by acetyl-CoA in the TCA cycle (Gaur *et al.*, 2022). Esterase are also responsible for PUR degradation. It was later demonstrated that the polyurethanase that was separated from Pseudomonas species—*Pseudomonas fluorescens, Pseudomonas chlororaphis, and Pseudomonas sp.* AKS31—was implicated in the breakdown of PUR(Saha *et al.*, 2022). Compared to PET and PUR, which contain hydrolyzable bonds, PE, PS, PP, and PVC are difficult to break due to the C-C backbone(Y. Zhang *et al.*, 2022). There are currently very few investigations on the characterization of particular enzymes for PE, PS, PP, and PVC. Oxidation and depolymerization of PE are carried out by enzymes such as alkane hydroxylase, laccase, manganese superoxide dismutase, and peroxidase (Chow *et al.*, 2023). According to data degradative pathways for PS involve enzymes such as oxidoreductases, laccase, & lipase (Lv *et al.*, 2024).

5. Bioremediation Mechanisms for Different Types of Plastics

5.1. Biodegradation of PET

By polycondensation of TPA and EG, PET is synthesized (Nisticò, 2020). PET is a colorless, semicrystalline, hygroscopic resin that has superior tensile strength, transparency, and resistance to wear and tear (Gao *et al.*, 2022). Because of their durability and thermostability, polyethylene terephthalate (PET)-related plastics stand out above other types of plastics and are frequently utilized in the packaging industry. PET is most commonly found in food jars, plastic films, and soft drink bottles(Koshti *et al.*, 2018). Due to the increased demand for PET-based plastics, particularly in the packaging industry, in 2015 out of a total of 269 million tons of Plastics, 18.8 million tons of PET plastic was produced and only 28.4% was recycled(Taniguchi *et al.*, 2019)Thus, the removal of PET will be a great challenge. One effective and environmentally benign method of combating the threat of plastic and PET contamination is biological recycling, which involves the enzymatic breakdown of plastic. PET is composed of ester linkages that can be broken down into its monomers TPA, EG, MHET, and BHET by a variety of hydrolytic enzymes(Maurya *et al.*, 2020). One important enzyme for PET breakdown is PETase, and the enzyme cutinase and PETase have the most similarities(Liu *et al.*, 2018). However, PETase has a larger active site than cutinase, which makes it a highly accommodating zone(Chen *et al.*, 2018). Yoshida and colleagues discovered a new bacteria called *Ideonella sakaiensis* 201-F6 that can use PET film with low crystallinity (1.9%) as a significant source of carbon and energy. PET-hydrolyzing enzymes (PETase and MHETase) are secreted by *I. sakaiensis* when it is cultured on PET, and they work together to hydrolyze PET polymers into monomeric forms(Yoshida *et al.*, 2016). Membrane-anchored PET esterase from the marine bacterium *Rhodococcus pyridinivorans* P23 is capable of PET biodegradation (Guo *et al.*, 2023) .Several researchers have also employed lipase to hydrolyze PET. Ma *et al.* (2012) demonstrated that lipase from *Candida cylindracea* and *Pseudomonas sp.* effectively degraded PET nanoparticles. Numerous microorganisms, including *Thermobifida, Saccharomonospora, Streptomyces, Thermobifida fusca,* & *Ideonella sakaiensis*, have been observed to degrade PET aerobically (Kushwaha *et al.*, 2023).

5.2. Biodegradation of PUR (Polyurethane)

Polyurethane is a synthetic polymer made up of repeating units of urethane (or carbamate) bonds (-NH-COO-) in their primary chains.PUR is derived from condensation of polyisocyanates and polyalcohols(Howard, 2002). With applications in coatings, adhesives, foams, shoe soles, synthetic leather, and bumpers, PUR is the sixth most popular class of heteropolymers(Vargas-Suárez *et al.*, 2021). Large amounts of PUR waste remain in the environment continuously and develop into potentially dangerous substances because of their strong resistance to degradation, endangering both human health and the integrity of different ecosystems(Geyer *et al.*, 2017). Microbes have been shown to exhibit esterase activity, both extracellular and membrane-bound, which enables them to use polyurethane. The hydrolysis of ester bonds by these esterase enzymes is thought to be the primary cause of the microbial degradation of polyester polyurethane(Howard, 2002). PUR (polyester- and polyether-PUR) can be broken down by *Serratia* sp. HY-72, which was isolated from the Asian mantis' gut. This results in modifications to the shape and structure of the PUR surface(Kim *et al.*, 2022). Microbial consortia from landfill leachate are capable of degrading PUR when enriched using PUR as the sole carbon source. Esterase and urease are estimated as key enzymes capable of inducing physical and chemical changes to the PUR Within 48 hours (Su *et al.*, 2023). When *B. subtilis* MZA-75 and *P. aeruginosa* MZA-85 are isolated from soil together as a consortium, they generate a higher amount of esterase to break down PUR(Shah *et al.*, 2016). The structural diversity of PUR results in the production of various metabolites and metabolic pathways. Depolymerization of PUR is carried out by enzymatic activities of polyesterases, urethanase, and peroxidases which break ester and urea bonds. Intermediates such as alkanes, 4,4′ -methylene-di-aniline (MDA), and polyols are generated by the depolymerization of PUR, which is further degraded in acid by hydrolases, dioxygenases, and dienoate hydrolase. To accomplish full metabolism, the PUR intermediates ultimately make their way into the TCA cycle and the fatty acid metabolic pathway (Lv *et al.*, 2024).

5.3. Biodegradation of PS (Polystyrene)

Styrene monomers combine to form the aromatic polymer known as polystyrene (PS). The blending of other polymers such as styrene acrylonitrile (SAN) and acrylonitrile butadiene styrene (ABS) improves the properties of PS (Lv *et al.*, 2024).PS, the third most significant petroleum-based plastic, which makes up around 7% of all plastics manufactured and is utilized for insulating materials, disposable cups, and packaging containers(Urbanek *et al.*, 2020). Because of aromatic rings in its polymer chain, PS becomes very difficult to biologically degrade(Kim *et al.*, 2021). PS biodegradation in soils, seas, landfill debris, activated sludge, and compost has been the subject of direct research. Certain microorganisms, including fungi and bacteria, can grow a biofilm on PS films and some of them can alter the chemical composition of PS(Kim *et al.*, 2021). The crucial enzymes involved in the early depolymerization have not been identified, even though numerous investigations have shown that different microorganisms biodegrade PS(Amobonye *et al.*, 2021b). Styrene monomers microbiologically degrade through two potential metabolic pathways : (i) styrene oxide–phenylacetaldehyde pathway: styrene converted into styrene oxide by enzyme styrene monooxygenase. After various steps end product 4-maleyl-acelo-acetate is produced by the action of the enzyme phenyl-acetaldehyde dehydrogenase, which undergoes a β-oxidation pathway and is converted into acetyl-CoA.(Mooney *et al.*, 2006);(Lv *et al.*, 2024) (ii) styrene cis-glycol–3-vinyl catechol pathway: Pathway initiated by hydroxylation of styrene into styrene cis-glycol by enzyme styrene dioxygenase and end product pyruvate produced by enzyme hydroxy-2-oxovalerate aldolase, which further enters to the TCA cycle (Mooney *et al.*, 2006); (Lv *et al.*, 2024). Other degradative pathways Polystyrene produced oligomers which after subsequent stages produce end product benzoyl Co-A by 4-hydroxybenzoyl-CoA reductase enzyme. In the TCA cycle complete mineralization of benzoyl, Co-A occurs (Lv *et al.*, 2024).

Moreover, some studies have shown the microbially driven Fenton reaction, which produces some of the natural hydroxyl radicals (OH), can also be capable of to breakdown of polystyrene microplastics (PS-MPs). However exact mechanism of the Fenton reaction is unknown(Yang *et al.*, 2022). It was stated that virgin PS may be converted to polyhydroxyalkanoate (PHA), a biodegradable thermoplastic, using a two-step pyrolysis and bacterial conversion combinational technique. This method involved pyrolyzing PS to styrene oil, which was subsequently used as the sole carbon source for *Pseudomonas putida* CA-3 growth and PHA buildup(Ward *et al.*, 2006). Through weight loss, FTIR, and microscopic analyses, several fungi, including

Aspergillus niger, Aspergillus flavus, Trichoderma sp., Pullularia pullulans, Penicillium sp., and Curvularia sp., have shown demonstrated PS-degradation capabilities(Y. Zhang *et al.*, 2022). Different insects were also investigated with promising PS degradation capacity. PS ingestion and biodegradation have also been studied in Achatina fulica, a type of widely distributed soil invertebrate. According to the results, each A. fulica consumed 18.5 ± 2.9 mg of PS foam over four weeks, and the feces with the micro-PS shape showed a notable mass reduction of 30.7% (Song *et al.*, 2020).

5.4. Biodegradation of PE (Polyethylene)

The most prevalent plastic, PE has a linear saturated C-C backbone (−[CH2-CH2]n−) and contributed almost 30% of all plastics in 2018. PE is classified as low-density PE (LDPE), linear LDPE (LLDPE), branching LDPE, medium-density PE (MDPE), and high-density PE (HDPE) according to its characteristics(Bardají *et al.*, 2020). PE is inert against the majority of chemical catalysts because its backbone is made up entirely of linear alkyl chains and lacks polar or hydrolyzable functional groups(Thiounn and Smith, 2020). Biodegradation of PE is done in three major steps: Depolymerization, assimilation, and mineralization (Y. Zhang *et al.*, 2022).

i. Depolymerization: In the first depolymerization step, a long polyethylene carbon chain is depolymerized into oligomers, dimers, and monomer mixtures (Elsamahy *et al.*, 2023). The key enzymes responsible for this are laccase and peroxidase (A. Zhang *et al.*, 2022);(Gao *et al.*, 2022).

ii. Assimilation: Formed dimers and oligomers are further assimilated into metabolic pathways. Produced alkane monomers assimilated through hydroxylation by terminal oxidation monooxygenase and subterminal oxidation monooxygenase. Fatty acids are formed through terminal hydroxylated alkanes by alcohol dehydrogenase and aldehyde dehydrogenase. Another hand subterminal hydroxylated alkane is converted into esters and produces fatty acids. (Y. Zhang *et al.*, 2022); (Zadjelovic *et al.*, 2022)

iii. Mineralization: both the fatty acids generated enter into a beta-oxidation pathway followed by a TCA cycle for final mineralization. (Elsamahy *et al.*, 2023); (Gao *et al.*, 2022)

5.5. Biodegradation of PVC (Polyvinylchloride)

One of the six common polymers that make up 10% of the world's plastic manufacturing is polyvinyl chloride (PVC), a synthetic plastic with a carbon backbone and chlorine(Xu *et al.*, 2023). Despite PVC's well-known resistance to biodegradation, there is evidence that some bacteria and fungi may be able to break down this stubborn polymer. Furthermore, a class of proteins known as hydrophobins is produced by fungi and is essential for promoting adhesion to hydrophobic surfaces(El-Dash *et al.*, 2023). In the process of bioremediation, this first attachment phase is crucial(Sánchez, 2020). Nevertheless, the lack of a hydrolyzable ester link in PVC makes its disintegration more challenging, leading to a poorly understood biodegradation pathway(Z. Zhang *et al.*, 2022). Depolymerization of PVC by enzymes catalase and peroxidase generates lower molecular grade polymers which are further transformed by enzymes monooxygenase, esterase, dioxygenase, aldehyde dehydrogenase, and dihydroxy-acid dehydratase to form fatty acids. Which ultimately through beta-oxidation enters into the TCA cycle(Peng *et al.*, 2020). Enzyme dehalogenase liberates HCL and other chlorinated compounds from PVC(Peng *et al.*, 2020). PVC can be depolymerized by the bacteria *Klebsiella* sp. EMBL-1 that was isolated from insect guts (Z. Zhang *et al.*, 2022). *Citrobacter koseri* isolated from the gut of superworm capable of degrading PVC (Nyamjav *et al.*, 2023).

5.6. Biodegradation of PP(Polypropylene)

After polyethylene, polypropylene is the second most widely used petroleum-based plastic. The substance has uses in practically every industry and is utilized in disposable goods, packaging, and film. However, due to its structural stability, it is difficult to degrade(Jeon *et al.*, 2021). Many microorganisms break down polyethylene, such as *Bacillus species, Brevibacillus borstelensis, Comamonas* species, *Delftia* species, *Penicillium simplicissimum, Pseudomonas* species, *Rhodococcus ruber*, and *Staphylococcus* species(Jeon *et al.*, 2021). Previous research indicates that the breakdown mechanism of PP and PE is comparable, meaning that increasing oxygenation functional groups, chain scission, and subsequent methyl ketones cause the polymer to break down into oligomers. Nevertheless, no enzymes that can effectively break down PP have been found, and not much is known about the process underlying PP breakdown(Yang *et al.*, 2021); (Lv *et al.*, 2024).

6. Microbial Diversity Involved in Plastic Remediation

Utilizing microorganisms in the clean-up process is known as bioremediation.

Post-consumer plastic pollution is seen in the air, on land, and in bodies of water including the sea and rivers. Microorganisms are good candidates for bioremediation because of their innate capacity to break down hydrocarbons, albeit slowly. Through biotechnology, scientists merely maximize and utilize these microbes' inherent potential (Idowu *et al.*, 2021).

6.1. Bacteria

Bacteria are the most significant and prevalent organisms in nature and are capable of degrading complex polymers including plastics. Plastic-degrading bacteria have been discovered in recent years in a variety of environments, including soil, compost, landfills, and marine environments(Lv *et al.*, 2024). Various bacterial species such *as Bacillus, Rhizopus, Streptomyces, Pseudomonas, Klebsiella, Halomonas, Enterobacter, Commamonas, and Alkanivorax* reported for Plastic bioremediation(Agarwal *et al.*, 2024).

Table 1 gives a brief detail about types of bacteria and their bioremediation capacity for different kinds of plastics.

Table 1. Bioremediation of Different Plastics by various Microorganisms

Microorganisms	Type of Plastic	Enzyme involved	Plastic degradation efficiency	Duration	References
***Exiguobacterim* species**					
Exiguobacterium sp. DR11, *Exiguobacterium* sp. DR14	Polystyrene chips	hydrolysing enzymes	8 % 8.8%	30 days	(Chauhan *et al.*, 2018)
Exiguobacterium sp. YT2	PS film		7.5%	60 days	(Yang *et al.*, 2015)
***Bacillus* species**					
Bacillus cereus	UV-treated PE microplastic	Cellular enzymes	1.6%	40 days	(Mohan *et al.*, 2016)
B. cereus	LDPE	-	38.88	30 days	(Ibrahim *et al.*, 2021)
Bacillus sp.	LDPE	-	11.8	40 days	(Wang *et al.*, 2022)
Bacillus sp.	PE	-	9.4%		
Bacillus mixture (*Bacillus megaterium Bacillus subtilis* & *Bacillus cereus*)	Photodegraded LDPE-iron	-	11.5%	90 days	(Abrusci *et al.*, 2013)
	Photodegraded LDPE-cobalt	-	41.1%		
Bacillus amyloliquefaciens	LDPE	-	18%	30 days	(Waqas *et al.*, 2021)
B.sporothermodurans	LDPE	-	36.54%	2 Months	(Aragaw, 2020)
B.coagulans			18.37%		
B. neidei			36.07%		
B.smithii			16.40%		
B.megaterium			34.48%		
***Enterobacter* species**					
Consortia of *Enterobacter sp*.nov.bt *E.cloacae* nov.bt DSCE02, and *Pseudomonas*	LDPE and PP	-	64.25% & 64% respectively	160 days	(Skariyachan *et al.*, 2021)

E. cloacae AKS7	LDPE	-	9%	45 days	(Sarker *et al.*, 2020)
E. hormaechei	PS	Peroxidase, alkyl hydroperoxide reductase	-	-	(Kang *et al.*, 2023)
Enterobacter sp. (with *Brevundi-monasdimina* & *Alcaligenes sp*)	PS	Hydrolyase enzymes	12.4%	30 days	(Sekhar *et al.*, 2016)
***Pseudomonas* species**					
Pseudomonas sp.	PS emulsion	Esterase	97%	30 days	(Mohan *et al.*, 2016)
Pseudomonas sp.	HDPE	Oxidoreductase	15%	30 days	(Balasubramanian *et al.*, 2010)
P. fluorescens	LDPE	Alkane monooxygenase	37%	1 month	(Lakshmi, 2020)
P. putida			40%		
Consortia of *Pseudomonas and Bacillus*	PET	Hydrolase, lipase and esterase	-	-	(Roberts *et al.*, 2020)
P.knackmussii *P.aerugenosa*	PE	PE degrading enzymes	5.95% and 3.62% respectively	8 weeks	(Hou *et al.*, 2022)
P. aeruginosa WD4 *P. putida*	LDPE	Alkane monooxygenase	9.2% and 6.5% respectively	100 days	(Shilpa *et al.*, 2024)
***Streptomyces* species**					
Streptomyces coelicolor	PET	BHETase/PETase enzymes	-	-	(Verschoor *et al.*, 2024)
Consortia of *Arthrobacter* sp. and *Streptomyces sp.*	PE	-	-	90 days	(Han *et al.*, 2020)
S. ardesiacus	PP	Lipase and esterase	17.52 %	90 days	(Wichatham *et al.*, 2024)
Streptomyces sp.	PET	PET hydrolases	-	-	(Carr *et al.*, 2023)
S. calvus	Semicrystalline PET and low crystalline PET	PET hyrolases	40% and 6% respectively	30 days	(Han *et al.*, 2024)

***Penicillium* species**					
Penicillium simplicissimum	LDPE	lipase	58%	150 days	(Ghosh & Pal, 2021)
Penicillium sp.	PUR	-	18%	90 days	(Miah *et al.*, 2023)
Penicillium fellutanum	Polyester vylone 200	lipase	49.65%	7 days	(Amin *et al.*, 2021)
Penicillium citrinum	LDPE	Laccase, lipase, esterase, and manganese peroxidase	47.22%	-	(Khan *et al.*, 2023)
Penicillium brevicompactum	Polyvinyl alcohol	lipase, laccase, and manganese peroxidase	81%	10 days	(Mohamed *et al.*, 2022)
P. brevicompactum	Epoxy resins microplastics	Lipase, protease and esterase	75%	14 days	(Ferreira-Filipe *et al.*, 2024)
***Aspergillus* Species**					
A. niger	PE	Lipase	3.8%	-	(Safdar *et al.*, 2024)
	PET		3.6%		
	PS		5%		
Consortia of *Aspergillus niger, Aspergillus flavus* and *Aspergillus oryzae*	LDPE		26.15%	55 days	(DSouza *et al.*, 2021)
A. flavus	PE microplastic	-	18.3	10 weeks	(Şimşek Uygun and Malkoç, 2024)
	PP microplastic		6.8		
	PS microplastic		1.9		
A. versicolor	PE microplastic		6.7		
	PP microplastic		5.1		
	PS microplastic		3.3		
Mix culture *of A. carbonarius* and *A. fumigatus*	Thermal treated LDPE	Laccases and manganese peroxidases	39.1%	16 weeks	(El-Sayed *et al.*, 2021)

6.2. Fungi

Fungi contribute to different biogeochemical processes and are known for decomposing various recalcitrant substrates involving plastics. Some well-known plastic degrading fungal species are *Aspergillus nidulans, Aspergillus flavus, A.glaucus, A. oryzae, A. nomius, penicillium griseofulvum, Bierkandera adusta, Cladosporium,* etc. and some other saprotrophic fungi, such as Pleurotus abalones, *P.ostreatus., Agaricus bisporus* and *Pleurotus eryngii* (Srikanth *et al.*, 2022). Detailed examples are illustrated in Table 1

7. Advances and Challenges in Plastic bioremediation

7.1. Omics Approaches (Metagenomic, Transcriptomics and Proteomics)

Because uncultured bacteria predominated, it was ineffective to identify and characterize plastic-degrading microbes and novel enzymes; as a result, metagenomic prospects have been chosen to understand the microbial community involved in plastic breakdown. Several researchers investigated the use of functional metagenomic techniques to extract PET hydrolase from a variety of microbial communities. They proposed that the producers of PET hydrolase are Actinobacteria in terrestrial ecosystems and Bacteroidetes in marine ecosystems (Purohit *et al.*, 2020). The structure of the plastic-degrading bacterial population at a municipal solid waste disposal site in Gujarat, India, was investigated using a metagenomic technique. The most common phyla found in compost and soil samples were Actinobacteria, Bacteroidetes, Firmicutes, and Proteobacteria; in contrast, leachate samples were primarily composed of Actinobacteria, Firmicutes, and Proteobacteria. Additionally, the functional analysis showed that genes were present engaged in the breakdown of certain polymeric polymers like PET, PS, & PE (Naresh Kumar *et al.*, 2021).

Metatranscriptomics and metaproteomics are examples of the omic method. mRNA transcripts that are directly extracted from natural samples are analyzed using the omic technique known as metatranscriptomics. This method uses direct mRNA samples, mRNA enrichment, and mRNA to cDNA conversion to obtain detailed information by accessing the codes through next-generation sequencing. It also aids in the direct identification of genes and their activities during environmental changes(Gaur *et al.*, 2022). A transcriptome method was used in a study by Gravouil *et al.* to determine the many pathways involved in the *Rhodococcus ruber C208* strain's consumption of polyethylene.

They discovered that there was overexpression of fatty acid degradation processes, which include B oxidation and alkane degradation(Gravouil *et al.*, 2017). To find the genes linked to plastic breakdown, Gao and Sun used transcriptome probing on samples taken from Huiquan Bay, China.

Halomonas sp., Exiguobacterium sp., and Ochrobactrum sp. were found to have upregulated genes linked to energy metabolism, citrate cycle, cell proliferation, and ribosome biosynthesis. Additionally, it was noted that some plastic-degrading enzymes, including lipase, hydrolase, esterase, and cutinase, were overexpressed(Gao and Sun, 2021). Using proteome profiling of *Pseudomonas pseudoalcaligenes*, a new esterase called PpEst was discovered that can hydrolyze co-aromatic-aliphatic polyester and polybutylene adipate terephthalate (PBAT)(Wallace *et al.*, 2017).

7.2. Metabolic Engineering

Advanced metabolic engineering and synthetic biology techniques have opened up numerous possibilities for reengineering and improving bacteria that can effectively break down plastic waste and use the resulting products for biomanufacturing. Numerous enzymes that break down plastic have been improved through the use of enzyme engineering approaches. Additionally, newer methods of protein engineering, such as AI-guided protein design, mutation, and direct evolution, may make it more likely that new enzymes will be produced that are resistant to impurities or inhibitors, stable, specific, and have higher activity. Other techniques compare plastic substrate topologies using protein alignment data(Agarwal *et al.*, 2024).

7.3. Site-directed Mutagenesis

Using in silico site-directed mutagenesis, a carboxylesterase from *Archaeoglobus fulgidus* was altered to yield a BTA-hydrolase from *Termobifda fusca* (Lameh *et al.*, 2022). After disulfide bridges were added to boost the thermostability of the provided PET and residues essential for substrate binding were mutated, the enzyme was able to depolymerize 90% of it in 10 hours(Khoonkari *et al.*, 2015). Analysis was done on Te TfCut2 from Termobifda fusca. Researchers increased the disintegration rate of PET film by a factor of 12.7 by employing site-directed mutagenesis to alter these residues on the chosen substrates and computational modeling to identify key residues(Furukawa *et al.*, 2019).

Challenges to overcome

- Specificity of microbes towards degradation of particular plastic polymers. The development of microbiological treatments that generically address the diversity of plastic materials is hampered by the wide range of plastic polymers. Moreover, microbial remediation of plastic is a slow process, increased degradation rate without affecting efficiency remains a great challenge for researchers (Agarwal *et al.*, 2024).

- The effectiveness of microbial bioremediation is further complicated by environmental parameters, which include temperature, pH, and the presence of other chemicals.
- A significant challenge is to optimize these parameters for broad applicability and scalability in various contexts(Mishra *et al.*, 2020).
- Engineering & logistical difficulties arise when microbial bioremediation is expanded from lab tests to large-scale, real-world applications. It takes creative thinking and a deep comprehension of the complexity involved to ensure the industrial viability of microbial processes while preserving cost-effectiveness. Furthermore, a thorough evaluation of the ecological effects of introducing particular microbes into ecosystems is necessary. Microbial bioremediation has potential, but any unforeseen environmental effects must be carefully considered through risk assessments (Agarwal *et al.*, 2024).

8. Polyhydroxyalkanoates (PHAs) - Bioplastic as an Eco-friendly alternate

It is strongly advised to use biodegradable polymers, or bioplastics, in place of biochemically inert ones. In light of this, industrial applications should expand the use of recently marketed polymers like polyhydroxyalkanoates. Because they share characteristics with traditional plastics, polyhydroxyalkanoates (PHAs) constitute an important class of biodegradable plastics. (Moshood *et al.*, 2022). PHAs are microbial polyesters that are mostly produced by prokaryotic organisms, including bacteria, and appear as 0.2–0.5 μm water-insoluble granules within the cytoplasm or cell under nutrient imbalanced conditions(Muneer *et al.*, 2020). PHAs are members of a broad class of polymers that contain more than 150 distinct kinds of monomers(Diniz *et al.*, 2023). PHAs have drawn a lot of interest recently as eco-friendly, non-toxic, and sustainable biomaterials because of their capacity to create biobased and biodegradable plastics with physiochemical characteristics that enable them to be sustainable bioproducts of the current day(Muneer *et al.*, 2020).

8.1. Mechanisms of Synthesis and Degradation of PHA

In a natural habitat, over 300 bacteria and archaea were reported to synthesize biopolymer PHA (Deng *et al.*, 2024).PHA synthesis is carried out by three major pathways (i)Glycolysis, (ii)de novo fatty acid synthesis, and(iii) fatty acid β-oxidation which relies on types of carbon sources. The production of the PHA biopolymer is aided by several distinct enzyme processes in pathways(Muneer *et al.*, 2020). One of the most significant precursor metabolites found in all pathways for PHA biosynthesis is the critical intermediate acetoacetyl-CoA, which is generated during the metabolism and glycolysis of sugar and fatty acid catalysis. Acetoacetyl-CoA polymerizes into PHA in which PHA synthase plays a key role(Zher Neoh *et al.*, 2022). Many unconventional carbon feed sources, including mixed carbon sources, organic wastes, methane, peanut oil, soybean oil, palmitate oil, waste frying oil, margarine waste, and glycerol, are employed to produce bacterial PHA(Meereboer *et al.*, 2020).

Because the metabolic pathways are linked to bio-assimilation, PHA production plays a significant role in biodegradation(Meereboer *et al.*, 2020). Using intracellular depolymerase enzymes, a variety of bacteria may break down PHAs and use the resulting carbon and energy(Shrivastav *et al.*, 2011). Extracellular enzymes that can break down PHAs, like PHA hydrolases and PHA depolymerases, are produced and released by a variety of microorganisms in the environment(Shimao, 2001). 695 microbial species can break down PHB, according to Mergaert and Swings(Mergaert and Swings, 1996). The process of hydrolyzing polymers by enzymes involves two steps: first, the enzyme attaches itself to the polymer substrate, and then it catalyzes intracellular and extracellular depolymerase in bacteria and fungi that break down PHA(Kumaravel *et al.*, 2010).

Intracellular degradation

The process by which the accumulating bacteria hydrolyze endogenous carbon storage is known as intracellular degradation. In this, PHA is broken down to 3-hydroxyalkanoic acid by PHA depolymerase and oligomer hydrolase(Yean Ong *et al.*, 2017). Generated 3-hydroxyalkanoic acid oxidized by NAD-specific dehydrogenase and produced acetoacetate which finally converted into acetyl-CoA by beta-ketothiolase. Acetoacetyl-CoA is therefore both a precursor of PHB synthesis and a byproduct of PHB breakdown(Fernandes *et al.*, 2020). Animals naturally produce the breakdown products of PHA. As a result, PHA is classified as biocompatible because its biodegradation in live cells does not result in the creation of hazardous chemicals(Yean Ong *et al.*, 2017).

Extracellular degradation

The term extracellular degradation describes the use of an external carbon source, though not always by the microbes that are accumulating. Microorganisms that break down polyhydroxyalkanoate (PHA) release PHA depolymerases, which hydrolyze the polymer extracellularly into water-soluble compounds and use those products as sources of carbon and energy for growth(Kumaravel *et al.*, 2010). Besides depolymerase, several lipases are also capable of hydrolyzing PHA by a similar mechanism of substrate hydrolysis (Yean Ong *et al.*, 2017).

9. Plastic valorization: Circular approach of bioplastic production through synthetic plastic degradation

Enzymatic breakdown of PET results in the formation of EG, MHET, and BHET.By action of MHETase MHET further degrades into TPA and EG and they are subsequently metabolized by specific microorganisms(Yoshida *et al.*, 2016). Three microorganisms—*Pseudomonas putida* GO16, *Pseudomonas putida* GO19, and *Pseudomonas frederiksbergensis* GO23—that could use TPA for both growth and accumulation of medium-chain length PHA (mclPHA) were initially discovered by Kenny *et al.* in 2008(Kenny *et al.*, 2008). Mutant *Pseudomonas putida* KT2440 was able to utilize EG as a sole carbon source for growth and PHA accumulation in nitrogen-limiting M9 medium(Franden *et al.*, 2018). Styrene, an aromatic ring monomer of PS utilized as a carbon source by various microbes to produce PHA(Ru *et al.*, 2020) In their shake flask tests, (Ward *et al.* (2006) employed styrene oil, which is the byproduct of pyrolyzing PS waste at 520°C in a fluidized bed reactor, as the only source of carbon and energy to support *P. putida* CA-3 growth and PHA buildup. 10% of PS waste was converted to PHA during a run(Ward *et al.*, 2006). pyrolytic hydrocarbons of PE applied as a sole source to produce Microbial PHA (Guzik *et al.*, 2014). *Pseudomonas aeruginosa* PAO-1 capable to accumulate 25% PHA by dry cell mass. *Ralstonia eutropha* is also capable of accumulating PHA when supplemented with PE pyrolytic product as a carbon source(Johnston *et al.*, 2017) PP could also be depolymerized into branched-chain fatty alcohols and alkenes by pyrolysis and strain *R. eutropha* H16 capable to convert these intermediates into PHA (Johnston *et al.*, 2019). Fig. 2 also gives a brief illustration of the circular economy.

10. Conclusion and Prospects

Despite being an essential everyday tool, plastic is becoming a threat to both human health and the environment. Microbial-based remediation of plastic pollutants has become the main emphasis due to the ineffectiveness of conventional treatment methods. The waste can be mineralized into simpler, non-toxic end products via microbial and enzymatic processes, which will lessen its toxicity. The increasing demand for plastic calls for more studies to improve the biomineralization of plastic contaminants and reduce their persistence on a worldwide scale. Given the hazardous effects of microplastics on human health, there is a global need to find more effective ways to separate and break down microplastics found in various environments. Research on plastic degradation may take a hopeful turn in the future with synergistic microbe-enzyme treatment. The microbial strain's capacity to break down microplastic particles offers a possible use for microplastic remediation. Along with microbial bioremediation, the Production of microbial biopolymers such as PHAs contributes to greener and sustainable environments.

In this field, little is known about creating microbial cell factories and altering enzymes through protein engineering. Internalization of microplastics into microbial cells, which necessitates a dedicated transport mechanism system, is another barrier to their breakdown. For such a transport system, certain transcripts encoding potential enzymes have not yet been described(Gaur *et al.*, 2022). A limited number of bacterial species exhibit many routes for the breakdown of plastic, and the full-fledged activation of those pathways may present additional challenges for researchers. Additionally, the circular economy would be improved by using synthetic biology to produce microorganisms that might transform plastics into value-added products such as PHAs. Products made of highly biodegradable plastics ought to be promoted on the international market. Implementing environmentally friendly regulations about the responsible usage and disposal of microplastic litter is necessary. Without compromising the standard of living for current and future generations, all of the gaps mentioned above should be filled to satisfy community demands (Khan *et al.*, 2021)

Declaration of competing interest

The authors declare that they have no known competing financial interests or personal relationships that could have appeared to influence the work reported in this paper.

Acknowledgement

The authors want to acknowledge the funding from SHODH - ScHeme of Developing High-Quality Research, Education Department, Government of Gujarat. Award Ref. number: 20220138100

References

Abrusci, C., Pablos, J. L., Marín, I., Espí, E., Corrales, T., & Catalina, F. (2013). Comparative effect of metal stearates as pro-oxidant additives on bacterial biodegradation of thermal- and photo-degraded low density polyethylene mulching films. International Biodeterioration and Biodegradation, 83, 25–32. https://doi.org/10.1016/j.ibiod.2013.04.002

Agarwal, T., Atray, N., & Sharma, J. G. (2024). A critical examination of advanced approaches in green chemistry: microbial bioremediation strategies for sustainable mitigation of plastic pollution. Future Journal of Pharmaceutical Sciences, 10(1). https://doi.org/10.1186/s43094-024-00645-x

Akhbarizadeh, R., Dobaradaran, S., Amouei Torkmahalleh, M., Saeedi, R., Aibaghi, R., & Faraji Ghasemi, F. (2021). Suspended fine particulate matter (PM2.5), microplastics (MPs), and polycyclic aromatic hydrocarbons (PAHs) in air: Their possible relationships and health implications. Environmental Research, 192. https://doi.org/10.1016/j.envres.2020.110339

Al-Jaibachi, R., Cuthbert, R. N., & Callaghan, A. (2018). Up and away: Ontogenic transference as a pathway for aerial dispersal of microplastics. Biology Letters, 14(9). https://doi.org/10.1098/rsbl.2018.0479

Amin, M., Bhatti, H. N., Sadaf, S., & Bilal, M. (2021). Optimization of Lipase Production by Response Surface Methodology and Its Application for Efficient Biodegradation of Polyester vylon-200. Catalysis Letters, 151(12), 3603–3616. https://doi.org/10.1007/s10562-021-03603-x

Amobonye, A., Bhagwat, P., Raveendran, S., Singh, S., & Pillai, S. (2021). Environmental Impacts of Microplastics and Nanoplastics: A Current Overview. Frontiers in Microbiology, 12. https://doi.org/10.3389/fmicb.2021.768297

Amobonye, A., Bhagwat, P., Singh, S., & Pillai, S. (2021). Plastic biodegradation: Frontline microbes and their enzymes. Science of the Total Environment, 759. https://doi.org/10.1016/j.scitotenv.2020.143536

Anuar Sharuddin, S. D., Abnisa, F., Wan Daud, W. M. A., & Aroua, M. K. (2016). A review on pyrolysis of plastic wastes. Energy Conversion and Management, 115, 308–326. https://doi.org/10.1016/j.enconman.2016.02.037

Anunobi, T. J. (2022). Hazardous effects of plastic wastes on land biodiversity: A review. Zoologist (The), 20(1), 80–86. https://doi.org/10.4314/tzool.v20i1.10

Aragaw, T. A. (2020). Surgical face masks as a potential source for microplastic pollution in the COVID-19 scenario. Marine Pollution Bulletin, 159. https://doi.org/10.1016/j.marpolbul.2020.111517

Asgher, M., Qamar, S. A., Bilal, M., & Iqbal, H. M. N. (2020). Bio-based active food packaging materials: Sustainable alternative to conventional petrochemical-based packaging materials. Food Research International, 137. https://doi.org/10.1016/j.foodres.2020.109625

Azevedo-Santos, V. M., Brito, M. F. G., Manoel, P. S., Perroca, J. F., Rodrigues-Filho, J. L., Paschoal, L. R. P., Gonçalves, G. R. L., Wolf, M. R., Blettler, M. C. M., Andrade, M. C., Nobile, A. B., Lima, F. P., Ruocco, A. M. C., Silva, C. V., Perbiche-Neves, G., Portinho, J. L., Giarrizzo, T., Arcifa, M. S., & Pelicice, F. M. (2021). Plastic pollution: A focus on freshwater biodiversity. Ambio, 50(7), 1313–1324. https://doi.org/10.1007/s13280-020-01496-5

Balasubramanian, V., Natarajan, K., Hemambika, B., Ramesh, N., Sumathi, C. S., Kottaimuthu, R., & Rajesh Kannan, V. (2010). High-density polyethylene (HDPE)-degrading potential bacteria from marine ecosystem of Gulf of Mannar, India. Letters in Applied Microbiology, 51(2), 205–211. https://doi.org/10.1111/j.1472-765X.2010.02883.x

Bardají, D. K. R., Moretto, J. A. S., Furlan, J. P. R., & Stehling, E. G. (2020). A mini-review: current advances in polyethylene biodegradation. World Journal of Microbiology and Biotechnology, 36(2). https://doi.org/10.1007/s11274-020-2808-5

Barker, D. J. P. (1995). Intrauterine programming of adult disease. Molecular Medicine Today, 1(9), 418–423. https://doi.org/10.1016/S1357-4310(95)90793-9

Biber, N. F. A., Foggo, A., & Thompson, R. C. (2019). Characterising the deterioration of different plastics in air and seawater. Marine Pollution Bulletin, 141, 595–602. https://doi.org/10.1016/j.marpolbul.2019.02.068

Blettler, M. C. M., & Mitchell, C. (2021). Dangerous traps: Macroplastic encounters affecting freshwater and terrestrial wildlife. Science of the Total Environment, 798. https://doi.org/10.1016/j.scitotenv.2021.149317

Brennecke, D., Duarte, B., Paiva, F., Caçador, I., & Canning-Clode, J. (2016). Microplastics as vector for heavy metal contamination from the marine environment. Estuarine, Coastal and Shelf Science, 178, 189–195. https://doi.org/10.1016/j.ecss.2015.12.003

Briassoulis, D., Hiskakis, M., & Babou, E. (2013). Technical specifications for mechanical recycling of agricultural plastic waste. Waste Management, 33(6), 1516–1530. https://doi.org/10.1016/j.wasman.2013.03.004

Browne, M. A., Dissanayake, A., Galloway, T. S., Lowe, D. M., & Thompson, R. C. (2008). Ingested microscopic plastic translocates to the circulatory system of the mussel, Mytilus edulis (L.). Environmental Science and Technology, 42(13), 5026–5031. https://doi.org/10.1021/es800249a

Campanale, C., Massarelli, C., Savino, I., Locaputo, V., & Uricchio, V. F. (2020). A detailed review study on potential effects of microplastics and additives of concern on human health. International Journal of Environmental Research and Public Health, 17(4). https://doi.org/10.3390/ijerph17041212

Carmen, S. (2021). Microbial capability for the degradation of chemical additives present in petroleum-based plastic products: A review on current status and perspectives. Journal of Hazardous Materials, 402. https://doi.org/10.1016/j.jhazmat.2020.123534

Carr, C. M., Keller, M. B., Paul, B., Schubert, S. W., Clausen, K. S. R., Jensen, K., Clarke, D. J., Westh, P., & Dobson, A. D. W. (2023). Purification and biochemical characterization of SM14est, a PET-hydrolyzing enzyme from the marine sponge-derived Streptomyces sp. SM14. Frontiers in Microbiology, 14. https://doi.org/10.3389/fmicb.2023.1170880

Chae, Y., Kim, D., Kim, S. W., & An, Y. J. (2018). Trophic transfer and individual impact of nano-sized polystyrene in a four-species freshwater food chain. Scientific Reports, 8(1). https://doi.org/10.1038/s41598-017-18849-y

Chauhan, D., Agrawal, G., Deshmukh, S., Roy, S. S., & Priyadarshini, R. (2018). Biofilm formation by Exiguobacterium sp. DR11 and DR14 alter polystyrene surface properties and initiate biodegradation. RSC Advances, 8(66), 37590–37599. https://doi.org/10.1039/C8RA06448B

Chen, C. C., Han, X., Ko, T. P., Liu, W., & Guo, R. T. (2018). Structural studies reveal the molecular mechanism of PETase. FEBS Journal, 285(20), 3717–3723. https://doi.org/10.1111/febs.14612

Chen, R., Li, Q., Xu, X., & Zhang, D. (2019). Comparative pyrolysis characteristics of representative commercial thermosetting plastic waste in inert and oxygenous atmosphere. Fuel, 246, 212–221. https://doi.org/10.1016/j.fuel.2019.02.129

Chow, J., Perez-Garcia, P., Dierkes, R., & Streit, W. R. (2023). Microbial enzymes will offer limited solutions to the global plastic pollution crisis. Microbial Biotechnology, 16(2), 195–217. https://doi.org/10.1111/1751-7915.14135

Deng, B., Rao, L., & Rodriguez-Freire, L. (2024). Evaluation and optimization of FTIR spectroscopy to quantify PHA production by municipal wastewater sludge. Spectrochimica Acta - Part A: Molecular and Biomolecular Spectroscopy, 312. https://doi.org/10.1016/j.saa.2024.124012

Diniz, M. S. da F., Mourão, M. M., Xavier, L. P., & Santos, A. V. (2023). Recent Biotechnological Applications of Polyhydroxyalkanoates (PHA) in the Biomedical Sector—A Review. Polymers, 15(22). https://doi.org/10.3390/polym15224405

DSouza, G. C., Sheriff, R. S., Ullanat, V., Shrikrishna, A., Joshi, A. V., Hiremath, L., & Entoori, K. (2021). Fungal biodegradation of low-density polyethylene using consortium of Aspergillus species under controlled conditions. Heliyon, 7(5). https://doi.org/10.1016/j.heliyon.2021.e07008

El-Dash, H. A., Yousef, N. E., Aboelazm, A. A., Awan, Z. A., Yahya, G., & El-Ganiny, A. M. (2023). Optimizing Eco-Friendly Degradation of Polyvinyl Chloride (PVC) Plastic Using Environmental Strains of Malassezia Species and Aspergillus fumigatus. International Journal of Molecular Sciences, 24(20). https://doi.org/10.3390/ijms242015452

Elsamahy, T., Sun, J., Elsilk, S. E., & Ali, S. S. (2023). Biodegradation of low-density polyethylene plastic waste by a constructed tri-culture yeast consortium from wood-feeding termite: Degradation mechanism and pathway. Journal of Hazardous Materials, 448. https://doi.org/10.1016/j.jhazmat.2023.130944

El-Sayed, M. T., Rabie, G. H., & Hamed, E. A. (2021). Biodegradation of low-density polyethylene (LDPE) using the mixed culture of Aspergillus carbonarius and A. fumigates. Environment, Development and Sustainability, 23(10), 14556–14584. https://doi.org/10.1007/s10668-021-01258-7

Enyoh, C. E., Verla, A. W., Verla, E. N., Ibe, F. C., & Amaobi, C. E. (2019). Airborne microplastics: a review study on method for analysis, occurrence, movement and risks. Environmental Monitoring and Assessment, 191(11). https://doi.org/10.1007/s10661-019-7842-0

Evode, N., Qamar, S. A., Bilal, M., Barceló, D., & Iqbal, H. M. N. (2021). Plastic waste and its management strategies for environmental sustainability. Case Studies in Chemical and Environmental Engineering, 4. https://doi.org/10.1016/j.cscee.2021.100142

Fernandes, M., Salvador, A., Alves, M. M., & Vicente, A. A. (2020). Factors affecting polyhydroxyalkanoates biodegradation in soil. Polymer Degradation and Stability, 182. https://doi.org/10.1016/j.polymdegradstab.2020.109408

Ferreira-Filipe, D. A., Oliveira, L., Paço, A., Fernandes, A. J. S., Costa, F. M., Duarte, A. C., Rocha-Santos, T., & Patrício Silva, A. L. (2024). Biodegradation of e-waste microplastics by Penicillium brevicompactum. Science of the Total Environment, 935. https://doi.org/10.1016/j.scitotenv.2024.173334

Fournier, S. B., D'Errico, J. N., Adler, D. S., Kollontzi, S., Goedken, M. J., Fabris, L., Yurkow, E. J., & Stapleton, P. A. (2020). Nanopolystyrene translocation and fetal deposition after acute lung exposure during late-stage pregnancy. Particle and Fibre Toxicology, 17(1). https://doi.org/10.1186/s12989-020-00385-9

Franden, M. A., Jayakody, L. N., Li, W. J., Wagner, N. J., Cleveland, N. S., Michener, W. E., Hauer, B., Blank, L. M., Wierckx, N., Klebensberger, J., & Beckham, G. T. (2018). Engineering Pseudomonas putida KT2440 for efficient ethylene glycol utilization. Metabolic Engineering, 48, 197–207. https://doi.org/10.1016/j.ymben.2018.06.003

Furukawa, M., Kawakami, N., Tomizawa, A., & Miyamoto, K. (2019). Efficient Degradation of Poly(ethylene terephthalate) with Thermobifida fusca Cutinase Exhibiting Improved Catalytic Activity Generated using Mutagenesis and Additive-based Approaches. Scientific Reports, 9(1). https://doi.org/10.1038/s41598-019-52379-z

Gao, R., & Sun, C. (2021). A marine bacterial community capable of degrading poly(ethylene terephthalate) and polyethylene. Journal of Hazardous Materials, 416. https://doi.org/10.1016/j.jhazmat.2021.125928

Gao, R., Liu, R., & Sun, C. (2022). A marine fungus Alternaria alternata FB1 efficiently degrades polyethylene. Journal of Hazardous Materials, 431. https://doi.org/10.1016/j.jhazmat.2022.128617

Gaur, V. K., Gupta, S., Sharma, P., Gupta, P., Varjani, S., Srivastava, J. K., Chang, J. S., & Bui, X. T. (2022). Metabolic Cascade for Remediation of Plastic Waste: a Case Study on Microplastic Degradation. Current Pollution Reports, 8(1), 30–50. https://doi.org/10.1007/s40726-021-00210-7

Geyer, R., Jambeck, J. R., & Law, K. L. (2017). Production, use, and fate of all plastics ever made. Science Advances, 3(7). https://doi.org/10.1126/sciadv.1700782

Ghosh, S. K., & Pal, S. (2021). De-polymerization of LDPE plastic by Penicillium simplicissimum isolated from municipality garbage plastic and identified by ITSs locus of rDNA. Vegetos, 34(1), 57–67. https://doi.org/10.1007/s42535-020-00176-9

Ghosh, S. K., Pal, S., & Ray, S. (2013). Study of microbes having potentiality for biodegradation of plastics. Environmental Science and Pollution Research International, 20(7), 4339–4355. https://doi.org/10.1007/s11356-013-1706-x

Goodman, K. E., Hare, J. T., Khamis, Z. I., Hua, T., & Sang, Q. X. A. (2021). Exposure of Human Lung Cells to Polystyrene Microplastics Significantly Retards Cell Proliferation and Triggers Morphological Changes. Chemical Research in Toxicology, 34(4), 1069–1081. https://doi.org/10.1021/acs.chemrestox.0c00486

Gravouil, K., Ferru-Clément, R., Colas, S., Helye, R., Kadri, L., Bourdeau, L., Moumen, B., Mercier, A., & Ferreira, T. (2017). Transcriptomics and Lipidomics of the Environmental Strain Rhodococcus ruber Point out Consumption Pathways and Potential Metabolic Bottlenecks for Polyethylene Degradation. Environmental Science and Technology, 51(9), 5172–5181. https://doi.org/10.1021/acs.est.7b00846

Guo, W., Duan, J., Shi, Z., Yu, X., & Shao, Z. (2023). Biodegradation of PET by the membrane-anchored PET esterase from the marine bacterium Rhodococcus pyridinivorans P23. Communications Biology, 6(1). https://doi.org/10.1038/s42003-023-05470-1

Guzik, M. W., Kenny, S. T., Duane, G. F., Casey, E., Woods, T., Babu, R. P., Nikodinovic-Runic, J., Murray, M., & O'Connor, K. E. (2014). Conversion of post consumer polyethylene to the biodegradable polymer polyhydroxyalkanoate. Applied Microbiology and Biotechnology, 98(9), 4223–4232. https://doi.org/10.1007/s00253-013-5489-2

Habib, S., Iruthayam, A., Shukor, M. Y. A., Alias, S. A., Smykla, J., & Yasid, N. A. (2020). Biodeterioration of untreated polypropylene microplastic particles by antarctic bacteria. Polymers, 12(11), 2616. https://doi.org/10.3390/polym12112616

Han, Y. N., Wei, M., Han, F., Fang, C., Wang, D., Zhong, Y. J., Guo, C. L., Shi, X. Y., Xie, Z. K., & Li, F. M. (2020). Greater biofilm formation and increased biodegradation of polyethylene film by a microbial consortium of arthrobacter sp. And streptomyces sp. Microorganisms, 8(12), 1–15. https://doi.org/10.3390/microorganisms8121979

Han, Z., Nina, M. R. H., Zhang, X., Huang, H., Fan, D., & Bai, Y. (2024). Discovery and characterization of two novel polyethylene terephthalate hydrolases: One from a bacterium identified in human feces and one from the Streptomyces genus. Journal of Hazardous Materials, 472. https://doi.org/10.1016/j.jhazmat.2024.134532

Haque, F., & Fan, C. (2023). Fate of microplastics under the influence of climate change. IScience, 26(9). https://doi.org/10.1016/j.isci.2023.107649

Hou, L., Xi, J., Liu, J., Wang, P., Xu, T., Liu, T., Qu, W., & Lin, Y. B. (2022). Biodegradability of polyethylene mulching film by two Pseudomonas bacteria and their potential degradation mechanism. Chemosphere, 286. https://doi.org/10.1016/j.chemosphere.2021.131758

Howard, G. T. (2002). Biodegradation of polyurethane: A review. International Biodeterioration and Biodegradation, 49(4), 245–252. https://doi.org/10.1016/S0964-8305(02)00051-3

Ibrahim, S., Gupta, R. K., War, A. R., Hussain, B., Kumar, A., Sofi, T., Noureldeen, A., & Darwish, H. (2021). Degradation of chlorpyriphos and polyethylene by endosymbiotic bacteria from citrus mealybug. Saudi Journal of Biological Sciences, 28(6), 3214–3224. https://doi.org/10.1016/j.sjbs.2021.03.058

Idowu, S. A., Arotupin, D. J., & Oladejo, S. O. (2021). Plastic Pollution and Climate Change: Role of Bioremediation as a Tool to Achieving Sustainability. African Handbook of Climate Change Adaptation: With 610 Fig.ures and 361 Tables, 1159–1168. https://doi.org/10.1007/978-3-030-45106-6_102

Jaiswal, S., Sharma, B., & Shukla, P. (2020). Integrated approaches in microbial degradation of plastics. Environmental Technology and Innovation, 17. https://doi.org/10.1016/j.eti.2019.100567

Jeon, J. M., Park, S. J., Choi, T. R., Park, J. H., Yang, Y. H., & Yoon, J. J. (2021). Biodegradation of polyethylene and polypropylene by Lysinibacillus species JJY0216 isolated from soil grove. Polymer Degradation and Stability, 191. https://doi.org/10.1016/j.polymdegradstab.2021.109662

Johnston, B., Jiang, G., Hill, D., Adamus, G., Kwiecień, I., Zięba, M., Sikorska, W., Green, M., Kowalczuk, M., & Radecka, I. (2017). The molecular level characterization of biodegradable polymers originated from polyethylene using non-oxygenated polyethylene wax as a carbon source for polyhydroxyalkanoate production. Bioengineering, 4(3). https://doi.org/10.3390/bioengineering4030073

Johnston, B., Radecka, I., Chiellini, E., Barsi, D., Ilieva, V. I., Sikorska, W., Musiol, M., Zięba, M., Chaber, P., Marek, A. A., Mendrek, B., Ekere, A. I., Adamus, G., & Kowalczuk, M. (2019). Mass spectrometry reveals molecular structure of polyhydroxyalkanoates attained by bioconversion of oxidized polypropylene waste fragments. Polymers, 11(10). https://doi.org/10.3390/polym11101580

Kang, M. G., Kwak, M. J., & Kim, Y. (2023). Polystyrene microplastics biodegradation by gut bacterial Enterobacter hormaechei from mealworms under anaerobic conditions: Anaerobic oxidation and depolymerization. Journal of Hazardous Materials, 459. https://doi.org/10.1016/j.jhazmat.2023.132045

Kenny, S. T., Runic, J. N., Kaminsky, W., Woods, T., Babu, R. P., Keely, C. M., Blau, W., & O'Connor, K. E. (2008). Up-cycling of PET (Polyethylene Terephthalate) to the biodegradable plastic PHA (Polyhydroxyalkanoate). Environmental Science and Technology, 42(20), 7696–7701. https://doi.org/10.1021/es801010e

Khan, M. J., Rai, A., Ahirwar, A., Sirotiya, V., Mourya, M., Mishra, S., Schoefs, B., Marchand, J., Bhatia, S. K., Varjani, S., & Vinayak, V. (2021). Diatom microalgae as smart nanocontainers for biosensing wastewater pollutants: recent trends and innovations. Bioengineered, 12(2), 9531–9549. https://doi.org/10.1080/21655979.2021.1996748

Khan, S., Ali, S. A., & Ali, A. S. (2023). Biodegradation of low density polyethylene (LDPE) by mesophilic fungus 'Penicillium citrinum' isolated from soils of plastic waste dump yard, Bhopal, India. Environmental Technology (United Kingdom), 44(15), 2300–2314. https://doi.org/10.1080/09593330.2022.2027025

Khoonkari, M., Haghighi, A. H., Sefidbakht, Y., Shekoohi, K., & Ghaderian, A. (2015). Chemical Recycling of PET Wastes with Different Catalysts. International Journal of Polymer Science, 2015. https://doi.org/10.1155/2015/124524

Kim, H. W., Jo, J. H., Kim, Y. Bin, Le, T. K., Cho, C. W., Yun, C. H., Chi, W. S., & Yeom, S. J. (2021). Biodegradation of polystyrene by bacteria from the soil in common environments. Journal of Hazardous Materials, 416. https://doi.org/10.1016/j.jhazmat.2021.126239

Kim, J. H., Choi, S. H., Park, M. G., Park, D. H., Son, K. H., & Park, H. Y. (2022). Polyurethane biodegradation by Serratia sp. HY-72 isolated from the intestine of the Asian mantis Hierodula patellifera. Frontiers in Microbiology, 13. https://doi.org/10.3389/fmicb.2022.1005415

Koshti, R., Mehta, L., & Samarth, N. (2018). Biological Recycling of Polyethylene Terephthalate: A Mini-Review. Journal of Polymers and the Environment, 26(8), 3520–3529. https://doi.org/10.1007/s10924-018-1214-7

Kumaravel, S., Hema, R., & Lakshmi, R. (2010). Production of polyhydroxybutyrate (Bioplastic) and its biodegradation by pseudomonas lemoignei and aspergillus niger. E-Journal of Chemistry, 7(SUPPL. 1). https://doi.org/10.1155/2010/148547

Kushwaha, A., Goswami, L., Singhvi, M., & Kim, B. S. (2023). Biodegradation of poly(ethylene terephthalate): Mechanistic insights, advances, and future innovative strategies. Chemical Engineering Journal, 457. https://doi.org/10.1016/j.cej.2022.141230

Lakshmi, P. J. (2020). Genetic analysis of low-density polyethylene degrading bacteria from plastic dump sites. Indian Journal of Science and Technology, 13(48), 4732–4738. https://doi.org/10.17485/ijst/v13i48.2066

Lameh, F., Baseer, A. Q., & Ashiru, A. G. (2022). Comparative molecular docking and molecular-dynamic simulation of wild-type- and mutant carboxylesterase with BTA-hydrolase for enhanced binding to plastic. Engineering in Life Sciences, 22(1), 13–29. https://doi.org/10.1002/elsc.202100083

Landrigan, P. J., Raps, H., Cropper, M., Bald, C., Brunner, M., Canonizado, E. M., Charles, D., Chiles, T. C., Donohue, M. J., Enck, J., Fenichel, P., Fleming, L. E., Ferrier-Pages, C., Fordham, R., Gozt, A., Griffin, C., Hahn, M. E., Haryanto, B., Hixson, R., … Dunlop, S. (2023). The Minderoo-Monaco Commission on Plastics and Human Health. Annals of Global Health, 89(1). https://doi.org/10.5334/aogh.4056

Liang, Y., Tan, Q., Song, Q., & Li, J. (2021). An analysis of the plastic waste trade and management in Asia. Waste Management, 119, 242–253. https://doi.org/10.1016/j.wasman.2020.09.049

Lin, D., Yang, G., Dou, P., Qian, S., Zhao, L., Yang, Y., & Fanin, N. (2020). Microplastics negatively affect soil fauna but stimulate microbial activity: insights from a field-based microplastic addition experiment. Proceedings of the Royal Society B: Biological Sciences, 287(1934). https://doi.org/10.1098/rspb.2020.1268

Liu, B., He, L., Wang, L., Li, T., Li, C., Liu, H., Luo, Y., & Bao, R. (2018). Protein crystallography and site-direct mutagenesis analysis of the poly(Ethylene terephthalate) hydrolase petase from Ideonella sakaiensis. ChemBioChem, 19(14), 1471–1475. https://doi.org/10.1002/cbic.201800097

Lokesh, P., Shobika, R., Omer, S. N., Reddy, M., Saravanan, P., Rajeshkannan, R., Saravanan, V., & Venkatkumar, S. (2023). Bioremediation of plastics by the help of microbial tool: A way for control of plastic pollution. Sustainable Chemistry for the Environment, 3. https://doi.org/10.1016/j.scenv.2023.100027

Lu, K., Zhan, D., Fang, Y., Li, L., Chen, G., Chen, S., & Wang, L. (2022). Microplastics, potential threat to patients with lung diseases. Frontiers in Toxicology, 4. https://doi.org/10.3389/ftox.2022.958414

Lv, S., Li, Y., Zhao, S., & Shao, Z. (2024). Biodegradation of Typical Plastics: From Microbial Diversity to Metabolic Mechanisms. International Journal of Molecular Sciences, 25(1). https://doi.org/10.3390/ijms25010593

Malankowska, M., Echaide-Gorriz, C., & Coronas, J. (2021). Microplastics in marine environment: A review on sources, classification, and potential remediation by membrane technology. Environmental Science: Water Research and Technology, 7(2), 243–258. https://doi.org/10.1039/d0ew00802h

Mao, Y., Li, H., Gu, W., Yang, G., Liu, Y., & He, Q. (2020). Distribution and characteristics of microplastics in the Yulin River, China: Role of environmental and spatial factors. Environmental Pollution, 265. https://doi.org/10.1016/j.envpol.2020.115033

Masnadi, M. S., El-Houjeiri, H. M., Schunack, D., Li, Y., Englander, J. G., Badahdah, A., Monfort, J. C., Anderson, J. E., Wallington, T. J., Bergerson, J. A., Gordon, D., Koomey, J., Przesmitzki, S., Azevedo, I. L., Bi, X. T., Duffy, J. E., Heath, G. A., Keoleian, G. A., McGlade, C., … Brandt, A. R. (2018). Global carbon intensity of crude oil production. Science, 361(6405), 851–853. https://doi.org/10.1126/science.aar6859

Massy, J. (2017). Thermoplastic and Thermosetting Polymers. SpringerBriefs in Materials, 19–26. https://doi.org/10.1007/978-3-319-54831-9_5

Maurya, A., Bhattacharya, A., & Khare, S. K. (2020). Enzymatic Remediation of Polyethylene Terephthalate (PET)–Based Polymers for Effective Management of Plastic Wastes: An

Overview. Frontiers in Bioengineering and Biotechnology, 8. https://doi.org/10.3389/fbioe.2020.602325

Meereboer, K. W., Misra, M., & Mohanty, A. K. (2020). Review of recent advances in the biodegradability of polyhydroxyalkanoate (PHA) bioplastics and their composites. Green Chemistry, 22(17), 5519–5558. https://doi.org/10.1039/d0gc01647k

Mergaert, J., & Swings, J. (1996). Biodiversity of microorganisms that degrade bacterial and synthetic polyesters. Journal of Industrial Microbiology and Biotechnology, 17(5–6), 463–469. https://doi.org/10.1007/bf01574777

Miah, M. J., Pei, J., Kim, H., Sharma, R., Jang, J. G., & Ahn, J. (2023). Property assessment of an eco-friendly mortar reinforced with recycled mask fiber derived from COVID-19 single-use face masks. Journal of Building Engineering, 66. https://doi.org/10.1016/j.jobe.2023.105885

Mishra, M., Singh, S. K., & Kumar, A. (2020). Environmental factors affecting the bioremediation potential of microbes. Microbe Mediated Remediation of Environmental Contaminants, 47–58. https://doi.org/10.1016/B978-0-12-821199-1.00005-5

Mohamed, H., Shah, A. M., Nazir, Y., Naz, T., Nosheen, S., & Song, Y. (2022). Biodegradation of poly (vinyl alcohol) by an orychophragmus rhizosphere-associated fungus Penicillium brevicompactum OVR-5, and its proposed PVA biodegradation pathway. World Journal of Microbiology and Biotechnology, 38(1). https://doi.org/10.1007/s11274-021-03197-x

Mohan, A. J., Sekhar, V. C., Bhaskar, T., & Nampoothiri, K. M. (2016). Microbial assisted High Impact Polystyrene (HIPS) degradation. Bioresource Technology, 213, 204–207. https://doi.org/10.1016/j.biortech.2016.03.021

Mooney, A., Ward, P. G., & O'Connor, K. E. (2006). Microbial degradation of styrene: Biochemistry, molecular genetics, and perspectives for biotechnological applications. Applied Microbiology and Biotechnology, 72(1), 1–10. https://doi.org/10.1007/s00253-006-0443-1

Moshood, T. D., Nawanir, G., Mahmud, F., Mohamad, F., Ahmad, M. H., & AbdulGhani, A. (2022). Sustainability of biodegradable plastics: New problem or solution to solve the global plastic pollution? Current Research in Green and Sustainable Chemistry, 5. https://doi.org/10.1016/j.crgsc.2022.100273

Muneer, F., Rasul, I., Azeem, F., Siddique, M. H., Zubair, M., & Nadeem, H. (2020). Microbial Polyhydroxyalkanoates (PHAs): Efficient Replacement of Synthetic Polymers. Journal of Polymers and the Environment, 28(9), 2301–2323. https://doi.org/10.1007/s10924-020-01772-1

Naresh Kumar, A., Kim, G. B., Muhorakeye, A., Varjani, S., & Kim, S. H. (2021). Biopolymer production using volatile fatty acids as resource: Effect of feast-famine strategy and lignin reinforcement. Bioresource Technology, 326. https://doi.org/10.1016/j.biortech.2021.124736

Nisticò, R. (2020). Polyethylene terephthalate (PET) in the packaging industry. Polymer Testing, 90. https://doi.org/10.1016/j.polymertesting.2020.106707

Nyamjav, I., Jang, Y., Lee, Y. E., & Lee, S. (2023). Biodegradation of polyvinyl chloride by Citrobacter koseri isolated from superworms (Zophobas atratus larvae). Frontiers in Microbiology, 14. https://doi.org/10.3389/fmicb.2023.1175249

Ong, S. Y., Chee, J. Y., & Sudesh, K. (2017). Degradation of Polyhydroxyalkanoate (PHA): a Review. Journal of Siberian Federal University. Biology, 10(2), 21–225. https://doi.org/10.17516/1997-1389-0024

Osman, A. I., Hosny, M., Eltaweil, A. S., Omar, S., Elgarahy, A. M., Farghali, M., Yap, P. S., Wu, Y. S., Nagandran, S., Batumalaie, K., Gopinath, S. C. B., John, O. D., Sekar, M., Saikia, T., Karunanithi, P., Hatta, M. H. M., & Akinyede, K. A. (2023). Microplastic sources, formation, toxicity and remediation: a review. Environmental Chemistry Letters, 21(4), 2129–2169. https://doi.org/10.1007/s10311-023-01593-3

Othman, A. R., Hasan, H. A., Muhamad, M. H., Ismail, N. 'Izzati, & Abdullah, S. R. S. (2021). Microbial degradation of microplastics by enzymatic processes: a review. Environmental Chemistry Letters, 19(4), 3057–3073. https://doi.org/10.1007/s10311-021-01197-9

Pan, D., Su, F., Liu, C., & Guo, Z. (2020). Research progress for plastic waste management and manufacture of value-added products. Advanced Composites and Hybrid Materials, 3(4), 443–461. https://doi.org/10.1007/s42114-020-00190-0

Pathak, V. M., & Navneet. (2017). Review on the current status of polymer degradation: a microbial approach. Bioresources and Bioprocessing, 4(1). https://doi.org/10.1186/s40643-017-0145-9

Peng, B. Y., Chen, Z., Chen, J., Yu, H., Zhou, X., Criddle, C. S., Wu, W. M., & Zhang, Y. (2020). Biodegradation of Polyvinyl Chloride (PVC) in Tenebrio molitor (Coleoptera: Tenebrionidae) larvae. Environment International, 145. https://doi.org/10.1016/j.envint.2020.106106

Prata, J. C., da Costa, J. P., Lopes, I., Duarte, A. C., & Rocha-Santos, T. (2020). Environmental exposure to microplastics: An overview on possible human health effects. Science of the Total Environment, 702. https://doi.org/10.1016/j.scitotenv.2019.134455

Prata, J. C., Reis, V., da Costa, J. P., Mouneyrac, C., Duarte, A. C., & Rocha-Santos, T. (2021). Contamination issues as a challenge in quality control and quality assurance in microplastics analytics. Journal of Hazardous Materials, 403. https://doi.org/10.1016/j.jhazmat.2020.123660

Prietl, B., Meindl, C., Roblegg, E., Pieber, T. R., Lanzer, G., & Fröhlich, E. (2014). Nano-sized and micro-sized polystyrene particles affect phagocyte function. Cell Biology and Toxicology, 30(1), 1–16. https://doi.org/10.1007/s10565-013-9265-y

Priyanka, N., & Archana, T. (2011). Biodegradability of Polythene and Plastic By The Help of Microorganism: A Way for Brighter Future. Journal of Environmental & Analytical Toxicology, 01(02). https://doi.org/10.4172/2161-0525.1000111

Purohit, J., Chattopadhyay, A., & Teli, B. (2020). Metagenomic Exploration of Plastic Degrading Microbes for Biotechnological Application. Current Genomics, 21(4), 253–270. https://doi.org/10.2174/1389202921999200525155711

Qamar, S. A., Qamar, M., Bilal, M., Bharagava, R. N., Ferreira, L. F. R., Sher, F., & Iqbal, H. M. N. (2021). Cellulose-deconstruction potential of nano-biocatalytic systems: A strategic drive from designing to sustainable applications of immobilized cellulases. International Journal of Biological Macromolecules, 185, 1–19. https://doi.org/10.1016/j.ijbiomac.2021.06.079

Ragusa, A., Notarstefano, V., Svelato, A., Belloni, A., Gioacchini, G., Blondeel, C., Zucchelli, E., De Luca, C., D'avino, S., Gulotta, A., Carnevali, O., & Giorgini, E. (2022). Raman Microspectroscopy Detection and Characterisation of Microplastics in Human Breastmilk. Polymers, 14(13). https://doi.org/10.3390/polym14132700

Revell, L. E., Kuma, P., Le Ru, E. C., Somerville, W. R. C., & Gaw, S. (2021). Direct radiative effects of airborne microplastics. Nature, 598(7881), 462–467. https://doi.org/10.1038/s41586-021-03864-x

Rillig, M. C., Leifheit, E., & Lehmann, J. (2021). Microplastic effects on carbon cycling processes in soils. PLoS Biology, 19(3). https://doi.org/10.1371/JOURNAL.PBIO.3001130

Roberts, C., Edwards, S., Vague, M., León-Zayas, R., Scheffer, H., Chan, G., Swartz, N. A., & Mellies, J. L. (2020). Environmental Consortium Containing Pseudomonas and Bacillus Species Synergistically Degrades Polyethylene Terephthalate Plastic . MSphere, 5(6). https://doi.org/10.1128/msphere.01151-20

Ru, J., Huo, Y., & Yang, Y. (2020). Microbial Degradation and Valorization of Plastic Wastes. Frontiers in Microbiology, 11. https://doi.org/10.3389/fmicb.2020.00442

Rubio, L., Barguilla, I., Domenech, J., Marcos, R., & Hernández, A. (2020). Biological effects, including oxidative stress and genotoxic damage, of polystyrene nanoparticles in different human hematopoietic cell lines. Journal of Hazardous Materials, 398. https://doi.org/10.1016/j.jhazmat.2020.122900

Safdar, A., Ismail, F., & Imran, M. (2024). Biodegradation of synthetic plastics by the extracellular lipase of Aspergillus niger. Environmental Advances, 17. https://doi.org/10.1016/j.envadv.2024.100563

Saha, M., Dutta, S. P., Mukherjee, G., Basu, A., Majumder, D., & Sil, A. K. (2022). Cloning, expression and characterization of PURase gene from Pseudomonas sp. AKS31. Archives of Microbiology, 204(8). https://doi.org/10.1007/s00203-022-03110-6

Sánchez, C. (2020). Fungal potential for the degradation of petroleum-based polymers: An overview of macro- and microplastics biodegradation. Biotechnology Advances, 40. https://doi.org/10.1016/j.biotechadv.2019.107501

Sarker, R. K., Chakraborty, P., Paul, P., Chatterjee, A., & Tribedi, P. (2020). Degradation of low-density poly ethylene (LDPE) by Enterobacter cloacae AKS7: a potential step towards sustainable environmental remediation. Archives of Microbiology, 202(8), 2117–2125. https://doi.org/10.1007/s00203-020-01926-8

Sekhar, V. C., Nampoothiri, K. M., Mohan, A. J., Nair, N. R., Bhaskar, T., & Pandey, A. (2016). Microbial degradation of high impact polystyrene (HIPS), an e-plastic with decabromodiphenyl oxide and antimony trioxide. Journal of Hazardous Materials, 318, 347–354. https://doi.org/10.1016/j.jhazmat.2016.07.008

Shah, Z., Gulzar, M., Hasan, F., & Shah, A. A. (2016). Degradation of polyester polyurethane by an indigenously developed consortium of Pseudomonas and Bacillus species isolated from soil. Polymer Degradation and Stability, 134, 349–356. https://doi.org/10.1016/j.polymdegradstab.2016.11.003

Sharma, B., Rawat, H., ja, P., & Sharma, R. (2017). Bioremediation - A Progressive Approach toward Reducing Plastic Wastes. International Journal of Current Microbiology and Applied Sciences, 6(12), 1116–1131. https://doi.org/10.20546/ijcmas.2017.612.126

Shen, M., Huang, W., Chen, M., Song, B., Zeng, G., & Zhang, Y. (2020). (Micro)plastic crisis: Un-ignorable contribution to global greenhouse gas emissions and climate change. Journal of Cleaner Production, 254. https://doi.org/10.1016/j.jclepro.2020.120138

Shilpa, Basak, N., & Meena, S. S. (2024). Biodegradation of low-density polythene (LDPE) by a novel strain of Pseudomonas aeruginosa WD4 isolated from plastic dumpsite. Biodegradation, 35(5), 641–655. https://doi.org/10.1007/s10532-023-10061-2

Shimao, M. (2001). Biodegradation of plastics. Current Opinion in Biotechnology, 12(3), 242–247. https://doi.org/10.1016/S0958-1669(00)00206-8

Show, P. L., Thangalazhy-Gopakumar, S., & Foo, D. C. Y. (2021). Sustainable technologies for waste reduction and pollutants removals. Clean Technologies and Environmental Policy, 23(1). https://doi.org/10.1007/s10098-020-02001-y

Shrivastav, A., Mishra, S. K., Pancha, I., Jain, D., Bhattacharya, S., Patel, S., & Mishra, S. (2011). Biodegradability studies of polyhydroxyalkanoate (PHA) film produced by a marine bacteria using Jatropha biodiesel byproduct as a substrate. World Journal of Microbiology and Biotechnology, 27(7), 1531–1541. https://doi.org/10.1007/s11274-010-0605-2

Şimşek Uygun, B., & Malkoç, S. (2024). Microplastics Biodegradation by Aspergillus flavus and Aspergillus versicolor. Eurasian Journal of Biological and Chemical Sciences, 7(1), 5–9. https://doi.org/10.46239/ejbcs.1374947

Singh, S., & Rawat, P. S. (2019). Biodegradation of Plastic. 435–461. https://doi.org/10.4018/978-1-5225-9452-9.ch022

Skariyachan, S., Taskeen, N., Kishore, A. P., Krishna, B. V., & Naidu, G. (2021). Novel consortia of enterobacter and pseudomonas formulated from cow dung exhibited enhanced biodegradation of polyethylene and polypropylene. Journal of Environmental Management, 284. https://doi.org/10.1016/j.jenvman.2021.112030

Song, Y., Qiu, R., Hu, J., Li, X., Zhang, X., Chen, Y., Wu, W. M., & He, D. (2020). Biodegradation and disintegration of expanded polystyrene by land snails Achatina fulica. Science of the Total Environment, 746. https://doi.org/10.1016/j.scitotenv.2020.141289

Sridharan, S., Kumar, M., Singh, L., Bolan, N. S., & Saha, M. (2021). Microplastics as an emerging source of particulate air pollution: A critical review. Journal of Hazardous Materials, 418. https://doi.org/10.1016/j.jhazmat.2021.126245

Srikanth, M., Sandeep, T. S. R. S., Sucharitha, K., & Godi, S. (2022). Biodegradation of plastic polymers by fungi: a brief review. Bioresources and Bioprocessing, 9(1). https://doi.org/10.1186/s40643-022-00532-4

Stanković, J., Milošević, D., Savić-Zdraković, D., Yalçın, G., Yildiz, D., Beklioğlu, M., & Jovanović, B. (2020). Exposure to a microplastic mixture is altering the life traits and is causing deformities in the non-biting midge Chironomus riparius Meigen (1804). Environmental Pollution, 262. https://doi.org/10.1016/j.envpol.2020.114248

Su, T., Zhang, T., Liu, P., Bian, J., Zheng, Y., Yuan, Y., Li, Q., Liang, Q., & Qi, Q. (2023). Biodegradation of polyurethane by the microbial consortia enriched from landfill. Applied Microbiology and Biotechnology, 107(5–6), 1983–1995. https://doi.org/10.1007/s00253-023-12418-2

Taniguchi, I., Yoshida, S., Hiraga, K., Miyamoto, K., Kimura, Y., & Oda, K. (2019). Biodegradation of PET: Current Status and Application Aspects. ACS Catalysis, 9(5), 4089–4105. https://doi.org/10.1021/acscatal.8b05171

Thiounn, T., & Smith, R. C. (2020). Advances and approaches for chemical recycling of plastic waste. Journal of Polymer Science, 58(10), 1347–1364. https://doi.org/10.1002/pol.20190261

Tokiwa, Y., & Suzuki, T. (1978). Hydrolysis of Polyesters by Rhizopus delemar Lipase. Agricultural and Biological Chemistry, 42(5), 1071–1072. https://doi.org/10.1271/bbb1961.42.1071

Tokiwa, Y., Calabia, B. P., Ugwu, C. U., & Aiba, S. (2009). Biodegradability of plastics. International Journal of Molecular Sciences, 10(9), 3722–3742. https://doi.org/10.3390/ijms10093722

Tsuji, H., & Miyauchi, S. (2001). Poly(L-lactide): VI. Effects of crystallinity on enzymatic hydrolysis of poly(L-lactide) without free amorphous region. Polymer Degradation and Stability, 71(3), 415–424. https://doi.org/10.1016/S0141-3910(00)00191-9

Urbanek, A. K., Rybak, J., Wróbel, M., Leluk, K., & Mirończuk, A. M. (2020). A comprehensive assessment of microbiome diversity in Tenebrio molitor fed with polystyrene waste. Environmental Pollution, 262. https://doi.org/10.1016/j.envpol.2020.114281

Urbanski, B. Q., Denadai, A. C., Azevedo-Santos, V. M., & Nogueira, M. G. (2020). First record of plastic ingestion by an important commercial native fish (Prochilodus lineatus) in the middle tietê river basin, southeast brazil. Biota Neotropica, 20(3), 1–6. https://doi.org/10.1590/1676-0611-bn-2020-1005

Vargas-Suárez, M., Savín-Gámez, A., Domínguez-Malfavón, L., Sánchez-Reyes, A., Quirasco-Baruch, M., & Loza-Tavera, H. (2021). Exploring the polyurethanolytic activity and microbial composition of landfill microbial communities. Applied Microbiology and Biotechnology, 105(20), 7969–7980. https://doi.org/10.1007/s00253-021-11571-w

Varjani, S. J., & Upasani, V. N. (2017). A new look on factors affecting microbial degradation of petroleum hydrocarbon pollutants. International Biodeterioration and Biodegradation, 120, 71–83. https://doi.org/10.1016/j.ibiod.2017.02.006

Verschoor, J. A., Croese, M. R. J., Lakemeier, S. E., Mugge, A., Burgers, C. M. C., Innocenti, P., Willemse, J., Crooijmans, M. E., van Wezel, G. P., Ram, A. F. J., & de Winde, J. H. (2024). Polyester degradation by soil bacteria: identification of conserved BHETase enzymes in Streptomyces. Communications Biology, 7(1). https://doi.org/10.1038/s42003-024-06414-z

Walker, T. R., & Fequet, L. (2023). Current trends of unsustainable plastic production and micro(nano)plastic pollution. TrAC - Trends in Analytical Chemistry, 160. https://doi.org/10.1016/j.trac.2023.116984

Wallace, P. W., Haernvall, K., Ribitsch, D., Zitzenbacher, S., Schittmayer, M., Steinkellner, G., Gruber, K., Guebitz, G. M., & Birner-Gruenberger, R. (2017). PpEst is a novel PBAT degrading polyesterase identified by proteomic screening of Pseudomonas pseudoalcaligenes. Applied Microbiology and Biotechnology, 101(6), 2291–2303. https://doi.org/10.1007/s00253-016-7992-8

Wang, J., Peng, C., Li, H., Zhang, P., & Liu, X. (2021). The impact of microplastic-microbe interactions on animal health and biogeochemical cycles: A mini-review. Science of the Total Environment, 773. https://doi.org/10.1016/j.scitotenv.2021.145697

Wang, P., Zhao, J., Ruan, Y., Cai, X., Li, J., Zhang, L., & Huang, H. (2022). Degradation of Polypropylene by the Pseudomonas aeruginosa Strains LICME WZH-4 and WGH-6. Journal of Polymers and the Environment, 30(9), 3949–3958. https://doi.org/10.1007/s10924-022-02480-8

Wang, Y. L., Lee, Y. H., Chiu, I. J., Lin, Y. F., & Chiu, H. W. (2020). Potent impact of plastic nanomaterials and micromaterials on the food chain and human health. International Journal of Molecular Sciences, 21(5). https://doi.org/10.3390/ijms21051727

Waqas, M., Haris, M., Asim, N., Islam, H. ul, Abdullah, A., Khan, A., Khattak, H., Waqas, M., & Ali, S. (2021). Biodegradable Potential of Bacillus amyloliquefaciens and Bacillus safensis Using Low Density Polyethylene Thermoplastic (LDPE) Substrate. European Journal of Environment and Public Health, 5(2), em0069. https://doi.org/10.21601/ejeph/9370

Ward, P. G., Goff, M., Donner, M., Kaminsky, W., & O'Connor, K. E. (2006). A two step chemo-biotechnological conversion of polystyrene to a biodegradable thermoplastic. Environmental Science and Technology, 40(7), 2433–2437. https://doi.org/10.1021/es0517668

Ware, R. L., Rowland, S. M., Rodgers, R. P., & Marshall, A. G. (2017). Advanced Chemical Characterization of Pyrolysis Oils from Landfill Waste, Recycled Plastics, and Forestry Residue. Energy and Fuels, 31(8), 8210–8216. https://doi.org/10.1021/acs.energyfuels.7b00865

Wichatham, K., Piyaviriyakul, P., Boontanon, N., Surinkul, N., Visvanathan, C., Fujii, S., & Boontanon, S. K. (2024). Biodegradation of polypropylene plastics in vitro and natural condition by Streptomyces sp. isolated from plastic-contaminated sites. Environmental Technology and Innovation, 35. https://doi.org/10.1016/j.eti.2024.103681

Xu, Y., Xian, Z. N., Yue, W., Yin, C. F., & Zhou, N. Y. (2023). Degradation of polyvinyl chloride by a bacterial consortium enriched from the gut of Tenebrio molitor larvae. Chemosphere, 318. https://doi.org/10.1016/j.chemosphere.2023.137944

Yang, S. S., Ding, M. Q., He, L., Zhang, C. H., Li, Q. X., Xing, D. F., Cao, G. L., Zhao, L., Ding, J., Ren, N. Q., & Wu, W. M. (2021). Biodegradation of polypropylene by yellow mealworms (Tenebrio molitor) and superworms (Zophobas atratus) via gut-microbe-dependent depolymerization. Science of the Total Environment, 756. https://doi.org/10.1016/j.scitotenv.2020.144087

Yang, Y., Chen, J., Chen, Z., Yu, Z., Xue, J., Luan, T., Chen, S., & Zhou, S. (2022). Mechanisms of polystyrene microplastic degradation by the microbially driven Fenton reaction. Water Research, 223. https://doi.org/10.1016/j.watres.2022.118979

Yang, Y., Yang, J., Wu, W. M., Zhao, J., Song, Y., Gao, L., Yang, R., & Jiang, L. (2015). Biodegradation and Mineralization of Polystyrene by Plastic-Eating Mealworms: Part 2. Role of Gut Microorganisms. Environmental Science and Technology, 49(20), 12087–12093. https://doi.org/10.1021/acs.est.5b02663

Yoshida, S., Hiraga, K., Takehana, T., Taniguchi, I., Yamaji, H., Maeda, Y., Toyohara, K., Miyamoto, K., Kimura, Y., & Oda, K. (2016). A bacterium that degrades and assimilates poly(ethylene terephthalate). Science, 351(6278), 1196–1199. https://doi.org/10.1126/science.aad6359

Zadjelovic, V., Erni-Cassola, G., Obrador-Viel, T., Lester, D., Eley, Y., Gibson, M. I., Dorador, C., Golyshin, P. N., Black, S., Wellington, E. M. H., & Christie-Oleza, J. A. (2022). A mechanistic understanding of polyethylene biodegradation by the marine bacterium Alcanivorax. Journal of Hazardous Materials, 436. https://doi.org/10.1016/j.jhazmat.2022.129278

Zhang, A., Hou, Y., Wang, Q., & Wang, Y. (2022). Characteristics and polyethylene biodegradation function of a novel cold-adapted bacterial laccase from Antarctic sea ice psychrophile Psychrobacter sp. NJ228. Journal of Hazardous Materials, 439. https://doi.org/10.1016/j.jhazmat.2022.129656

Zhang, Y., Pedersen, J. N., Eser, B. E., & Guo, Z. (2022). Biodegradation of polyethylene and polystyrene: From microbial deterioration to enzyme discovery. Biotechnology Advances, 60. https://doi.org/10.1016/j.biotechadv.2022.107991

Zhang, Z., Peng, H., Yang, D., Zhang, G., Zhang, J., & Ju, F. (2022). Polyvinyl chloride degradation by a bacterium isolated from the gut of insect larvae. Nature Communications, 13(1). https://doi.org/10.1038/s41467-022-32903-y

Zhao, X., Zhou, Y., Liang, C., Song, J., Yu, S., Liao, G., Zou, P., Tang, K. H. D., & Wu, C. (2023). Airborne microplastics: Occurrence, sources, fate, risks and mitigation. Science of the Total Environment, 858. https://doi.org/10.1016/j.scitotenv.2022.159943

Zher Neoh, S., Fey Chek, M., Tiang Tan, H., Linares-Pastén, J. A., Nandakumar, A., Hakoshima, T., & Sudesh, K. (2022). Polyhydroxyalkanoate synthase (PhaC): The key enzyme for biopolyester synthesis. Current Research in Biotechnology, 4, 87–101. https://doi.org/10.1016/j.crbiot.2022.01.002

Zhou, C., Fang, W., Xu, W., Cao, A., & Wang, R. (2014). Characteristics and the recovery potential of plastic wastes obtained from landfill mining. Journal of Cleaner Production, 80, 80–86. https://doi.org/10.1016/j.jclepro.2014.05.083

Zhu, L., Zhu, J., Zuo, R., Xu, Q., Qian, Y., & AN, L. (2023). Identification of microplastics in human placenta using laser direct infrared spectroscopy. Science of the Total Environment, 856. https://doi.org/10.1016/j.scitotenv.2022.159060

Ziajahromi, S., Kumar, A., Neale, P. A., & Leusch, F. D. L. (2018). Environmentally relevant concentrations of polyethylene microplastics negatively impact the survival, growth and emergence of sediment-dwelling invertebrates. Environmental Pollution, 236, 425–431. https://doi.org/10.1016/j.envpol.2018.01.094

16

Bioremediation of Plastics for Sustainable Environment

S. Abirami

Department of Microbiology, Madras Christian College, Chennai

**Corresponding Author Email: abirami@mcc.edu.in*

Abstract

Due to the widespread usage of plastics and the ensuing MPs pollution, there have been detrimental effects on the environment. Polyethylene(PE), Polyvinylchloride(PVC), Polyethene terephthalate (PET), Polypropylene (PP), Polystyrene (PS), and Polyurethane(PU) are the six main polymers that are frequently used in plastic products. These wastes has its impacts on all forms of life. As a result, creating efficient and ecologically beneficial methods for breaking down these plastics has gained importance. Plastics are prone to both abiotic and biotic degradation processes, which encompass environmental chemical, physical, and biological responses.

Microorganisms can absorb and mineralize degradation products when plastics undergo advanced deterioration because of their low molecular weight. It has been claimed that a number of microbes possess the ability to degrade plastic; however, their slow pace of disintegration and incapacity to overcome resistance provided by some compounds restrict its use on a broad scale. The process of genetic modification aids in improves the way that polymers degrade biologically. Microorganisms are thought to be innate decomposers of natural polymers, converting them into value added compounds through bioconversion.

Keywords: *Plastic degradation, Biodeterioration, Biofragmentation, Mineralization*

1. Introduction

Plastic pollution is a global problem that requires immediate attention due to its negative impacts on ecosystems, infrastructure, (MacLeod, M., Arp *et al.* 2021) society, and the economy (Honingh, D. *et al.*,2020).

In terms of absolute emissions, we find that South-eastern part of Asia, and Sub-Saharan part of Africa possess the greatest quantities of plastic pollution, with India emitting the most (9.3 Mt year−1), or about one-fifth of the world's total plastic pollution (Joshua W. Cottom *et al.*).

Future regulations pertaining to the manufacturer, usage, and getting rid of plastics will be influenced by the discussions for a worldwide plastic pollution treaty. A baseline with great resolution of waste flows and sources of plastic emissions will be useful for the stakeholders in order to identify pollution hotspots and the reasons behind them (United Nations Environment Programme). Furthermore, in addition to the production of municipal waste, the pandemic of coronavirus illness has raised the usage of disposable polymer products, masks, tissues, gloves, and additional personal protection equipment, which made the plastic pollution disaster worse (De-la-Torre GE *et al.,* 2021).

Wasted plastics can break down and fragment into three different sizes: 1 mm to 5 mm big microplastics, 1 μm to 1 mm microplastics, and 1 nm to 1 μm nano plastics, (Waller *et al.,* 2017; De-la-Torre 2020; Atugoda *et al.,* 2022) being exposed to ultraviolet (UV) light and other processes that cause environmental damage

1.1. Types of Plastics

The creation of different thermoplastic materials is the main focus of industrial-level plastic production because of their necessity for numerous commodities. Polystyrene (PS), polypropylene (PP), polyvinyl chloride (PVC), low-density polyethylene (LDPE), high-density polyethylene (HDPE), and polyethylene terephthalate (PET), and polyurethane (PU) are the six main categories of plastics.

Polyethylene is one of the most commonly produced plastics and comes in two varieties: LDPE and HDPE. Due to its hardness and brittleness, HDPE is excellent for both domestic items (like toys and utensils) and construction (such pipes and window frames). The most common uses for LDPE, however, are in the production of plastic bags, packaging, containers, etc. (Kaushal J *et al.*, Singh N, Hui D *et al.,* 2021). Another kind of plastic called PP is used to make single-use straws and bottle caps. It is the least expensive, fireproof, and densest of all the polymers (Alsabri A, Tahir F *et al.*). PET manufacture is widely utilized in single-use plastic bottles (PET market volume worldwide Statista) and because of its mechanical strength, PVC is used as the foundation material to produce home goods including floor coverings, shower curtains, pipes, and window frames . (Ru J, Huo Y, *et al.,* 2020). In general, comprehending the properties of plastic polymers, as seen in table 1, requires a grasp of their classification.

Table 1. Major types of commercially available plastics

Plastic types		Back bone	Density	Melting temperature	Plastic utilization (%)
Polyethylene (PE)	HDPE	C-C (Non Hydrolysable)	0.94-0.97g/ cm^3	130-140°C	36.30%
	LDPE	C-C (Non Hydrolysable)	0.917-0.940g/ cm^3	85-125°C	
Polypropylene (PP)		C-C (Non Hydrolysable)	0.9-0.91g/cm^3	165-175°C	21%
Polyvinyl chloride (PVC)		C-C (Non Hydrolysable)	1.3-1.7g/cm^3	100-260°C	11.80%
Polystyrene (PS)		C-C (Non Hydrolysable)	1.04-1.05g/ cm^3	240°C	7.60%
Polyethylene terephthalate (PET)		C-O (Hydrolysable)	1.3-1.4g/cm^3	245-265°C	10.20%
Polyurethane(PU)		C-O (Hydrolysable)	1.23-1.25g/ cm^3	65-121°C	8.20%

2. Detrimental Impact of Plastic Waste on the Environment

Numerous dangerous gaseous chemicals, such as Dioxins, hydrogen cyanide, carbon monoxide, and nitrogen oxides are discharged into the atmosphere throughout the plastics manufacturing process. These pollutants are extremely dangerous to the environment and human health (S.S. Ali, T. Elsamahy, E. Koutra *et al.*,2020). Low density polyethylene (LDPE) produces trace gases that rise with incubation time (212 days) to methane at 5.8 nmol/g/day and ethylene at 14.5 nmol/g/day. As more plastic is made and deposited in the environment, trace gas emissions are predicted to rise, suggesting that plastics may be an unidentified contributor to climate change (Royer *et al.,*2018). Plastic products are now widely found in all habitats worldwide. because of their low biodegradability and persistent character, they will remain in these surroundings for a long time (Andrady 2017; Aragaw 2020;). Because plastic trash takes 500 years to fully decompose, it damages the microbiological content of soil and causes infertility when it is dumped into landfills (Chamasetal., 2020).

Organic pollutants, chemicals, heavy metals, and pathogens can all be transported by plastic garbage (Cregut *et al.,* 2013; Galloway and Lewis, 2016; Chamas *et al.,* 2020; Tang *et al.,* 2021). Additionally, highly toxic compounds are released during the abiotic degradation of plastic (Chen *et al.,* 2019), and plants and soil organism activity can move this fragmented MP deeper into the soil, (Zhu *et al.,* 2018) contaminating ground water and degrading soil qualities (Scheurer and Bigalke, 2018).

Microplastics can inadvertently infiltrate the food chain since the majority of them possess a range of sizes that is comparable for the foods that zooplankton often eat (Gregory 2009). The choice of microplastics over food particles may result in sublethal impacts on the reproductive pattern of the species as well as a loss of energy resources (Enders *et al.*, 2015). Furthermore, intestinal obstruction caused by the plastic that is consumed may change the balance of hormones and reduce the absorption of nutrients (Derraik, 2002).

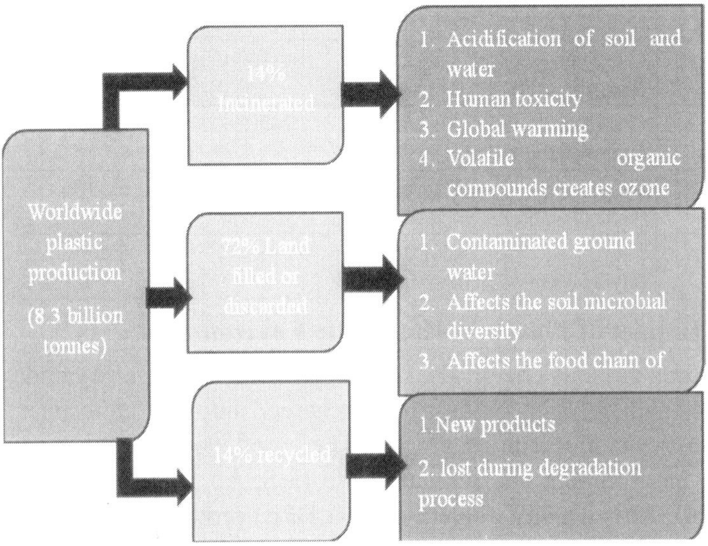

Fig. 1: Detrimental impacts of plastic waste

3. Molecular Mechanism of Plastic Degradation

3.1. Abiotic Activity

When polymeric materials are exposed to external factors including weather, aging, and burial, they may experience more or less significant changes in their mechanical, light, thermal, and chemical properties. The polymeric materials' capacity to biodegrade is altered by this exposure. Abiotic factors typically help to degrade the polymeric structure, which favours undesired changes (Helbling *et al.*, 2006; Ipekoglu *et al.*, 2007). The abiotic characteristics might be used to start the biodegradation process or as a synergistic component. To better estimate the endurance of polymeric materials, it is vital to investigate the role of abiotic environments (Nathalie Lucasa *et al.* 2008).

3.1.1. Light Degradation

There are a number of photosensitive materials. Photon energy has the ability to induce unstable states in a variety of compounds. Thermal radiation, fluorescence, photoionization, and luminescence can all be used to transmit energy. Occasionally, contaminants found in manufactured goods might unintentionally alter the material's resistance. In other instances, light-induced macromolecular degradation is induced by the deliberate addition (i.e., copolymerization or simple addition) of photosensitive molecular structures (For instance, prooxidant compounds that can be activated based on exposure duration and light intensity) into the polymer framework (Kounty *et al.*, 2006; Wiles and Scott, 2006). Manufacturers of polyolefins employ this tactic to improve the biodegradability of agricultural films, plastic bags, and packaging, and other materials (Weiland *et al.*, 1995; Schyichuk *et al.*, 2001).

3.1.2. Thermal Degradation

Depending on its molecular weight, thermoplastic polymers undergo thermal degradation at the melting point when they change from solid to liquid (for example:159–178°C for L-PLA, 137–169°Cfor P(HB/HV) (poly hydroxybutyrate-co hydroxy valerate) (Ojumu *et al.*, 2004). The melting point of thermoplastic polymers is typically higher than the ambient temperature. Nonetheless, melting temperatures for certain PCL and other thermoplastic polymers (tm 60 C), are close to ambient conditions. For the thermophile stage of composting, this is true.

3.1.3. Chemical Degradation

The other crucial factor in abiotic deterioration is chemical transformation. Polymers may interact with agrochemicals and atmospheric contaminants to alter the characteristics of the macromolecule (Briassoulis, 2005). Oxygen is the most potent of the substances that cause materials to degrade. Free radicals are created when oxygen is in its atmospheric form, such as O2 or O3, breaks the covalent bonds. The polymer structure—such as unsaturated bonds and branching chains—determines the oxidative breakdown (Duval, 2004). To create free radicals, these oxidations may occur concurrently or in concert with light degradation.

3.1.4. Mechanical Deterioration

Shear, tension, and/or compression stresses can all result in mechanical degeneration. Several elements, such as restrictions during the installation of materials, load-induced aging, turbulence in water and air, decreased pressure, and minor harm, are among the many causes of these forces. Accordingly, in field conditions, thermoplastic films (such as mulches, low-tunnel films, etc.) can experience a variety of mechanical degradations (Briassoulis, 2004, 2006, 2007).

Physicochemical pre-treatment, which includes agents that oxidize chemicals, thermo oxidation, and UV exposure of microplastics, is the primary degradation process. This process reduces the length of the polymer and alters the microplastics functional group, increasing their susceptibility of microbial enzyme activity.

3.2. Biotic Activity

There are five primary processes by which microbes break down plastic waste: colonization, biodeterioration, bio fragmentation, absorption, and mineralization.

3.2.1. Colonization

In the initial phase, Microbial species' colonization of the plastic surface is a component of the degradation mechanism. Microbes create biofilms and consortia during this stage, seriously harming the polymer surface (Gu, 2003). Microbes stick to the polymer surface by producing various proteins and polysaccharides that enter the material pores and change the size of those pores. Therefore, the biodegradation step is caused by microbial adherence and the plastic surface's activity (Capitelli *et al.*, 2006). Surface deterioration during this phase modifies the plastic polymer's physicochemical characteristics.

3.2.2. Biodeterioration

Biodeterioration is the process by which different microorganisms, filamentous fungi, and other microbes utilize their mycelial to pierce the polymer, increase the size of the pores, and cause fractures to form. This reduces the durability and resistance of the polymer (Bonhomme *et al.*, 2003). Additionally, the microbial extracellular release molecules influence phases that are hydrophilic and hydrophobic can increase the rate of microbial penetration. The extracellular chemicals promote the buildup of pollutants, which in turn promotes microbial growth and the pace of biodeterioration Zanardini *et al.*, (2000).

Physical biodeterioration is caused by filamentous microorganisms, such as fungi and bacteria, which can infiltrate the polymer material and enlarge its holes and fissures (Hakkarainen *et al.*, 2000).

Oxidation-reductions and chemical biodeterioration are also possible in certain microorganisms, Iron ($Fe3+$) and/or manganese cations can be absorbed by chemolithotrophic bacteria via certain proteins in their cell membrane (Pelmont, 2005).

Peroxidases are among the several extracellular enzymes implicated in enzymatic biodeterioration (Otsuka *et al.*, 2003). In few plastic polymers, like PU and PVC, are not susceptible to deterioration (Shah *et al.*, 2008). In order to overcome the crystallinity of these polymers, certain bacteria have a tendency to create enzyme groups, including proteases, ureases, lipases, and esterases (Ameen *et al.,* 2015; Yoshida *et al.,* 2016).

3.2.3. Fragmentation

The breakdown into monomers, dimers, and/or oligomers from polymers depends on the lytic process of fragmentation. Microorganisms cleave polymers by a variety of methods, such as the release of certain enzymes like hydrolases and oxidoreductases and/or free radicals. The monomers that make up plastic are joined by chemical linkages that resemble those in lignocellulosic compounds and other natural polymers. Thus, it makes sense to believe that the majority of the enzymes that break down lignocelluloses may also break down plastic polymers (Chen *et al.,* 2020).

This process involves the microbial endo- and exoenzymes, which are not created instantly. The creation of some enzymes requires time to initiate the cell machinery Lucasa *et al.,* (2008). Furthermore, the duration required for microorganisms to break down plastic is explained by the fact that the concentration of enzymes rises with time and that the activity ends when the substrate is exhausted.

Electron, photonic, and/or polarization microscopy can be used to study this stage.

3.2.4. Assimilation

Regarding the absorption of the generated simpler molecules, the cell of microbes use these integrated monomers as energy sources, which enables the growth of the microbial cell. Some monomers come in the cell via individually particular membrane carriers. The permeability of membranes of cells prevents assimilation of other molecules.

The microbiological cell may then employ the non-assimilated monomers through the process of biotransformation, which includes conversion that is catalysed by enzymes (Lucasa *et al.,* 2008; Al-Tohamy *et al.,* 2020). Aerobic respiration, an aerobic respiration, and fermentation are the three pathways by which the microbial catabolic pathways oxidize the broken-down monomers inside the microbial cell to create Adenosine triphosphate (ATP) and other molecules that carry energy in cell organelles, in order to maintain cellular activity, structure, and reproduction as well as to produce energy (Hong and Gu, 2009; Al-Tohamy *et al.,* 2020b; Ali *et al.,* 2020c, 2020d).

TGA, UV-visible spectroscopy, High performance liquid chromatography (HPLC) and/or Fourier-transform infrared spectroscopy (FT-IR) can all be employed to confirm the assimilation stage (Sameh S. Alia,b, 2021 V.Mahalakshmi., 2014)

3.2.5. Mineralization

3.2.5.1. Aerobic Respiration

Many microorganisms can use oxygen as the last electron acceptor through aerobic respiration. These microbes require oxidized substrates that enter the cell. First, a limited amount of energy is produced via fundamental catabolic pathways, such as purine and pyrimidine catabolism, amino acid catabolic processes, glycolysis, and β-oxidation. Second, electron transport mechanisms that convert oxygen to water realize oxidative phosphorylation, which subsequently generate more energy (Moussard, 2006).

3.2.5.2. Anaerobic Respiration

Oxygen cannot be used as the final electron acceptor by a number of microbes. They can, however, attain complete oxidation in membrane systems through altered electron transport. Other than oxygen, the last electron acceptors include fumarate, NO3, SO2, CO2, and Fe3+ Brock and Madigan (1991). Moreover, more ATP molecules are synthesized as a result compared to a partial oxidation.

3.2.5.3. Fermentation

Electron transport mechanisms are absent in certain microbes. They have evolved to exploit exogenous mineral molecules, such as oxygen, as ultimate electron acceptors. Their only means of energy production is fermentation, an imperfect oxidation route. As ultimate electron acceptors, the cell uses its own endogenous organic compounds. Organic and/or mineral substances discharged into the environment, including butanediol, ethanol, lactate, acetate, and CO2, are examples of fermentation products (Regnault, 1990; Brock and Madigan, 1991; Alcamo, 1998).

Since these molecules still have the ability to reduce, other species can often exploit them as sources of carbon. Since microorganisms follow the biogeochemical cycles, the mineral compounds they produce often do not pose an ecotoxicity risk. On the other hand, under some circumstances and to varying degrees, microbial organic compounds that are expelled or altered may pose ecotoxic risks. Since these molecules still possess a reduction power, they can often be utilized by other species as sources of carbon. Due to their adherence to the biogeochemical cycles, mineral compounds produced by microbes often do not pose a danger of ecotoxicity. Contrarily, under certain circumstances and at varying concentrations, microbially produced organic compounds may provide ecotoxic risks.

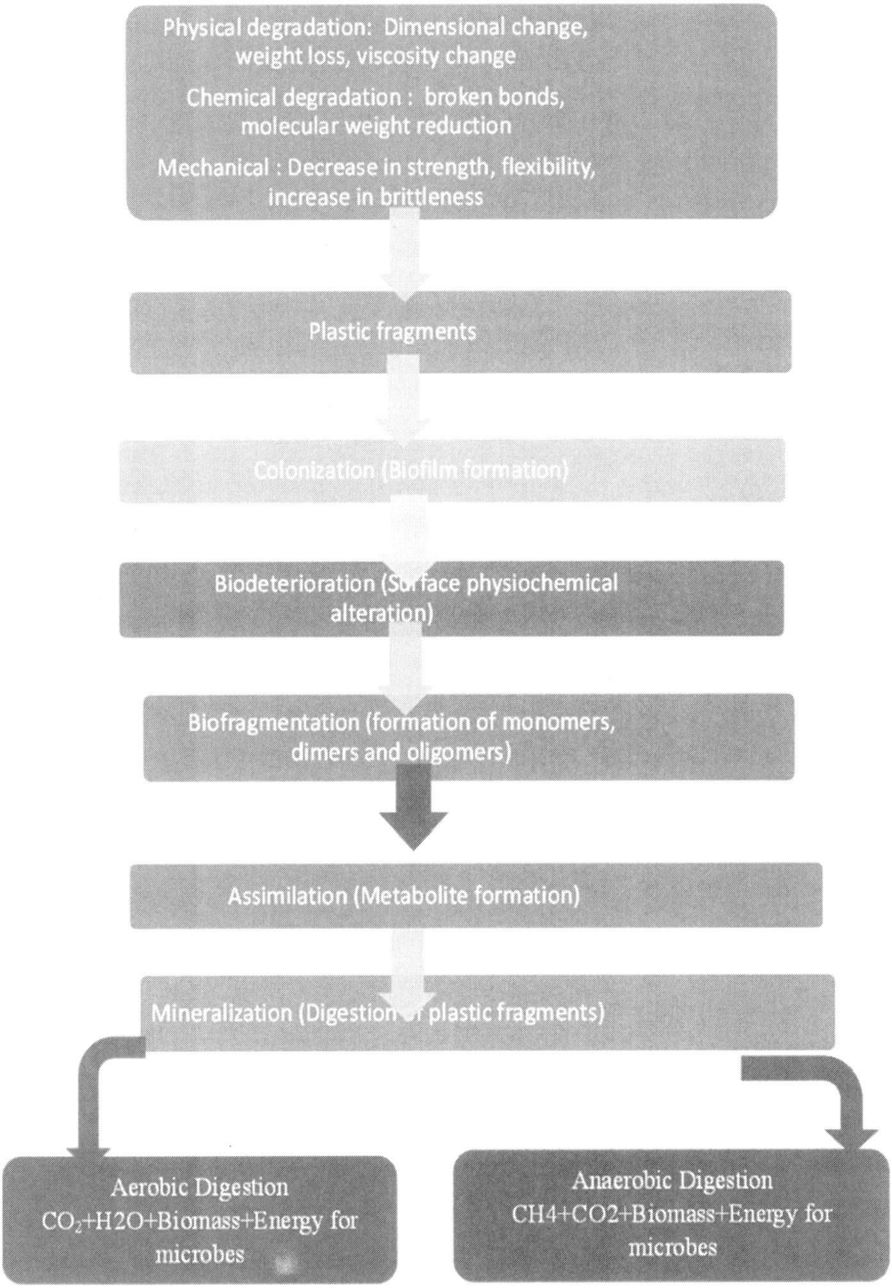

Fig. 2: Molecular Mechanisms involved in Plastic Degradation

4. Microbial Degradation of Plastics

Over 90 distinct kinds of Aerobes, anaerobes, photosynthetic bacteria, and other microorganisms archaebacteria, together with lower eukaryotic microbes, are in charge of bioplastics' catabolism and biodegradation. Soil and compost materials are rich sources of these microorganisms (Lee *et al.,* 2005; Kumaravel *et al.,* 2010; Accinelli *et al.,* 2012). When bacteria or fungi break down bioplastics, a clear- zone forms around growth in a plate with the bio-plastic as sole sources of carbon. After that, the diameter is considered for the extension of biodegradation (Tezuka *et al.,* 2004; Lee *et al.,* 2005; Ishii *et al.,* 2008; Kumaravel *et al.,* 2010).

Scanning electron microscopy (SEM) examinations are widely used by researchers to determine how biological activity alters polymer structure. Furthermore, the change in the strength of the bond, which was caused by microbial deterioration, and the Fourier Transform Infrared spectroscopy - technique was employed to find it (Wu, 2009, 2011, 2014; Phukon *et al.,* 2012).

Through active involvement of intracellular and extracellular enzymes, microorganisms primarily facilitate the deterioration of plastics, which leads to their mineralization.

4.1. Degradation of Plastics by Algae

Microalgae, together with their toxins and enzymes, can aid in the biological degradation of polymeric substances (Manzi *et al.,* 2022). Their main benefit is that they can adapt to a variety of conditions, which is where most microplastics are located, and they do not require a high carbon supply to flourish, unlike the system of bacteria, microalgae, which are recognized for colonizing surfaces made of plastic in wastewater streams, initiates degradation of plastic by producing enzymes that break down lignin and exopolysaccharides (Yan *et al.,* 2016). The main function among these polymers is to supply carbon, which increases the amount of carbs and proteins inside cells and promotes faster expanding. Only lately the exterior degradation or disintegration of polyethylene with a low density sheets caused by algal colonization has been found. using SEM (Sanniyasi *et al.,* 2021).

Numerous factories of genetically altered microalgal cells that can generate and secrete the biocatalyst required for breakdown of plastic have been established thanks to recent advancements in a variety of biotechnological techniques (Shen *et al.,* 2019). The green microalgae-Chlamydomonas reinhardtii underwent genetic modification to create polyethylene terephthalate hydrolase biocatalyst, which degrades polyethylene terephthalate films and terephthalic acid (Kim *et al.,* 2020). A comparable change was successfully accomplished inside *P. tricornutum* by generating polyethylene terephthalate hydrolase as well as demonstrating catalytic activity against the copolymer polyethylene terephthalate glycol and polyethylene terephthalate itself (Moog *et al.,* 2019).

4.2. Degradation of Plastics by fungi

Saprotrophs, which are obligatory or opportunistic parasites, comprise the majority of the fungus organisms. Because of their remarkable adaptability, they could thrive in a wide range of both land and water habitats with a variety of environmental circumstances. They generate a range comprising natural biosurfactants such hydrophobins and extracellular enzymes, that may break down intricate polymers into more basic monomers, and they are resistant to a number of dangerous chemicals and metals. Additionally, they provide microorganisms with carbons and electrons, which aid in the breakdown and mineralization of complicated contaminants (Olicón-Hernández *et al.,* 2017).

Polyethylene terephthalate, polypropylene, and polyethylene are among few polymers that are degraded by the primary genus *Aspergillus niger, Cladosporium, Penicillium simplicissimum,* and *Zalerion maritimum.* (Paço *et al.,* 2017; de Oliveira *et al.,* 2020a, b; Devi *et al.,* 2015). Comparable polyurethane degradation was demonstrated by isolates of *Pestalotiopsis microspora, Fusarium solani, Penicillium chrysogenum, Aspergillus fumigatus, Aspergillus tubingensis,* and *Cladosporium pseudo cladosporioides.* (Khan *et al.,* 2017; Álvarez-Barragán *et al.,* 2016; Magnin *et al.,* 2020; 2015; Russell *et al.,* 2011)

4.3. Degradation of Plastics by Bacteria

Several studies using microorganisms to degrade microplastics have been conducted. Microplastics may be broken down by bacteria found in a range of settings, includes sludge compost, wastewater, and municipal dumps, and contaminated sediments (Mehmood *et al.,* 2016; Awasthi *et al.,* 2020). Additionally, microbes that degrade plastics have additionally been identified in the stomach microbiota of earth-worms. It's well known the microorganisms living in polluted environments have the capacity to turn on the enzyme mechanism in charge of plastics' breakdown.

The breakdown of microplastics can be accomplished using bacterial consortiums or pure cultures. Conversely, Microbial consortium provide a number of advantages in the degradation process and offer a useful tool to investigate the metabolic processes at play. Additionally, it's simpler to monitor the consequences of external factors on temperature, pH, and substrate characteristics, and the surfactants that affect the breakdown process (Janssen *et al.,* 2002). However, the extremely slow rate of disintegration is the main disadvantage. Consequently, to optimize circumstances, more innovative approaches are needed. It is often desirable to use a consortium of bacteria because biodegradation by a single bacterium frequently produces harmful end products, which, in a stable microbial ecosystem, can be effectively eliminated (Singh and Wahid 2015).

Within 60 days, the dry mass of microplastic particles decreased by 14.7% by a bacterial consortium that included *Bacillus* and *Paenibacillus* species (Park & Kim 2019). Additionally, low-density polyethylene is broken down by earthworms (Huerta Lwanga *et al.,* (2018). In addition to being examined independently, the intestinal isolates from the genus Firmicutes and Actinobacteria were found to be capable of breaking down low-density polyethylene microplastics and releasing eicosane, docosane, and tricosane, among other volatile chemicals, the Cow manure contains the mixture of Enterobacter and Pseudomonas. promoted reduction in weight by 14% in 110 days (Skariyachan *et al.,* 2021).

5. Biotechnological Methods to Enhance the Breakdown of Plastics

With development of various methods of genetic engineering, it is possible to alter the genetic composition of microorganisms to improve their ability to decompose. Numerous methods, including as genetic alteration, cloning, and rDNA technologies, have been used to boost the microorganisms' potential for bio-remediation in the presence of different hydrocarbons and heavy metals (Kumar *et al.,* 2020). But up until now, not much research has been done on using genetic engineering to develop a superior strain for plastic decomposition.

These techniques can alter the specificity and affinity of enzymes for different microplastics and are used to develop new pathways. For gene editing to be successful, the right genes needed to break down and metabolize microplastics must be identified, as well as the appropriate host organisms, such as E. coli, that express these genes.

Site-directed mutagenesis, anti-sense ribonucleic acid (RNA) technology, and polymerase chain reaction are the primary procedures involved. The ability of synthetic antisense RNA to efficiently control gene expression in host cells has made antisense RNA technology a novel genetic editing tool. However, site directed mutagenesis can also change the function of genes linked to the decomposition of plastics.

Furthermore, genome editing makes it easier to alter a gene of interest, enabling studies that alter the expression of many genes through function loss and gain.

This method can successfully incorporate genes that encode enzymes like as laccase, esterase, depolymerase, dehalogenase, and polyethylene terephthalate hydrolase that are involved in the degradation of microplastics. However, biosafety concerns now restrict the application of genetically modified bacteria in on-site research, which restricts the actual assessment of the efficacy of gene editing in microplastic bioremediation.

5. Conclusion

To properly evaluate the potential use of microbes for polymer removal, more investigation is needed. Using microbes to biodegrade plastics has several limitations and challenges, but these is surmountable by various genetic changes. Also, there are few research on the effectiveness of genetically modified microbes in rural areas, and most have only undergone validation in lab settings. Recent developments in uncultivated microbial community engineering and metagenomic analysis, collected from polluted areas (Schloss and Handelsman, 2005), can help build novel bioremediation strategies. The gradual ecosystem consequences must be assessed through long-term, coordinated remediation efforts.

References

Accinelli, C., Saccà, M.L., Mencarelli, M., Vicari, A., 2012. Deterioration of bioplastic carrier bags in the environment and assessment of a new recycling alternative. Chemosphere 89, 136–143.

Ali, S.S., Kornaros, M., Manni, A., Sun, J., El-Shanshoury, A.E.R., Kenawy, E.-R., Khalil, M.A., 2020c. Enhancedanaerobic digestion performance by two artificially constructed mi crobial consortia capable of woody biomass degradation and chlorophenols detoxifi cation. J. Hazard. Mater. 389, 122076. https://doi.org/10.1016/j.jhazmat.2020.122076.

Ali, S.S., Mustafa, A.M., Kornaros, M., Manni, A., Sun, J., Khalil, M.A., 2020d. Construction of novel microbial consortia CS-5 and BC-4 valued for the degradation of catalpa saw dust and chlorophenols simultaneously with enhancing methane production. Bioresour. Technol. 301, 122720. https://doi.org/10.1016/j.biortech.2019.122720.

Alsabri A, Tahir F, Al-Ghamdi SG.(2022) Environmental impacts of polypropylene (PP) production and prospects of its recycling in the GCC region. Mater Today Proc. 2022;56:2245–51.

Al-Tohamy, R., Kenawy, E.R., Sun, J., Ali, S.S., 2020b. Performance of a newly isolated salt-tolerant yeast strain Sterigmatomyces halophilus SSA-1575 for azo dye decol orization and detoxification. Front. Microbiol. 11, 1163. https://doi.org/10.3389/fmicb.2020.01163.

Álvarez-Barragán J, Domínguez-Malfavón L, Vargas-Suárez M, González-Hernández R, Aguilar-Osorio G, Loza-Tavera H (2016) Biodegradative Activities of Selected Environmental Fungi on a Polyester Polyurethane Varnish and Polyether Pol yurethane Foams. Appl Environ Microbiol 82(17):5225–5235. https://doi.org/10.1128/AEM.01344-16

Andrady AL (2017) The plastic in microplastics: a review. Mar Pollut Bull 119:12–22. https://doi. org/ 10. 1016/j. marpolbul. 2017. 01. 082

Aragaw TA (2020) Surgical face masks as a potential source for microplastic pollution in the COVID-19 scenario. Mar Pollut Bull 159:111517. https:// doi. org/ 10. 1016/j. marpolbul. 2020. 111517

Awasthi AK, Tan Q, Li J (2020) Biotechnological potential for micro plastic waste. Trends Biotechnol 38(11):1196–1199. https:// doi. org/10.1016/j.tibtech.2020.03.002

Besseling E, Wegner A, Foekema EM, Van Den Heuvel-Greve MJ, Koelmans AA (2013) Effects of microplastic on fitness and PCB bioaccumulation by the lugworm Arenicola marina (L). Environ Sci Technol 47(1):593–600. https://doi.org/10.1021/es302763x

Bonhomme, S., Cuer, A., Delort, A.-M., Lemaire, J., Sancelme, M., Scott, G., 2003. Environ mental biodegradation of polyethylene. Polym. Degrad. Stab. 81, 441–452. https:// doi. org/10.1016/S0141-3910(03)00129-

Briassoulis, D., 2005. The effects of tensile stress and the agrochemicals Vapam on the ageing of low density polyethylene (LDPE) agricultural films. Part I. Mechanical behaviour. Polym. Degrad. Stab. 86, 489–503.

Briassoulis, D., 2006. Mechanical behaviour of biodegradable agricultural films under real field conditions. Polym. Degrad. Stab. 91, 1256–1272. Briassoulis, D., 2007. Analysis of the mechanical and degradation performances of optimised agricultural biodegradable films. Polym. Degrad. Stab. 92, 1115 1132.

Brock, T.D., Madigan, M.T., 1991. Biology of Microorganisms. Prentice-Hall, Inc. Call, H.P., Mücke, I., 1997. History, overview and applications of mediated lignolytic systems, especially laccase–mediator-systems (Lignozym process). J. Biotechnol. 53, 163–202.

Bugoni L, Krause L, Petry MV (2001) Marine debris and human impacts on sea turtles in southern Brazil. Mar Pollut Bull 42(12):1330–1334. https:// doi. org/ 10. 1016/ S0025-326X(01) 00147-3

Capitelli, F., Principi, P., Sorlini, C., 2006. Biodeterioration of modern materials incontemporary collections: can biotechnology help? Trends Biotechnol. 24 (8), 350–354. https://doi.org/10.1016/j.tibtech.2006.06.001.

Chamas, A., Moon, H., Zheng, J., Qiu, Y., Tabassum, T., Jang, J.H., Abu-Omar, M., Scott, S.L., Suh, S., 2020. Degradation rates of plastics in the environment. ACS Sustain. Chem. Eng. 8, 3494–3511. https://doi.org/10.1021/acssuschemeng.9b06635.

Chen, C.-C., Dai, L., Ma, L., Guo, R.-T., 2020. Enzymatic degradation of plant biomass and synthetic polymers. Nat. Rev. Chem. 4, 114–126. https://doi.org/10.1038/s41570 020-0163-6.

Chen, Q., Allgeier, A., Yin, D., Hollert, H., 2019. Leaching of endocrine disrupting chemicals from marine microplastics and meso plastics under common life stress conditions. Environ. Int. 130, 104938. https://doi.org/10.1016/j.envint.2019.104938.

Cregut, M., Bedas, M., Durand, M.-J., Thouand, G., 2013. New insights into polyurethane biodegradation and realistic prospects for the development of a sustainable waste recycling process. Biotechnol. Adv. 31 (8), 1634–1647. https://doi.org/10.1016/j.biotechadv.2013.08.011.

De-la-Torre GE, Dioses-Salinas DC, Pérez-Baca BL, Millones Cumpa LA, Pizarro- Ortega CI, Torres FG, Gonzales KN, Santillán L (2021) Marine macroinvertebrates inhabiting plastic litter in Peru. Mar Pollut Bull 167:112296. https:// doi. org/ 10. 1016/j.marpolbul.2021.112296

Derraik JGB (2002) The pollution of the marine environment by plastic debris: a review. Mar Pollut Bull 44:842–852. https:// doi. org/ 10. 1016/S0025-326X(02)00220-5

Devi RS, Kannan VR, Nivas D, Kannan K, Chandru S, Antony AR (2015) Biodegradation of HDPE by Aspergillus spp. from marine ecosystem of Gulf of Mannar, India. Mar Pollut Bull 96(1–2):32 40. https://doi.org/10.1016/j.marpolbul.2015.05.050

Duval, C., 2004. Matériaux dégradables In Matières plastiques et environnement Recyclage, valorisation, biodégradabilité, écoconception. DUNOD, Paris.

Enders K, Lenz R, Stedmon CA, Nielsen TG (2015) Abundance, size and polymer composition of marine microplastics≥ 10 μm in the Atlantic Ocean and their modelled vertical distribution. Mar Pollut Bull 100(1):70–81. https:// doi. org/ 10. 1016/j. marpo lbul. 2015.09.027

Galloway, T.S., Lewis, C.N., 2016. Marine microplastics spell big problems for future generations. Proc. Natl. Acad. Sci. 113, 2331–2333. https://doi.org/10.1073/ pnas.1600715113.

Gregory MR (2009) Environmental implications of plastic debris in marine settings–entanglement, ingestion, smothering, hangers on, hitch-hiking and alien invasions. Philos Trans R Soc Lond B Biol Sci 364(1526):2013–2025. https:// doi. org/ 10. 1098/ rstb. 2008.0265

Gu, J.-D., 2003. Microbiological deterioration and degradation of synthetic polymeric ma terials: recent research advances. Int. Biodeterior. Biodegradation 52, 69–91. https:// doi.org/10.1016/S0964-8305(02)00177-4.

Hakkarainen, M., Karlsson, S., Albertsson, A.-C., 2000. Rapid (bio)degradation of polylactide by mixed culture of compost microorganisms– low molecular weight products and matrix changes. Polymer 41, 2331–2338. https://doi.org/10.1016/ S0032-3861(99)00393-6.

Helbling, C., Abanilla, M., Lee, L., Karbhari, V.M., 2006. Issues of variability and durability under synergistic exposure conditions related to advanced polymer composites in civil infrastructure. Compos. Part A-Appl. Sci. 37 (8), 1102–1110.

Hong, Y., Gu, J.-D., 2009. Bacterial anaerobic respiration and electron transfer relevant to the biotransformation of pollutants. Int Biodeterior. Biodegrad. 63 (8), 973–980.https://doi. org/10.1016/j.ibiod.2009.08.001.

Honingh, D. *et al.*, (2020), Urban river water level increase through plastic waste accumulation at a rack structure. Front. Earth Sci. 8, 28.

Huerta Lwanga E, Thapa B, Yang X, Gertsen H, Salánki T, Geissen V, Garbeva P (2018) Decay of low-density polyethylene by bacteria extracted from earthworm's guts: a potential for soil restoration. Sci Total Environ 624:753–757. https:// doi. org/ 10. 1016/j. scito tenv.2017.12.144

Ipekoglu, B., Böke, H., Cizer, O., 2007. Assessment of material use in relation to climate in historical buildings. Build. Environ. 42, 970–978.

Ishii, N., Inoue, Y., Tagaya, T., Mitomo, H., Nagai, D., Kasuya, K., (2008). Isolation and characterization of poly(butylene succinate)-degrading fungi. Polym. Degrad. Stab. 93, 883–888.

Jakubowicz, I., Yarahmadi, N., Petersen, H., 2006. Evaluation of the rate of abiotic degradation of biodegradable polyethylene in various environments. Polym. Degrad. Stab. 91 (6), 1556–1562.

Janssen PH, Yates PS, Grinton BE, Taylor PM, Sait M (2002) Acido bacteria Actinobacteria Proteobacteria Verrucomicrobia. Appl Environ Microbiol 68(5):2391–2396. https:// doi. org/ 10. 1128/ AEM.68.5.2391-2396.2002

Joshua W. Cottom1, Ed Cook1 & Costas A. Velis A local-to-global emissions inventory of macroplastic pollution, Nature Vol 633, 5 September 2024

Kaushal J, Khatri M, Arya SK. (2021) Recent insight into enzymatic degradation of plastics prevalent in the environment: A mini - review. Cleaner Engineering and Technology. 2021;2:100083.

Khan S, Nadir S, Shah ZU, Shah AA, Karunarathna SC, Xu J, Hasan F (2017) Biodegradation of polyester polyurethane by Aspergil lus tubingensis. Environ Pollut 225:469–480. https://doi.org/10. 1016/j.envpol.2017.03.012

Kim, M.-N., Park, S.T., 2010. Degradation of poly(L-lactide) by a mesophilic bacterium. J. Appl. Polym. Sci. 117, 67–74.

Kounty, M., Lemaire, J., Delort, A.-M., 2006. Biodegradation of polyethylene films with prooxidant additives. Chemosphere 64, 1243–1252.

Kumar M, Xiong X, He M, Tsang DC, Gupta J, Khan E, Bolan NS (2020) Microplastics as pollutants in agricultural soils. Environ Pollut 265:114980. https:// doi. org/ 10. 1016/j. envpol. 2020. 114980

Kumaravel, S., Hema, R., Lakshmi, R., 2010. Production of polyhydroxybutyrate (bioplastic) and its biodegradation by Pseudomonas lemoignei and Aspergillus niger. E– J. Chem. 7, S536–S542.

Lee, K.-M., Gimore, D.F., Huss, M.J., 2005. Fungal degradation of the bioplastic PHB (poly-3-hydroxy- butyric acid). J. Polym. Environ. 13, 213–219.

Lucasa, N., Bienaimeb, C., Belloyc, C., Queneudeca, M., Silvestred, F., Nava-Saucedob, J.-E., 2008. Polymer biodegradation: mechanisms and estimation techniques. Chemosphere 73, 429–442. https://doi.org/10.1016/j.chemosphere.2008.06.064.

MacLeod, M., Arp, H. P. H., Tekman, M. B. & Jahnke, (2021), A. The global threat from plastic pollution. Science 373, 61–65

Manzi HP, Abou-Shanab RA, Jeon BH, Wang J, Salama ES (2022) Algae: a frontline photosynthetic organism in the microplastic catastrophe. Trends Plant Sci S1360–1385(22):00157–00161. https://doi.org/10.1016/j.tplants.2022.06.005

Mehmood CT, Qazi IA, Hashmi I, Bhargava S, Deepa S (2016) Bio degradation of low density polyethylene (LDPE) modified with dye sensitized titania and starch blend using Stenotrophomonas pavanii. Int Biodeterior Biodegrad 113:276–286. https:// doi. org/ 10.1016/j.ibiod.2016.01.025

Moog D, Schmitt J, Senger J, Zarzycki J, Rexer KH, Linne U, Erb T, Maier UG (2019) Using a marine microalga as a chassis for polyethylene terephthalate (PET) degradation. Microb Cell Fact 18(1):1–15. https://doi.org/10.1186/s12934-019-1220-z

Moussard, C., 2006. Biochimie Structurale et Métabolique, 3ème ed. De Boeck & Larcier, Bruxelles.

Nathalie Lucasa, Christophe Bienaimeb, Christian Belloyc, Michèle Queneudeca, FrançoiseSilvestred, José-Edmundo Nava-Saucedob, 2008 Polymer biodegradation: Mechanisms and estimation techniques doi:10.1016/j.chemosphere.2008.06.064

Ojumu, T.V., Yu, J., Solomon, B.O., 2004. Production of polyhydroxyalkanoates, a bacterial biodegradable polymer. Afr. J. Biotechnol. 3 (1), 18–24.

Olicón-Hernández DR, González-López J, Aranda E (2017) Overview on the biochemical potential of filamentous fungi to degrade pharmaceutical compounds. Front Microbiol 8:1792. https:// doi.org/10.3389/fmicb.2017.01792

Oliveira J, Belchior A, da Silva VD, Rotter A, Petrovski Ž, Almeida PL, Gaudêncio SP (2020b) Marine environmental plastic pol lution: mitigation by microorganism degradation and recycling valorization. Front Mar Sci 7:567126. https:// doi. org/ 10. 3389/ fmars.2020.567126

Otsuka, Y., Sonoki, T., Ikeda, S., Kajita, S., Nakamura, M., Katayama, Y., 2003. Detection and characterization of a novel extracellular fungal enzyme that catalyses the specificand hydrolytic cleavage of lignin guaiacylglycerol b-aryl ether linkages. Eur. J. Biochem. 270, 2353–2362. https://doi.org/10.1046/j.1432-1033.2003.03545.x.

Paço A, Duarte K, da Costa JP, Santos PS, Pereira R, Pereira ME, Frei tas AC, Duarte AC, Rocha-Santos TA (2017) Biodegradation of polyethylene microplastics by the marine fungus Zalerion mar itimum. Sci Total Environ 586:10–15. https://doi.org/10.1016/j. scitotenv.2017.02.017

Park SY, Kim CG (2019) Biodegradation of micro-polyethylene par ticles by bacterial colonization of a mixed microbial consortium isolated from a landfill site. Chemosphere 222:527–533. https:// doi.org/10.1016/j.chemosphere.2019.01.159

Pelmont, J., 2005. Biodégradations et métabolismes. Les bactéries pour les technologies de l'environnement. EDP Sciences, Les Ulis.

PET market volume worldwide 2015-2030 (2023)| Statista. https://www.stati sta.com/statistics/1245264/polyethylene-terephthalate-market-vol ume-worldwide/. Accessed 2 Dec 2023.

Phukon, P., Saikia, J.P., Konwar, B.K., (2012). Bio-plastic (P-3HB-co-3HV) from Bacillus circulans (MTCC 8167) and its biodegradation. Colloids Surf. B Biointerf. 92, 30 34.

Regnault, J.-P., 1990. Microbiologie Générale. Décarie Editeur Inc., Montreal. Reid, B.J., MacLeod, J.A., Lee, P.H., Morriss, W.J., Stockes, J.D., Semple, K.T., 2001. A simple 14C-respirometric method for assessing microbial catabolic potential and contaminant bioavailability. FEMS Microbial. Lett. 196, 141–146.

Royer, S.-J., Ferrón, S., Wilson, S.T., Karl, D.M., 2018. Production of methane and ethylene from plastic in the environment. PLoS One 13 (8),e0200574. https://doi.org/10.1371/journal.pone.0200574.

Ru J, Huo Y, Yang Y. (2020) Microbial degradation and valorization of plastic wastes. Front Microbiol. 2020;11:507487.

Russell JR, Huang J, Anand P, Kucera K, Sandoval AG, Dantzler KW, Hickman DS, Jee J, Kimovec FM, Koppstein D, Marks DH, Mit termiller PA, Nunez SJ, Santiago M, Townes MA, Vishnevetsky M, Williams NE, Nunez Vargas MP, Boulanger LA, Slack CB, Strobel SA (2011) Biodegradation of polyester polyurethane by endophytic fungi. Appl Environ Microbiol 77(17):6076–6084. https://doi.org/10.1128/AEM.00521-11

S. Mehdi Emadian, Turgut T. Onay, Burak Demirel (2016) Biodegradation of bioplastics in natural environments http://dx.doi.org/10.1016/j.https://www.vecteezy.com/vector-art/697403-flat-sales-funnel-diagram-template

Sameh S. Alia,b, Tamer Elsamahya, Rania Al-Tohamya, Daochen Zhua, Yehia A.-G. Mahmoudb, Eleni Koutrac,d, Metwally A. Metwallyb, Michael Kornarosc,d, Jianzhong Suna, Plastic wastes biodegradation: Mechanisms, challenges and future prospects https://doi.org/10.1016/j.scitotenv.2021.146590

Sameh Samir Alia, Tamer Elsamahya, Eleni Koutrac, Michael Kornarosc, Mostafa El-Sheekhb, Esraa A. Abdelkarima, Daochen Zhua, Jianzhong Sun (2020) Degradation of conventional plastic wastes in the environment: A review on current status of knowledge and future perspectives of disposal https://doi.org/10.1016/j.scitotenv.2020.144719

Sanniyasi E, Gopal RK, Gunasekar DK, Raj PP (2021) Biodegrada tion of low-density polyethylene (LDPE) sheet by microalga, Uronema Africanum Borge. Sci Rep 11(1):1–33. https:// doi. org/ 10.1038/s41598-021-96315-6

Scheurer, M., Bigalke, M., 2018. Microplastics in Swiss floodplain soils. Environ. Sci. Technol. 52, 3591–3598. https://doi.org/10.1021/acs.est.7b06003.

Schloss PD, Handelsman J (2005) Metagenomics for studying uncul turable microorganisms: cutting the Gordian knot. Genome biol 6(8):1–4. https://doi.org/10.1186/gb-2005-6-8-229

Schyichuk, A.V., Stavychna, D.Y., White, J.R., 2001. Effect of tensile stress on chain scission and crosslinking during photo-oxidation of polypropylene. Polym. Degrad. Stab. 72, 279–285.

Shah, A.A., Hasan, F., Hameed, A., Ahmed, S., 2008. Biological degradation of plastics: a comprehensive review. Biotechnol. Adv. 26 (3), 246–265. https://doi.org/10.1016/j.biotechadv.2007.12.005.

Shen M, Zeng G, Zhang Y, Wen X, Song B, Tang W (2019) Can bio technology strategies effectively manage environmental (micro) plastics? Sci Total Environ 697:134200. https:// doi. org/ 10. 1016/j.scitotenv.2019.134200

Singh L, Wahid ZA (2015) Methods for enhancing bio-hydrogen production from biological process: a review. J Ind Eng Chem 21:70–80. https://doi.org/10.1016/j.jiec.2014.05.035

Singh N, Hui D, Singh R, Ahuja IPS, Feo L, Fraternali F. (2017) Recycling of plastic solid waste: A state of art review and future applications. Compos B Eng. 2017;115:409–22.

Skariyachan S, Taskeen N, Kishore AP, Krishna BV, Naidu G (2021) Novel consortia of Enterobacter and Pseudomonas formulated from cow dung exhibited enhanced biodegradation of polyethyl ene and polypropylene. J Environ Manage 284:112030. https:// doi.org/10.1016/j.jenvman.2021.112030

Tang, Y., Liu, Y., Chen, Y., Zhang, W., Zha, J., He, S., Yang, C., Zhang, T., Tang, C., Zhang, C., Yang, Z., 2021. A review: research progress on microplastic pollutants in aquatic environments. Sci. Total Environ. https://doi.org/10.1016/j.scitotenv.2020.142572.

Tezuka, Y., Ishii, N., Kasuya, K.I., Mitomo, H., (2004). Degradation of poly(ethylene succinate) by mesophilic bacteria. Polym. Degrad. Stab. 84, 115–121.

United Nations Environment Programme. Zero draft text of the international legally binding instrument on plastic pollution, including in the marine environment. (2023), United Nations Environment Assembly of the United Nations Environment Programme. https:// wedocs.unep.org/bitstream/handle/20.500.11822/43239/ZERODRAFT.pdf.

Uttpal Anand· Satarupa Dey · Elza Bontempi · Serena Ducoli· A. Dick Vethaak· Abhijit Dey· Stefania Federici (2024) - Biotechnological methods to remove microplastics: a review https://doi.org/10.1007/s10311-022-01552-4

V. Mahalakshmi (2014), Evaluation pf Biodegradation of plastic. International journal of innovative research and development, ISSN 2278-0211, 185-190

Waller CL, Griffiths HJ, Waluda CM, Thorpe SE, Loaiza I, Moreno B, Hughes KA (2017) Microplastics in the Antarctic marine system: an emerging area of research. Sci Total Environ 598:220–227. https://doi.org/10.1016/j.scitotenv.2017.03.283

Weiland, M., Daro, D., David, C., 1995. Biodegradation of thermally oxidized polyethylene. Polym. Degrad. Stab. 48, 275–289.

Wiles, D.M., Scott, G., 2006. Polyolefins with controlled environmental degradability. Polym. Degrad. Stab. 91, 1581–1592.

Wright SL, Thompson RC, Galloway TS (2013b) The physical impacts of microplastics on marine organisms: a review. Environ Pollut 178:483–492. https://doi.org/10.1016/j.envpol.2013.02.031

Wu, C.-S., (2009). Renewable resource-based composites of recycled natural fibers and maleated polylactide bioplastic: characterization and biodegradability. Polym. Degrad. Stab. 94, 1076–1084.

Wu, C.S., (2012). Preparation, characterization and biodegradability of renewable resource-based composites from recycled polylactide bioplastic and ssal fibers. J. Appl. Polym. Sci. 123, 347–355.

Wu,C.S., (2012a). Characterization and biodegradability of polyester bioplastic-based green renewable composites from agricultural residues. Polym. Degrad. Stab. 97, 64–71.

Yan N, Fan C, Chen Y, Hu Z (2016) The potential for microalgae as bioreactors to produce pharmaceuticals. Int J Mol Sci 17(6):962. https://doi.org/10.3390/ijms17060962

Yousif, E., Haddad, R. (2013). Photodegradation and photo stabilization of polymers, especially polystyrene: review. Springer Plus 2, 398 https://doi.org/10.1186/2193-1801-2-398

Yulong Niu, Fengying Pan, Kailiang Shen, Xingfeng Yang, Shiyu Niu, Xinyan Xu, Hongyi Zhou, Qiang Fu andXiaowei Li (2024) Sustainability, Status and Enhancement

Techniques of Plastic Waste Degradation in the Environment: A Review https://doi.org/10.3390/su16219395

Zanardini, E., Abbruscato, P., Ghedini, N., Realini, M., Sorlini, C., 2000. Influence of atmo spheric pollutants on the biodeterioration of stone. Int. Biodeterior. Biodegradation 45, 35–42. https://doi.org/10.1016/S0964-8305(00)00043-3.

Zhu, D., Chen, Q.L., An, X.L., Yang, X.R., Christie, P., Ke, X., Wu, L.H., Zhu, Y.G., 2018. Expo sure of soil collembolans to microplastics perturbs their gut microbiota and alters their isotopic composition. Soil Biol. Biochem. 116, 302–310. https://doi.org/ 10.1016/j.soilbio.2017.10.027

17

Bioremediation of Biomedical Waste

Vartika Jain[1] and Deeksha Dave[1,2*]

[1]*Dept. of Botany, Government Meera Girls College, Udaipur - 313001, Rajasthan, India*

[2]*Environmental Studies, School of Interdisciplinary and Trans-disciplinary Studies, IGNOU, New Delhi, India*

[*]*Corresponding Author Email: deekshadave@ignou.ac.in*

Abstract

Biomedical waste poses significant hazards due to its potential carcinogenic, genotoxic, and mutagenic properties, along with other toxic characteristics. Current disposal methods for biomedical waste are often expensive and merely convert pollutants from one toxic form to another, rather than eliminating them. In contrast, bioremediation utilizes naturally occurring microorganisms and plants to detoxify anthropogenic pollutants into benign products, simultaneously enhancing soil fertility. Recent advancements have led to the development of genetically engineered microorganisms capable of remediating environmental pollutants, including radionuclides. Phytoremediation represents another sustainable approach, effectively addressing heavy metals and radioactive waste generated by healthcare facilities. Microbial degradation has emerged as a sustainable and efficient strategy for treating diverse waste types, such as plastics, human anatomical waste, laboratory waste, expired pharmaceuticals, contaminated water, and chemical residues. However, it is crucial to acknowledge that no existing practices or methods for medical waste treatment can guarantee sustainability. Therefore, the most effective solution to the waste management crisis is to minimize waste generation. The chapter will examine the application of bioremediation techniques, emphasizing the role of microorganisms in the removal of various pollutants in biomedical waste. The chapter will also explore the future developments in biomedical waste management strategies.

***Keywords**: Bioremediation, toxicity, biomedical waste, sustainability, microbes*

1. Introduction to Biomedical Waste (BMW)

The waste produced at different medical facilities, including hospitals, laboratories, research centers, mortuaries, and autopsy centers is considered as

Biomedical Waste (BMW). Some of the BMW is hazardous whereas some are non-hazardous in nature. Hazardous waste includes, Infectious, Pathological, Microbiological, Pharmaceutical, Chemical, Cytotoxic and Radioactive waste material as described by World Health Organization (Fig. 1). Healthcare personnel and the general public are at risk from physical, chemical, and microbiological effects of hazardous waste. On the other hand, the non-hazardous BMW which are not harmful, include packaging material made up of plastic & thermocol, cardboards, paper waste etc. (Capoor & Parida, 2021).

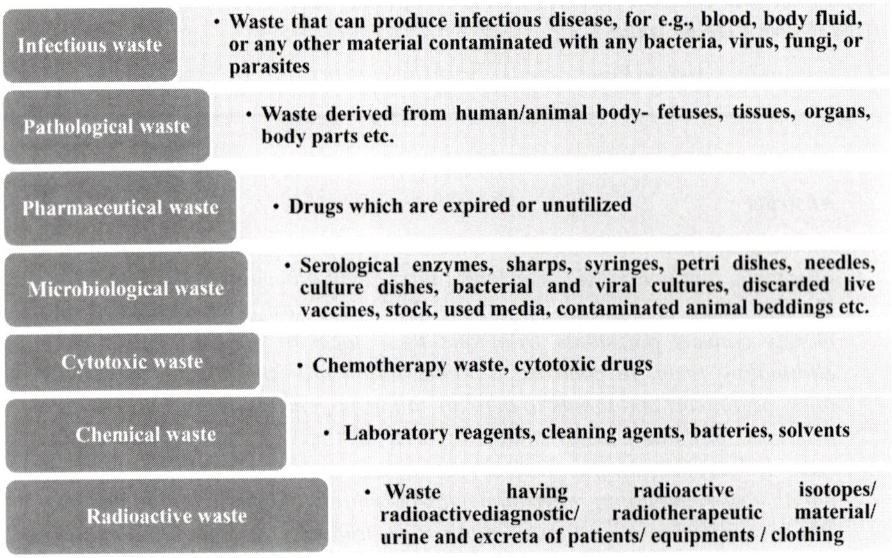

Fig. 1: Types of hazardous biomedical waste material

2. Hazards of Biomedical Waste

Persons who are directly involved in patient care such as health care professional as well as who are indirectly involved in handling of BMW for example, sanitation workers, labors, waste management personnel etc. are at high risk of getting different type of bacterial and viral infections (Porta *et al.*, 2009) which could even lead to epidemics. Healthcare institutions use a wide range of chemical substances which can enter a person's body through ingestion, skin/mucous membrane contact, gas, vapour droplets, or inhalation etc. Most of these are harmful to human health by having the ability to cause cancer, corrosion, reprotoxicity, irritation, sensitization etc. (Bhusnure *et al.*, 2018). Similarly, improper disposal of needles can cause injuries and infections in the health care workers (Sharma *et al.*, 2013). Radioactive BMW is quite dangerous for environment as well as for human beings (Abdulla *et al.*, 2008). Hence, proper management of BMW is essential.

Pathogens or hazardous BMW may contaminate water and soil as a result of low-quality wastewater, sewage sludge, or by-products generated during industrial wastewater sewage treatment. Moreover, septic tanks and biological sewage treatment facilities may malfunction by inflow of chemical and pharmaceutical waste, for example antibiotics and their byproducts. These could contaminate water supplies and overall, the biosphere. Further, the increase in antibiotic resistant bacteria is one of the grave situations developing rapidly due to improper disposal and overuse (Makowska *et al.*, 2021; Sartelli *et al.*, 2023).

Besides, improper incineration of the BMW can seriously damage the environment. Moreover, the human health is also jeopardized by toxic chemicals for example, dioxins, furans, coplanar polychlorinated biphenyls etc. released through inappropriate combustion of BMW (Srogi, 2008). Landfills and dump sites have the potential to contaminate soil, surface water, and groundwater with heavy metals, lead, and coliforms. These unplanned dumping sites are not only polluting environment but also risky for human health. Even mere exposure to these sites can cause serious health issues (Morita *et al.*, 2021). Investigations also show the effects of exposure to bioaerosols in particular work settings. Bronchitis is more common in workers exposed to bioaerosols at processing plants, storage facilities, and landfills (Kim *et al.*, 2017).

3. Current Disposal Techniques

Different types of techniques are employed for disposal of BMW according to the nature of the waste. However, some practices are quite common. For example incineration, that is considered good in sterilization of infectious wastes and turning toxic wastes, like animal carcasses into harmless ash and also recovering waste heat for further utilization (Giakoumakis *et al.*, 2021). Use of autoclave for moist heat sterilization of equipments is quite common in hospitals and research laboratories. Though, high cost, use of electricity, long time consuming process and inability to dispose chemical wastes are its limitations (Kumar and Rana, 2021). Chemical treatment for disinfection is another technique used to manage the infectious biomedical wastes as it kills or retards the growth of infectious agents. Several disinfectants are used for this purpose for example, Sodium hypochlorite, Ethanol, Isopropanol, Hydrogen peroxide, Formaldehyde, Iodine, Triclosan, Benzalkonium chloride etc. However, the safety issues associated with chemical disinfectants are many and concerning. Hence, adequate care is required while handling these chemicals regarding their concentrations, contact duration etc. (Dindarloo *et al.*, 2020).

The use of landfills is quite common method for waste management. The waste which is dumped in open sites in an uncontrolled manner is a hazardous method which can spread the transmission of infectious pathogens, fires, pollution without any restriction to scavenger, birds or animals. Sanitary landfills are better options over open dump sites as these reduce the risk of environment pollution and disease spread to a greater extent. These landfills require daily monitoring from efficient personnel for effectively managing the BMW (Idris *et al.*, 2004; Kulkarni, 2020).

Several mechanical techniques are also employed for treatment of BMW, for example, shredding, granulation, grinding, pulverization, mixing, crushing, agitation etc. (Hooshmand *et al.*, 2020) Shredding is a valuable technique which is used to make small fragments of BMW. Mostly, syringes and needles are shredded through shredding machines which operate through electricity. Unfortunately, if the waste is not disinfected before shredding, it may lead to spread of pathogens to persons working near the shredding machines and pose threat to their health (Sponza and Ağdağ, 2005). Encapsulation of pharmaceutical wastes, ash produced in incinerators and sharps in a specific composition of cement, lime/clay and water and their final disposition to any waste pit or sanitary landfill is a comparatively safer and cost-effective technique. However, it makes the waste material bigger which require larger space to dispose off (Sabbas *et al.*, 2023). Inertization of heavy metal containing ash or pharmaceutical waste by combining with cement, lime or water is another affordable technique which reduce the leaching of toxins into the ground or surface water (Nyekwere, 2021).

In India, the Bio-Medical Waste Management Rules were amended in 2016 which have classified the BMW on the basis of its risk profile into four categories as shown in Fig. 2. These rules suggest that BMW must be classified at the production stage and stored in appropriate color containers. Transportation of those containers should be done in special vehicles without any chance of leaking out. The regulations also specify the obligations of those who produce biomedical waste, including teaching employees on how to handle biomedical waste and maintaining track of the trash produced and disposed of.

Among the treatment techniques include incineration, chemical treatment, autoclaving, and use of microwaves. The kind and volume of waste produced determine the treatment procedure that is chosen. The 2016 regulations also mandate that processed waste be disposed of in an incinerator or landfill that satisfies the required requirements and is scientifically constructed (Datta *et al.*, 2018; Singh *et al.*, 2024).

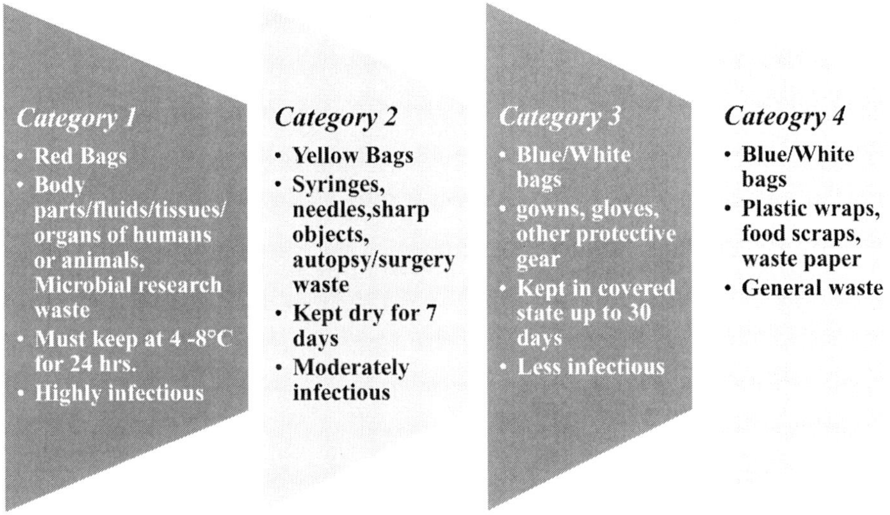

Fig. 2: Four categories of BMW in India

In recent years, Biomedical waste production has significantly increased and whatever may be the technique utilized for management of BMW, the associated disadvantages are many. Hence, an effective, environment friendly and rather inexpensive technique is required to get rid of BMW in an efficient manner. In this regard, Bioremediation is a one such promising technique. It utilizes some microbes or plants and converts the toxic waste into non-toxic or lesser toxic forms. Moreover, the by-products generated during such biological treatment are also not harmful. Therefore, there is an urgent need to follow this sustainable technique in management of BMW (Khan, 2024).

4. Biological Treatment of Biomedical Waste

Biomedical waste is disposed of using a variety of methods, including incineration, autoclaving, microwaving, and landfilling. All these methods have their own advantages and disadvantages. The current common treatment methods such as incineration and landfills are costly and need well-developed infrastructure and technological support. In these methods, in many cases, the pollutants are transformed into other toxic forms of pollution. However, since Biomedical waste generated from hospital is an accumulation of liquid or solid materials comprising high amounts of organic pollutants, it could be treated biologically.

The biological disposal of biomedical waste involves the utilization of living organisms to treat and dispose the waste through various methods. An emerging system called "Bio-converter" is widely used to treat and dispose the biomedical waste. It uses a solution of enzyme to decontaminate medical waste, and the resulting sludge is put through an extruder used to remove water for sewage disposal and the remaining solid waste is sent to landfill.

Another method of biological biomedical waste disposal is by using biodegradable plastics (Datta *et al.*, 2018). Bioplastics like polylactic acid (PLA), polyhydroxyalkanoates (PHAs), and polycaprolactone (PCL) are most suitable materials for biomedical waste disposal due to their biodegradable nature. It allows the bioplastics to break down into harmless components under specific conditions, potentially reducing the environmental impact of biomedical waste.

Many biomedical embeds made up of biodegradable plastics undergo biological degradation with microbial enzymes. These microbes utilize these biodegradable polymers as substrate in the absence of food and unavailability of suitable substrate. Further research needs to be done for large-scale manufacture of biodegradable plastics.

5. Role of Microbes in Managing Biomedical Waste

Biomedical waste can be harmful due to its potential genotoxicity, carcinogenicity, mutagenicity, and other toxicities. Microbial degradation can be considered an environmentally friendly and sustainable treatment option without causing any adverse impact (Krueger *et al.*, 2015). Microbial degradation of the plastic involves steps like bio-deterioration, bio fragmentation, assimilation and mineralization.

In most cases, bacteria-mediated degradation is very slow, and significant changes can only be visible after 2–3 months. Therefore, more innovative methods are required to optimize cultural conditions and shorten the degradation process. Several types of pre-treatments like UV treatment and HNO_3 treatment are recommended to enhance the degradation of plastic polymers (Dey, *et al.*, 2023).

One of the encouraging biological treatments is bioremediation (Dixit, 2015). Around 20% of the biomedical waste is treated through bioremediation technique (Yadav *et al.*, 2024). This technique is considered as an alternative to currently used remediation technologies which include harmful chemicals and physical processes as it is cheaper and more sustainable method (Raklami *et al.*, 2022). It is being used to treat polluted soil and water by changing the environmental conditions that facilitate active growth of plants or microorganisms towards degradation of certain pollutants especially biomedical wastes (Rajan *et al.*, 2019).

Large reservoirs containing liquid biomedical waste contain remains of varied types of organic matters which are found to be rich in bacteria. Ethica *et al.*, (2018) reviewed the exploitation of hydrolytic bacteria in remediating the biomedical wastes. The use of hydrolytic bacteria as bioremediation agents could be an encouraging strategy to resolve the problem of disposal of liquid biomedical waste. (Rao and Wankhede, 2012).

Till now, eradication of soils contaminated with hospital solid waste especially radionuclides is done by digging and transporting it to a distant waste disposal site, but it is very costly (Lloyd and Renshaw, 2005). Bioremediation via microorganisms could be a remarkable alternative to revive contaminated soil (Saha *et al.*, 2021). Microorganisms such as *Desulfuromus aferrireducens* and *Rhodanobacter sp.* are found to be interacting with the contaminants of biomedical waste (Green, *et al.*, 2012).

6. Biotechnological Interventions

Biotechnology can aid in the management of medical waste. Biotechnology can help to dispose pharmaceutical, chemical, radioactive and infectious waste. These methods reduce the bad effect on environment and human health by preventing the emission of harmful chemicals and gases. Recombinant enzymes can be used to treat biomedical waste (Humer *et al.,* 2020, Kumar *et al.,* 2016). Horseradish peroxidase is recombinant enzyme that can be utilized for the treatment of biomedical waste. Lipases, proteases and cellulose are other recombinant enzymes that can be utilized for the bioremediation as biomedical waste as they can be broken down into lipids, proteins and cellulose, respectively. By the combination of these enzymes also, a variety of biomedical waste can be bioremediated (Yadav *et al.,* 2022).

7. Composting

Medical waste can be managed naturally and sustainably via composting. Composing could be an effective way to treat biomedical waste generated at the health care centers, hospitals, etc., including food waste from hospital canteens, clinics, and labs, as well as outdated pharmaceuticals and other organic waste products (Li *et al.*, 2023). The fungus *Aspergillus niger,* is frequently employed in the pharmaceutical and food industries due to its capacity to manufacture enzymes. *A. niger* can aid in the composting process by reducing complex organic molecules like lignin and cellulose into less complex ones that are more easily employed by other bacteria in the compost pile (Searca *et al.*, 2022). Composting medical waste helps to reduce the amount of the garbage that is dumped in landfills and creates an important soil improvement that can be utilized to enhance the fertility and health of the soil. When done properly, it is safe, toxin free and organic method of treating medical waste that can help to check the release of dangerous toxins into the environment (Taiwo, 2011). Cellulase is one type of recombinant enzyme that can be utilized in the composting of medical waste including bandages, gauze, and cotton swabs which are cellulose, and can be broken down by the enzyme cellulase. It intensifies the decomposition process, reducing waste volume, minimizes environmental pollution and leads to the production of nutrient-rich compost in the process. In addition to lipase and cellulase, protease can be employed to digest proteins in medical waste, including biological fluids like blood, lowering the risk of infection (Khan *et al.*, 2016, Patel *et al.*, 2019).

8. Phytoremediation

Hazardous chemicals, germs and other pollutants in the biomedical waste if not effectively controlled, may be dangerous to human health and the environment (Ansari *et al.*, 2020). To clean-up the medical waste, phytoremediation can provide a safe, affordable, and sustainable alternative. Vetiver grass (*Chrysopogon zizanioides*) is a type of phytoremediation agent that has been used to treat the medical waste. It has been demonstrated that Vetiver grass is good at removing pathogens, organic pollutants, and heavy metals from medical waste. Its extensive root system enables it to absorb and store contaminants in its tissues, delaying the rate of evaporation. Additionally, Vetiver grass might encourage the development of advantageous soil microbes, further accelerating the clean-up procedure (Liu and Tran, 2021). Willow (*Salix spp.*) is also a phytoremediation agent that is widely used in the remediation of medical waste. It is also used to clean up polluted air, water, and soil. It can effectively absorb and store toxins in its tissues and has a high tolerance for them. Additionally, willow can encourage the development of advantageous soil microbes, which can accelerate the detoxification process (Jiang *et al.*, 2015).

9. Genetic Engineering

Researchers have attempted to create native or genetically engineered microbes for the remediation of hospital waste contaminants including radionuclide (Raklami *et al.,* 2022). It has been found that remediation by microbes can affect the solubility, bioavailability, and mobility of radionuclides (Ayilara *et al.,* 2020). Genetic engineering utilizes microorganisms to transform contaminants, especially medical waste, into non-toxic chemicals. This method can be applied to specific contaminants, is economical, and is favourable to the environment (Kaur and Sodhi, 2022). The biodegradation of polyethylene terephthalate (PET), a common plastic utilized for medical purposes, is an example of how genetic engineering is applied in the bioremediation of medical waste. As PET is resistant to biodegradation, it can withstand the environmental stressors for a long period of time. However, the scientists have discovered a bacterium called *Ideonella sakaiensis* that can degrade PET. Researchers have improved this bacterium's capacity to break down PET by tinkering with its DNA, making it a more useful bioremediation tool (Ebah *et al.*, 2022).

10. Waste to Energy (WTE) Conversion of BioMedical Waste

In the waste-to-energy (WTE) conversion of medical waste, the waste is converted into a renewable energy source using a variety of technologies. WTE conversion is an efficient means of getting rid of medical waste. During the incineration process, the biomedical waste is heated in the absence of oxygen in a process known as gasification, which produces a gas that can be utilized to produce power (Manegdeg *et al.*, 2022).

The conversion of medical waste using WTE has a number of advantages. First of all, it lessens the amount of waste that is dumped in landfills, lowering the chance of pollution and environmental degradation. WTE conversion of medical waste has its own challenges. One of the risks is the potential release of hazardous chemicals during the combustion or gasification process, which poses threat to the health of the local population. The maintenance of the infrastructure and technology is also expensive and is also a bottleneck (Chen *et al.*, 2022).

11. Challenges and Future Directions

The healthcare industry faces a serious issue in managing medical waste since inappropriate disposal poses serious health and environmental dangers. Environmental biotechnology provides creative ways to manage medical waste sustainably, but there are many obstacles that must be overcome before it can be put into practice successfully (Mishra *et al.*, 2016). The lack of knowledge and comprehension of the dangers posed by medical waste among healthcare professionals, waste handlers, and the general public is one of the main issues. Due to inappropriate handling and disposal procedures, diseases may spread and the environment may get contaminated. To raise awareness and encourage safe handling and disposal methods, it is, therefore, necessary to implement effective communication and teaching programs (Mugabi *et al.*, 2018; Thakur and Ramesh, 2015). The complex nature and heterogeneity of biomedical waste presents another difficulty. Biomedical waste contains radioactive, toxic, infectious, non-infectious substances that must be handled and disposed of according to specific rules. To ensure adequate management of the various waste streams, efficient segregation, collection, and treatment techniques are required (Windfeld and Brooks, 2015). Furthermore, implementing sustainable medical waste management practices can be expensive, which can be particularly difficult for healthcare organizations with limited resources. As a result, innovative, long lasting, simple to use and durable solutions are required (Lee and Lee, 2022). The development of accessible, sustainable technologies that can address the challenges and difficulties is a future research area for managing medical waste in the field of environmental biotechnology. This may entail treating and managing medical waste using cutting-edge technologies like plasma gasification, microbial degradation, and waste-to-energy conversion (Singh *et al.*, 2022). The management of medical waste must also be incorporated into the larger healthcare sustainability framework, which emphasizes resource efficiency, community engagement, and a reduction in the environmental impact of healthcare facilities (Lee and Lee, 2022).

12. Conclusion

Environmental biotechnology offers a promising way to minimize the harm that medical waste causes to the environment. A variety of technologies, including microbial treatment, composting, and phytoremediation, can effectively treat medical waste and lessen its impact on the environment. Further enhancing the efficacy of environmental biotechnology in waste management is the use of biodegradable materials and the adoption of sustainable practices in healthcare facilities. Stakeholders in the healthcare sector, decision-makers, and researchers must work together to develop more creative and long-lasting solutions to the problem of managing medical waste. The negative effects of BMW on the environment can be minimized to safeguard public health by using environmental biotechnology solutions.

References

Abdulla, F., Abu Qdais, H., & Rabi, A. (2008). Site investigation on medical waste management practices in northern Jordan. Waste management (New York, N.Y.), 28(2), 450–458. https://doi.org/10.1016/j.wasman.2007.02.035

Ansari, A. A., Naeem, M., Gill, S. S., & Al Zuaibr, F. M. (2020). Phytoremediation of contaminated waters: An eco-friendly technology based on aquatic macrophytes application. The Egyptian Journal of Aquatic Research, 46(4), 371-376.

Ayilara, M. S., Olanrewaju, O. S., Babalola, O. O., & Odeyemi, O. (2020). Waste management through composting: Challenges and potentials. Sustainability, 12(11), 4456.

Bhusnure, O.G., Dongare, R.B., Gholve, S.B., Giram, P.S. (2018). Chemical hazards and safety management in pharmaceutical industry. Journal of Pharmacy Research, 12(3), 357–69.

Capoor, M., & Parida, A. (2021). Biomedical Waste and Solid Waste Management in the Time of COVID-19: A Comprehensive Review of the National and International Scenario and Guidelines. Journal of Laboratory Physicians, 13, 175 - 182.

Chen, H., Li, J., Li, T., Xu, G., Jin, X., Wang, M., & Liu, T. (2022). Performance assessment of a novel medical-waste-to-energy design based on plasma gasification and integrated with a municipal solid waste incineration plant. Energy, 245, 123156.

Datta, P., Mohi, G., Chander, J. (2018). Biomedical waste management in India: critical appraisal. Journal of Laboratory Physicians, 10(01), 6–14.

Dey, S., Anand, U., Kumar, V., Kumar, S., Ghorai, M., Ghosh, A., & Dey, A. (2023). Microbial strategies for degradation of microplastics generated from COVID-19 healthcare waste. Environmental Research, 216, 114438.

Dindarloo, K., Aghamolaei, T., Ghanbarnejad, A., Turki, H., Hoseinvandtabar, S., Pasalari, H., & Ghaffari, H. R. (2020). Pattern of disinfectants use and their adverse effects on the consumers after COVID-19 outbreak. Journal of environmental health science & engineering, 18(2), 1301–1310. https://doi.org/10.1007/s40201-020-00548-y

Dixit, R., Wasiullah, X., Malaviya, D., Pandiyan, K., Singh, U. B., Sahu, A., ... & Paul, D. (2015). Bioremediation of heavy metals from soil and aquatic environment: an overview of principles and criteria of fundamental processes. Sustainability, 7(2), 2189-2212.

Ebah, E. E., Yange, I. A., Ohie, I. R., & Inya, O. J. (2022). Application of Genetically Modified Organisms in Waste Management– A Review. Stamford Journal of Microbiology, 12(1), 15-20.

Ethica, S. N., Saptaningtyas, R., Muchlissin, S. I., & Sabdono, A. (2018). The development method of bioremediation of hospital biomedical waste using hydrolytic bacteria. Health and Technology, 8, 239-254.

Giakoumakis, G., Politi, D., Sidiras, D. (2021). Medical waste treatment technologies for energy, fuels, and materials production: a review. Energies, 14(23), 8065.

Green, S. J., Prakash, O., Jasrotia, P., Overholt, W. A., Cardenas, E., Hubbard, D., ... & Kostka, J. E. (2012). Denitrifying bacteria from the genus Rhodanobacter dominate bacterial communities in the highly contaminated subsurface of a nuclear legacy waste site. Applied and environmental microbiology, 78(4), 1039-1047.

Hooshmand, S., Kargozar, S., Ghorbani, A., Darroudi, M., Keshavarz, M., Baino, F., & Kim, H. W. (2020). Biomedical Waste Management by Using Nanophotocatalysts: The Need for New Options. Materials (Basel, Switzerland), 13(16), 3511. https://doi.org/10.3390/ma13163511

Humer, D., Ebner, J., & Spadiut, O. (2020). Scalable high-performance production of recombinant horseradish peroxidase from E. coli inclusion bodies. International Journal of Molecular Sciences, 21(13), 4625.

Idris, A., Inanc, B., Hassan, M.N. (2004). Overview of waste disposal and landfills/dumps in Asian countries. Journal of Material Cycles and Waste Management, 6, 104–10.

Jiang, Y., Lei, M., Duan, L., & Longhurst, P. (2015). Integrating phytoremediation with biomass valorisation and critical element recovery: A UK contaminated land perspective. Biomass and Bioenergy, 83, 328-339.

Kaur, M., & Sodhi, H. S. (2022). Genetically engineered microorganisms for bioremediation processes. In Microbial Bioremediation: Sustainable Management of Environmental Contamination (pp. 91-107). Cham: Springer International Publishing.

Khan, M. N., Luna, I. Z., Islam, M. M., Sharmeen, S., Salem, K. S., Rashid, T. U., ... & Rahman, M. M. (2016). Cellulase in waste management applications. In New and future developments in microbial biotechnology and bioengineering (pp. 237-256). Elsevier.

Khan, M.S.A. (2024). Applications of Bioremediation in Biomedical Waste Management: Current and Future Prospects. Brazilian Archives of Biology and Technology, 67, e24230161.

Kim, K. H., Kabir, E., & Jahan, S. A. (2018). Airborne bioaerosols and their impact on human health. Journal of environmental sciences (China), 67, 23–35.

Krueger, M. C., Harms, H., & Schlosser, D. (2015). Prospects for microbiological solutions to environmental pollution with plastics. Applied microbiology and biotechnology, 99, 8857-8874.

Kulkarni, B.N. (2020). Environmental sustainability assessment of land disposal of municipal solid waste generated in Indian cities–a review. Environmental Development, 33, 100490.

Kumar, R. & Rana, A. (2021). Basic concepts of sterilization techniques. Research Journal of Pharmacology and Pharmacodynamics, 13(4), 155–61.

Kumar, V., Misra, N., Goel, N. K., Thakar, R., Gupta, J., & Varshney, L. (2016). A horseradish peroxidase immobilized radiation grafted polymer matrix: a biocatalytic system for dye waste water treatment. RSC Advances, 6(4), 2974-2981.

Lee, S. M., & Lee, D. (2022). Effective medical waste management for sustainable green healthcare. International Journal of Environmental Research and Public Health, 19(22), 14820.

Li, H., He, Y., Yan, Z., Yang, Z., Tian, F., Liu, X., & Wu, Z. (2023). Insight into the microbial mechanisms for the improvement of spent mushroom substrate composting efficiency driven by phosphate-solubilizing Bacillus subtilis. Journal of Environmental Management, 336, 117561.

Liu, Z., & Tran, K. Q. (2021). A review on disposal and utilization of phytoremediation plants containing heavy metals. Ecotoxicology and Environmental Safety, 226, 112821.

Lloyd, J. R., & Renshaw, J. C. (2005). Microbial transformations of radionuclides: fundamental mechanisms and biogeochemical implications. Metal Ions In Biological Systems, Volume 44, 205-240.

Makowska, N., Bresa, K., Koczura, R., Philips, A., Nowis, K., & Mokracka, J. (2021). Urban wastewater as a conduit for pathogenic Gram-positive bacteria and genes encoding resistance to β-lactams and glycopeptides. The Science of the total environment, 765, 144176. https://doi.org/10.1016/j.scitotenv.2020.144176

Manegdeg, R. F., Rollon, A., Ballesteros Jr, F., Magdaluyo Jr, E., De Sales-Papa, L., Clemente, E., ... & Cervera, R. B. (2022). Multi-Attribute Assessment of Waste-to-Energy Technologies for Medical, Industrial, and Electronic Residual Wastes. Journal of Energy Resources Technology, 144(7), 070908.

Mishra, K., Sharma, A., Sarita, A. S., & Ayub, S. (2016). A study: biomedical waste management in India. IOSR Journal of Environment Science, Technology and Food Technology, 10(5), 64-67.

Morita, A. K. M., Ibelli-Bianco, C., Anache, J. A. A., Coutinho, J. V., Pelinson, N. S., Nobrega, J., Rosalem, L. M. P., Leite, C. M. C., Niviadonski, L. M., Manastella, C., & Wendland, E. (2021). Pollution threat to water and soil quality by dumpsites and non-sanitary landfills in Brazil: A review. Waste management (New York, N.Y.), 131, 163–176.

Mugabi, B., Hattingh, S., & Chima, S. C. (2018). Assessing knowledge, attitudes, and practices of healthcare workers regarding medical waste management at a tertiary hospital in Botswana: A cross-sectional quantitative study. Nigerian Journal of clinical practice, 21(12), 1627-1638.

Nyekwere, E.H. (2021). A legal assessment of the impact of the improper disposal of medical waste in Nigeria. Nnamdi Azikiwe University Journal of International Law and Jurisprudence, 12(1), 25–45.

Patel, N., Rai, D., Shivam, Shahane, S., & Mishra, U. (2019). Lipases: sources, production, purification, and applications. Recent patents on biotechnology, 13(1), 45-56.

Porta, D., Milani, S., Lazzarino, A. I., Perucci, C. A., & Forastiere, F. (2009). Systematic review of epidemiological studies on health effects associated with management of solid waste. Environmental health, 8, 60. https://doi.org/10.1186/1476-069X-8-60

Rajan, R., Robin, D. T., & Vandanarani, M. (2019). Biomedical waste management in Ayurveda hospitals–current practices and future prospectives. Journal of Ayurveda and integrative medicine, 10(3), 214-221.

Raklami, A., Meddich, A., Oufdou, K., & Baslam, M. (2022). Plants—Microorganisms-based bioremediation for heavy metal cleanup: Recent developments, phytoremediation techniques, regulation mechanisms, and molecular responses. International Journal of Molecular Sciences, 23(9), 5031.

Rao, M. S., & Wankhede, S. (2012). Management of Hospital Wastes with Potential Pathogenic Microbes. Microorganisms in Environmental Management: Microbes and Environment, 365-401.

Sabbas, T., Polettini, A., Pomi, R., Astrup, T., Hjelmar, O., Mostbauer, P., Cappai, G., Magel, G., Salhofer, S., Speiser, C., Heuss-Assbichler, S., Klein, R., Lechner, P., & pHOENIX Working Group on Management of MSWI Residues (2003). Management of municipal solid waste incineration residues. Waste management (New York, N.Y.), 23(1), 61–88. https://doi.org/10.1016/S0956-053X(02)00161-7

Saha, L., Tiwari, J., Bauddh, K., & Ma, Y. (2021). Recent developments in microbe–plant-based bioremediation for tackling heavy metal-polluted soils. Frontiers in Microbiology, 12, 731723.

Sartelli, M., Bartoli, S., Borghi, F., Busani, S., Carsetti, A., Catena, F., Cillara, N., Coccolini, F., Cortegiani, A., Cortese, F., Fabbri, E., Foghetti, D., Forfori, F., Giarratano, A., Labricciosa, F. M., Marini, P., Mastroianni, C., Pan, A., Pasero, D., Scatizzi, M., … Moro, M. L. (2023). Implementation Strategies for Preventing Healthcare-Associated Infections across the Surgical Pathway: An Italian Multisociety Document. Antibiotics (Basel, Switzerland), 12(3), 521. https://doi.org/10.3390/antibiotics12030521

Searca, R., Dela Cruz, J. P., Ballesteros, F. C., Undan, J. R., Lou, L., Lopez, M. A., Eloisa, M., Cortez, F. C., & Porciuncula, F. L. (2022). Assessment of Aspergillus niger and Aspergillus flavus in Inoculant Induced Composting to Remove Antibiotics in Poultry Manure. Eco. Env. & Cons., 28(2), 677–684. https://doi.org/10.53550/EEC.2022.v28i02.015

Sharma, A., Sharma, V., Sharma, S., & Singh, P. (2013). Awareness of biomedical waste management among health care personnel in jaipur, India. Oral Health and Dental Management, 12(1), 32–40.

Singh, H., YT, K., Mishra, A. K., Singh, M., Mohanto, S., Ghumra, S., ... & Thangadurai, D. (2024). Harnessing the foundation of biomedical waste management for fostering public health: strategies and policies for a clean and safer environment. Discover Applied Sciences, 6(3), 89.

Singh, J. S., Abhilash, P. C., Singh, H. B., Singh, R. P., & Singh, D. P. (2011). Genetically engineered bacteria: an emerging tool for environmental remediation and future research perspectives. Gene, 480(1-2), 1-9.

Singh, N., Ogunseitan, O. A., & Tang, Y. (2022). Medical waste: Current challenges and future opportunities for sustainable management. Critical Reviews in Environmental Science and Technology, 52(11), 2000-2022.

Sponza, D.T. and Ağdağ, O.N. (2005). Effects of shredding of wastes on the treatment of municipal solid wastes (MSWs) in simulated anaerobic recycled reactors. Enzyme and Microbial Technology, 36(1), 25–33.

Srogi, K. (2008). Levels and congener distributions of PCDDs, PCDFs and dioxin-like PCBs in environmental and human samples: a review. Environmental Chemistry Letters, 6, 1–28.

Taiwo, A.M. (2011). Composting as a Sustainable Waste Management Technique in Developing Countries. Journal of Environmental Science and Technology, 4, 93-102.

Thakur, V., & Ramesh, A. (2015). Healthcare waste management research: A structured analysis and review (2005–2014). Waste Management & Research, 33(10), 855-870.

Windfeld, E. S., & Brooks, M. S. L. (2015). Medical waste management–A review. Journal of environmental management, 163, 98-108.

Yadav, A. N., Suyal, D. C., Kour, D., Rajput, V. D., Rastegari, A. A., & Singh, J. (2022). Bioremediation and waste management for environmental sustainability. Journal of Applied Biology and Biotechnology, 10(2), 1-5.

Yadav, C. S., Awasthi, K. K., & Sankhla, M. S. (Eds.). (2024). Bioremediation of Environmental Toxicants: Toxicants, Sources, Mechanism, Impact on Human Health, and Bioremediation Approaches. CRC Press.

18

Prospects of Nanobioremediation as an Ecofriendly Approach

Sonika Kashyap, Riddhi Garg, Vandana Jhalora, Imran Khan and Renu Bist**[*]**

Department of Zoology, University of Rajasthan, Jaipur - 302004, India

[*]Corresponding Author Email: renu_bisht22@yahoo.co.in

Abstract

Nanobioremediation is an emerging field which amalgamates the approaches of nanotechnology and traditional bioremedial measures, thereby leading to an increase in degradation potential of pollutants. Bioremediation involves microbial and enzymatic processes that break down contaminants, yet various factors can influence their efficiency, including environmental conditions and the nature of pollutants. Nanotechnology plays a major role in advancing bioremediation by utilizing nanomaterials, such as metallic nanoparticles (e.g., silver, gold, iron), carbon-based nanomaterials (e.g., graphene, carbon nanotubes), and other functionalized nanomaterials. The nanoparticles deploy unique mechanisms (such as increase in reactivity and surface area) in degradation of pollutants. The utilisation of nanoparticles becomes much important due to their ability to address specific environmental challenges, such as soil, water, heavy metals, and organic pollutants remediation. Despite these benefits, there are certain challenges, including toxicity concerns and the potential environmental impact of nanomaterials. Therefore, the extensive research is cardinal towards the synthesis and characterization of nanomaterials, alongside the improved understanding of their environmental interactions, which could be beneficial to more effective nanobioremediation applications.

Keywords: *Bioremediation, Nanobioremediation, Nanotechnology, Pollutants, Nanomaterials*

1. Introduction

The environment and human health are seriously threatened by contaminants, both organic and inorganic, whose prevalence and persistence have been steadily rising in recent years. In addition to propelling economic and scientific advancement, the industrial revolutions of the 19th and 20th centuries have also led to widespread pollution (Wuana & Okieimen, 2011; Singh *et al.*, 2013). Industrial activity has increased trace metal concentrations in the environment, with sediments acting as key sources and reservoirs for heavy metals. These metals build up in soils and aquatic bodies as a result of scouring, erosion, rainfall leaching, wastewater discharge, and atmospheric deposition (Alonso Castillo *et al.*, 2013). Additionally, toxic contaminants like phenols, industrial effluents, domestic wastewater, agricultural runoff, fly ash, fertilizers, and waste from insecticides, pesticides, and pharmaceutical industries infiltrate the environment and food chain via mechanisms such as bioaccumulation, biomagnification, and bioconcentration (Arora & Bae, 2014; Azubuike *et al.*, 2016).

Bioremediation is an eco-friendly and sustainable approach that uses microorganisms to degrade and neutralize pollutants, providing an alternative to conventional physical and chemical remediation methods (Azubuike *et al.*, 2016; Ajona & Vasanthi, 2021). It is essential for eliminating environmental toxins, restoring ecosystems, and reducing contamination risks (Berillo *et al.*, 2021). This technique efficiently removes contaminants from industrial and agricultural processes, including heavy metals, pesticides, industrial effluents, and chlorinated chemicals (Sharma, 2020). Recent advancements highlight its efficacy in reducing the effects of human-made pollutants (Devarajan *et al.*, 2015). The chemical structure and toxicity of pollutants have a considerable impact on the biodegradation process. A key factor in the effectiveness of bioremediation is the surrounding environment (Azad *et al.*, 2024). It is necessary to optimise factors including pH, temperature, oxygen availability, and nutrient levels in order to stimulate microbial activity.

Bioremediation offers a versatile and effective solution for addressing various pollutants, but its efficiency can decline when dealing with high pollutant concentrations or stubborn compounds like xenobiotics (Mapelli *et al.*, 2017). To overcome these challenges, the advent of nanotechnology has opened exciting new possibilities. Incorporating nanomaterials (NMs), particularly nanoparticles (NPs), into bioremediation enhances its effectiveness by overcoming current limitations, paving the way for innovative and sustainable environmental restoration strategies. This integrated approach, known as nano-bioremediation, holds significant potential for improving remediation efficiency (Rizwan *et al.*, 2014). NMs can be applied directly or in combination with microbes or microbial enzymes to enhance the remediation process.

2. Types of Bioremediation

Bioremediation applications are divided into two primary areas: *in situ* or *ex situ.*

2.1. In-situ Bioremediation

In-situ bioremediation refers to the process of treating pollution directly at its source, offering a less invasive and more environmentally friendly alternative to traditional cleanup methods (Fernandez-Lopez *et al.*, 2022). This technique treats contaminated groundwater and soil without requiring extensive excavation or expensive removal procedures. In-situ bioremediation minimises environmental disturbance while efficiently eliminating chemical pollutants by employing natural microbes, plants, and animals. Key benefits include reduced physical labor, lower transportation costs, and the use of natural processes. Additionally, because contaminants remain confined, smaller amounts of NMs can treat large areas or volumes. However, the effectiveness of in-situ bioremediation depends on environmental conditions such as soil type, seasonal temperature changes, pH levels, microbial gene expression, nutrition availability, and oxygen levels (Romantschuk *et al.*, 2023).

2.2. Ex-situ Bioremediation

Ex-situ bioremediation involves excavating and transporting contaminated materials from polluted sites to another location for treatment. This approach is highly efficient against a wide variety of hydrocarbon-based contaminants, comprising carbon compounds, aliphatic chains, lubricating oils, ketones, cresols, phenols, semi-volatile organic compounds (SVOCs), and polycyclic aromatic hydrocarbons. Ex-situ bioremediation is influenced by parameters like the treatment costs, type and concentration of contaminants, geographical location, and topography (Philp & Atlas, 2005). While more resource-intensive than in-situ methods, ex-situ techniques offer faster treatment, greater control, and the ability to address a broader variety of contaminants and soil types (Prokop, 2000). This makes it a reliable option for complex or highly polluted sites.

3. Overview of Bioremediation Techniques

Bioremediation, supported by new technologies, offers a hopeful solution to important environmental problems (Agrawal *et al.*, 2024). Table 1 shows different bioremediation techniques.

3.1 Microbial Process

Microbial bioremediation is a natural and efficient way to reduce environmental contamination which uses the aerobic or anaerobic activity of microorganisms.

This biotechnological approach employs living organisms, notably microbes and plants, to remove or immobilise toxins, thereby aiding in environmental restoration (Srivastava *et al.*, 2019). In addition to conventional sites such as soil, water, and sediments, new areas emerged as possible microbial activity zones. Broadening the scope of bioremediation initiatives, these include marshland subsurfaces, aquatic shorelines, plant rhizospheres, gravel-associated communities, and worn rocks (Boopathy, 2000). Also, microbes thrive across a wide range of conditions and can tolerate even high concentrations of chemical contaminants, making them highly adaptable for bioremediation purposes.

Table 1. Recent advancements in microbial bioremediation techniques

Advancement	Description	References
Biostimulation	Enhances the activity of native microbes by providing nutrients or stimulants.	Omokhagbor Adams *et al.*, 2020
Bioaugmentation	Increases the population of viable microbes by introducing specialized strains.	Muter, 2023
Bioaccumulation	Microbes store toxic or non-toxic elements within their cells.	Fernández *et al.*, 2018
Biosorption	Adsorption of environmental elements onto microbial cell surfaces.	Yaashikaa *et al.*, 2024
Biofilms	Use of microbial biofilms for efficient pollutant degradation.	Zapata-Peñasco *et al.*, 2024

3.2. Enzymatic Process

Enzyme-mediated bioremediation utilizes naturally occurring enzymes from microorganisms or plants to break down and neutralize harmful and persistent environmental pollutants, aiding in site cleanup. Numerous enzymes derived from bacteria, fungi, and plants have been identified as effective in degrading toxic organic compounds. This process relies on microbial enzymes to transform pollutants into harmless byproducts (Karigar & Rao, 2011). Enzyme-mediated bioremediation is an eco-friendly approach with the advantage of highly specific degradation capabilities, particularly for targeted compounds. The smaller size of enzymes compared to microbial cells allows them to easily interact with contaminants, enhancing mobility and contact with pollutants. This leads to faster, more efficient, and more precise degradation, transforming harmful substances into less toxic or acceptable forms (Karigar & Rao, 2011).

4. Factors Influencing Bioremediation Efficiency

According to a study, a number of factors affect the physical and chemical interactions between NMs, organisms, and contaminants. These factors include the size and shape of NMs, their surface coating, the chemical makeup of the NMs and contaminants, as well as the organism type, medium, pH, and temperature (Tan *et al.*, 2018). The choice of NM depends on the type of contaminant present. For instance, the magnetic characteristics of magnetite, a nanoiron substance, are utilised to extract heavy metals from water or soil. Additionally, carbon-based NMs are utilised to capture heavy metals or organic contaminants from the air, soil, or water. Photocatalytic reactions may destroy pollutants, depending on the type of NMs. The biotic systems may further biotransform the resulting products, which would lower the concentration of pollutants in the media. Furthermore, a number of pollutants may be broken down by certain enzymes produced by organisms (Peixoto *et al.*, 2011).

5. Nanotechnology in Environmental Remediation

Exponentially increasing global population and rapid industrialization have pushed environmental pollutants levels to a very alarming stage. Pollutants like Particulate matter ($PM_{2.5}$, PM_{10}), heavy metals like Cadmium (Cd), Chromium (Cr), Arsenic (As), Lead (Pb), etc. (Huang *et al.*, 2004), agricultural pollutants like pesticides, herbicides, fertilizers, fungicides etc. (Tudi *et al.*, 2021), oil spills, toxic gases, effluents of industries, dyes from textile industries (Singh & Chandra, 2019; Goswami *et al.*,2024), sewage, and organic compounds are contaminants that are threatening health by hindering clean ambient air, supply of nutritional food, potable water, and safe shelter. Various remediation techniques based on physical, chemical and biological treatments have evolved over the last couple of decades to address air, water and soil contaminants (Rachna *et al.*, 2024). Sources and concentrations of these pollutants need to be identified and technologies like nanotechnology that safeguard precise, efficient and sustaining mitigation is need of the hour.

5.1. Introduction to Nanotechnology

"Nanotechnology is a branch of science which deals with the design, production and application of structures, devices and systems by controlling shape and size at the nanoscale" (European Commission). 'Nano' in term nanotechnology means one billionth of a meter (1×10^{-9}). Nanostructures or devices have at least one dimension on a nanometre scale. Nanostructured materials have discrete physicochemical properties like large surface area to volume ratio, distinct structure, electrical and optical properties, etc., rendering them most suitable and efficient for fields like medicine, electronics, food, agriculture and environmental remediation. Nanotechnology in environmental remediation is contributing remarkably (Hazrati Saadabadi *et al.*, 2024; Guan *et al.*, 2024). Nanomaterials utilized in environmental remediation are typically categorized into three types: inorganic, carbon-based, and polymer-based. Inorganic nanomaterials like titanium oxides (TiO_2) NPs, Silver nanoparticles (AgNPs), silica gel, Modified jacobsite ($MnFe_2O_4$), Maghemite ($-Fe_2O_3$) Magnetite nanoparticles modified with di-mercaptosuccinic acid (Guerra *et al.*, 2018; Bhatti *et al.*, 2022), Carbon-based nanomaterials, such as carbon nanohorns, carbyne, graphene, fullerenes, single-walled carbon nanotubes (SWCNTs), nanodiamonds, carbon fibers, multi-walled carbon nanotubes (MWCNTs), hybrid carbon nanotubes (HCNTs), carbon beads, and nanoporous carbon, along with polymer-based nanomaterials like Fe^0/Pd combined with polyvinyl alcohol (PVA) or polyacrylic acid (PAA), TiO_2 with polyaniline, Fe^0 with alginate or carboxymethyl cellulose, Cu^0 with chitosan, and Ni/Fe with cellulose acetate, have been extensively studied. These materials have shown great potential for applications such as wastewater treatment, air purification, disinfection, and the removal of metals and organic dyes. (Gerasimenko *et al.*, 2022; Sacco & Vollebregt, 2023).

5.2. Synthesis and Characterization of Nanomaterials

5.2.1. Synthesis of Nanomaterials

Top-down nanofabrication techniques, such as laser ablation and mechanical milling, reduce bulk materials to nanoscale particles. While these methods are relatively simple, they can potentially induce structural and chemical modifications to the resulting nanoparticles. Bottom-up approaches, including biological and chemical synthesis, offer precise control over nanoparticle synthesis. Solid-state, sol-gel, and gas-phase methods provide alternative approaches with varying degrees of control, cost-effectiveness, and environmental impact (Paramasivam *et al.*, 2021). Green synthesis emerges as a promising approach for nanoparticle production, leveraging biological resources to overcome the limitations of conventional physical and chemical methods. This eco-friendly technique offers several advantages, including reduced environmental impact, cost-effectiveness, and scalability for large-scale applications. Biological agents, such as plant extracts, fungi, and bacteria, can act as both reducing and capping agents, facilitating the synthesis of stable nanoparticles with unique properties. However, challenges remain in terms of reproducibility, standardization, and scale-up, requiring further research and optimization. Green synthesis is achieved by reducing metal solution using cellular extractants from bacteria viruses, fungi, flowers, roots and leaves of plants (Huston *et al.*, 2021; Singh *et al.*, 2023). These extractants have wide varieties of bio compounds like intracellular and extracellular enzymes proteins, coenzymes, surfactants, alkaloids, flavonoids, tannins, monoterpenes, carotenoids, saponins and phenolics etc. Some nanoparticles using green synthesis are shown in Table 2.

Table 2. Microbe based Green synthesised metal nanoparticles

S. No.	Microbe	Nanoparticles	Method of synthesis	Size in nm	references
1.	*Rhizopus stolonifera*	Gold	Culture filtrate	9	AbdelRahim *et al.*, 2017
2.	*Aspergillus niger*	Silver	Extracellular	20	Gade *et al.*, 2008
3.	*Fusarium oxysporum*	Silver	Cell free filtrate	24	Hamedi *et al.*, 2017
4.	*Sargassum tenerrimum*	Gold	Cell extract	5-45	Ramakrishna *et al.*, 2016)
5.	*Sargassum muticum*	Silver	Cell extract	5-15	Azizi *et al.*, 2013
6.	*Bacillus amyloliquefaciens*	Titanium Dioxide	Cell culture	22-97	Khan and Fulekar 2016
7.	*Lactobacillus species*	Titanium	Cell culture	40-60	Prasad *et al.*, 2007
8.	*Bacillus methylotrophicus*	Silver	Extracellular	10-30	Wang *et al.*, 2016
9.	*M13 bacteriophage*	ZnS, CdS	Inorganic synthesis	560 - 20 nm Qdots & wires	Mao *et al.*, 2003

5.2.2. Methods of Characterization

The size and shape of nanoparticles are key criteria for their characterization, but additional parameters such as surface charge, surface area, elemental and chemical composition, crystal structure, oxidation state, and optical and magnetic properties are also evaluated (Saleh & Hassan, 2023). Some important techniques used for these purposes are X-ray absorption spectroscopy (XAS), X-ray diffraction (XRD), High-resolution transmission electron microscope (HRTEM), scanning electron microscope (SEM), *X-ray* absorption fine structure (EXAFS), Dynamic light scattering (DLS), Small-angle X-ray scattering (SAXS), Atomic force microscopy (AFM), Nanoparticle tracking analysis (NTA), X-ray absorption near edge structure (XANES), UV-visible (UV-Vis) spectroscopy, ferromagnetic resonance (FMR), Nuclear Magnetic Resonance (NMR) spectroscopy, Fourier transform infrared (FTIR spectroscopy, Matrix-assisted laser desorption/ionization (MALDI), etc. (Lin *et al.*, 2014; Jayawardena *et al.*, 2021).

5.3. Mechanisms of Nanoparticle Action in Pollutant Degradation

Environmental remediation fundamentally relies on approaches like physical, chemical, mechanical, thermal and biological treatments. Each technique employs distinct reaction mechanisms to restrain pollution. Physical remediation techniques leverage adsorption dynamics and radiative processes to mitigate contaminants. Adsorption is an excellent technique that involves the eradication of pollutants by adsorbing them on adsorbent materials like activated carbon, molecular sieves, nano-SiO- AgNPs, zeolites and Mesoporous TiO, etc (Dhaka & Chattopadhyay, 2021; Ojija, 2024). In the nanofiltration technique nanomaterial-based membrane filter selectively allows particles to pass through its nanopore based on particle's hydrophillic or hydrophobic nature, shape, size and charge. Nanomaterials like Ag, TiO, Zno, and carbon nanotubes are added with polymers to develop desired membrane.

The photocatalysis technique has emerged as a very cost-effective and environmentally friendly approach. Homogenous photocatalysis also known as Photo-Fenton reaction uses 600 nm wavelength light while heterogenous photocatalysis (Fenton reaction) does not need light irradiation. OH• and $\bullet O_2$ – like radicals are used as oxidizing agents in this process (Chen *et al.*, 2024; Bopape *et al.*, 2024). Chemical methodologies utilize oxidative-reductive reactions to transform or neutralize pollutants. In contrast, biological interventions employ microbial consortia—both aerobic and anaerobic—for biocatalytic disinfection and degradation, complemented by enzymatic pathways that facilitate the biochemical transformation of hazardous substances (Singh *et al.*, 2023) (Fig. 1).

6. Significance of using NPs in Bioremediation

NPs are increasingly recognized as powerful tools in bioremediation due to their unique properties, such as a high surface area-to-volume ratio, reactivity, and customizable surface chemistry. NPs can adsorb and degrade heavy metals, organic contaminants, and pathogens, offering targeted and efficient solutions for environmental cleanup (Ningthoujam *et al.*, 2022). For instance, metal oxide NPs like titanium dioxide (TiO_2) have photocatalytic properties, breaking down organic pollutants under light (Anucha *et al.*, 2022), while functionalized carbon-based NPs are effective in adsorbing organic compounds (Krishna *et al.*, 2023). NPs can also be combined with microorganisms, creating "nano-bio hybrids" (Liang & Liang, 2023) that enhance pollutant degradation through synergistic effects, improving both the speed and efficiency of remediation processes (Shen *et al.*, 2017). These features allow NPs to effectively target and degrade pollutants in hard-to-reach areas like soil pores and water bodies, where traditional methods are less efficient (Singh & Saxena, 2022).

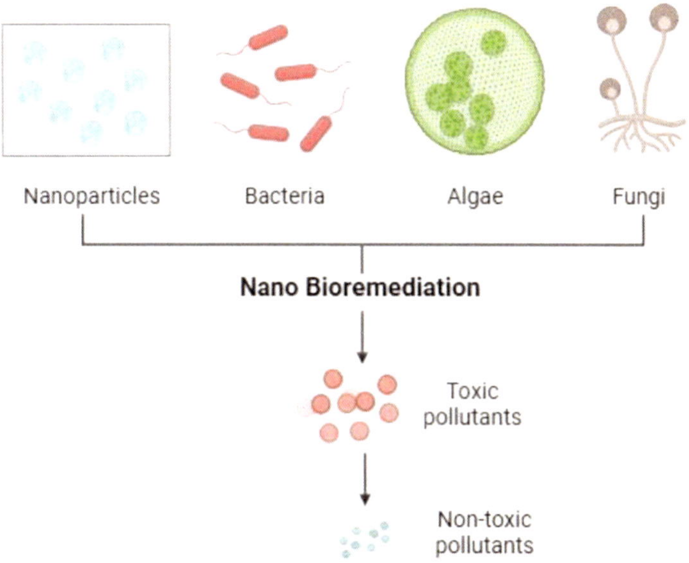

Fig. 1. The Fig. shows nanoparticles interacting with microbial cells to facilitate the degradation of environmental pollutants, such as heavy metals, organic compounds, and dyes.

6.1. Metallic NPs

Metallic NPs are increasingly recognized as potent agents in bioremediation due to their unique physicochemical properties, including high reactivity, large surface area-to-volume ratio, and strong catalytic abilities. Their applications in environmental cleanup focus primarily on the removal, degradation, or neutralization of contaminants such as heavy metals, organic pollutants, dyes, and pathogens. Silver NPs (AgNPs) play a significant role in the bioremediation of heavy metals by interacting with microorganisms to enhance the removal and detoxification of toxic metals. AgNPs can adsorb heavy metals like lead, mercury, cadmium, and chromium onto their surface due to their high surface area and reactivity, reducing the mobility and toxicity of these metals in the environment (Tao *et al.*, 2016). Zero-valent AgNPs synthesized from the leaf extract of the *Ficus Benjamina* have shown exceptional efficiency in removing cadmium (Al-Qahtani, 2017). Likewise, AgNPs derived from the leaves of *Piliostigma thonningii* have demonstrated significant heavy metal removal from waste water (Shittu & Ihebunna, 2017). Additionally, AgNPs can act as carriers for metal-reducing enzymes, facilitating their delivery to contaminated sites for more efficient bioremediation. However, the potential toxicity of AgNPs, particularly through the release of silver ions, requires careful management to avoid inhibiting beneficial microbes essential for bioremediation (Ihtisham *et al.*, 2021).

Gold NPs (AuNPs) are highly stable and can be easily functionalized with various surface ligands, allowing them to target specific pollutants. While AuNPs are less reactive in directly degrading pollutants, their primary role in bioremediation lies in pollutant detection and sensing (Kumalasari *et al.*, 2024). AuNP-based biosensors are widely used due to their stability, ease of functionalization, and unique optical properties, enabling the detection of trace levels of heavy metals and organic contaminants (Akintelu *et al.*, 2021).

Iron NPs (FeNPs), particularly zero-valent iron (nZVI) and iron oxide (e.g., magnetite, Fe_3O_4), are extensively studied for their effectiveness in bioremediation. The nZVI particles are especially known for their ability to degrade chlorinated organic compounds through reductive dechlorination, where nZVI effectively broke down pollutants like trichloroethylene (TCE) and dichloroethane into non-toxic byproducts (Galdames *et al.*, 2020). FeNPs also possess magnetic properties, allowing for easy separation of pollutants in waste water treatment (Keshta *et al.*, 2024).

Zinc Oxide NPs (ZnO NPs) are photocatalytic and highly effective in the degradation of organic contaminants, especially when exposed to UV light (Mousa *et al.*, 2024). Under UV radiation, ZnO NPs generate reactive oxygen species (ROS) that break down a variety of organic pollutants, including dyes, pharmaceuticals, and pesticides, making them ideal for solar-driven remediation processes. Similarly, TiO_2 NPs are also highly effective in photocatalytic degradation, making them valuable for both water and air bioremediation (Palajonnala Narasaiah *et al.*, 2022).

6.2. Carbon-Based NMs in Bioremediation

Carbon-based NMs are increasingly recognized for their bioremediation potential, addressing various contaminants due to their high surface area, mechanical strength, and chemical stability. Research highlights the effectiveness of carbon nanotubes (CNTs) in adsorbing pollutants due to their extensive surface area and unique structure (Sajid *et al.*, 2022). CNTs can effectively adsorb aromatic compounds from contaminated water through hydrophobic interactions (Pan & Xing, 2008). Furthermore, functionalized CNTs with carboxyl or hydroxyl groups significantly enhance metal adsorption capacity, particularly for heavy metals like lead, cadmium, and arsenic (Ganzoury *et al.*, 2020). Additionally, to mitigate CNT retrieval challenges post-application, researchers have developed magnetic CNT composites that enable easy recovery using magnetic fields, thereby improving the practicality of CNTs in repeated bioremediation applications (Masotti & Caporali, 2013).

Graphene and graphene oxide (GO) also demonstrate significant potential in bioremediation applications. Due to its large surface area and high conductivity, graphene is particularly effective at adsorbing organic pollutants, including dyes, pesticides, and pharmaceuticals, through π-π interactions (Thakur *et al.*, 2024). Zhang *et al.* (2014) reported that GO can effectively adsorb various contaminants from wastewater, suggesting its utility in water treatment (Anegbe *et al.*, 2024). Furthermore, oxygen-containing functional groups of GO confer a high affinity for heavy metals. It was found that GO could remove up to 98% of lead ions from water, indicating its robustness for metal remediation (Bloor *et al.*, 2021).

Fullerenes (C60), which are spherical molecules composed entirely of carbon, exhibit both hydrophobic and hydrophilic characteristics, allowing them to interact effectively with a broad spectrum of pollutants. Their surface allows them to adsorb and degrade organic contaminants like hydrocarbons and phenolic compounds in water (Ballesteros *et al.*, 2000). Notably, fullerenes have photocatalytic properties as fullerene-based nanocomposites can act as photosensitizers to generate reactive oxygen species (ROS) under light, facilitating the oxidative breakdown of organic contaminants such as dyes and pesticides (Hamblin, 2018).

Activated carbon, a widely used porous carbon material, is known for its high adsorption capacity for organic pollutants and heavy metals (Wang *et al.*, 2023). In its nanostructured form, activated carbon exhibits enhanced surface area and porosity, further improving its efficiency in contaminant adsorption. Studies have demonstrated its capacity for removing pollutants such as mercury, lead, benzene, and toluene from wastewater (Jafari *et al.*, 2023).

7. Applications of Nanobioremediation

In recent years, rapid population growth, technological advancements, and persistent use of inorganic or organic contaminants and pollutants are posing a concern to the environment and human health (Vázquez-Núñez *et al.*, 2020; Ningombam *et al.*, 2024). A cost-effective approach of combining nanotechnology with bioremediation using microbes can combat this environmental contamination. The application of NMs have gained attention in various fields due to their unique properties of being smaller in size and having a large surface area (Araújo *et al.*, 2015; Singh & Saxena, 2022). Quantum effects and Surface Plasmon Resonance phenomenon exhibited by NMs are used for the detection of toxic chemicals and environmental clean-up (Rizwan *et al.*, 2014).

7.1. Soil Remediation

Agriculture, an important human activity contaminated soil health and fertility by adding potentially toxic chemicals. Soil pollution is caused by pesticides, industrial activity and wastewater, which persists in soil and imbalances the nutrient cycle ultimately harming human health (Rajput *et al.*, 2022; Aliyari Rad *et al.*, 2023). Natural soil has a higher concentration of toxic heavy metals like Copper (Cu), Zinc (Zn), Lead (Pb), and Nickel (Ni). Bioaccumulation ability of NPs of *Noaea Mucronata* accumulates Pb and also Zn, Cu, and Ni, *Reseda lutea* NPs best accumulate Fe, and Cd is best collected by *Marrubium vulgare* plant species and the amount of heavy metals decreases many folds combating soil pollution (Mohsenzadeh & Rad, 2012). Bioremediation of lindane pesticide using laccase immobilized from *Myceliophthora thermophilia* on nano-silica particles (Bebić *et al.*, 2020).

7.2. Water and Organic Pollutants Remediation

Water pollutants P-nitrophenol and Acid orange 7 are efficiently degraded enzymatically by bacterially overexpressed organophosphorus hydrolase using Zinc Sulphide (ZnS) nanocrystals using glutathione or cysteine (Torres-Martínez *et al.*, 2001). Organic water pollutants Polybrominated diphenyl ethers (PDBEs) are effectively treated through reductive debromination using nanoscale zero-valent iron (nZVI) followed by biological oxidation with *Sphingomonas* sp (Kim *et al.*, 2012). Industrial waste water cyanide compounds, methylene blue dye were removed by synthesizing nano-adsorbents using copper NPs with biological extract of *Cynomorium coccineum* (Sebeia *et al.*, 2020). Decomposition of Azo dye direct yellow 86 from water using iron-oxide nanopowder and soil microorganisms by fentom method (Katsumata *et al.*, 2010) and *Bacillus subtilis* immobilized on FeO (Nadi *et al.*, 2019).

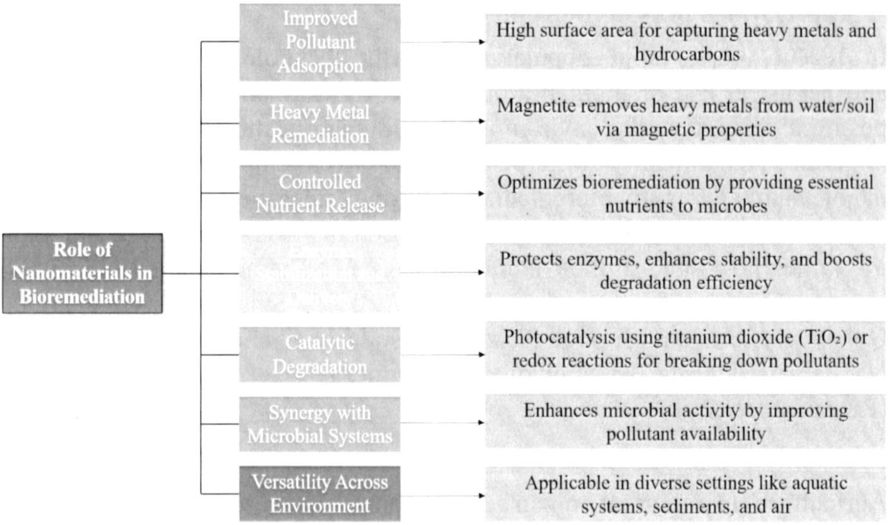

Fig. 2: Role of nanomaterials in bioremediation.

7.3. Heavy Metal Remediation

Industrial processing has caused heavy metal contamination in water and soil. Accumulation of heavy metals causes harms to plants, animals and human life. Due to their highly toxic nature and non-biodegradability, soil productivity and fertility is reduced affecting plant's physiology and metabolic activities (Bai *et al.*, 2019). In the bioremediation techniques, potential microorganisms are consumed to convert highly toxic forms of heavy metal into low toxic forms to manage and combat soil and water pollution. NPs of iron oxide (FeO) were used in removal of various toxic heavy metals and textile effluents from industrial wastewater (Mahanty *et al.*, 2020; Fouda *et al.*, 2021). Bio-organic NP' used for the removal of soil heavy metals like AgNPs produced in *Morganella psychrotolerans* (Arif *et al.*, 2016; Enez *et al.*, 2018) Rice paddy soil is contaminated by Cu and Pb whose bioavailability is significantly decreased by isolated *Bacillus subtilis* strain by biosorption and bioaccumulation (Bai *et al.*, 2019), magnetic NPs of Fe3O4 immobilized with phthalic acid treated with *Staphylococcus aureus* for bioremediation (Mahmoud *et al.*, 2016). Heavy metal Pb adsorption has been increased considerably by synthesizing a novel nitrogen and phosphorus-enriched activated carbon biochar co-pyrolysis with *Camellia oleifera* shells (COS's) with different mass ratio of ammonium polyphosphate (APP@AC's). as the mass ratio of APP to COS's has increased, the amount of N and P on biochar also increase which beneficially adsorb Pb by chemisorption (Fan *et al.*, 2020). Plant Growth Promoting Rhizobacteria (PGPR) have been used for field treatment to maintain soil productivity, health, disease suppression and nutrient bioavailability. These soil bacteria tolerate, degrade and remediate soil pollutants (Mahapatra *et al.*, 2022; Gupta *et al.*, 2024). Sharkawy in their studies use nano-Magnesium Oxide (nano-MgO) in combination with *R. nepotum* and *G. halophytocola* bacterial strains in arsenic (As) contaminated soil to significantly decrease As uptake in maize plants. Uranium (U) mines have As as a co-contaminant which exploits the bioremediation process. Bacterial isolates with *Microbacterium*, *Shinella*, and *Bacillus* have As oxidation potential indicated by environmental application to remove harmful heavy metals from contaminated sites (Bhakat *et al.*, 2019; El Sharkawy *et al.*, 2024) (Fig. 2).

8. Challenges and Limitations in Nanobioremediation

Nanotechnology is revolutionizing bioremediation processes by offering innovative solutions to challenging environmental problems providing many significant benefits through the use of highly efficient and adaptable NMs. An innovative approach to nanobioremediation holds immense potential for addressing pollution and environmental cleanup with unparalleled precision and efficiency. Despite its enormous potential, various challenges remain constant like high production costs, NP toxicity, scalability issues and environmental persistence. NMs pose significant challenges due to their unique characteristics fundamentally differing in size, shape, composition, and solubility. Synthesizing NPs with such precise properties is expensive and involves complex chemical processes. These materials in their nano forms, can undergo transformations, altering properties and interaction within the environment. Their potential toxic effects on microbes and plants can disrupt cellular membranes, inhibit enzymatic activity and induce oxidative stress affecting bioremediation capabilities. Environmental matrices may also influence the performance of nanobioremediation where NPs may aggregate reducing their bioavailability and reactivity. However, lack of interactions between NPs and biological systems as well as insufficient research between nanotechnology and bioremediation slow down the development of practical solutions. Therefore, with more research and development nanoremediation may develop into effective and efficient method for cleaning up the environment (Ramírez-García *et al.*, 2019; Patra Shahi *et al.*, 2021; G. Pandey *et al.*, 2022).

9. Future Perspectives in Nanobioremediation

The rapid advancements and progress in the field of nanotechnology have significantly enhanced the economic impact by unlocking its full potential as a sustainable solution for environmental cleanup. NPs are becoming a viable option for environmentally friendly approaches. It is applied in various domains to treat and remediate the environment, sensor development, and eco-friendly green synthesis, giving new avenues for research development. Innovations in green synthesis for the treatment and elimination of harmful and toxic pollutants using biological agents like plants and microbes revolutionize the large-scale production of NPs. The large-scale productivity of biologically synthesized NMs makes them cost-effective and environmentally safe, playing a vital role in remediating contamination. The efficiency of the remediation process can be achieved by integrating artificial intelligence to optimize NP design and predict their interactions with contaminants, which is crucial for achieving nanoremediation of harmful toxicants. Alongside this, integrating nanobioremediation with phytoremediation and photocatalysis might create hybrid approaches for tackling complicated pollution scenarios. Thus, nanobioremediation holds the potential to become a key component of sustainable environmental management by facilitating the efficient production of advanced NMs for the eradication of pollutants without endangering the environment. This can be achieved by the promotion of interdisciplinary research, global collaborations, green synthesis techniques and circular economic principles.

References

AbdelRahim, K., Mahmoud, S. Y., Ali, A. M., Almaary, K. S., Mustafa, A. E., & Husseiny, S. M. (2017). Extracellular biosynthesis of silver nanoparticles using Rhizopus stolonifer. Saudi journal of biological sciences, 24(1), 208–216. https://doi.org/10.1016/j.sjbs.2016.02.025

Agrawal, K., Ruhil, T., Gupta, V. K., & Verma, P. (2024). Microbial assisted multifaceted amelioration processes of heavy-metal remediation: A clean perspective toward sustainable and greener future. Critical Reviews in Biotechnology, 44(3), 429–447. https://doi.org/10.1080/07388551.2023.2170862

Ajona, M., & Vasanthi, P. (2021). Bioremediation of petroleum contaminated soils–A review. Materials Today: Proceedings, 45, 7117–7122. https://www.sciencedirect.com/science/article/pii/S2214785321010531

Akintelu, S. A., Yao, B., & Folorunso, A. S. (2021). Bioremediation and pharmacological applications of gold nanoparticles synthesized from plant materials. Heliyon, 7(3), e06591. https://doi.org/10.1016/j.heliyon.2021.e06591

Aliyari Rad, S., Nobaharan, K., Pashapoor, N., Pandey, J., Dehghanian, Z., Senapathi, V., Minkina, T., Ren, W., Rajput, V. D., & Asgari Lajayer, B. (2023). Nano-Microbial Remediation of Polluted Soil: A Brief Insight. Sustainability, 15(1), Article 1. https://doi.org/10.3390/su15010876

Alonso Castillo, M. L., Sánchez Trujillo, I., Vereda Alonso, E., García de Torres, A., & Cano Pavón, J. M. (2013). Bioavailability of heavy metals in water and sediments from a typical Mediterranean Bay (Málaga Bay, Region of Andalucía, Southern Spain). Marine Pollution Bulletin, 76(1–2), 427–434. https://doi.org/10.1016/j.marpolbul.2013.08.031

Al-Qahtani, K. M. (2017). Cadmium removal from aqueous solution by green synthesis zero valent silver nanoparticles with Benjamina leaves extract. Egyptian Journal of Aquatic Research, 43(4), 269–274. https://doi.org/10.1016/j.ejar.2017.10.003

Anegbe, B., Ifijen, I. H., Maliki, M., Uwidia, I. E., & Aigbodion, A. I. (2024). Graphene oxide synthesis and applications in emerging contaminant removal: A comprehensive review. Environmental Sciences Europe, 36(1), Article 1. https://doi.org/10.1186/s12302-023-00814-4

Anucha, C. B., Altin, I., Bacaksiz, E., & Stathopoulos, V. N. (2022). Titanium dioxide (TiO_2)-based photocatalyst materials activity enhancement for contaminants of emerging concern (CECs) degradation: In the light of modification strategies. Chemical Engineering Journal Advances, 10, 100262. https://doi.org/10.1016/j.ceja.2022.100262

Araújo, R., Castro, A. C. M., & Fiúza, A. (2015). The Use of Nanoparticles in Soil and Water Remediation Processes. Materials Today: Proceedings, 2(1), 315–320. https://doi.org/10.1016/j.matpr.2015.04.055

Arif, N., Yadav, V., Singh, S., Singh, S., Ahmad, P., Mishra, R. K., Sharma, S., Tripathi, D. K., Dubey, N. K., & Chauhan, D. K. (2016). Influence of High and Low Levels of Plant-Beneficial Heavy Metal Ions on Plant Growth and Development. Frontiers in Environmental Science, 4. https://doi.org/10.3389/fenvs.2016.00069

Arora, P. K., & Bae, H. (2014). Bacterial degradation of chlorophenols and their derivatives. Microbial Cell Factories, 13(1), 31. https://doi.org/10.1186/1475-2859-13-31

Azad, A., Sheikh, M. N., & Hai, F. I. (2024). A critical review of the mechanisms, factors, and performance of pervious concrete to remove contaminants from stormwater runoff. Water Research, 251, 121101. https://doi.org/10.1016/j.watres.2024.121101

Azizi, S., Namvar, F., Mahdavi, M., Ahmad, M. B., & Mohamad, R. (2013). Biosynthesis of Silver Nanoparticles Using Brown Marine Macroalga, Sargassum Muticum Aqueous

Extract. Materials (Basel, Switzerland), 6(12), 5942–5950. https://doi.org/10.3390/ma6125942

Azubuike, C. C., Chikere, C. B., & Okpokwasili, G. C. (2016). Bioremediation techniques-classification based on site of application: Principles, advantages, limitations and prospects. World Journal of Microbiology & Biotechnology, 32(11), 180. https://doi.org/10.1007/s11274-016-2137-x

Bai, J., Chao, Y., Chen, Y., Wang, S., & Qiu, R. (2019). The effect of interaction between Bacillus subtilis DBM and soil minerals on Cu(II) and Pb(II) adsorption. Journal of Environmental Sciences, 78, 328–337. https://doi.org/10.1016/j.jes.2018.11.012

Ballesteros, E., Gallego, M., & Valcárcel, M. (2000). Analytical potential of fullerene as adsorbent for organic and organometallic compounds from aqueous solutions. Journal of Chromatography. A, 869, 101–110. https://doi.org/10.1016/S0021-9673(99)01050-X

Bebić, J., Banjanac, K., Ćorović, M., Milivojević, A., Simović, M., Marinković, A., & Bezbradica, D. (2020). Immobilization of laccase from Myceliophthora thermophila on functionalized silica nanoparticles: Optimization and application in lindane degradation. Chinese Journal of Chemical Engineering, 28(4), 1136–1144. https://doi.org/10.1016/j.cjche.2019.12.025

Berillo, D., Yeskendir, A., Zharkinbekov, Z., Raziyeva, K., & Saparov, A. (2021). Peptide-Based Drug Delivery Systems. Medicina (Kaunas, Lithuania), 57(11), 1209. https://doi.org/10.3390/medicina57111209

Bhakat, K., Chakraborty, A., & Islam, E. (2019). Characterization of arsenic oxidation and uranium bioremediation potential of arsenic resistant bacteria isolated from uranium ore. Environmental Science and Pollution Research, 26(13), 12907–12919. https://doi.org/10.1007/s11356-019-04827-6

Bhatti, R., Shakeel, H., Malik, K., Qasim, M., Khan, M. A., Ahmed, N., & Jabeen, S. (2022). Inorganic Nanoparticles: Toxic Effects, Mechanisms of Cytotoxicity and Phytochemical Interactions. Advanced pharmaceutical bulletin, 12(4), 757–762. https://doi.org/10.34172/apb.2022.077

Bloor, J. M., Handy, R. D., Awan, S. A., & Jenkins, D. F. L. (2021). Graphene oxide biopolymer aerogels for the removal of lead from drinking water using a novel nano-enhanced ion exchange cascade. Ecotoxicology and Environmental Safety, 208, 111422. https://doi.org/10.1016/j.ecoenv.2020.111422

Boopathy, R. (2000). Factors limiting bioremediation technologies. Bioresource Technology, 74(1), 63–67. https://doi.org/10.1016/S0960-8524(99)00144-3

Bopape, D. A., Ntsendwana, B., & Mabasa, F. D. (2024). Photocatalysis as a pre-discharge treatment to improve the effect of textile dyes on human health: A critical review. Heliyon, 10(20), e39316. https://doi.org/10.1016/j.heliyon.2024.e39316

Chen, C. X., Yang, S. S., Pang, J. W., He, L., Zang, Y. N., Ding, L., Ren, N. Q., & Ding, J. (2024). Anthraquinones-based photocatalysis: A comprehensive review. Environmental science and ecotechnology, 22, 100449. https://doi.org/10.1016/j.ese.2024.100449

Devarajan, R., Hanumappa, K. R., & Kuppan, N. (2015). The study of change in physicochemical properties of soil due to cement dust pollutionan hazardous terrorization to ecosystem. Canadian Journal of Pure and Applied Sciences, 9(1), 68–67. https://www.researchgate.net/profile/Raajasubramanian-Devarajan/publication/335202234

Dhaka, A., & Chattopadhyay, P. (2021). A review on physical remediation techniques for treatment of marine oil spills. Journal of environmental management, 288, 112428. https://doi.org/10.1016/j.jenvman.2021.112428

El Sharkawy, M., AL-Huqail, A. A., Aljuaid, A. M., Kamal, N., Mahmoud, E., Omara, A. E.-D., El-Kader, N. A., Li, J., Mahmoud, N. N., El Baroudy, A. A., Ghoneim, A. M., &

Ismail, S. M. (2024). Nano-Bioremediation of Arsenic and Its Effect on the Biological Activity and Growth of Maize Plants Grown in Highly Arsenic-Contaminated Soil. Nanomaterials, 14(13), 1164. https://doi.org/10.3390/nano14131164

Enez, A., Hudek, L., & Bräu, L. (2018). Reduction in Trace Element Mediated Oxidative Stress towards Cropped Plants via Beneficial Microbes in Irrigated Cropping Systems: A Review. Applied Sciences, 8(10), 1953. https://doi.org/10.3390/app8101953

Fan, Y., Wang, H., Deng, L., Wang, Y., Kang, D., Li, C., & Chen, H. (2020). Enhanced adsorption of Pb(II) by nitrogen and phosphorus co-doped biochar derived from Camellia oleifera shells. Environmental Research, 191, 110030. https://doi.org/10.1016/j.envres.2020.110030

Fernández, P. M., Viñarta, S. C., Bernal, A. R., Cruz, E. L., & Fig.ueroa, L. I. C. (2018). Bioremediation strategies for chromium removal: Current research, scale-up approach and future perspectives. Chemosphere, 208, 139–148. https://doi.org/10.1016/j.chemosphere.2018.05.166

Fernandez-Lopez, C., Posada-Baquero, R., & Ortega-Calvo, J.-J. (2022). Nature-based approaches to reducing the environmental risk of organic contaminants resulting from military activities. The Science of the Total Environment, 843, 157007. https://doi.org/10.1016/j.scitotenv.2022.157007

Ferreyra-Suarez, D., García-Depraect, O., & Castro-Muñoz, R. (2024). A review on fungal-based biopesticides and biofertilizers production. Ecotoxicology and environmental safety, 283, 116945. https://doi.org/10.1016/j.ecoenv.2024.116945

Fouda, A., Hassan, S. E.-D., Abdel-Rahman, M. A., Farag, M. M. S., Shehal-deen, A., Mohamed, A. A., Alsharif, S. M., Saied, E., Moghanim, S. A., & Azab, M. S. (2021). Catalytic degradation of wastewater from the textile and tannery industries by green synthesized hematite (α-Fe2O3) and magnesium oxide (MgO) nanoparticles. Current Research in Biotechnology, 3, 29–41. https://doi.org/10.1016/j.crbiot.2021.01.004

G. Pandey, S. Bajpai, S. Tripathi, Chauhan, R., & Sharma, P. (2022). Approaches, challenges, and prospects of life cycle assessment applied in nano bioremediation of contaminated sites. Rasayan Journal of Chemistry, 15(02), 1108–1125. https://doi.org/10.31788/RJC.2022.1526904

Gade, A. K., Bonde, P., Ingle, A. P., Marcato, P. D., Duran, N., & Rai, M. K. (2008). Exploitation of Aspergillus niger for synthesis of silver nanoparticles. Journal of Biobased Materials and Bioenergy, 2(3), 243-247.

Galdames, A., Ruiz-Rubio, L., Orueta, M., Sánchez-Arzalluz, M., & Vilas-Vilela, J. L. (2020). Zero-Valent Iron Nanoparticles for Soil and Groundwater Remediation. International Journal of Environmental Research and Public Health, 17(16), 5817. https://doi.org/10.3390/ijerph17165817

Ganzoury, M. A., Chidiac, C., Kurtz, J., & de Lannoy, C.-F. (2020). CNT-sorbents for heavy metals: Electrochemical regeneration and closed-loop recycling. Journal of Hazardous Materials, 393, 122432. https://doi.org/10.1016/j.jhazmat.2020.122432

Gerasimenko, A. Y., Kuksin, A. V., Shaman, Y. P., Kitsyuk, E. P., Fedorova, Y. O., Murashko, D. T., Shamanaev, A. A., Eganova, E. M., Sysa, A. V., Savelyev, M. S., Telyshev, D. V., Pavlov, A. A., & Glukhova, O. E. (2022). Hybrid Carbon Nanotubes-Graphene Nanostructures: Modeling, Formation, Characterization. Nanomaterials (Basel, Switzerland), 12(16), 2812. https://doi.org/10.3390/nano12162812

Goswami, D., Mukherjee, J., Mondal, C., & Bhunia, B. (2024). Bioremediation of azo dye: A review on strategies, toxicity assessment, mechanisms, bottlenecks and prospects. The Science of the total environment, 954, 176426. https://doi.org/10.1016/j.scitotenv.2024.176426

Guan, C., Han, Y., Ling, Z., Meng, X., Zhang, B., Dong, W., Zhang, D., & Chen, K. (2024). Nanomaterials: breaking the bottleneck of breast cancer drug resistance. Frontiers in immunology, 15, 1492546. https://doi.org/10.3389/fimmu.2024.1492546

Guerra, F. D., Attia, M. F., Whitehead, D. C., & Alexis, F. (2018). Nanotechnology for Environmental Remediation: Materials and Applications. Molecules (Basel, Switzerland), 23(7), 1760. https://doi.org/10.3390/molecules23071760

Gupta, R., Khan, F., Alqahtani, F. M., Hashem, M., & Ahmad, F. (2024). Plant Growth–Promoting Rhizobacteria (PGPR) Assisted Bioremediation of Heavy Metal Toxicity. Applied Biochemistry and Biotechnology, 196(5), 2928–2956. https://doi.org/10.1007/s12010-023-04545-3

Hamblin, M. R. (2018). Fullerenes as Photosensitizers in Photodynamic Therapy: Pros and Cons. Photochemical & Photobiological Sciences: Official Journal of the European Photochemistry Association and the European Society for Photobiology, 17(11), 1515. https://doi.org/10.1039/c8pp00195b

Hamedi, S., Ghaseminezhad, M., Shokrollahzadeh, S., & Shojaosadati, S. A. (2017). Controlled biosynthesis of silver nanoparticles using nitrate reductase enzyme induction of filamentous fungus and their antibacterial evaluation. Artificial cells, nanomedicine, and biotechnology,45(8),1588–1596.https://doi.org/10.1080/21691401.2016.1267011Artif. Cells Nanomed. Biotechnol. 45, 1588–1596. doi: 10.1080/21691401.2016. 1267011

Hazrati Saadabadi, R., Shariatmadar Tehrani, F., Sabouri, Z., & Darroudi, M. (2024). Photocatalytic activity and anticancer properties of green synthesized ZnO-MgO-Mn_2O_3 nanocomposite via Ocimum basilicum L. seed extract. Scientific reports, 14(1), 29812. https://doi.org/10.1038/s41598-024-81692-5

Huang, J., Mao, J., Liu, H., Li, Z., Liang, G., Zhang, D., Yang, J., Qin, W., Wen, P., Jiang, Y., & Mo, Z. (2024). Association between exposure to arsenic, cadmium, and lead and chronic kidney disease: evidence from four practical statistical models. Environmental geochemistry and health, 47(1), 6. https://doi.org/10.1007/s10653-024-02318-3

Huston, M., DeBella, M., DiBella, M., & Gupta, A. (2021). Green Synthesis of Nanomaterials. Nanomaterials (Basel, Switzerland), 11(8), 2130. https://doi.org/10.3390/nano11082130

Ihtisham, M., Noori, A., Yadav, S., Sarraf, M., Kumari, P., Brestic, M., Imran, M., Jiang, F., Yan, X., & Rastogi, A. (2021). Silver Nanoparticle's Toxicological Effects and Phytoremediation. Nanomaterials, 11(9), 2164. https://doi.org/10.3390/nano11092164

Jafari, D., Esfandyari, M., & Mojahed, M. (2023). Optimization of removal of toluene from industrial wastewater using RSM Box–Behnken experimental design. Sustainable Environment Research, 33(1), Article 1. https://doi.org/10.1186/s42834-023-00189-4

Jayawardena, H. S. N., Liyanage, S. H., Rathnayake, K., Patel, U., & Yan, M. (2021). Analytical Methods for Characterization of Nanomaterial Surfaces. Analytical chemistry, 93(4), 1889–1911. https://doi.org/10.1021/acs.analchem.0c05208

Karigar, C. S., & Rao, S. S. (2011). Role of microbial enzymes in the bioremediation of pollutants: A review. Enzyme Research, 2011, 805187. https://doi.org/10.4061/2011/805187

Katsumata, H., Koike, S., Kaneco, S., Suzuki, T., & Ohta, K. (2010). Degradation of Reactive Yellow 86 with photo-Fenton process driven by solar light. Journal of Environmental Sciences, 22(9), 1455–1461. https://doi.org/10.1016/S1001-0742(09)60275-8

Keshta, B. E., Gemeay, A. H., Kumar Sinha, D., Elsharkawy, S., Hassan, F., Rai, N., & Arora, C. (2024). State of the art on the magnetic iron oxide Nanoparticles: Synthesis, Functionalization, and applications in wastewater treatment. Results in Chemistry, 7, 101388. https://doi.org/10.1016/j.rechem.2024.101388

Khan, R., & Fulekar, M. H. (2016). Biosynthesis of titanium dioxide nanoparticles using Bacillus amyloliquefaciens culture and enhancement of its photocatalytic activity for the degradation of a sulfonated textile dye Reactive Red 31. Journal of colloid and interface science, 475, 184–191. https://doi.org/10.1016/j.jcis.2016.05.001

Kim, Y., Murugesan, K., Chang, Y., Kim, E., & Chang, Y. (2012). Degradation of polybrominated diphenyl ethers by a sequential treatment with nanoscale zero valent iron and aerobic biodegradation. Journal of Chemical Technology & Biotechnology, 87(2), 216–224. https://doi.org/10.1002/jctb.2699

Krishna, R. H., Chandraprabha, M. N., Samrat, K., Krishna Murthy, T. P., Manjunatha, C., & Kumar, S. G. (2023). Carbon nanotubes and graphene-based materials for adsorptive removal of metal ions – A review on surface functionalization and related adsorption mechanism. Applied Surface Science Advances, 16, 100431. https://doi.org/10.1016/j.apsadv.2023.100431

Kumalasari, M. R., Alfanaar, R., & Andreani, A. S. (2024). Gold nanoparticles (AuNPs): A versatile material for biosensor application. Talanta Open, 9, 100327. https://doi.org/10.1016/j.talo.2024.100327

Liang, J., & Liang, K. (2023). Nanobiohybrids: Synthesis strategies and environmental applications from micropollutants sensing and removal to global warming mitigation. Environmental Research, 232, 116317. https://doi.org/10.1016/j.envres.2023.116317

Lin, P. C., Lin, S., Wang, P. C., & Sridhar, R. (2014). Techniques for physicochemical characterization of nanomaterials. Biotechnology advances, 32(4), 711–726. https://doi.org/10.1016/j.biotechadv.2013.11.006

Mahanty, S., Chatterjee, S., Ghosh, S., Tudu, P., Gaine, T., Bakshi, M., Das, S., Das, P., Bhattacharyya, S., Bandyopadhyay, S., & Chaudhuri, P. (2020). Synergistic approach towards the sustainable management of heavy metals in wastewater using mycosynthesized iron oxide nanoparticles: Biofabrication, adsorptive dynamics and chemometric modeling study. Journal of Water Process Engineering, 37, 101426. https://doi.org/10.1016/j.jwpe.2020.101426

Mahapatra, S., Yadav, R., & Ramakrishna, W. (2022). Bacillus subtilis impact on plant growth, soil health and environment: Dr. Jekyll and Mr. Hyde. Journal of Applied Microbiology, 132(5), 3543–3562. https://doi.org/10.1111/jam.15480

Mahmoud, M. E., Abdou, A. E. H., Mohamed, S. M. S., & Osman, M. M. (2016). Engineered staphylococcus aureus via immobilization on magnetic Fe3O4-phthalate nanoparticles for biosorption of divalent ions from aqueous solutions. Journal of Environmental Chemical Engineering, 4(4), 3810–3824. https://doi.org/10.1016/j.jece.2016.08.022

Mao, C., Flynn, C. E., Hayhurst, A., Sweeney, R., Qi, J., Georgiou, G., Iverson, B., & Belcher, A. M. (2003). Viral assembly of oriented quantum dot nanowires. Proceedings of the National Academy of Sciences of the United States of America, 100(12), 6946–6951. https://doi.org/10.1073/pnas.0832310100

Mapelli, F., Scoma, A., Michoud, G., Aulenta, F., Boon, N., Borin, S., Kalogerakis, N., & Daffonchio, D. (2017). Biotechnologies for Marine Oil Spill Cleanup: Indissoluble Ties with Microorganisms. Trends in Biotechnology, 35(9), 860–870. https://doi.org/10.1016/j.tibtech.2017.04.003

Masotti, A., & Caporali, A. (2013). Preparation of Magnetic Carbon Nanotubes (Mag-CNTs) for Biomedical and Biotechnological Applications. International Journal of Molecular Sciences, 14(12), 24619. https://doi.org/10.3390/ijms141224619

Mohsenzadeh, F., & Rad, A. C. (2012). Bioremediation of Heavy Metal Pollution by Nano-Particles of Noaea Mucronata. International Journal of Bioscience, Biochemistry and Bioinformatics, 85–89. https://doi.org/10.7763/IJBBB.2012.V2.77

Mousa, S. A., Wissa, D. A., Hassan, H. H., Ebnalwaled, A. A., & Khairy, S. A. (2024). Enhanced photocatalytic activity of green synthesized zinc oxide nanoparticles using low-cost plant extracts. Scientific Reports, 14(1), 16713. https://doi.org/10.1038/s41598-024-66975-1

Muter, O. (2023). Current Trends in Bioaugmentation Tools for Bioremediation: A Critical Review of Advances and Knowledge Gaps. Microorganisms, 11(3), 710. https://doi.org/10.3390/microorganisms11030710

Nadi, A., Boyer, D., Charbonnel, N., Boukhriss, A., Forestier, C., & Gmouh, S. (2019). Immobilisation of bacteria onto magnetic nanoparticles for the decolorisation and degradation of azo dyes. IET Nanobiotechnology, 13(2), 144–149. https://doi.org/10.1049/iet-nbt.2018.5026

Ningombam, L., Mana, T., Apum, G., Ningthoujam, R., & Disco Singh, Y. (2024). Nano-bioremediation: A prospective approach for environmental decontamination in focus to soil, water and heavy metals. Environmental Nanotechnology, Monitoring & Management, 21, 100931. https://doi.org/10.1016/j.enmm.2024.100931

Ningthoujam, R., Singh, Y. D., Babu, P. J., Tirkey, A., Pradhan, S., & Sarma, M. (2022). Nanocatalyst in remediating environmental pollutants. Chemical Physics Impact, 4, 100064. https://doi.org/10.1016/j.chphi.2022.100064

Ojija F. (2024). Emerging environmental contaminants: Sources, effects on biodiversity and humans, remediation, and conservation implications. Science progress, 107(2), 368504241253720. https://doi.org/10.1177/00368504241253720

Omokhagbor Adams, G., Tawari Fufeyin, P., Eruke Okoro, S., & Ehinomen, I. (2020). Bioremediation, Biostimulation and Bioaugmention: A Review. International Journal of Environmental Bioremediation & Biodegradation, 3(1), 28–39. https://doi.org/10.12691/ijebb-3-1-5

Palajonnala Narasaiah, B., Banoth, P., Bustamante Dominguez, A. G., Mandal, B. K., Kumar, C. K., Barnes, C. H. W., De Los Santos Valladares, L., & Kollu, P. (2022). Biogenic Photo-Catalyst TiO2 Nanoparticles for Remediation of Environment Pollutants. ACS Omega, 7(30), 26174–26189. https://doi.org/10.1021/acsomega.2c01763

Pan, B., & Xing, B. (2008). Adsorption Mechanisms of Organic Chemicals on Carbon Nanotubes. Environmental Science & Technology, 42(24), 9005–9013. https://doi.org/10.1021/es801777n

Paramasivam, G., Palem, V. V., Sundaram, T., Sundaram, V., Kishore, S. C., & Bellucci, S. (2021). Nanomaterials: Synthesis and Applications in Theranostics. Nanomaterials (Basel, Switzerland), 11(12), 3228. https://doi.org/10.3390/nano11123228

Patra Shahi, M., Kumari, P., Mahobiya, D., & Kumar Shahi, S. (2021). Nano-bioremediation of environmental contaminants: Applications, challenges, and future prospects. In Bioremediation for Environmental Sustainability (pp. 83–98). Elsevier. https://doi.org/10.1016/B978-0-12-820318-7.00004-6

Peixoto, R. S., Vermelho, A. B., & Rosado, A. S. (2011). Petroleum-degrading enzymes: Bioremediation and new prospects. Enzyme Research, 2011, 475193. https://doi.org/10.4061/2011/475193

Philp, J. C., & Atlas, R. M. (2005). Bioremediation of Contaminated Soils and Aquifers. In Bioremediation (pp. 139–236). John Wiley & Sons, Ltd. https://doi.org/10.1128/9781555817596.ch5

Prasad, K., Jha, A. K., & Kulkarni, A. R. (2007). Lactobacillus assisted synthesis of titanium nanoparticles. Nanoscale Research Letters, 2, 248-250.

Prokop, G. (2000). Management of contaminated sites in Western Europe. Office for official publications of the European communities. https://ketlib.lib.unipi.gr/xmlui/

bitstream/handle/ket/1036/Management%20of%20contaminated%20sites%20in%20Western%20Europe.pdf.pdf?sequence=2

Rachna, Singh, M. P., Goswami, S., & Singh, U. K. (2024). Pesticide pollution: toxicity, sources and advanced remediation approaches. Environmental science and pollution research international, 10.1007/s11356-024-35502-0. Advance online publication. https://doi.org/10.1007/s11356-024-35502-0

Rajput, V. D., Minkina, T., Upadhyay, S. K., Kumari, A., Ranjan, A., Mandzhieva, S., Sushkova, S., Singh, R. K., & Verma, K. K. (2022). Nanotechnology in the Restoration of Polluted Soil. Nanomaterials, 12(5), Article 5. https://doi.org/10.3390/nano12050769

Ramakrishna, M., Rajesh Babu, D., Gengan, R. M., Chandra, S., & Nageswara Rao, G. (2016). Green synthesis of gold nanoparticles using marine algae and evaluation of their catalytic activity. Journal of Nanostructure in Chemistry, 6, 1-13.

Ramírez-García, R., Gohil, N., & Singh, V. (2019). Recent Advances, Challenges, and Opportunities in Bioremediation of Hazardous Materials. In Phytomanagement of Polluted Sites (pp. 517–568). Elsevier. https://doi.org/10.1016/B978-0-12-813912-7.00021-1

Rizwan, Md., Singh, M., Mitra, C. K., & Morve, R. K. (2014). Ecofriendly Application of Nanomaterials: Nanobioremediation. Journal of Nanoparticles, 2014(1), 431787. https://doi.org/10.1155/2014/431787

Romantschuk, M., Lahti-Leikas, K., Kontro, M., Galitskaya, P., Talvenmäki, H., Simpanen, S., Allen, J. A., & Sinkkonen, A. (2023). Bioremediation of contaminated soil and groundwater by in situ biostimulation. Frontiers in Microbiology, 14, 1258148. https://doi.org/10.3389/fmicb.2023.1258148

Sacco, L. N., & Vollebregt, S. (2023). Overview of Engineering Carbon Nanomaterials Such As Carbon Nanotubes (CNTs), Carbon Nanofibers (CNFs), Graphene and Nanodiamonds and Other Carbon Allotropes inside Porous Anodic Alumina (PAA) Templates. Nanomaterials (Basel, Switzerland), 13(2), 260. https://doi.org/10.3390/nano13020260

Sajid, M., Asif, M., Baig, N., Kabeer, M., Ihsanullah, I., & Mohammad, A. W. (2022). Carbon nanotubes-based adsorbents: Properties, functionalization, interaction mechanisms, and applications in water purification. Journal of Water Process Engineering, 47, 102815. https://doi.org/10.1016/j.jwpe.2022.102815

Saleh, H. M., & Hassan, A. I. (2023). Synthesis and Characterization of Nanomaterials for Application in Cost-Effective Electrochemical Devices. Sustainability, 15(14), 10891. https://doi.org/10.3390/su151410891

Sebeia, N., Jabli, M., Ghith, A., & Saleh, T. A. (2020). Eco-friendly synthesis of Cynomorium coccineum extract for controlled production of copper nanoparticles for sorption of methylene blue dye. Arabian Journal of Chemistry, 13(2), 4263–4274. https://doi.org/10.1016/j.arabjc.2019.07.007

Sharma, I. (2020). Bioremediation techniques for polluted environment: Concept, advantages, limitations, and prospects. In Trace metals in the environment-new approaches and recent advances. IntechOpen. https://books.google.com/books

Shen, L., Ziheng, J., Wang, D., Wang, Y., & Lu, Y. (2017). Enhance wastewater biological treatment through the bacteria induced graphene oxide hydrogel. Chemosphere, 190. https://doi.org/10.1016/j.chemosphere.2017.09.105

Shittu, K. O., & Ihebunna, O. (2017). Purification of simulated waste water using green synthesized silver nanoparticles of Piliostigma thonningii aqueous leave extract. Advances in Natural Sciences: Nanoscience and Nanotechnology, 8(4), 045003. https://doi.org/10.1088/2043-6254/aa8536

Singh, A. K., & Chandra, R. (2019). Pollutants released from the pulp paper industry: Aquatic toxicity and their health hazards. Aquatic toxicology (Amsterdam, Netherlands), 211, 202–216. https://doi.org/10.1016/j.aquatox.2019.04.007

Singh, A., Tyagi, P., Ranjan, R., Sushkova, S. N., Minkina, T., Burachevskaya, M., & Rajput, V. D. (2023). Bioremediation of Hazardous Wastes Using Green Synthesis of Nanoparticles. Processes, 11(1), 141. https://doi.org/10.3390/pr11010141

Singh, R., Manickam, N., Mudiam, M. K. R., Murthy, R. C., & Misra, V. (2013). An integrated (nano-bio) technique for degradation of γ-HCH contaminated soil. Journal of Hazardous Materials, 258–259, 35–41. https://doi.org/10.1016/j.jhazmat.2013.04.016

Singh, Y., & Saxena, M. K. (2022). Insights into the recent advances in nano-bioremediation of pesticides from the contaminated soil. Frontiers in Microbiology, 13, 982611. https://doi.org/10.3389/fmicb.2022.982611

Srivastava, A., Srivastava, P., Pandey, A., Khanna, V. K., & Pant, A. B. (2019). Phytomedicine: A potential alternative medicine in controlling neurological disorders. In New look to phytomedicine (pp. 625–655). Elsevier. https://www.sciencedirect.com/science/article/pii/B9780128146194000252

Tan, W., Peralta-Videa, J. R., & Gardea-Torresdey, J. L. (2018). Interaction of titanium dioxide nanoparticles with soil components and plants: Current knowledge and future research needs – a critical review. Environmental Science: Nano, 5(2), 257–278. https://doi.org/10.1039/C7EN00985B

Tao, W., Chen, G., Zeng, G., Yan, M., Chen, A., Guo, Z., Huang, Z., he, K., Hu, L., & Wang, L. (2016). Influence of silver nanoparticles on heavy metals of pore water in contaminated river sediments. Chemosphere, 162, 117–124. https://doi.org/10.1016/j.chemosphere.2016.07.043

Thakur, S., Bi, A., Mahmood, S., Samriti, Ruzimuradov, O., Gupta, R., Cho, J., & Prakash, J. (2024). Graphene oxide as an emerging sole adsorbent and photocatalyst: Chemistry of synthesis and tailoring properties for removal of emerging contaminants. Chemosphere, 352, 141483. https://doi.org/10.1016/j.chemosphere.2024.141483

Torres-Martínez, C. L., Kho, R., Mian, O. I., & Mehra, R. K. (2001). Efficient Photocatalytic Degradation of Environmental Pollutants with Mass-Produced ZnS Nanocrystals. Journal of Colloid and Interface Science, 240(2), 525–532. https://doi.org/10.1006/jcis.2001.7684

Tudi, M., Daniel Ruan, H., Wang, L., Lyu, J., Sadler, R., Connell, D., Chu, C., & Phung, D. T. (2021). Agriculture Development, Pesticide Application and Its Impact on the Environment. International journal of environmental research and public health, 18(3), 1112. https://doi.org/10.3390/ijerph18031112

Vázquez-Núñez, E., Molina-Guerrero, C. E., Peña-Castro, J. M., Fernández-Luqueño, F., & de la Rosa-Álvarez, M. G. (2020). Use of Nanotechnology for the Bioremediation of Contaminants: A Review. Processes, 8(7), Article 7. https://doi.org/10.3390/pr8070826

Wang, B., Lan, J., Bo, C., Gong, B., & Ou, J. (2023). Adsorption of heavy metal onto biomass-derived activated carbon: Review. RSC Advances, 13(7), 4275. https://doi.org/10.1039/d2ra07911a

Wang, C., Kim, Y. J., Singh, P., Mathiyalagan, R., Jin, Y., & Yang, D. C. (2016). Green synthesis of silver nanoparticles by Bacillus methylotrophicus, and their antimicrobial activity. Artificial cells, nanomedicine, and biotechnology, 44(4), 1127–1132. https://doi.org/10.3109/21691401.2015.1011805.

Wuana, R. A., & Okieimen, F. E. (2011). Heavy Metals in Contaminated Soils: A Review of Sources, Chemistry, Risks and Best Available Strategies for Remediation. International Scholarly Research Notices, 2011(1), 402647. https://doi.org/10.5402/2011/402647

Yaashikaa, P. R., Palanivelu, J., & Hemavathy, R. V. (2024). Sustainable approaches for removing toxic heavy metal from contaminated water: A comprehensive review of bioremediation and biosorption techniques. Chemosphere, 357, 141933. https://doi.org/10.1016/j.chemosphere.2024.141933

Zapata-Peñasco, I., Avelino-Jiménez, I. A., Mendoza-Pérez, J., Vázquez Guevara, M., Gutiérrez-Ladrón de Guevara, M., Valadez-Martínez, M., Hernández-Maya, L., Garibay-Febles, V., Fregoso-Aguilar, T., & Fonseca-Campos, J. (2024). Environmental stressor assessment of hydrocarbonoclastic bacteria biofilms from a marine oil spill. Biotechnology Reports (Amsterdam, Netherlands), 42, e00834. https://doi.org/10.1016/j.btre.2024.e00834

19

Nano-bioremediation of Radioactive Waste: Challenges and Opportunities

***Satyendra Barber*[1] *Girima Nagda*[2] *and Chhavi Bhalothia*[2]**

[1]*Dept. of Zoology, Guru Nanak Girls P.G. College, Udaipur - 313001, Rajasthan, India*

[2]*Dept. of Zoology, Mohanlal Sukhadia University, Udaipur - 313001, Rajasthan, India*

**Corresponding Author Email: barbersatyen@gmail.com*

Abstract

Waste that emits harmful levels of nuclear radiation is radioactive waste. None of the places on the earth is free from natural radioactive background. Nano-bioremediation takes a center stage while addressing the case of radioactive waste as it integrates nanotechnology with microbial and plant-based bioremediation. Microbes like Bacillus sphaericus, exhibit unique capabilities by using their surface layer proteins for binding and sequestering radionuclides such as uranium via biomineralization and biosorption. To enhance the efficacy of these methods carbon-based nanomaterial (CBN) like graphene sheets and nanotubes can be used to increase the surface area scaffolds for redox transformations and microbial adhesion. Despite its promising potential challenges exist in scalability and mitigating risks associated with ecosystems. The viability of hybrid nano-bioremediation is demonstrated by field studies at U.S. DOE sites (e.g., Hanford, Oak Ridge). Nuclear waste management and disposal is an intricate and time-consuming process that requires thoughtful coordination planning, and creativity. Ensuring security is critical, necessitating strict guidelines, legal frameworks, supervision, and reliable contingency protocols. Several technical and regulatory barriers can be overcome by interdisciplinary collaborations. Nano bioremediation should focus on the combined use of nanoparticles, genetically modified microbes, and plant to design eco-friendly, cost – effective, robust, and sustainable remediation strategies.

Keywords: *Nano bioremediation, Radionuclides, Nanomaterials, Bio reduction, Bio sorption, Bioaccumulation, Phytoremediation.*

1. Introduction

It is the need of the hour that disposal of radioactive wastes be taken seriously and given appropriate attention. We house around one million gallons of liquid radioactive waste awaiting remediation at U.S. Department of Energy's Hanford Reservation alone (Parshley & McDermott, 2021). Not only the numbers are disturbing but the increasing demand of nuclear energy paves the way for generation of even more radioactive waste in future. The conventional methods have certain limitations. Ion-exchange resins have reported to be degraded under high radiation and chemical precipitation has proven to be ineffective in treating low concentration of radionuclides found in groundwater, posing risk for centuries (Zhang & Liu, 2020). Remediation methods including soil washing and vitrification are costly as estimates for cleanup in U.S alone are up to hundreds of millions of dollars (L. Liu *et al.*, 2018). Even after we bear this cost most of these methods prove to be disruptive to the ecosystem and overall failures in long run. Major nuclear accidents such as Chernobyl (1986) and Fukushima (2011) have further increased the burden of global nuclear waste. Cesium-137 and strontium-90 dispersed during these disasters continue to pollute the soil and ground water till date and will sustain for centuries if not treated effectively (Kyne & Bolin, 2016). India faces a compounded challenge in terms of radioactive waste management as its expanding nuclear power and uranium mining add to its burden. However, in Indian nuclear fuel cycle, used material undergoes refinement to extract plutonium along with remaining uranium. But this is just a band aid on a gushing wound because as we expand into new horizons of nuclear energy, we need sustainable bioremediation practices in place which can effectively tackle the radioactive waste (Raj *et al.*, 2006).

Key methods for handling radioactive materials involve compression, combustion, and high-temperature processing. Additionally, containment approaches include solidifying with cement, thermal decomposition, and vitrification (R. Randive *et al.*, 2023). These methods are not effective nor sustainable for instance incineration of radioactive solid waste disperses toxic particulates in air, escalating the forever increasing air pollution. Solidifying radioactive waste using cement is no relief either as it posses' risks of leaching under acidic conditions, as evidenced by a Mayak Production Association site in Russia (Rozhkova *et al.*, 2021).These methods are ineffective for heterogenous waste matrices and often lack selectivity and remove beneficial compounds along with non-desirable ones and fail to address subsurface plumes effectively.

In response, sustainable bioremediation strategies have gained prominence. These methods leverage natural processes, such as microbial metabolism and plant hyperaccumulation, to detoxify pollutants. For example, *Arabidopsis thaliana* genetically modified with phytochelatin synthase has shown enhanced uranium uptake in laboratory trials (Berthet *et al.*, 2018) . Unlike energy-intensive approaches, bioremediation operates at ambient temperatures, reducing carbon footprints and aligning with the United Nations' Sustainable Development Goals for clean water and responsible consumption. The integration of nanotechnology and bioremediation-termed nano-bioremediation-further enhances efficiency by targeting subsurface contaminants and enabling the recovery and potential reuse of valuable radionuclides (Singh *et al.*, 2020).

Nanotechnology has contributed majorly to a variety of fields such as healthcare, apparel, pharmaceuticals, electronics, beauty products, optical advancements, athletics, and much more. A ton of published articles and cutting-edge research in nanotechnology evidences its role in efforts to clean up the environment (Ramezani *et al.*, 2021; Singh *et al.*, 2020). This technology has demonstrated potential in tackling restoration challenges efficiently.

Recently the strategies for sustainable strategies to cleanup environment have gained popularity, emphasising on decreasing the levels of contamination below threshold levels long with lowering associated ecological damage. Recent strategies have unified systems by combining various approaches which ensure timely site purification, cost effectivity alongside recovery of ecosystem. Biological restoration is the only viable technique amongst various restoration techniques as it employs microbes to decontaminate water and soil environments from pollutants. Also, it is economical and nature-friendly (Ramezani *et al.*, 2021; Srinath *et al.*, 2024).

To the EPA, remediation that is biological makes use of natural occurring microorganisms that transform unsafe chemicals into either non-toxic or safer equivalents. Although there are benefits for this process to include precision, it is cheaper, energy friendly, and possess minimal resource demand, it also has limitations-is lengthy, for it takes several months to several years to degenerate contaminants. Lastly, massive levels of contamination compromised of substances of high toxicity could be hindering to this effectiveness (Nair & Abraham, 2018).

All remediation approaches have their merits and demerits. A promising solution to address challenges associated with decontamination of environment is the unification of various strategies used for remediation. One such approach that has gained significant attention in recent years is nano-bioremediation. This method harnesses the benefits of both nanotechnology and bioremediation toto effectively remove pollutants from the environment.

This chapter offers a concise overview of microbial mechanisms in removal of radioactive waste, nanotechnology, exploring a range of nanoscale materials utilized for extracting organic and inorganic pollutants from environmental systems. Additionally, it provides an in-depth discussion on the nano-bioremediation process, covering its applications, challenges, and future perspectives. The overview of the nano bioremediation is shown in Fig. 1.

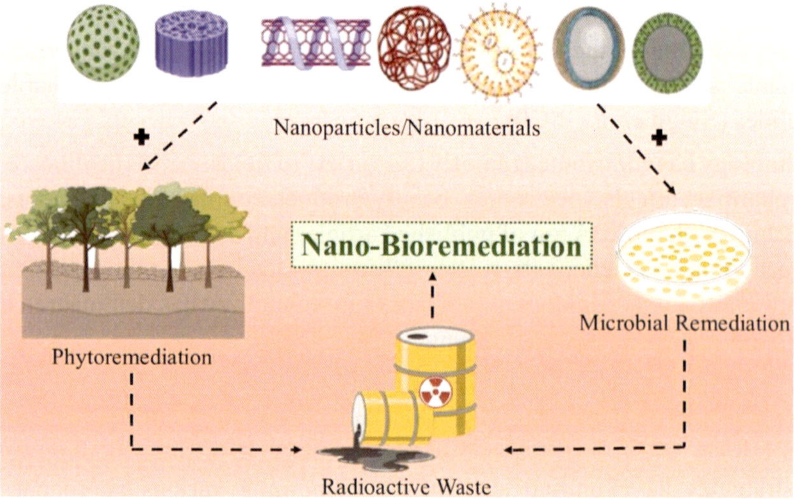

Fig. 1: An outline of Nano-Bioremediation

2. Microbial Mechanisms in Radioactive Waste Remediation

Microbial remediation of nuclear waste takes advantage of the extraordinary metabolic diversity and versatility of some bacteria and fungi to immobilize, convert, or detoxify radionuclides in affected environments. Of central importance in this regard are biosorption and biomineralization processes, where microbial cell walls or extracellular matrices serve as reactive surfaces for radionuclide adsorption. For instance, *Bacillus sphaericus* has surface-layer (S-layer) proteins containing high levels of phosphoryl and carboxyl groups, capable of chelating uranium and other actinides with good specificity and efficacy, particularly at slightly acidic conditions. This metabolically independent and passive uptake or biosorption occurs fast and can work even in the case of non-viable biomass (Velásquez & Dussan, 2009).

Biomineralization, but not the natural abiotic, involves the energetic conversion of soluble radionuclides into insoluble phases of minerals. Enzyme-mediated reduction by metal-reducing bacteria. *Sheewanella oneidensis* and *Geobacter sulffurreducens* can decrease soluble U(IV) enzymatically to precipitate insoluble U(IV) as nanocrystals of uraninite. Not only does this entomb uranium from migrating in groundwater, but it also produces the stable mineral from that will linger in the environment. Microorganisms such as *Aspergillus niger* help by excreting organic acids and phosphatases that immobilize uranium as a stable phosphate mineral in the fungal matrix(Jin *et al.*, 2019; J. Liu *et al.*, 2015; Marshall *et al.*, 2006).

Microbial reduction of radionuclides such as technetium and chromium is done via a unique reductase enzyme. For example, nitrate reductase is utilised by *Desulfovibrio desulfuricans* to change Tc(VII) to less mobile Tc (IV). The effectiveness of these microbial remediation methods is challenges by environmental factors like pH, presence of competing ions and redox potential (Lloyd *et al.*, 1999).

Technical innovations in omics technologies, more specifically metagenomics and transcriptomics, have transformed the view of microbial consortia in irradiated environments. These techniques exhibit the genetic capacities and adaptive measures of native microbes, for instance, up-regulation of efflux pumps, siderophone production, and stress response machinery under radionuclide pressure. These reveal the possibilities that allow for rationale design of designed strains or consortia with extended remediation traits. For example, genetic engineering can introduce radiation-resistance or metal-binding genes into hardy hosts such as *Deinococcus radiodurans*, equipping them to survive in extreme conditions and actively engage in immobilization of radionuclide (Lloyd *et al.*, 1999; Tang *et al.*, 2024).

In general, microbial processes provide a sustainable, low-energy, and self-sustaining means for bioremediation of radioactive wastes, with the further advantage associated with adaptation to varying and fluctuating environmental conditions. Further various microbial species and their remediation mechanisms are listed in Table 1.

Table 1. Microbial Species and Their Remediation Mechanisms

Microorganism	Mechanism	Target Radionuclide	Key Feature	Reference
Bacillus sphaericus	S-layer biosorption	Uranium (U)	High-affinity, rapid binding	(Velásquez & Dussan, 2009)
Shewanella oneidensis	Enzymatic reduction, biomineralization	U	Nanoparticle precipitation	(J. Liu *et al.*, 2015; Marshall *et al.*, 2006)
Geobacter sulfurreducens	Reductive precipitation	U	Coupled to energy metabolism	(Marshall *et al.*, 2006)
Desulfovibrio desulfuricans	Enzymatic reduction	Technetium (Tc)	Nitrate reductase pathway	(Lloyd *et al.*, 1999)
Aspergillus niger	Phosphate biomineralization	U	Organic acid/ phosphatase secretion	(Jin *et al.*, 2019)
Pseudomonas spp.	Biosorption, bioaccumulation	Lead (Pb), Cadmium (Cd), U	High biosorption capacity; multiple functional groups on cell wall	(Gul *et al.*, 2024)
Paenibacilus jamilae	Biosorption, biomineralization	Pb, Cd	Native soil bacteria; robust durability and efficiency	(H. Liu *et al.*, 2022)
Chlamydomonas reinhardtii (Microalgae)	Bioaccumulation, biosorption	Pb, Cd	High photosynthetic efficiency; effective metal uptake	(Flouty & Estephane, 2012)
Bacillus cereus	Bioaugmentation, plant-microbe interaction	Cd	Enhances plant remediation potential; reshapes microbial community	(Tang *et al.*, 2024)
Shinella and *Ensifer* spp.	Biotransformation (oxidation, reduction)	Antimony (Sb)	Oxidize or reduce antimony; reduce toxicity in soil	(H. Liu *et al.*, 2022)

3. Phytoremediation of Radioactive Waste

Phytoremediation has emerged as a sustainable and cost-effective alternative to conventional methods for remediating radioactive contamination in soils and water. Unlike traditional ex-situ techniques, which often involve excavation, chemical leaching, or thermal treatments-each carrying high financial and environmental costs-phytoremediation harnesses the natural abilities of green plants to extract, stabilize, or degrade contaminants directly *in situ*. The underlying principle is elegantly simple: certain plant species, through their anatomical and physiological adaptations, can absorb radionuclides such as uranium (U-238), thorium (Th-232), and radium (Ra-226) from contaminated substrates, translocate them into their shoots, leaves, or roots, and in some cases, even volatilize or transform them into less harmful forms (Malhotra *et al.*, 2014). The process not only reduces the bioavailability of hazardous substances but also preserves soil structure and fertility, making it especially suitable for large, low-to-moderate contamination sites where traditional remediation would be impractical or prohibitively expensive (Jhilta *et al.*, 2022). Notably, the roots, stems, and leaves of these plants act as natural bio-pumps, drawing up contaminants in a manner that is both environmentally benign and visually unobtrusive.

The effectiveness of phytoremediation depends on a complex interplay of factors, including plant species, the chemical nature and bioavailability of radionuclides, soil properties, and environmental conditions such as pH, temperature, and humidity (Jhilta *et al.*, 2022; Malhotra *et al.*, 2014). For example, uranium is most readily absorbed by plants when present in the soil as uranyl cations or uranyl-carbonate complexes, while radium, being chemically similar to calcium, is easily taken up and incorporated into plant tissues through physiological processes akin to those used for essential nutrients (Abreu *et al.*, 2014; Yan & Luo, 2016). The selection of plant species is critical: hyperaccumulators like Indian mustard (*Brassica juncea*) and sunflower (*Helianthus annuus*) have demonstrated exceptional abilities to extract cesium, strontium, and uranium from contaminated soils and waters, as evidenced by their successful deployment in post-Chernobyl remediation projects (Ehlken & Kirchner, 2002; Malhotra *et al.*, 2014). Meanwhile, aquatic and wetland species such as *Juncus effusus* and *Caltha palustris* have shown promise for remediating radionuclide-contaminated sediments and tailings, particularly in mining-impacted landscapes (Hemming *et al.*, 2023). The efficiency of uptake is often quantified using transfer factors (TF), bioconcentration factors (BF), and phytoremediation factors (PF), which collectively assess a plant's capacity to absorb, translocate, and accumulate radionuclides relative to their concentrations in the surrounding environment (Abreu *et al.*, 2014).

Table 2. Plant species for phytoremediation of radionuclides

Plant Species	Target Radionuclide(s)	Uptake Mechanism	Notable Features/Performance	References
Brassica juncea	Cs, Sr, U	Phytoextraction	High uptake; used at Chernobyl; fast-growing	(Jhilta *et al.*, 2022)
Helianthus annuus	Cs, Sr, U	Rhizofiltration, Extraction	Used for water and soil; effective at Chernobyl	(Dubchak & Bondar, 2019)
Juncus effusus	Ra, U	Root accumulation	High Ra transfer factor in mine tailings	(Ehlken & Kirchner, 2002)
Caltha palustris	U, Ra, Th	Root accumulation	Moderate uptake; suited for wetlands	(Hemming *et al.*, 2023)
Pokeberry (*Phytolacca americana*)	U, Th, Ra, K	Hyperaccumulation	Highest phytoremediation factor for multiple radionuclides	(Yan & Luo, 2016)
Acacia nilotica	U	Root accumulation	Accumulates up to 73% of U in roots; suited for dry soils	(Fathi *et al.*, 2014)
Spinach (*Spinacia oleracea*)	Ra, U	Rhizofiltration	Effective in water-based systems	(Malhotra *et al.*, 2014)
Populus spp.	Hg, Se (organics)	Phytovolatilization	Used for organics; limited for radionuclides	(Li *et al.*, 2024)

Despite its many advantages, phytoremediation is not without limitations. The process is inherently slow, often requiring several growing seasons to achieve significant reductions in contaminant levels and is generally limited to the rooting depth of the selected plant species (Malhotra *et al.*, 2014). Additionally, the survival and growth of plants can be adversely affected by the toxicity of highly contaminated soils, nutrient deficiencies, or unfavourable climatic conditions. Another challenge is the safe disposal of radioactive biomass generated during the process; harvested plant material enriched with radionuclides must be managed as radioactive waste, necessitating appropriate containment or incineration strategies (Ehlken & Kirchner, 2002; Lee *et al.*, 2003). There is also the risk of contaminants entering the food chain if phytoremediation sites are not properly managed, particularly when edible or invasive species are used (Yan & Luo, 2016). Nevertheless, ongoing research into genetic engineering and the use of chelating agents or soil amendments continues to expand the range of suitable plant species and enhance the efficiency of radionuclide uptake (Jhilta *et al.*, 2022).

Phytoremediation stands out as a green technology that leverages the natural abilities of plants to address the persistent challenge of radioactive contamination. Its greatest strengths lie in its environmental compatibility, cost-effectiveness, and potential for large-scale application, particularly in areas where traditional remediation is unfeasible. However, its success is contingent upon careful species selection, site assessment, and long-term management of radioactive biomass. As scientific understanding of plant-radionuclide interactions deepens and novel biotechnological tools become available, phytoremediation is poised to play an increasingly central role in the sustainable management of radioactive waste and the restoration of contaminated landscapes.

4. Nanomaterials in Bioremediation: Synthesis and Applications

Nanomaterials have transformed environmental remediation through the provision of distinctive physicochemical properties, such as tunable surface chemistry, high surface area, and increased reactivity, which are specifically beneficial for overcoming the obstacles posed by radioactive waste. Carbon-based nanomaterials (CBNs) such as carbon nanotubes (CNTs) and graphene oxide (GO) are among the leading contenders due to their high adsorption capabilities and potential to enable electron transfer processes (Asghar *et al.*, 2024).

Graphene oxide's rich oxygen-functional groups allow functionalization with amines, thiols, or other ligands to preconcentrate selectively uranium and other radionuclides. The GO have about 2,630 m^2/g of surface area which is very high this helps in fast and efficient adsorption, which enables it to achieve removal rates higher than 98% in studies conducted in laboratory (Nanda *et al.*, 2015). Metal-reducing bacteria like *Shewanella* and *Geobacter have greatly* benifited from the use of carbon nanotubes as they act as conductive bridges when introduced in microbial systems, enhancing electron transfer in extracellular region. This synergy accelerates the reduction of soluble radionuclides to insoluble forms, thereby improving the overall efficiency of bioremediation (J. Liu *et al.*, 2015; Marshall *et al.*, 2006).

Magnetic nanocomposites, such as Fe_3O_4-GO hybrids, combine the adsorption power of graphene with the magnetic properties of iron oxide, allowing for facile separation and recovery of the nanomaterials after remediation through the application of an external magnetic field. This reduces the risk of secondary contamination and facilitates the reuse or safe disposal of spent adsorbents(Donga *et al.*, 2024).

Bionanocomposites represent a convergence of biological and nanomaterial strategies. For example, chitosan-nanocellulose sponges embedded with microbial biofilms can simultaneously adsorb and biologically transform radionuclides, achieving high removal efficiencies for strontium and cesium (Al-Gethami *et al.*, 2024). Layer-by-layer assembly of polyelectrolytes and nanoparticles can create membranes that mimic biological ion channels, enabling selective separation of specific radionuclides from complex waste streams. Three-dimensional printed scaffolds further enhance the stability and longevity of these systems in variable environmental conditions (Ali *et al.*, 2008).

Synthesis methods for nanomaterials are increasingly shifting toward green chemistry approaches. Biogenic synthesis using plant extracts or microbial processes avoids toxic solvents and reduces energy consumption, producing nanoparticles with controlled size and surface properties(Gour *et al.*, 2019). Plasma-assisted chemical vapor deposition (PACVD) offers another route to defect-free graphene at lower temperatures, further minimizing the environmental footprint (Kotbi *et al.*, 2025).

The integration of nanomaterials with biological systems not only enhances the kinetics and selectivity of radioactive waste remediation but also opens avenues for resource recovery and circular economy applications, such as the reuse of recovered radionuclides in industry or medicine.

Table 3. Nanomaterials in Radioactive Waste Remediation

Nanomaterial	Application	Advantage	Limitation	References
Graphene oxide (GO)	Uranium adsorption	High surface area, tunable chemistry	pH-dependent efficiency	(H. Liu *et al.*, 2021; Nanda *et al.*, 2015)
Carbon nanotubes (CNTs)	Electron transfer enhancement	Boosts microbial reduction rates	Aggregation in saline environments	(Asghar *et al.*, 2024)
Magnetic Fe_3O_4-GO composite	Magnetic recovery, adsorption	Easy separation, reusability	Complex synthesis	(Donga *et al.*, 2024)
Chitosan-nanocellulose sponge	Biohybrid sorption/ transformation	Biocompatible, high efficiency	Stability in harsh conditions	(Al-Gethami *et al.*, 2024)
Layered nanoparticle membranes	Selective ion separation	Mimics biological selectivity	Scalability, fouling	(Ali *et al.*, 2008)

5. Case Studies and Field Applications

Field-scale applications of nano-bioremediation have demonstrated the potential of integrating microbial and nanomaterial strategies for radioactive waste management. At the U.S. Department of Energy's Hanford Site, pilot studies have employed *Bacillus* S-layer proteins immobilized on biochar to selectively capture uranium from groundwater, achieving removal efficiencies comparable to conventional ion exchange resins (Wood *et al.*, 2001). Similarly, Oak Ridge National Laboratory has tested bioaugmentation with *Shewanella oneidensis* and graphene-based nanocomposites for the reduction and immobilization of U(VI), with promising results in both lab and field trials (Marshall *et al.*, 2006).

Another notable example is the use of nano-enhanced microbial consortia for hydrocarbon degradation in oil waste, where carbon nanotubes (CNTs) have been shown to accelerate electron transfer and improve the metabolic activity of indigenous bacteria (X.-W. Liu *et al.*, 2014). These case studies highlight the adaptability of nano-bioremediation to diverse contaminants and environmental conditions.

But scaling-up from the laboratory to full-scale field deployment is fraught with challenges. Site heterogeneity, varying environmental conditions, and regulatory issues can put a damper on the effectiveness of these technologies. For example, aggregation of nanomaterials in saltwater or high-ionic-strength groundwater decreases their adsorption capacity and mobility (Petersen & Henry, 2012). Further, long-term stability and toxicity of engineered nanomaterials in natural systems are concerns, and therefore risk assessments and monitoring protocols need to be carried out comprehensively.

Despite these obstacles, field studies have provided valuable insights into the practical feasibility of nano-bioremediation. Ongoing research aims to optimize material synthesis, improve delivery systems, and enhance the resilience of microbial consortia to ensure reliable performance in real-world settings (Roy *et al.*, 2021).

Table 4. Notable Field Applications of Nano-Bioremediation

Site/Project	Technology Used	Target Contaminant	Key Results/ Challenges	References
Hanford Site, USA	*Bacillus* S-layer proteins on biochar	Uranium	High removal efficiency, scalability	(Wood *et al.*, 2001)
Oak Ridge National Laboratory	*Shewanella oneidensis* + graphene nanocomposite	Uranium	Lab and field success, stability	(Marshall *et al.*, 2006)
Oil Waste Sites	CNT-enhanced microbial consortia	Hydrocarbons	Accelerated degradation, aggregation	(X.-W. Liu *et al.*, 2014)

6. Challenges in Nano-Bioremediation

Even though nano-bioremediation offers distinct merits, various environmental, technical, and regulatory challenges must be addressed to optimize its potential to fullest. One of the major risks amongst the potential environmental risks is the release of engineered nanoparticles in the environment. Unintended ecological damage may be caused by nanoparticles such as graphene oxide and carbon nanotubes as they may exhibit toxicity to aquatic organisms and soil microbiota, and their persistence in ecosystem may cause bioaccumulation in future (Petersen & Henry, 2012). Long-term monitoring and ecotoxicological studies are essential to assess these risks and develop mitigation strategies.

Technical Challenges also involve the survival and viability of introduced microbial strains for a short period in harsh or fluctuating field conditions. Major factors such as competition with native microbes, nutrient availability, and exposure to high levels of radiation may inhibit the efficacy of bioaugmentation process (Vandana *et al.*, 2021). Also, scaling up nanomaterial synthesis and economy of bulk production are also key challenges.

The adoption of nano-bioremediation is further complicated by economic and regulatory challenges. The lack of designated guidelines about the use of nanomaterials in environmental remediation in current regulatory framework causes uncertainty and delays in approval of projects (Roy *et al.*, 2021). Cost-benefit analyses must also consider the trade-offs between initial investment, operational costs, and long-term environmental benefits.

Addressing these challenges will require interdisciplinary collaboration, advances in material science and biotechnology, and the development of robust risk assessment protocols. Only through a comprehensive and cautious approach can the promise of nano-bioremediation be fully realized.

Table 5. Key Challenges in Nano-Bioremediation

Challenge Type	Description	Potential Solutions	References
Environmental Risk	Toxicity and persistence of nanomaterials in ecosystems	Ecotoxicology studies, safe-by-design	(Petersen & Henry, 2012)
Technical Barriers	Microbial survival, nanomaterial stability, scalability	Strain engineering, green synthesis	(Roy *et al.*, 2021)
Regulatory Hurdles	Lack of guidelines for nanomaterial use in remediation	Policy development, stakeholder engagement	(Sethi *et al.*, 2024)
Economic Factors	High costs of nanomaterial synthesis and field deployment	Cost-benefit analysis, value recovery	(Patra Shahi *et al.*, 2021)

7. Future Directions and Opportunities

Looking forward, there are some exciting possibilities that may expand the nano-bioremediation field. One is a circular economy based on recovered radionuclides being used in industrial or medicinal purposes. An example of this is the possible utilization of extracted uranium and rare earth elements from polluted environments to use them in nuclear fuel cycles, in medicine as medical isotopes, or in innovative materials (Roy *et al.*, 2021). The creation of bio nanocomposites with designed functionality would make it possible to recover and reuse valuable metals selectively, closing the cycle in resource use.

Artificial intelligence (AI) and machine learning have the potential to play a revolutionary role in the optimization of nano-bioremediation processes. These technologies can simulate intricate microbial-nanomaterial interactions, forecast system performance under different conditions, and inform the design of future remediation strategies (Patowary *et al.*, 2023). Approaches driven by AI can also increase the efficacy of monitoring and decision-making abilities. This could inspire real time alterations which can maximize the efficacy of bioremediation and minimize the damage on ecosystem.

For standardization of protocols and acceleration of innovation it is necessary that the knowledge sharing and collaborations are done on a global scale. International initiatives such as the IAEA's Environmental Remediation Program provide platforms for exchanging best practices and fostering partnerships between academia, industry, and government (Ramamoorthy, 2024).

In conclusion, the future of nano-bioremediation is integration with advanced materials, biotechnology, and information tools, augmented by strong policy frameworks and collaboration at the global level. In confronting existing problems and adopting innovative opportunities, the field can greatly contribute to environmentally sustainable management and global security.

References

Abreu, M. M., Lopes, J., Santos, E. S., & Magalhães, M. C. F. (2014). Ecotoxicity evaluation of an amended soil contaminated with uranium and radium using sensitive plants. Journal of Geochemical Exploration, 142, 112–121.

Al-Gethami, W., Qamar, M. A., Shariq, M., -Nasser, A., Alaghaz, M. A., Farhan, A., Areshi, A. A., & Hisham Alnasir, M. (2024). Emerging environmentally friendly bio-based nanocomposites for the efficient removal of dyes and micropollutants from wastewater by adsorption: a comprehensive. Pubs.Rsc.Org. https://doi.org/10.1039/d3ra06501d

Ali, M., Yameen, B., Cervera, J., Ramírez, P., Neumann, R., Ensinger, W., Knoll, W., & Azzaroni, O. (2008). Composite layer-by-layer (LBL) assembly with inorganic nanoparticles and nanowires. ACS Publications, 41(12), 1831–1841. https://doi.org/10.1021/AR8001377

Asghar, N., Hussain, A., Nguyen, D. A., Ali, S., Hussain, I., Junejo, A., & Ali, A. (2024). Advancement in nanomaterials for environmental pollutants remediation: a systematic review on bibliometrics analysis, material types, synthesis pathways, and related mechanisms. Journal of Nanobiotechnology, 22(1), 26.

Berthet, S., Villiers, F., Alban, C., Serre, N. B. C., Martin-Laffon, J., Fig.uet, S., Boisson, A., Bligny, R., Kuntz, M., & Finazzi, G. (2018). Arabidopsis thaliana plants challenged with uranium reveal new insights into iron and phosphate homeostasis. New Phytologist, 217(2), 657–670.

Donga, C., Mishra, S. B., Ndlovu, L. N., Abd-El-Aziz, A. S., Kuvarega, A. T., & Mishra, A. K. (2024). Magnetic Magnetite-Graphene Oxide (Fe3O4-GO) Nanocomposites for Removal of Dyes from Aqueous Solution. SpringerC Donga, SB Mishra, LN Ndlovu, AS Abd-El-Aziz, AT Kuvarega, AK MishraJournal of Inorganic and Organometallic Polymers and Materials, 2024•Springer, 34, 4192–4202. https://doi.org/10.1007/S10904-024-03077-5

Dubchak, S., & Bondar, O. (2019). Bioremediation and phytoremediation: Best approach for rehabilitation of soils for future use. Remediation Measures for Radioactively Contaminated Areas, 201–221.

Ehlken, S., & Kirchner, G. (2002). Environmental processes affecting plant root uptake of radioactive trace elements and variability of transfer factor data: a review. Journal of Environmental Radioactivity, 58(2–3), 97–112.

Fathi, R. A., Godbold, D. L., Al-Salih, H. S., & Jones, D. (2014). Potential of phytoremediation to clean up uranium-contaminated soil with Acacia species.

Flouty, R., & Estephane, G. (2012). Bioaccumulation and biosorption of copper and lead by a unicellular algae Chlamydomonas reinhardtii in single and binary metal systems: a comparative study. Journal of Environmental Management, 111, 106–114.

Gour, A., cells, N. J.-A., nanomedicine, undefined, biotechnology, and, & 2019, undefined. (2019). Advances in green synthesis of nanoparticles. Taylor & Francis, 47(1), 844–851. https://doi.org/10.1080/21691401.2019.1577878

Gul, I., Adil, M., Lv, F., Li, T., Chen, Y., Lu, H., Ahamad, M. I., Lu, S., & Feng, W. (2024). Microbial strategies for lead remediation in agricultural soils and wastewater: mechanisms, applications, and future directions. Frontiers in Microbiology, 15, 1434921.

Hemming, S. D., Purkis, J. M., Warwick, P. E., & Cundy, A. B. (2023). Current and emerging technologies for the remediation of difficult-to-measure radionuclides at nuclear sites. Environmental Science: Processes & Impacts, 25(12), 1909–1925.

Jhilta, P., Dipta, B., & Rana, A. (2022). Phytoremediation of heavy metals and radionuclides: sustainable approach to environmental management. In Phytoremediation for environmental sustainability (pp. 83–111). Springer.

Jin, M., Xiao, A., Zhu, L., Zhang, Z., Huang, H., & Jiang, L. (2019). The diversity and commonalities of the radiation-resistance mechanisms of Deinococcus and its up-to-date applications. AMB Express, 9, 1–12.

Kotbi, A., Lejeune, M., Oughaddou, H., Rajput, N., Zhang, X., Belhadi, J., Zakaria, Y., Richert, D., Piquemal, F., Morán-Meza, J., El Marssi, M., & Jouiad, M. (2025). Eco-Friendly Deposition of Catalyst-Free Graphene on Diverse Substrates. Advanced Sustainable Systems, 9(4), 2500105. https://doi.org/10.1002/ADSU.202500105

Kyne, D., & Bolin, B. (2016). Emerging environmental justice issues in nuclear power and radioactive contamination. International Journal of Environmental Research and Public Health, 13(7), 700.

Lee, S., Moon, J. S., Ko, T.-S., Petros, D., Goldsbrough, P. B., & Korban, S. S. (2003). Overexpression of Arabidopsis phytochelatin synthase paradoxically leads to hypersensitivity to cadmium stress. Plant Physiology, 131(2), 656–663.

Li, M., Heng, Q., Hu, C., Wang, Z., Jiang, Y., Wang, X., He, X., Yong, J. W. H., Dawoud, T. M., & Rahman, S. U. (2024). Phytoremediation efficiency of poplar hybrid varieties with diverse genetic backgrounds in soil contaminated by multiple toxic metals (Cd, Hg, Pb, and As). Ecotoxicology and Environmental Safety, 283, 116843.

Liu, H., Manufacturing, Y. M.-E. M. &, & 2021, undefined. (2021). Graphene oxide-based nanomaterials for uranium adsorptive uptake. Espublisher.Com, 13, 3–22. https://doi.org/10.30919/esmm5f453

Liu, H., Sun, W., & Häggblom, M. M. (2022). Microbial transformations of antimony. In Microbial metabolism of metals and metalloids (pp. 223–254). Springer.

Liu, J., Xiong, F., Rong, L., Li, S., & Xie, S. (2015). A reduction mechanism of U () on Shewanella oneidensis by spectral analysis. 2015 International Forum on Energy, Environment Science and Materials, 1398–1403.

Liu, L., Li, W., Song, W., & Guo, M. (2018). Remediation techniques for heavy metal-contaminated soils: Principles and applicability. Science of the Total Environment, 633, 206–219.

Liu, X.-W., Chen, J.-J., Huang, Y.-X., Sun, X.-F., Sheng, G.-P., Li, D.-B., Xiong, L., Zhang, Y.-Y., Zhao, F., & Yu, H.-Q. (2014). Experimental and theoretical demonstrations for the mechanism behind enhanced microbial electron transfer by CNT network. Nature. ComXW Liu, JJ Chen, YX Huang, XF Sun, GP Sheng, DB Li, L Xiong, YY Zhang, F Zhao, HQ YuScientific Reports, 2014•nature.Com. https://doi.org/10.1038/srep03732

Lloyd, J. R., Ridley, J., Khizniak, T., Lyalikova, N. N., & Macaskie, L. E. (1999). Reduction of technetium by Desulfovibrio desulfuricans: biocatalyst characterization and use in a flowthrough bioreactor. Applied and Environmental Microbiology, 65(6), 2691–2696.

Malhotra, R., Agarwal, S., & Gupta, P. (2014). Phytoremediation of radioactive metals. J Civ Eng Environ Technol, 1, 75–79.

Marshall, M. J., Beliaev, A. S., Dohnalkova, A. C., Kennedy, D. W., Shi, L., Wang, Z., Boyanov, M. I., Lai, B., Kemner, K. M., & McLean, J. S. (2006). c-Type cytochrome-dependent formation of U (IV) nanoparticles by Shewanella oneidensis. PLoS Biology, 4(8), e268.

Nair, S., & Abraham, J. (2018). Hazardous waste management with special reference to biological treatment. Handbook of Environmental Materials Management, 1–27.

Nanda, S. S., Papaefthymiou, G. C., & Yi, D. K. (2015). Functionalization of graphene oxide and its biomedical applications. Taylor & Francis, 40(5), 291–315. https://doi.org/10.1080/10408436.2014.1002604

Parshley, L., & McDermott, S. (2021). Cold War, Hot Mess: After decades of mismanaging its nuclear waste, the US Department of Energy wrestles with its toxic legacy. Virginia Quarterly Review, 97(3), 46–71.

Patowary, R., Devi, A., & Mukherjee, A. K. (2023). Advanced bioremediation by an amalgamation of nanotechnology and modern artificial intelligence for efficient restoration of crude petroleum oil-contaminated sites: a prospective study. Environmental Science and Pollution Research 2023 30:30, 30(30), 74459–74484. https://doi.org/10.1007/S11356-023-27698-4

Patra Shahi, M., Kumari, P., Mahobiya, D., & Kumar Shahi, S. (2021). Nano-bioremediation of environmental contaminants: applications, challenges, and future prospects. Bioremediation for Environmental Sustainability: Approaches to Tackle Pollution for Cleaner and Greener Society, 83–98. https://doi.org/10.1016/B978-0-12-820318-7.00004-6

Petersen, E. J., & Henry, T. B. (2012). Methodological considerations for testing the ecotoxicity of carbon nanotubes and fullerenes. Academic.Oup.ComEJ Petersen, TB HenryEnvironmental Toxicology and Chemistry, 2012•academic.Oup.Com, 31(1), 60–72. https://doi.org/10.1002/ETC.710

R. Randive, K., Godbole, P., Jawadand, S., Chopra, V., Dora, M. L., & Dhoble, S. J. (2023). 12 - Radioactive waste management in India: present status and future perspectives. In N. A. Raut, D. M. Kokare, B. A. Bhanvase, K. R. Randive, & S. J. Dhoble (Eds.), 360-Degree Waste Management, Volume 2 (pp. 273–298). Elsevier. https://doi.org/https://doi.org/10.1016/B978-0-323-90909-9.00005-8

Raj, K., Prasad, K. K., & Bansal, N. K. (2006). Radioactive waste management practices in India. Nuclear Engineering and Design, 236(7–8), 914–930.

Ramamoorthy, N. (2024). Harnessing Science-Diplomacy Synergy (SDS): Key feature of the IAEA roles and success since inception. https://inis.iaea.org/records/4ggyz-qc624

Ramezani, M., Rad, F. A., Ghahari, S., Ghahari, S., & Ramezani, M. (2021). Nano-bioremediation application for environment contamination by microorganism. Microbial Rejuvenation of Polluted Environment: Volume 2, 349–378.

Roy, A., Sharma, A., Yadav, S., Jule, L. T., & Krishnaraj, R. (2021). Nanotechnology for environmental remediation: materials and applications. Mdpi.Com, 2021. https://doi.org/10.1155/2021/1764647

Rozhkova, A. K., Kuzmenkova, N. V, Pryakhin, E. A., Mokrov, Y. G., & Kalmykov, S. N. (2021). Artificial radionuclides association with bottom sediment components from Mayak Production Association industrial reservoirs. Journal of Environmental Radioactivity, 232, 106569.

Sethi, S., Bharshankh, A., Sahu, R., & Biswas, R. (2024). Nano-bioremediation for Sustainable Treatment of Waste: Applications, Advancements, and Challenges. Environmental Engineering and Waste Management, 539–575. https://doi.org/10.1007/978-3-031-58441-1_19

Singh, R., Behera, M., & Kumar, S. (2020). Nano-bioremediation: An innovative remediation technology for treatment and management of contaminated sites. Bioremediation of Industrial Waste for Environmental Safety: Volume II: Biological Agents and Methods for Industrial Waste Management, 165–182.

Srinath, N., Vivek, P., Rajakumari, K., Meenambiga, S. S., & Yuvaraj, D. (2024). Biotechnological Innovations in Radioactive Waste Management. Radioactive Pollutant, 137.

Tang, H., Xiang, G., Xiao, W., Yang, Z., & Zhao, B. (2024). Microbial mediated remediation of heavy metals toxicity: mechanisms and future prospects. Frontiers in Plant Science, 15, 1420408.

Vandana, U. K., Gulzar, A. B. M., Laskar, I. H., Meitei, L. R., & Mazumder, P. B. (2021). Role of microbes in bioremediation of radioactive waste. Springer, 25, 329–352. https://doi.org/10.1007/978-981-15-7447-4_13

Velásquez, L., & Dussan, J. (2009). Biosorption and bioaccumulation of heavy metals on dead and living biomass of Bacillus sphaericus. Journal of Hazardous Materials, 167(1–3), 713–716.

Wood, T., Johnson, W., Kreid, D., & Walton, T. (2001). Hanford Site Cleanup Challenges and Opportunities for Science and Technology--A Strategic Assessment. https://www.osti.gov/biblio/965746

Yan, X., & Luo, X. G. (2016). Uptake of uranium, thorium, radium and potassium by four kinds of dominant plants grown in uranium mill tailing soils from the southern part of China. Radioprotection, 51(2), 141–144.

Zhang, X., & Liu, Y. (2020). Nanomaterials for radioactive wastewater decontamination. Environmental Science: Nano, 7(4), 1008–1040.

Index